Ecology of Tidal Freshwater Forested Wetlands
of the Southeastern United States

Ecology of Tidal Freshwater Forested Wetlands of the Southeastern United States

Edited by

William H. Conner
Clemson University
Georgetown, SC
USA

Thomas W. Doyle
US Geological Survey
Lafayette, LA
USA

and

Ken W. Krauss
US Geological Survey
Lafayette, LA
USA

A C.I.P. Catalogue record for this book is available from the Library of Congress.

ISBN 978-1-4020-5094-7 (HB)
ISBN 978-1-4020-5095-4 (e-book)

Published by Springer,
P.O. Box 17, 3300 AA Dordrecht, The Netherlands.

www.springer.com

Printed on acid-free paper

Cover Pictures from upper left clockwise:

Barnacles growing on base of baldcypress tree in Louisiana.
Hermit crab on base of baldcypress tree in South Carolina.
Tidal freshwater baldcypress stand on Turkey Creek, South Carolina.
(photos by William Conner).

All Rights Reserved
© 2007 Springer
No part of this work may be reproduced, stored in a retrieval system, or transmitted in
any form or by any means, electronic, mechanical, photocopying, microfilming, recording
or otherwise, without written permission from the Publisher, with the exception
of any material supplied specifically for the purpose of being entered
and executed on a computer system, for exclusive use by the purchaser of the work.

Contributors

T. Alphin, C.J. Anderson, G.B. Avery, A.H. Baldwin, J.L. Chambers, W.H. Conner, D. Creech, M.R. Darst, J.W. Day, R.H. Day, T.W. Doyle, J. Duberstein, R.S. Effler, S.P. Faulkner, A.S. From, E.S. Gardiner, R.A. Goyer, C.T. Hackney, S.S. Hoeppner, M.S. Hughes, C.R. Hupp, L.W. Inabinette, B.D. Keeland, R.F. Keim, S.L. King, W. Kitchens, K.W. Krauss, D.E. Kroes, L.A. Leonard, H.M. Light, B.G. Lockaby, R.A. Mattson, J.W. McCoy, M. MacDonald, K.W. McLeod, K. McPherson, M.P.V. Melder, C.A. Miller, T.H. Mirti, G.B. Noe, C.P. O'Neil, J.A. Nyman, M.M. Palta, M. Posey, M. Ratard, R.D. Rheinhardt, G.P. Shaffer, C.M. Swarzenski, K. Williams, T.M. Williams

Contents

Preface   xi

Chapter 1 - Tidal Freshwater Swamps of the Southeastern United States: Effects of Land Use, Hurricanes, Sea-level Rise, and Climate Change   1
Thomas W. Doyle, Calvin P. O'Neil, Marcus P.V. Melder, Andrew S. From, and Monica M. Palta

Chapter 2 - Hydrology of Tidal Freshwater Forested Wetlands of the Southeastern United States   29
Richard H. Day, Thomas M. Williams, and Christopher M. Swarzenski

Chapter 3 - Soils and Biogeochemistry of Tidal Freshwater Forested Wetlands   65
Christopher J. Anderson and B. Graeme Lockaby

Chapter 4 - Plant Community Composition of a Tidally Influenced, Remnant Atlantic White Cedar Stand in Mississippi   89
Bobby D. Keeland and John W. McCoy

Chapter 5 - Sediment, Nutrient, and Vegetation Trends Along the Tidal, Forested Pocomoke River, Maryland   113
Daniel E. Kroes, Cliff R. Hupp, and Gregory B. Noe

Chapter 6 - Vegetation and Seed Bank Studies of Salt-Pulsed Swamps of the Nanticoke River, Chesapeake Bay   139
Andrew H. Baldwin

Chapter 7 - Tidal Freshwater Swamps of a Lower Chesapeake Bay Subestuary   161
Richard D. Rheinhardt

**Chapter 8 - Biological, Chemical, and Physical
Characteristics of Tidal Freshwater Swamp Forests
of the Lower Cape Fear River/Estuary, North Carolina**   183
Courtney T. Hackney, G. Brooks Avery, Lynn A. Leonard,
Martin Posey, and Troy Alphin

**Chapter 9 - Ecology of Tidal Freshwater Forests in
Coastal Deltaic Louisiana and Northeastern South Carolina**   223
William H. Conner, Ken W. Krauss, and Thomas W. Doyle

**Chapter 10 - Ecology of the Coastal Edge of Hydric
Hammocks on the Gulf Coast of Florida**   255
Kimberlyn Williams, Michelina MacDonald, Kelly McPherson,
and Thomas H. Mirti

**Chapter 11 - Ecological Characteristics of Tidal Freshwater
Forests Along the Lower Suwannee River, Florida**   291
Helen M. Light, Melanie R. Darst, and Robert A. Mattson

**Chapter 12 - Community Composition of Select Areas
of Tidal Freshwater Forest Along the Savannah River**   321
Jamie Duberstein and Wiley Kitchens

**Chapter 13 - Ecology of the Maurepas Swamp:
Effects of Salinity, Nutrients, and Insect Defoliation**   349
Rebecca S. Effler, Gary P. Shaffer, Susanne S. Hoeppner,
and Richard A. Goyer

**Chapter 14 - Selection for Salt Tolerance in Tidal
Freshwater Swamp Species: Advances Using
Baldcypress as a Model for Restoration**   385
Ken W. Krauss, Jim L. Chambers, and David Creech

**Chapter 15 - Assessing the Impact of Tidal Flooding
and Salinity on Long-term Growth of Baldcypress
Under Changing Climate and Riverflow**   411
Thomas W. Doyle, William H. Conner, Marceau Ratard,
and L. Wayne Inabinette

**Chapter 16 - Conservation and Use of Coastal
Wetland Forests in Louisiana** 447
Stephen P. Faulkner, Jim L. Chambers, William H.
Conner, Richard F. Keim, John W. Day, Emile S. Gardiner,
Melinda S. Hughes, Sammy L. King, Kenneth W. McLeod,
Craig A. Miller, J. Andrew Nyman, and Gary P. Shaffer

**Chapter 17 - Tidal Freshwater Forested Wetlands:
Future Research Needs and an Overview of Restoration** 461
William H. Conner, Courtney T. Hackney,
Ken W. Krauss, and John W. Day, Jr.

Appendix 1 489

Index 497

Preface

Tidal freshwater forested wetlands represent an intriguing and understudied type of ecosystem in the southeastern United States. The physiographic position of tidal freshwater forested wetlands in occupying low lying, coastal areas makes them susceptible to upland runoff, tidal flooding, saltwater intrusion, and other global climate change phenomena. While information on them is rather sparse in the scientific literature, these ecosystems are among the most sensitive to sea-level rise and increased drought or flood frequency. Tidal freshwater forested wetlands are readily impacted by acute and chronic exposure to even low levels of salinity. The combined stress of flooding and salinity may compound the threat in these systems such that the margin for survival and compensation to changing climate is much less than for other coastal habitats. In this book, we bring together principal investigators whose research focus has targeted the hydrology, biogeochemistry, community ecology, forestry, stress physiology, and restoration of tidal freshwater forested wetlands in the southeastern United States. It is our foremost intent to develop an up-to-date treatise that includes not only peer-reviewed journal articles but also the dispersive grey literature on the topic in order to spark future research interest in tidal freshwater forested wetlands and to provide land managers with a concise overview of research findings. We have thus formalized all scientific and common names into the standard of ITIS (Integrated Taxonomic Information System, http://www.itis.gov, January 2007; for a complete listing of scientific and common names of all plants used in book, see Appendix 1).

This book resulted from several research projects that we were conducting under the auspices of the U.S. Geological Survey's Global Change Research Program. In particular, as we began to investigate carbon cycling in tidal freshwater forested wetlands impacted by salinity in South Carolina, we noticed that there were few applicable studies in the scientific literature. However, distribution of these forests along the tremendously altered coastlines is often in the form of remnant patches in need of major attention by restorationists. Without concise documentation of the many services provided by these ecosystems and of how those services shift with climate change, restoration is likely to be stalled considerably in light of the many natural and human impacts to tidal freshwater forested wetlands in the Southeast.

Tidal forests are certainly not new to the literature, but most attention has been devoted to mangrove wetlands in which tidal flooding, salinity, and nutrient gradients interact in complex ways. This literature has so dominated tidal forested wetland research that the terms "tidal" and "salinity" have become tightly linked, and some have even suggested that any mention of "tidal" and "freshwater" together is contradictory. We submit that this is not the case either etymologically or in natural systems. These wetlands – which make up at least 200,000 ha in the southeastern United States (Field et al. 1991) – can be periodically exposed to salinity either by large tides, droughts, or storm events, but their general function and species assemblage are those of a freshwater ecosystem.

In the first part of this book (Chapters 1–3), we describe land-use history in the southeastern United States that has led to the restricted distribution of tidal freshwater forested wetlands. We then describe what is known about the function of these systems by dedicated chapters on hydrology, soils, and biogeochemistry. The second part of this book (Chapters 4–13) describes specific tidal freshwater forested wetlands as case studies that detail specific hydrologic cycles, salinity regimes, floristics, disturbances, and discoveries made in particular tidal forests. Tidal freshwater forested wetlands can occur just about anywhere a river meets the sea, but they are more likely to occur where larger tidal ranges persist (Figure P.1). This section omits many tidal freshwater forested wetlands for which little information exists, but generalities can be made from focal systems. Specifically, we present information from a broad range of sites in the southeastern United States, including the Nanticoke River (Delaware/Maryland), Pocomoke River (Maryland), Lower Chesapeake Bay subestuary (Virginia), Cape Fear River (North Carolina), Waccamaw River (South Carolina), Savannah River (South Carolina/Georgia), Waccasassa Bay (Florida), Suwannee River (Florida), Escatawpa River (Mississippi), Lower Mississippi River Deltaic Plain (Louisiana), and Maurepas Swamp (Louisiana). In the final part of this book (Chapters 14–17), we present some of the research that has been conducted on restoration, and in assessing past

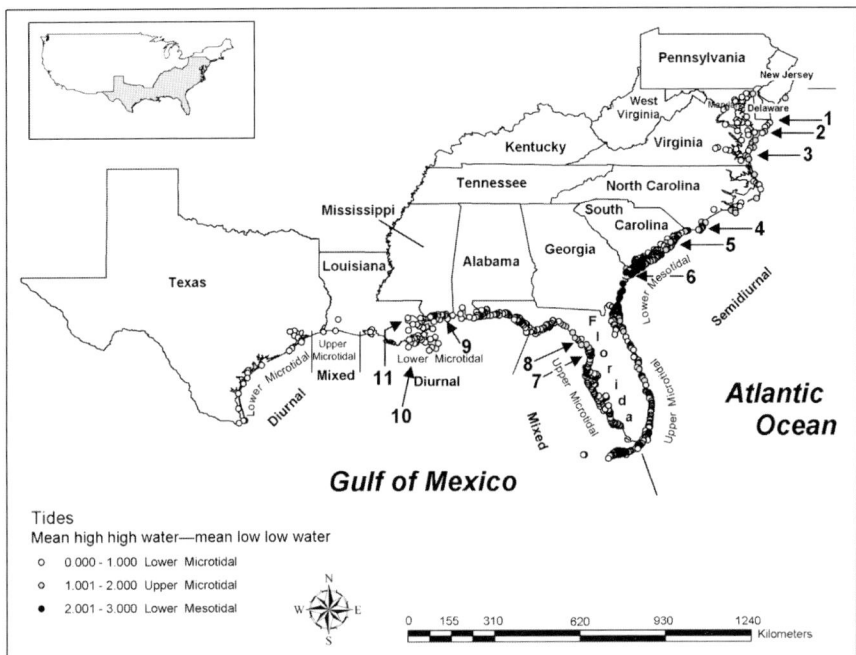

Fig. P.1. Map of the southeastern United States showing the location of tidal freshwater wetlands discussed in this book. 1=Nanticoke River (Delaware/Maryland), 2 =Pocomoke River (Maryland), 3=Lower Chesapeake Bay subestuary (Virginia), 4=Cape Fear River (North Carolina), 5=Waccamaw River (South Carolina), 6=Savannah River (South Carolina/Georgia), 7=Waccasassa Bay (Florida), 8=Suwannee River (Florida), 9=Escatawpa River (Mississippi), 10=Lower Mississippi River (Louisiana), and 11=Maurepas Swamp (Louisiana).

effects of climate change on tidal freshwater forested wetlands. We end with our thoughts on what some of the future research topics should include, and we highlight the importance of collaborative research and interactions with land managers in this endeavor.

We are very appreciative of the many peer and editorial reviewers who contributed to ensuring that these chapters were scientifically accurate and informative to a broad range of professionals. These reviewers include James A. Allen (Northern Arizona University), Christopher J. Anderson (Auburn University), Andrew H. Baldwin (University of Maryland), Mark M. Brinson (East Carolina University), Jim L. Chambers (Louisiana State University), Jarita Davis (IAP World Services, Inc.), John W. Day, Jr. (Louisiana State University), Richard H. Day (U.S. Geological Survey),

Diane DeSteven (USDA Forest Service), Jamie Duberstein (Clemson University), Rebecca S. Effler (University of Georgia), Katherine C. Ewel (USDA Forest Service, retired), Emile S. Gardiner (USDA Forest Service), Graeme B. Lockaby (Auburn University), Courtney Hackney (University of North Carolina at Wilmington), Bobby D. Keeland (U.S. Geological Survey), Richard Keim (Louisiana State University), Cheryl Kelley (University of Missouri-Columbia), James O. Luken (Coastal Carolina University), Thomas McGinnis II (IAP World Services, Inc.), Beth A. Middleton (U.S. Geological Survey), William J. Mitsch (Ohio State University), Leonard G. Pearlstine (University of Florida), Brian C. Perez (U.S. Geological Survey), Jack Putz (University of Florida), Gary P. Shaffer (Southeastern Louisiana University), Rebecca R. Sharitz (University of Georgia), Erik Shilling (National Council for Air and Stream Improvement, Inc.), Thomas J. Smith III (U.S. Geological Survey), David W. Stahle (University of Arkansas), Michael Stine (Louisiana State University), Robert R. Twilley (Louisiana State University), Beth A. Vairin (U.S. Geological Survey), Jos T.A. Verhoeven (Utrecht University), Michael G. Waldon (U.S. Fish and Wildlife Service), and Dennis F. Whigham (Smithsonian Environmental Research Center).

William H. Conner, Thomas W. Doyle, and Ken W. Krauss

Reference

Field DW, Reyer A, Genovese P, Shearer B (1991) Coastal wetlands of the United States-An accounting of a valuable national resource. Strategic Assessment Branch, Ocean Assessments Division, Office of Oceanography and Marine Assessment, National Ocean Service, National Oceanic and Atmospheric Administration, Rockville

Chapter 1 - Tidal Freshwater Swamps of the Southeastern United States: Effects of Land Use, Hurricanes, Sea-level Rise, and Climate Change*

Thomas W. Doyle[1], Calvin P. O'Neil[1], Marcus P.V. Melder[2], Andrew S. From[2], and Monica M. Palta[3]

[1]*U.S. Geological Survey, National Wetlands Research Center, 700 Cajundome Blvd., Lafayette, LA 70506*
[2]*IAP World Services, Inc., 700 Cajundome Blvd., Lafayette, LA 70506*
[3]*Rutgers University, Department of Ecology, Evolution and Natural Resources, New Brunswick, NJ 08901*

1.1. Introduction

Tidal freshwater wetlands are found worldwide at the outlets of coastal rivers with low gradient and low topographic relief at or near sea level. In the United States, they commonly occur in the lower Coastal Plain ecoregion along the Atlantic coast and Gulf of Mexico coast stretching from Maryland to Texas (Odum 1988; Mitsch and Gosselink 2000). These wetlands include both marsh and forest cover at or above mean sea level within the local tide range but also receive sufficient freshwater flows to generally keep surface water salinities less than 0.5 parts per thousand (ppt) (Cowardin et al. 1979). During the normal high tide cycle, tidal freshwater systems experience increased stage with the potential for reverse riverflow and salinity incursions from tidal forcing extending many kilometers upstream from the coast. Tidal freshwater systems typically share unique characteristics and biota of both riverine and estuarine systems, acting as a transitional habitat that occupies a substantial area of the coastal zone. The dynamic nature of tides and riverflow that intermix at the estuary interface within the tidal freshwater zone have made classification of these systems somewhat problematic.

---

* The U.S. Government's right to retain a non-exclusive, royalty-free licence in and to any copyright is acknowledged.

The actual extent of tidal freshwater swamps within the southeastern United States or even within a given river system is technically difficult to determine and therefore not well known. Conservative estimates indicate that there are approximately 200,000 ha of tidal freshwater swamps along the coast of the southeastern United States (Field et al. 1991). These systems are important breeding and wintering grounds for migratory birds and coastal fisheries, and they account for millions of dollars in commercial goods and ecological services (Costanza et al. 1998). Several factors contribute to the lack of knowledge and research attention given to tidal freshwater swamps. First, human intervention has resulted in the wholesale clearing of these forests for agriculture and timber products and in the alteration of river hydrology for navigation, hydropower, and flood control. Second, the original characteristics of these systems have not often returned after human intervention. Forest regeneration is often halted or hampered in former backswamp settings where controlled impoundments are still maintained for agricultural and recreational purposes. Conditions needed for favorable seedling germination and growth may be compromised by altered hydrology, subsidence, or salinization. Reforestation of some of these lands may be even beyond the reach of artificial techniques (Chapter 16).

In this chapter, we introduce landscape scale perspectives and controlling factors that shape the general characteristics and ecological dynamics of tidal freshwater swamps found in the southeastern United States. By virtue of their proximity to the coast, tidal freshwater swamps have been severely impacted by human regulation and contamination of streamflow, sea-level rise, hurricanes, and climate change. Major rivers in the southeastern Coastal Plain with associated tidal freshwater swamps are identified, and their history and characteristics in relation to human intervention and climatic change, past and future, are discussed. Subsequent chapters in this volume elucidate the hydrology, soils, and ecology that distinguish these systems from other forested wetland ecosystems. Case study chapters provide a comprehensive collection of known research studies on tidal freshwater swamps within various states and coastal reaches of the southeastern United States. Concluding chapters discuss relevant policy and research issues of wetland restoration and climate change impacts on future conservation and management of tidal freshwater swamps.

1.2. Coastal plain river systems

Tidal freshwater forests exist at the coastal terminus where rivers are subjected to tides. The geomorphology and gradient of the river outlet at the estuarine interface determine how the mix of riverflow and tides couple to affect river stage and salinity upstream. Higher tides and higher river discharge will generally equate to larger areas of tidally influenced freshwater zones and habitat.

Drainage basin areas and river discharge rates for coastal rivers of the southeastern United States vary from nearly insignificant streams to the massive Mississippi River that drains more than a third of the continental United States and nourishes half the coastal wetlands in the lower 48 states (Benke and Cushing 2005). Five river systems have drainage basins exceeding 100,000 km^2 (Table 1.1): the Mississippi, Rio Grande, Brazos, Mobile, and Colorado Rivers, and all drain into the Gulf of Mexico. Except for some Texas rivers of the western Gulf of Mexico, where arid conditions, low precipitation, and dewatering for municipal and agricultural needs reduce riverflow, mean discharge is positively correlated with basin area (Figure 1.1). Comparatively, these western rivers in Texas have some of the largest drainage basins yet the lowest mean discharge rates, about half or less of the expected outflow by drainage size. In contrast, the Atchafalaya River basin is a distributary of the Mississippi River and receives a controlled allotment of nearly one third of the Mississippi River and Red River discharge. Forest cover in most river basins has been reduced for agricultural purposes, and this reduction has impacted surface runoff rates and sediment loads. The Mississippi River drainage basin has been entrained within levees to prevent both overbank flow and coupling with backswamp settings. Most of the Mississippi River freshwater and sediment runoff is regulated upstream or shunted into the Gulf of Mexico beyond the continental shelf with little interaction with the remnant tidal swamps of Louisiana's deltaic wetlands.

1.2.1. Upstream river alterations

Humans have extensively altered most river systems in the United States since the earliest colonial period. Harbor towns and rivers were the lifeblood of the early settlers for travel, trade, and subsistence. Small creeks and rivers were commonly dammed for various agriculture, sawmill, and grist mill needs. The modern industrial period of the last century has spurred the construction of dams, reservoirs, and levees for navigation, hydropower, and flood control on larger rivers.

Altering natural flow regimes has had profound impacts on the timing and pattern of streamflow that in turn has altered the movement of sediment, nutrients, and biota in downstream receiving ecosystems, including tidal freshwater swamps. Few rivers today have not been altered by impoundments of one kind or another. Approximately 90 hydroelectric dams

Table 1.1. Major coastal rivers of the southeastern United States and basin characteristics related to drainage size, flow, climate, and land use (Benke and Cushing 2005).

River System	Drainage Basin Area (km²)	Mean Flow (m³s⁻¹)	Mean Precipitation (cm)	Land-use Forest Cover
Gulf of Mexico Coast				
Rio Grande, TX	870,000	37	21	14
Nueces, TX	43,512	20	61	0
Guadalupe, TX	26,231	79	81	0
Colorado, TX	103,341	75	82	0
Brazos, TX	115,566	249	81	3
Trinity, TX	46,540	222	115	35
Neches, TX	25,929	179	136	65
Sabine, TX, LA	25,268	238	127	67
Atchafalaya, LA	8,345	5,178	153	34
Mississippi, LA	3,270,000	18,400	94	16
Pearl, LA, MS	21,999	373	142	58
Pascagoula, MS	24,599	432	156	66
Mobile, AL	111,369	1,914	128	68
Escambia, AL	10,936	196	164	72
Apalachicola, FL	50,688	759	128	55
Suwannee, FL	24,967	294	134	38
Atlantic Ocean Coast				
St. Johns, FL	22,539	222	131	45
Satilla, GA	9,143	65	126	72
Altamaha, GA	37,600	393	130	64
Ogeechee, GA	13,500	115	113	54
Savannah, GA, SC	27,414	319	114	65
Santee, SC	39,500	434	125	64
Pee Dee, SC	27,560	371	111	58
Cape Fear, NC	24,150	217	119	56
Roanoke, NC	25,326	232	108	68
James, VA	26,164	227	108	71
York, VA	6,892	45	108	73
Potomac, VA, MD	37,995	320	99	58
Susquehanna, MD	71,432	1,153	98	63

and 120 impoundments are located along Coastal Plain rivers in the mid-Atlantic states alone (Schneider et al. 1989). Regulated streamflow can result in drastic changes in the geomorphology of river floodplain systems, altering the magnitude, frequency, duration, timing, and sediment loads of floods that shape floodplain features and functions (Poff et al. 1997). Similarly, changes in the hydrologic conditions of rivers, particularly changes in the size and timing of annual maximum peak discharge, have been implicated in alterations of growth and regeneration patterns in riparian tree populations (Sharitz and Lee 1985; Kozlowski 2002).

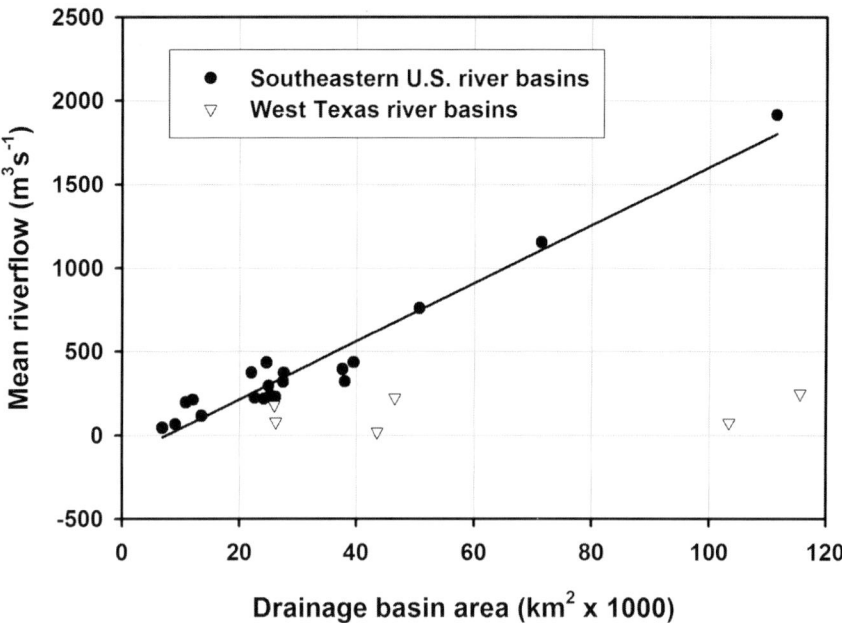

Fig. 1.1. Relationship of drainage basin area and mean discharge of major coastal rivers of the southeastern United States. Texas rivers are marked by open triangle symbol and show reduced riverflow in relation to basin area. Trend line is composed of all other non-Texan rivers listed in Table 1.1 excluding the Atchafalaya and Mississippi Rivers which have flow and basin areas beyond chart dimensions.

1.2.2. Water source and quality

Coastal Plain streams of the southeastern United States differ in water quality depending on whether they originate in the Coastal Plain or Piedmont region or are spring-fed rivers from surficial aquifers (Hupp 2000). The elemental and nutrient differences between rivers originating in the Piedmont or Coastal Plain are broadly defined by classifying these streams as redwater or blackwater. Redwater streams drain the mountainous regions of the Piedmont above the fall line and are generally sediment-laden, more alkaline, and nutrient rich. Blackwater streams, which are of low gradient and originate in the Coastal Plain, are distinguished by moderately acidic, low nutrient waters with high dissolved organic carbon content from humic soil pedogenesis. Spring-fed rivers of the Big Bend region of Florida are unique from black or redwater streams in that they are sterile, except for high carbonate concentrations caused by dissolved limestone from lateral flow through shallow surficial aquifers. Spring-fed rivers also demonstrate more uniform daily and seasonal flow than do alluvial streams, which are subject to flash floods.

1.3. Tidal forcing and hydrology

1.3.1. Tidal pattern

Tidal freshwater swamps are unique in their hydrologic and salinity characteristics. They are flooded and drained regularly by freshwater overflow attributed to local high tides, but they are also prone to saltwater influx during low flow periods (during drought) or from high storm tides (usually attributed to hurricanes). Normal astronomical forces combined with local meteorological conditions act to control river stage and salinities across the tidal freshwater zone more so than does freshwater outflow. Moreover, tides behave differently in different oceanic basins and coastal embayments. Tide patterns vary across the coastal zone of the southeastern United States from diurnal, semidiurnal, or mixed, and are of different amplitudes and ranges (Figure 1.2). The western Gulf Coast from Texas to the Florida panhandle exhibits a diurnal tide pattern of one high and one low tide each day, except for a mixed tide zone at the Texas-Louisiana border. The eastern Gulf Coast of the Florida peninsula exhibits a mixed tide pattern of two unequal high and low tides per day in contrast to the semidiurnal tides of the Atlantic coast which produce two similar high and low tides each day. Tide types mostly affect the frequency pattern of daily and seasonal stages and the probability or distance upstream of salinity exposure.

1.3.2. Diurnal range

The diurnal tide range is typically defined as the difference between the highest tide (mhhw) and lowest tide (mllw). The highest range of tides within the southeast coastal region is found from southern North Carolina to northeast Florida, achieving mesotidal (2-4 m) classification. Tides in the western Gulf of Mexico from Apalachicola Bay in Florida to the southern tip of Texas are mostly lower-microtidal (< 1m) in diurnal range. Tide range for most of the Florida peninsula and mid-Atlantic states is intermediate and classified as upper-microtidal (1-2 m) (Figure 1.2). Meteorological conditions can add or subtract from the predicted astronomical tide, particularly in microtidal and low gradient settings. The higher the

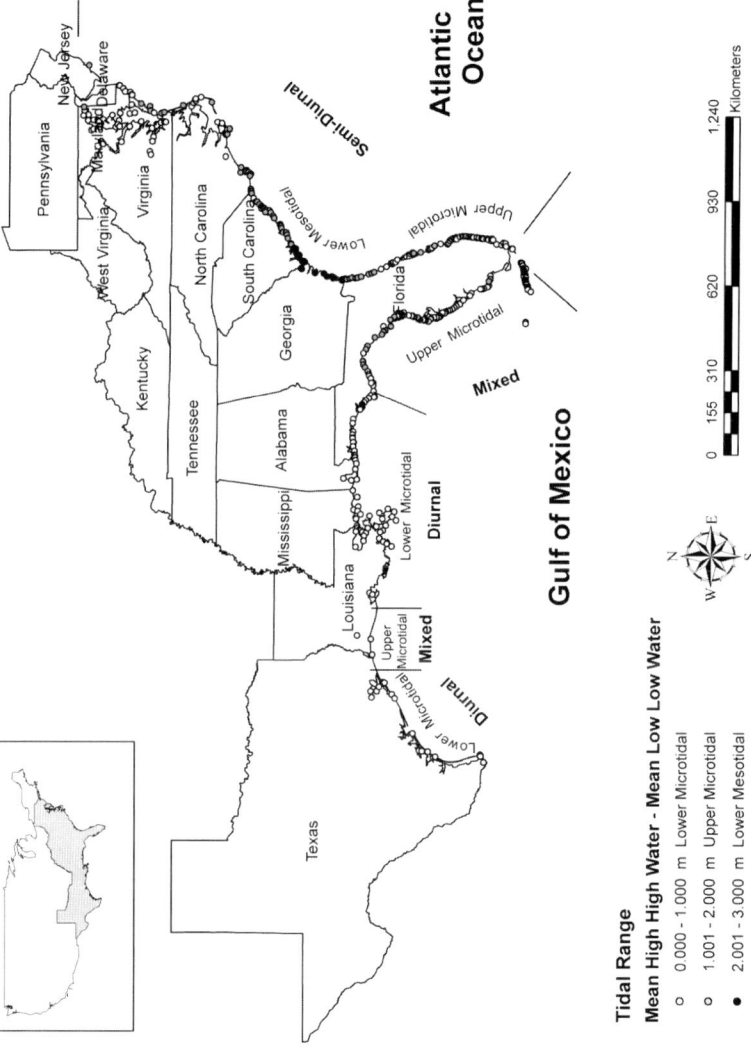

Fig. 1.2. Distribution of National Oceanic and Atmospheric Administration (NOAA) tide gages and classification of tide ranges and patterns for the southeastern coastal region and Gulf of Mexico coasts from Texas to Maryland.

tide range, in this case mesotidal, the higher the flood stage and the further inland saltwater intrusion can be expected for equal gradient and riverflow.

1.3.3. Seasonality

Tides also have a seasonal pattern that varies by oceanic basin and with meteorological conditions; but more importantly, seasonal patterns in tides may interact with seasonal patterns in riverflow to raise or lower stage and salinity at critical periods of forest growth. Seasonal patterns of tide and river stage in coastal rivers of the Southeast generally reflect low riverflow and above normal tides in the summer and fall, contrasted with higher riverflow and lower tides in winter and spring (Figure 1.3). This seasonal effect of inversely correlated streamflow and tide stage provides more freshwater flow when trees are mostly dormant and greater saltwater exchange during the active growing season. The seasonally driven effects of salinity on tree growth may play a greater role in controlling the growth and succession of tidal freshwater forests than what has been documented to date.

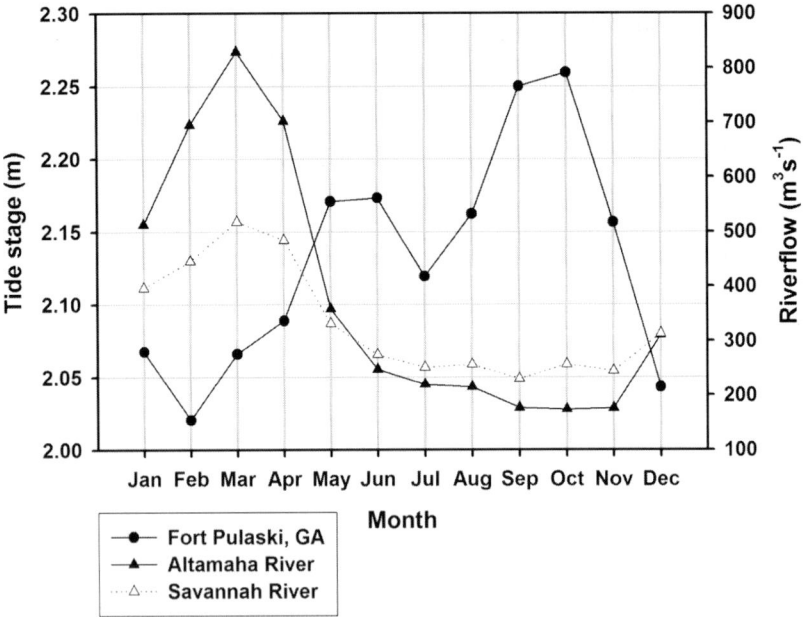

Fig. 1.3. Seasonal pattern of mean monthly tide levels (mhhw) at Fort Pulaski, Georgia and riverflow of the Savannah (regulated) and Altamaha (unregulated) Rivers in coastal Georgia.

1.4. Inventory of tidal freshwater forests

The extent of tidal freshwater hydrology and habitat is related to both the tidal range and to the geomorphology and gradient of the coastal reach. While coastal Louisiana has the lowest tidal range for the region, it also has low-gradient deltaic wetlands that account for its expansive coastal marsh complex of salt and freshwater wetlands. The Everglades of Florida similarly have a low-gradient shelf of coastal wetlands that support the largest and most contiguous intertidal mangrove ecosystem found anywhere. The actual criteria that are used to distinguish tidal freshwater forest and marshes vary with context and application. Currently, there are no reliable estimates of the real extent or potential natural vegetation zone of tidal freshwater swamps because of a lack of research and established criteria for a system that is both hydrologically complex and highly altered.

Field et al. (1991) conducted a nationwide coastal county survey of the U.S. Fish and Wildlife's National Wetlands Inventory to delineate acreages of wetland habitats, which included a category for tidal freshwater forests and marshes. Their work accounted for over 200,000 ha of tidal freshwater forests from Maryland to Texas (Figure 1.4). South Carolina has the largest distribution of tidal forests with over 40,000 ha. These

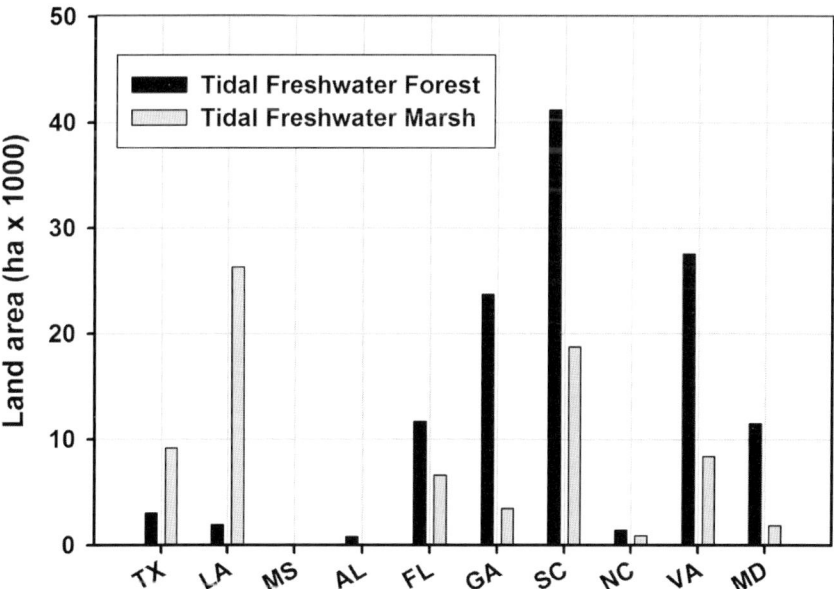

Fig. 1.4. Land area of tidal freshwater swamps by state in the southeastern United States based on the National Wetlands Inventory (Field et al. 1991).

estimates are conservative in terms of potential total habitat acreage because of the restricted definition applied to tidal freshwater habitats, the limited coastal county inventory, and the cumulative loss of tidal swamps to agriculture, logging, and sea-level rise since European settlement.

1.4.1. Land-use and forest history

Seaports and rivers along the coast were the highways of travel and trade of early colonists from the 1600s to mid-1800s before roads and rail eventually replaced steamboats (Rodgers 1970). Nearby tidal freshwater swamps provided the early capital for construction, trade, fuelwood, and agriculture. Early land grants were deeded along river corridors as a practical necessity for access and transportation of goods. Forest clearing on a local scale was necessary to establish a homestead and to grow crops. The plantation and mechanized eras were characterized by successive assaults on tidal freshwater swamps for agricultural and timber production; these activities effectively reduced the expanse of native forests and potential recovery of converted swamplands. The fate of tidal freshwater swamps has varied slightly among southeastern states, but nowhere have these swamps been more extensively altered than in South Carolina and Louisiana. Many of the case study chapters in this book include land-use history details that are specific to different areas, but all echo a similar exploitation that exhibits the resourcefulness of humans in reaping the rich, natural resources of tidal swamp systems.

According to historic plat surveys and aerial mapping of abandoned rice fields, it is estimated that more than 40,000 ha of tidal swamps were in rice cultivation at one time or another in coastal South Carolina; much of this land remains in marsh or pond today (Doar 1936; Wilms 1972; Gresham and Hook 1982; Smith 1985). An extensive patchwork of remnant rice-fields can be readily observed in coastal floodplains that were once cleared of tidal freshwater forests and are currently in various stages of marsh and forest regrowth (Figure 1.5). Swamp clearings for rice and indigo cultivation were selectively initiated in the head waters of tributaries and creeks during the early colonial period to create rainwater reservoirs and to avoid tidal inundation (Doar 1936). Impoundments were constructed in the upper reaches to hold water for irrigating the bottoms as crop development and climate dictated. Extended droughts interrupted the supply of rainwater and runoff that replenished reservoirs and prompted planters to exploit the abundant supply of freshwater along tidal rivers. The discovery of Carolina Gold rice and its preference for tidal waters under a dike and trunk delivery system driven by mesotidal conditions gave way to wholesale clearing

Fig. 1.5. Aerial photograph of Santee River floodplain in coastal South Carolina showing the patchwork and vast acreage of abandoned rice impoundments in the tidal freshwater zone. Darker ricefields are burned, moist, soil impoundments maintained for waterfowl use and hunting.

of tidal freshwater swamps for plantation agriculture and commercial enterprise along the southeast coastal region (Doar 1936; Porcher 1995). The higher tide ranges near or above 2 m provided a hydraulic head sufficient to support a passive irrigation system of controlled freshwater exchange into diked fields on both high and low tides. The success of plantation rice culture in Georgia and the Carolina colonies over many decades reduced the expanse of tidal swamps considerably, much of which has been retained in marsh cover as moist soil impoundments for wildlife and waterfowl management (Gresham and Hook 1982).

1.4.2. Logging practices of the Mississippi River delta region

The extensive tidal freshwater forests of Louisiana's delta region were exploited despite the soft and saturated organic substrate unfit for oxen and wagon. In 1850, Congress granted over 4 million ha of swampland unfit for cultivation to the state of Louisiana, much of this land was surveyed as cypress swamp and coastal wetland (Norgress 1947). Virgin baldcypress forests in tidal environments produced slow-growth, millennial-aged wood that was prized for its durability, stability, utility, and beauty. It was highly sought after even under the most adverse of swamp settings.

Prior to the invention of the steam engine in the mid 19th century, swamp trees were girdled at the base and left to die for a season before being felled and floated by river or bayou during spring high water. Eventually, railways and "pull boats" became the preferred means to harvest trees year round and sustain mill operations on a continual basis (Norgress 1947; Mancil 1969; Fetters 1990). In both Louisiana and South Carolina, thousands of kilometers of railway lines were constructed to gain access into swamp areas. The invention of the pull boat allowed deeper access into the swamps beyond natural waterways by dredging canals and skidding trees to the boat by overhead cable lines. Many logging companies in Louisiana maintained their own dredges to prevent delays in digging access canals (Davis 1975). Canals 3-12 m wide and 2-3 m deep resulted in partial drainage of many swamps (Mancil 1969, 1980). With the use of pull boats, felled trees could be reeled in from as far as 1,000 m from the canal through runs spaced in a fan-shaped pattern (Figure 1.6a). This skidding of timber across the swamp floor uprooted young saplings, disturbed the soil integrity for regrowth, and etched a permanent landscape scar in a distinctive wagon wheel-shaped pattern (Mancil 1980) that can still be seen today in aerial photographs of Louisiana and South Carolina swamps (Figure 1.6b). By 1934, pullboat and railway logging contributed to the clearcutting of nearly 650,000 ha of cypress swamps in Louisiana (Norgress 1947), an area equivalent to the size of Delaware. What proportion of this land was tidal swamp is unknown, but much of the cutover tidal forests have not regenerated and have regressed into freshwater marsh systems as a result of soil disturbance and relative sea-level rise.

1.4.3. Abandoned agriculture impoundments and dredged access canals

The low microtides of Louisiana and the advent of mechanical pumps prompted entrepreneurs of the early 20th century to drain tidal and nontidal swamps for agricultural enterprises (Okey 1918a, 1918b). Crop production in the absence of any major floods was generally good for a few years after initial drawdown but problematic in the long-term: rapid soil oxidation and subsidence brought on more sustained flooding and the need for continuous pumping at great expense and maintenance. Agricultural attempts of this kind were abandoned in tidal areas following the 1927 flood and sub-

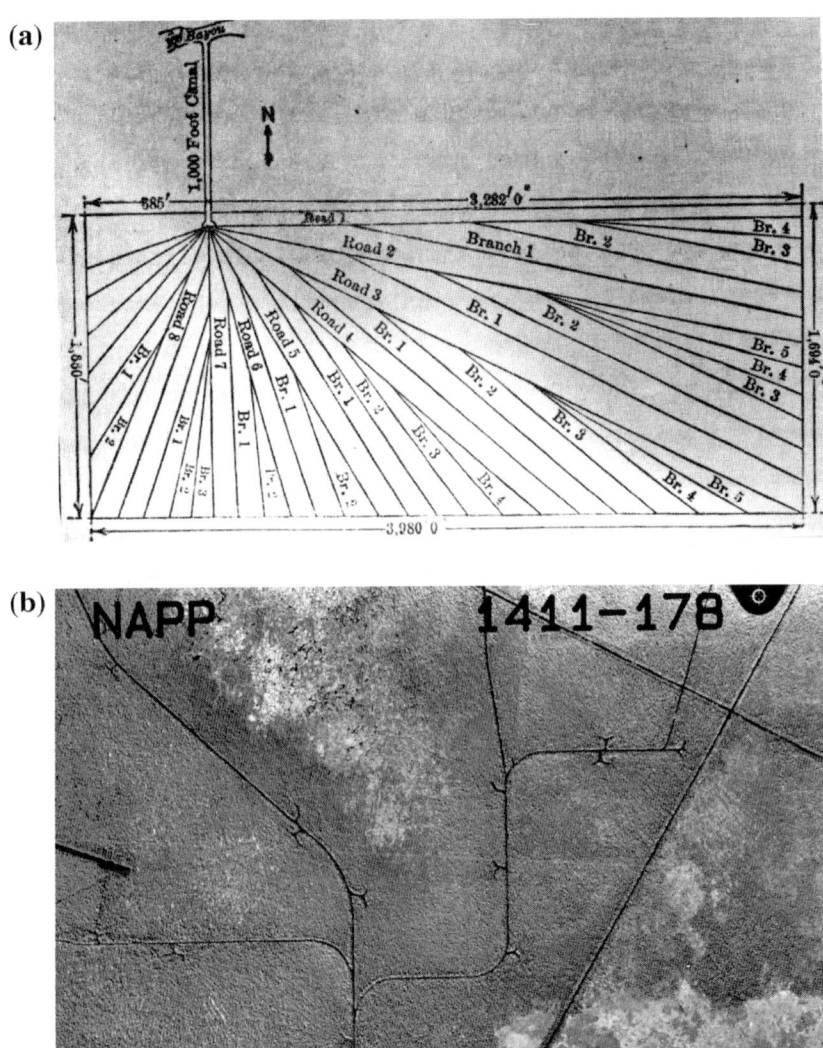

Fig. 1.6. (a) Schematic of pull boat logging runs and layout for cabling felled baldcypress to pull boat canal spur. (b) Aerial photograph of logged swamplands in the lower Atchafalaya River basin near Morgan City, Louisiana, where both pull boat and railway logging operations were employed in tidal freshwater swamps. Fan-shaped cable runs at docking spurs inflicted permanent scars on the landscape that are still visible after 100 years.

sequent floods over the decades that destroyed crops, infrastructure, levees, and the will to reinvest and replant (Trepangier et al. 1995). Following abandonment, a legacy of former tidal swamps and marsh that have reverted to shallow lakes and impoundments have been left behind. These remnants can be seen as distinctive rectangular features across the Louisiana coastal zone, such as "The Pen" near the town of Lafitte (Figure 1.7).

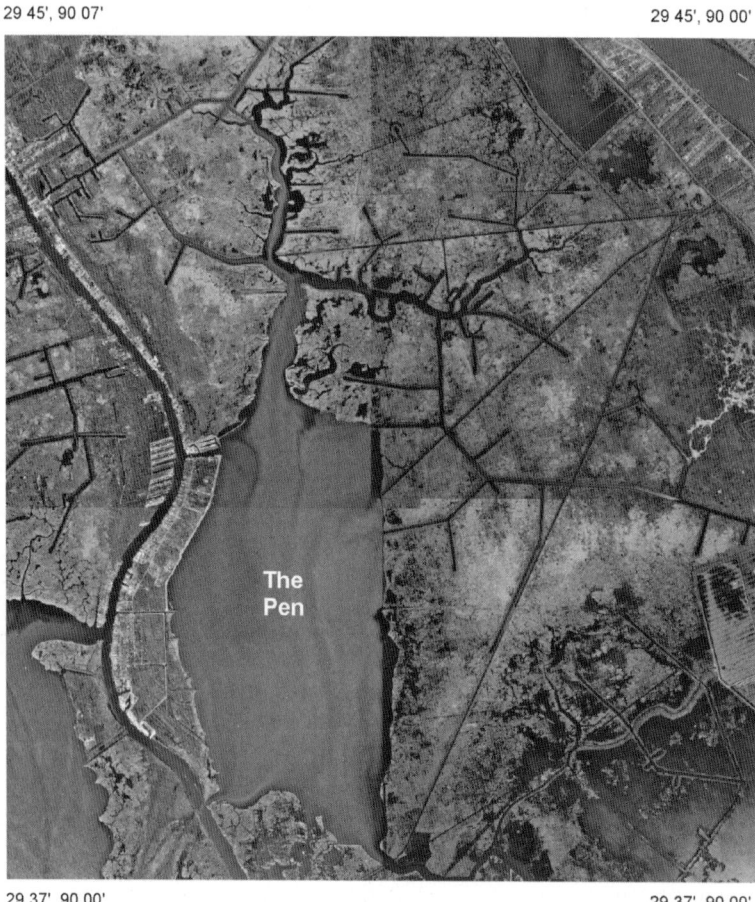

Fig. 1.7. Aerial view of "The Pen," an open water body, 1-2 m deep, in the center of the photo that was once a planned subdivision and agricultural truck farm. It was abandoned after the 1927 flood destroyed levees and reclaimed logged swampland on the eastern flank of the Barataria Pass and town of Lafitte, Louisiana. Other abandoned agricultural impoundments are visible in the lower right quadrant as distinctive rectangular features with remnant levee outlines are common in the Louisiana coastal zone.

Oil and gas access canals, shipping channels, and pipelines crisscross the coastal zone of Louisiana, intersecting saltwater and freshwater habitats. The orientation and depth of deep water shipping channels linking inland coastal communities for barge traffic and commerce have accelerated and exacerbated saltwater intrusion into freshwater zones (Swarzenski 2003). Straight canals of all kinds are efficient conduits for wind-generated tides and storm surges from tropical storms and hurricanes. Even during relatively mild frontal passages and normal drought periods, substantial saltwater pulses can penetrate far inland into tidal freshwater swamps that historically were isolated from saltwater effects. Tidal freshwater swamps on the flanks of shipping channels connected to the Gulf of Mexico exhibit signs of saltwater stress and forest dieback so severe that complete dieoff areas are known as "ghost forests." These former tidal baldcypress forests are composed of bare-barked tree snags of adequate density to deserve their own classification in the National Wetlands Inventory (NWI). Figure 1.8 shows a NWI habitat map of the Houma-Dulac area in Terrebonne Parish, Louisiana along the Houma Navigation Canal dredged in the 1960s and the surrounding tidal swamplands and ghost forests. Numerous studies have documented the general process and conditions that foster forest dieback from elevated saltwater concentrations (DeLaune et al. 1987; Conner and Day 1988; Conner and Brody 1989; Pezeshki et al. 1990; Conner 1993; Allen et al. 1994; Krauss et al. 2000).

1.5. Hurricanes

Forest systems of all kinds are subject to natural disturbances that shape their structure and biodiversity. Tidal freshwater swamps are vulnerable to recurring North Atlantic tropical storms and hurricanes that generate winds and surge high enough to snap or topple trees and deliver saltwater and elevated floodwater far inland. Field studies of coastal and inland forests after major Hurricanes Hugo (1989) and Andrew (1992) demonstrated that baldcypress-dominated tidal swamps were highly resistant to windthrow and crown damage relative to other hardwood species and forest types (Gresham et al. 1991; Hook et al. 1991; Doyle et al. 1995). Recent experiences with Hurricanes Katrina and Rita of the 2005 hurricane season across Louisiana affirmed the light to moderate damage to tidal swamps from high winds and the wide-scale saltwater overwash into freshwater zones from storm surge (Chapter 9; Doyle et al. in press). Porewater soil salinities in several tidal freshwater swamps commonly experience elevated salt concentrations following surge overwash of landfalling hurri-

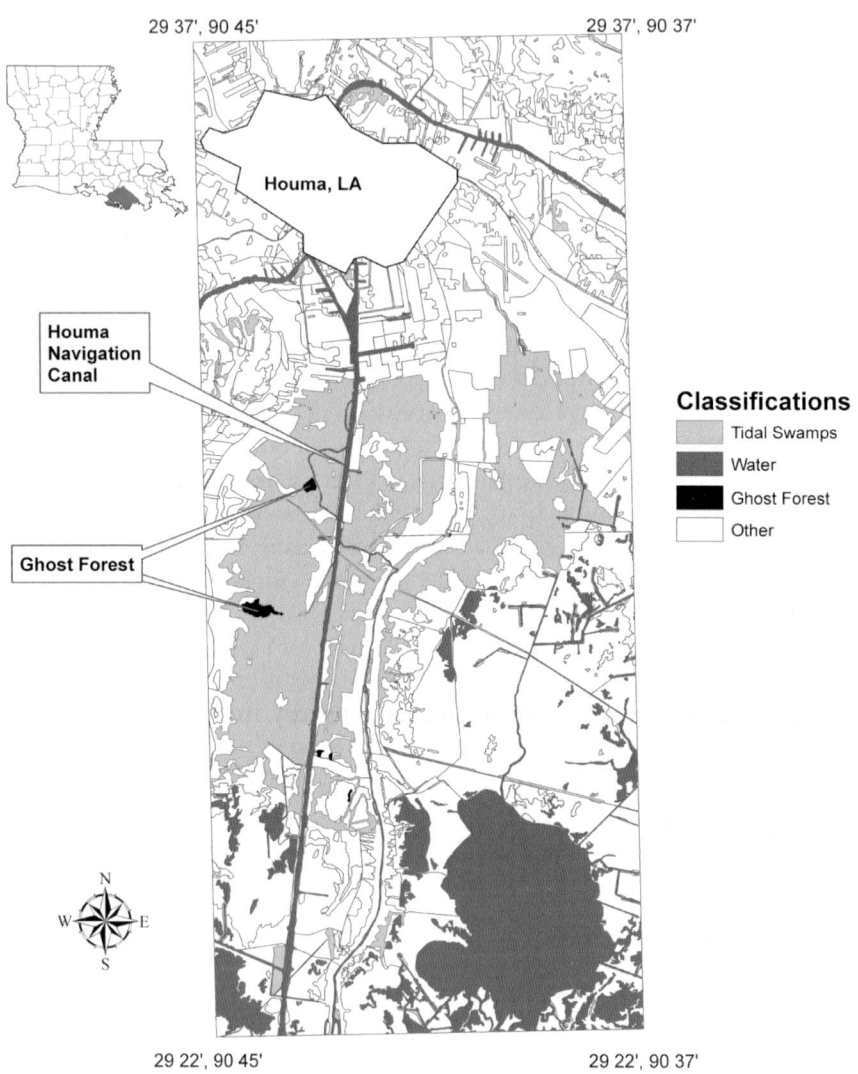

Fig. 1.8. Habitat map of the Houma-Dulac quadrangles in Terrebonne Parish, Louisiana along the Houma Navigation Canal dredged in the 1960s and the habitat designations including tidal swampland and ghost forest.

canes (Gardner et al. 1991; Gresham 1993; Rybczyk et al. 1995). Hurricanes can also bring heavy and sustained rain showers that increase coastal flooding from rainfall and storm tides, and increase flooding within the interior watersheds of coastal rivers from runoff as storms move overland.

Different coastal reaches of the southeastern United States have experienced higher or lower frequency of hurricane strikes potentially influencing the structure and composition of coastal forests depending on species' susceptibility to damage. Historically, the Atlantic coast of south Florida is known for its frequency of major hurricane strikes, followed by Louisiana and South Carolina (Simpson and Riehl 1981; Elsner and Kara 1999). The annual recurrence interval for tropical storms and hurricanes describes the average number of years between hurricane strikes calibrated for a given period of record. Recurring storm strikes increase the potential for tree mortality and turnover of branch debris. Coastal areas with higher storm surge frequencies, however, may experience chronically elevated soil salinities over time, which increases the probability of forest stress and dieback. Onshore surges are associated with the right quadrant of hurricane circulation and may be higher or lower depending on the angle of landfall relative to the coast.

A hurricane simulation model, HURASIM (Doyle and Girod 1997; Doyle et al. 2003), was used to determine the recurrence interval of category 1 storms or greater over the hurricane history from 1851 to 2004 for select coastal communities along the southeastern coastal region and Gulf Coast. The model accounts for storm and site orientation with respect to predicting maximum windspeeds and surge potential. Results show that south Florida and the central Gulf Coast have the highest incidence of hurricane landfalls for the period of record, and the Big Bend region and Spring Coast on Florida's western coast had the lowest frequency of storms and surge events (Figure 1.9). Miami, Florida recorded the greatest number of surge events by virtue of the highest incidence of storm activity; about 1 hurricane every 4 years. The Capes of North Carolina jut seaward and captures a high incidence of storms traveling north despite the latitudinal decline in storm strength and frequency for Virginia and Maryland. Coastal reaches and rivers along the Atlantic coast may experience more frequent storm tides than predicted by virtue of offshore storms moving parallel to the shoreline on northerly tracks. The western Gulf locations in Texas exhibit a higher than average number of surge events for the moderate number of storm events that occurred in the last 150 years.

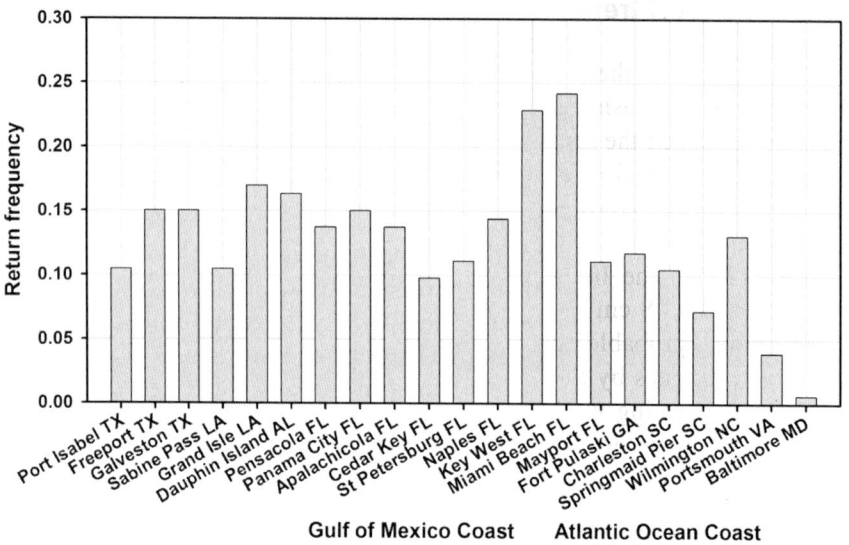

Fig. 1.9. Annual recurrence interval calculated for wind and surge events of hurricanes for select coastal communities along the southeastern coastal region and Gulf of Mexico coasts.

1.6. Climate change

Tidal freshwater swamps are expected to be among the most sensitive ecosystems to climate change and variability because of all the associated climate factors that impinge their growth and survival. Because of their physiographic position of occupying low lying coastal areas, tidal swamps are affected by streamflow, floods, droughts, tides, sea level, and storms. Climate change may be broadly defined as the pattern and range of climate variability for any processes or factors that change over time. More narrowly defined, climate change refers to historic and future changes in climate related to the consumption of fossil fuels since modern industrialization. Fossil fuel consumption has increased CO_2 concentrations and other "greenhouse" gas emissions, contributing to contemporary warming effects and trends. While the degree to which recent climate changes are attributed to natural or human causes is debatable, increasing sea level and tropical storm activity will directly and negatively impact tidal freshwater swamps of the southeastern United States. Tidal freshwater wetlands are readily impacted by acute and chronic exposure to even low levels of salinity. The combined stress of flooding and salinity may compound the threat in some systems, such that the margin for survival and compensation to changing climate is much less than for other coastal and upland habitats.

1.6.1. Sea-level trends

Coastal areas of the United States and worldwide are slowly being inundated by an increasing sea level. Warming of our global environment threatens to speed the rate of current sea-level rise and perhaps further amplify the detrimental effects of tropical storms, droughts, and record rainfall. Sea level has reportedly been rising since the last ice age (15,000 BP) and over the last century by as much as 1-2 mm/year (Douglas 1991, 1997; Gornitz 1995). The Intergovernmental Panel on Climate Change (2001) has projected a 48 cm rise in average global eustatic sea level by the year 2100 within a probable range of 9 - 88 cm (Watson et al. 1996). Other conservative estimates by the U.S. Environmental Protection Agency indicate that global warming will likely raise sea level by at least 42 cm by 2100 (4.2 mm/yr, Titus and Narayanan 1995). If realized, these moderate or "best estimate" projections will more than double the rate of sea-level rise over the past century. Thus, sea-level rise is expected to have a significant, sustained impact on future coastal evolution and tidal freshwater forests.

The process and data for calculating a historic rate of eustatic sea-level rise of 1-2 mm/year are derived largely by analyzing long-term tide records at select global stations. While the longest tide records from northern Europe exhibit slightly increasing sea levels during the 20th century (IPCC 2001), there is yet insufficient evidence of any significant acceleration trends related to greenhouse gas emissions (Douglas 1997). Satellite altimeter observations over the last decade have demonstrated the variability of sea-level changes associated with El Nina/Southern Oscillation and a short-term trend of rising sea level of 2.4 mm/year (Nerem et al. 1999). The historical local rate of relative sea-level rise is a process and product of several components, including eustatic change caused by deglaciation and thermal expansion, isostatic change from glacial rebound and tectonic uplift, and surficial subsidence caused by compaction and minerals/fluid extraction.

1.6.2. Geologic subsidence rates

Long-term tide gage records are among the most reliable measures of local and regional subsidence. However, tide records also include the long-term trend of eustatic sea-level change which over the last century has been estimated at 1.7-1.8 mm/yr on a global basis (Douglas 1991, 1997; IPCC 2001; Holgate and Woodworth 2004). Sea-level trends have been calculated by the National Oceanic and Atmospheric Administration from long-term tide gage records along the southeastern coastal region and

Atlantic and Gulf Coasts. These trends range from slightly accreting embayments of the Atchafalaya River delta in Louisiana and Apalachicola River delta in Florida to the rapidly subsiding deltaic region of the Mississippi River (Figure 1.10). Several studies have extracted subsidence rates from these and other tide gages within the Atlantic basin with some variability in rate estimates and methodology that mostly reaffirm regional patterns of generally high or low subsidence trends (Swanson and Thurlow 1973; Penland and Ramsey 1990; Paine 1993; Zervas 2001; Shingle and Dokka 2004). Gulf Coast wetlands, in particular, have shown high rates of land subsidence attributed to deltaic soil decomposition and compaction (Cahoon et al. 1998), dewatering effects (Burkett et al. 2002), deep fluid extraction (White and Morton 1997; Morton et al. 2001, 2002), and the lack of overbank floods and sediment deposition by leveeing the Mississippi River. Subsidence rates across a broad region like the Gulf Coast are highly variable on a local scale even within a representative coastal landform such as the Mississippi River Deltaic Plain or Chenier Plain. Some areas within the coastal zone of Louisiana have subsidence rates exceeding 20 mm/yr, demonstrating the potential range and variability within a subregion (Shingle and Dokka 2004).

1.6.3. Projected sea level under changing climate

Sea-surface temperatures are expected to keep pace with atmospheric warming over the next century and lead to warmer seas, thermal expansion, and thus, accelerated sea-level rise (IPCC 2001). There are many other complicated factors and circumstances beyond the scope of this chapter of glacier and polar ice melts that may ameliorate or exacerbate the rate of sea-level rise. Nevertheless, sea-level rise of any degree, large or small, will impact tidal freshwater swamps by increasing submergence and salinity over time. As sea level creeps upslope, tidal freshwater swamps at lower elevations near the estuary interface will experience salinity conditions that will compromise growth and survival and initiate forest dieback. Assuming global eustatic sea-level change will be nearly the same along the East and Gulf Coasts, the rates of local or regional subsidence will determine which forests will undergo stress and decline first and foremost. Worst-case scenarios of eustatic sea-level rise for representative coastal sites of high, low, and intermediate subsidence rates for the southeastern coastal region and Gulf Coast project potential sea-level change of 1-2 m by the year 2100 (Figure 1.11). These sea-level curves are composed of the historic monthly mean high water (mhw) records for select tide stations rectified to the North American Vertical Datum, 1988, and wrapped into

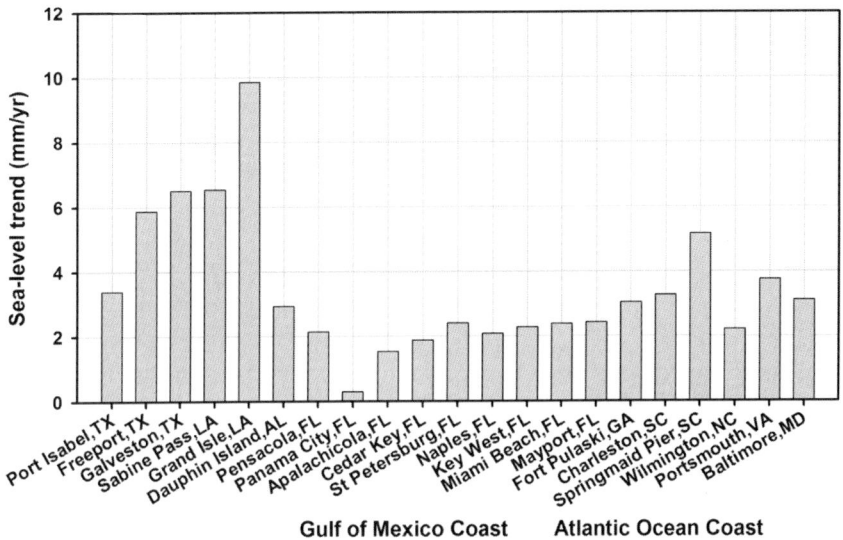

Fig. 1.10. Relative sea-level trends for representative NOAA tide gages based on long-term records for coastal cities across the Gulf of Mexico and south Atlantic Ocean coasts from Texas to Maryland.

the future with the addition of a eustatic sea-level projection based on the Goddard Fluid Dynamic Lab R15 climate model for the A1F1 emission scenario explained in the IPCC (2001) climate change report. By the year 2100, residents of Louisiana's delta region can expect sea level to rise by as much as 2 m. Coastal reaches with lower subsidence rates outside the Louisiana delta and Texas Chenier Plain can expect sea-level changes of over 1 m. Under these projections, tidal freshwater swamps will be severely threatened by the magnitude of tidal stage change and salinity increase that is likely to displace much of the tidal freshwater forest zone as we know it today.

1.7. Conclusion

Tidal freshwater swamps along the Gulf of Mexico and southeastern coastal region are unique ecosystems sharing characteristics and biota of both riverine and estuarine systems. These systems have been difficult to define and delineate because of complex daily and seasonal interactions of tide and riverflow. Their proximity to coastal waters, valued timber resources, and cultivation qualities and suitability for passive irrigation ac-

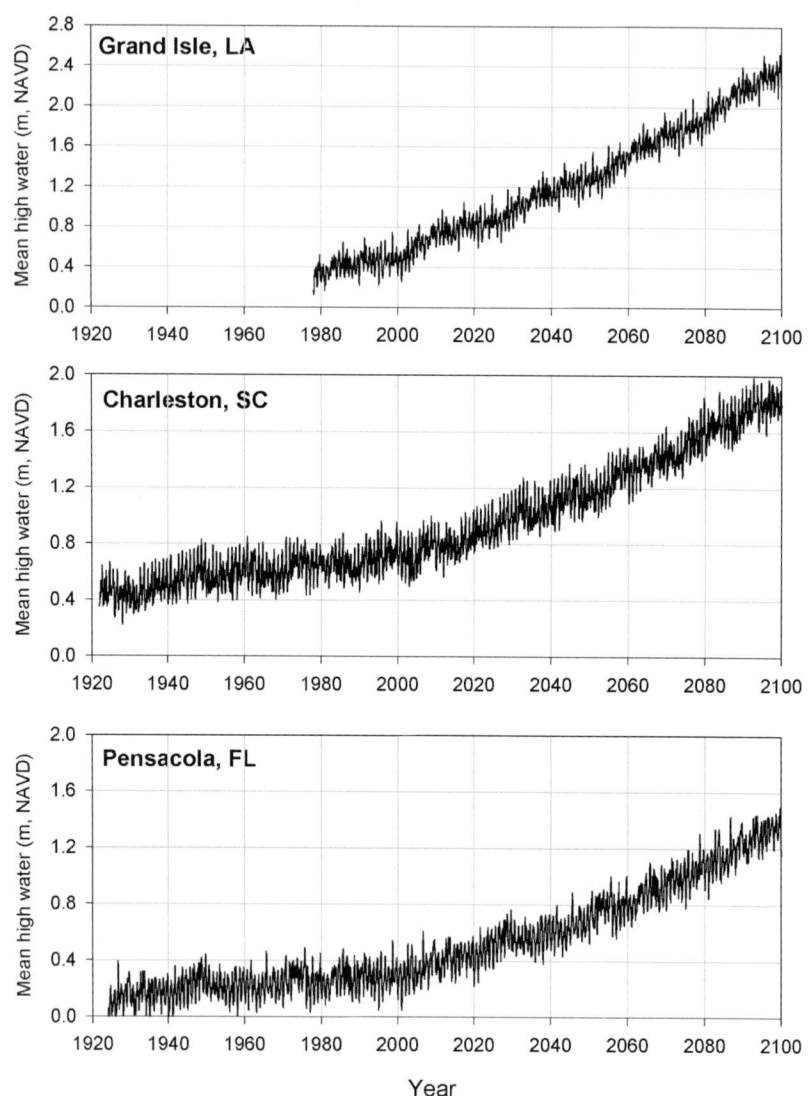

Fig. 1.11. Historic and projected sea-level rise for Gulf of Mexico coast and Atlantic Ocean Coast National Oceanic and Atmospheric Administration tide gage stations at Grand Isle, Louisiana, Pensacola, Florida, and Charleston, South Carolina based on mean monthly high water (mhw) records (North American Vertical Datum) with added curvilinear trend of increasing eustatic levels through the year 2100 under climate change. These stations are representative of high, low, and moderate subsidence rates and regions. Projected eustatic sea-level rise expected in the future under climate change was based on results from the Goddard Fluid Dynamic R15 climate model and A1F1 emission scenario (IPCC 2001).

counted for the success and wealth of early colonists, antebellum planters, and lumbermen alike. Human exploitation of these forests continued generation after generation to the point where the actual expanse of potential natural vegetation of this type is difficult to reconstruct. Remnant second and third growth forests comprise more than 200,000 ha of tidal freshwater swamps by conservative estimates for all states in the southeastern United States. Altered riverflow and vestiges of diked agricultural impoundments are problematic conditions that may limit regrowth or restoration of this forest type where warranted or needed. Reduced riverflow due to interbasin transfer, agricultural and municipal withdrawals, and dam operations, in concert with rising sea levels, have many conservationists concerned for the future of these forests under a changing climate. Tidal freshwater swamps of the Gulf and Atlantic regions persist under different hydrogeomorphic settings, tidal amplitudes, drought and hurricane frequencies, subsidence rates, and streamflow volume, which in part accounts for varying degrees of salinity exposure and dieback conditions on a local and regional basis. The frequency and severity of droughts and hurricanes are other major natural factors that influence the extent and concentration of saltwater distribution that contributes to forest dieback in coastal zones. Rapid subsidence rates in the Mississippi River delta of Louisiana have led to the drowning of distributary ridges and the subsequent demise of tidal swamp forests. Projected sea-level rise and changing climate are expected to accelerate the process and extent of saltwater intrusion, further impacting freshwater swamp habitats and restoration efforts in the absence of adaptive coastal management. In this chapter we have described the various physical elements and land-use history of coastal environments that control where tidal freshwater swamps may be found and function on the landscape. The chapters that follow will provide details of the hydrology, soils, and ecology of tidal freshwater swamps and identify which factors and relationships define their existence, uniqueness, and potential for restoration.

1.8. Acknowledgments

This material was based on work supported by the USGS Global Climate Change Program. The authors thank Drs. William Mitsch and John Day for their helpful comments on this manuscript.

References

Allen JA, Chambers JL, McKinney D (1994) Intra-specific variation in the response of *Taxodium distichum* seedlings to salinity. For Ecol Manage 70:203-214

Benke AC, Cushing CE (eds) (2005) Rivers of North America. Academic/Elsevier, Amsterdam/Boston

Burkett VB, Zilkowski DB, Hart DA (2002) Sea-level rise and subsidence: Implications for flooding in New Orleans, Louisiana. In: Prince KR, Galloway DL (eds) Subsidence interest group conference, proceedings of the technical meeting. USGS Water Resources Division Open-File Report Series 03-308. U.S. Geological Survey, Austin, pp 63-70

Cahoon DR, Day JW Jr, Reed D, Young R (1998) Global climate change and sea-level rise: Estimating the potential for submergence of coastal wetlands. In: Guntenspergen GR, Vairin BA (eds) Vulnerability of coastal wetlands in the southeastern United States: Climate change research results, 1992-97. Biological Science Report USGS/BRD/BSR-1998-0002. U.S. Geological Survey, Biological Resources Division, pp 19-32

Conner WH (1993) Artificial regeneration of baldcypress in three South Carolina forested wetland areas after Hurricane Hugo. In: Brissette JC (ed) Proceedings of the seventh biennial southern silvicultural research conference. General Technical Report SO-93. U.S. Department of Agriculture, Forest Service, Southern Forest Experiment Station, New Orleans, pp 185-188

Conner WH, Brody M (1989) Rising water levels and the future of southeastern Louisiana swamp forests. Estuaries 12(4):318-323

Conner WH, Day JW Jr (1988) Rising water levels in coastal Louisiana: implications for two forested wetland areas in Louisiana. J Coast Res 4:589-596

Costanza R, d'Arge R, de Groot R, Farber S, Grasso M, Hannon B, Limburg K, Naeem S, O'Neill RV, Paruelo J, Raskin RG, Sutton P, van den Belt M (1998) The value of the world's ecosystem services and natural capital. Ecol Econ 25:3-15

Cowardin LM, Carter V, Golet FC, LaRoe ET (1979) Classification of wetlands and deepwater habitats of the United States. U.S. Government Printing Office, Washington

Davis DW (1975) Logging canals: a distinctive pattern of the swamp landscape in south Louisiana. For People 25:14-17, 33-35

DeLaune RD, Pezeshki SR, Patrick WH Jr (1987) Response of coastal plants to increase in submergence and salinity. J Coast Res 3:535-546

Doar D (1936) Rice and rice planting in the South Carolina low country. Historic Contribution VIII. Charleston Museum, Charleston

Douglas BC (1991) Global sea level rise. J Geophys Res 96:6981–6992

Douglas BC (1997) Global sea rise: a redetermination. Surv Geo-phys 18:279-292

Doyle TW, Girod GF (1997) The frequency and intensity of Atlantic hurricanes and their influence on the structure of south Florida mangrove communities. In: Diaz H, Pulwarty R (eds) Hurricanes, climatic change and socioeconomic impacts: A current perspective. Westview Press, New York, pp 111-128

Doyle TW, Keeland BD, Gorham LE, Johnson DJ (1995) Structural impact of Hurricane Andrew on forested wetlands of the Atchafalaya Basin in coastal Louisiana. J Coast Res 18:354-364

Doyle TW, Girod GF, Books MA (2003) Modeling mangrove forest migration along the southwest coast of Florida under climate change. In: Ning ZH, Turner RE, Doyle T, Abdollahi K (eds) Integrated assessment of the climate change impacts on the Gulf coast region. GRCCC and LSU Graphic Services, Baton Rouge, pp 211-221

Doyle TW, Conner WH, Day RH, Krauss KW, Swarzenski CM (in press) Wind damage and salinity effects of Hurricanes Katrina and Rita (2005) on coastal baldcypress forests of Louisiana. In: Farris GS, Smith GJ, Crane MP, Demas CR, Robbins LL, Lavoie DL (eds) Science and the storms: The USGS response to the hurricanes of 2005. Hurricane Circular: Science and Storms. U.S. Geological Survey, Lafayette

Elsner JB, Kara AB (1999) Hurricanes of the North Atlantic. Oxford University Press, New York

Fetters T (1990) Logging railroads of South Carolina. Heimburger House Publishing Company, Forest Park

Field DW, Reyer A, Genovese P, Shearer B (1991) Coastal wetlands of the United States-an accounting of a valuable national resource. Strategic Assessment Branch, Ocean Assessments Division, Office of Oceanography and Marine Assessment, National Ocean Service, National Oceanic and Atmospheric Administration, Rockville

Gardner LR, Michener WK, Blood ER, Williams TM, Lipscomb DJ, Jefferson WH (1991) Ecological impact of hurricane Hugo – salinization of a coastal forest. J Coast Res SI 8:301-317

Gornitz V (1995) Sea-level rise: a review of recent past and near-future trends. Earth Surface Processes and Landforms 20:7-20

Gresham CA (1993) Changes in baldcypress-swamp tupelo wetland soil chemistry caused by Hurricane Hugo induced saltwater inundation.

In: Brissette JC (ed) Proceedings of the seventh biennial southern silvicultural research conference. General Technical Report SO-93. U.S. Department of Agriculture, Forest Service, Southern Research Station, New Orleans, pp 171-175

Gresham CA, Hook DD (1982) Rice fields of South Carolina: a resource inventory and management policy evaluation. Coast Zone Manage J 9:183-203

Gresham CA, Williams TM, Lipscomb DJ (1991) Hurricane Hugo wind damage to southeastern U.S. coastal forest tree species. Biotropica 23: 420-426

Holgate SJ, Woodworth PL (2004) Evidence for enhanced coastal sea level rise during the 1990s. Geophys Res Lett 31:L07305

Hook DD, Buford MA, Williams TM (1991) Impact of Hurricane Hugo on the South Carolina coastal plain forest. J Coast Res SI 8:291-300

Hupp CR (2000) Hydrology, geomorphology, and vegetation of Coastal Plain rivers in the southeastern USA. Hydrol Proc 14:2991-3010.

IPCC (Intergovernmental Panel on Climate Change) (2001) Climate change 2001: Impacts, adaptation, and vulnerability. McCarthy JJ, Canziani OF, Leary NA, et al. (eds) Contribution of Working Group II to the third assessment report of the Intergovernmental Panel on Climate Change. Cambridge University Press, New York

Kozlowski TT (2002) Physiological-ecological impacts of flooding on riparian forest ecosystems. Wetlands 22:550-561

Krauss KW, Chambers JL, Allen JA, Soileau DM Jr, DeBosier AS (2000) Growth and nutrition of baldcypress families planted under varying salinity regimes in Louisiana, USA. J Coast Res 16:153-163

Mancil E (1969) Some historical and geographical notes on the cypress lumbering industry in Louisiana. La Studies 8:14-25

Mancil E (1980) Pullboat logging. J For Hist 24:135-141

Mitsch WJ, Gosselink JG (2000) Wetlands, 3rd edn. John Wiley and Sons, New York

Morton RA, Purcell NA, Peterson RL (2001) Field evidence of subsidence and faulting induced by hydrocarbon production in coastal southeast Texas. Gulf Coast Assoc Geogr Soc Trans 51:239-248

Morton RA, Buster NA, Krohn MD (2002) Subsurface controls on historical subsidence rates and associated wetland loss in southcentral Louisiana. Gulf Coast Assoc Geogr Soc Trans 52:767-778

Nerem RS, Chambers DP, Leuliette EW, Mitchum GT, Giese BS (1999) Variations in global mean sea level associated with the 1997-98 ENSO event: Implications for measuring long term sea level change. Geophys Res Lett 26(19):3005-3008

Norgress RE (1947) History of the cypress lumber industry in Louisiana. La Hist Quart 30(3):46-61

Odum WE (1988) Comparative ecology of tidal freshwater and salt marshes. Ann Rev Ecol Syst 19:147-176

Okey CW (1918a) The wetlands of southern Louisiana and their drainage. Bulletin 6552. U.S. Department of Agriculture, Washington

Okey CW (1918b) The subsidence of muck and peat soils in southern Louisiana and Florida. Am Soc Civil Engineers 82:396-432

Paine JG (1993) Subsidence of the Texas coast-inferences from historical and late Pleistocene sea levels. Tectonophysics 222:445–458

Penland S, Ramsey K (1990) Relative sea-level rise in Louisiana and the Gulf of Mexico: 1908-1988. J Coast Res 6:323-342

Pezeshki SR, DeLaune RD, Patrick WH Jr (1990) Flooding and saltwater intrusion: potential effects on survival and productivity of wetland forests along the U.S. Gulf Coast. For Ecol Manage 33/34:287-301

Poff NL, Allan JD, Bain MB, Karr JR, Prestegaard KL, Richter BD, Sparks RE, Stromberg JC (1997) The natural flow regime. Bioscience 47:769-784

Porcher RD (1995) Wildflowers of the Carolina lowcountry and lower Pee Dee. University of South Carolina Press, Columbia

Rodgers GC Jr (1970) The history of Georgetown County, South Carolina. The University of South Carolina Press, Columbia (reprinted 2002 by The Reprint Company, Publishers, Spartanburg)

Rybczyk JM, Zhang X, Day J, Hesse I, Feagley S (1995) The impact of hurricane Andrew on tree mortality, litterfall, nutrient flux, and water quality in a Louisiana coastal swamp forest. J Coast Res 21:340-353

Schneider RL, Martin NE, Sharitz RR (1989) Impact of dam operations on hydrology and associated floodplain forests of southeastern rivers. In: Sharitz RR, Gibbons JW (eds) Freshwater wetlands and wildlife. U.S. Department of Energy, Office of Scientific and Technical Information, Oak Ridge, pp 1113-1122

Sharitz RR, Lee LC (1985) Limits on regeneration processes in southeastern riverine wetlands. In: Riparian ecosystems and their management: reconciling conflicting uses. U.S. Department of Agriculture, Forest Service, Rocky Mountain Forest and Range Experiment Station, Ft. Collins, pp 139-160

Shingle KD, Dokka RK (2004) Rates of vertical displacement at benchmarks in the Lower Mississippi valley and the northern Gulf Coast. NOAA Technical Report NOS/NGS 50. U.S. Department of Commerce

Simpson RH, Riehl H (1981) The hurricane and its impact. Louisiana State University Press, Baton Rouge

Smith, JF (1985) Slavery and rice culture in low country Georgia, 1750-1860. University of Tennessee Press, Knoxville

Swanson, RL, Thurlow CI (1973) Recent subsidence rates along the Texas and Louisiana coasts as determined from tide measurements. J Geophys Res 78:2665-2671

Swarzenski CM (2003) Surface-water hydrology of the Gulf Intracoastal Waterway in South-Central Louisiana, 1996-1999. Professional Paper 1672. U.S. Geological Survey, Denver

Titus JG, Narayanan VK (1995) The probability of sea level rise. EPA 230-R-95-008. U.S. Environmental Protection Agency, Office of Policy, Planning, and Evaluation (2122), Washington

Trepangier CM, Kogas MA, Turner RE (1995) Evaluation of wetland gain and loss of abandoned agricultural impoundments in South Louisiana, 1978-1988. Restor Ecol 3:299-303

Watson RT, Zinyowera MC, Moss RH (1996) Climate change 1995 - impacts, adaptations and mitigation of climate change: Scientific-technical analyses. Cambridge University Press, New York

White WA, Morton RA (1997) Wetland losses related to fault movement and hydrocarbon production, southeastern Texas coast. J Coast Res 13:1305-1320

Wilms DC (1972) The development of rice culture in 18th century Georgia. Southeast Geogr 12:45-57

Zervas C (2001) Sea level variations of the United States 1854-1999. NOAA Technical Report NOS CO-OPS 36. National Oceanic and Atmospheric Administration, Silver Spring

Chapter 2 - Hydrology of Tidal Freshwater Forested Wetlands of the Southeastern United States*

Richard H. Day[1], Thomas M. Williams[2], and Christopher M. Swarzenski[3]

[1]*U.S. Geological Survey, National Wetlands Research Center, 700 Cajundome Blvd., Lafayette, LA 70506*
[2]*Clemson University, Belle W. Baruch Institute of Coastal Ecology and Forest Science, PO Box 596, Georgetown, SC 29442*
[3]*U.S. Geological Survey, Louisiana Water Science Center, 3535 South Sherwood Forest Blvd., Suite 120, Baton Rouge, LA 70816*

2.1. Introduction

The downstream, coastward boundary of tidal freshwater forests is easy to define. There is usually an abrupt border from forest to fresh, brackish, or salt marsh as tidal effect and salinity increase. Upstream, however, there is a continuum within the forested community as salinity and tidal signature diminish. The boundary conditions can be defined by salinity ranges, species distributions, or hydraulic relations (Cowardin et al. 1979; Light et al. 2002; Hicks et al. 2000). Unfortunately, each of these criteria leads to a different spatial and functional delineation of the tidal freshwater zone. Odum et al. (1984) summarized the existing literature at that time on the ecology of tidal freshwater marshes, but little attention was given to analyzing the hydrology of the upper boundary of these systems, the forested edge.

We propose that tidal freshwater forested wetlands are coastal forested wetlands where the hydrology is affected by astronomical tides, and salinity does not normally exceed the tolerance of nonhalophytic tree species. This definition provides a comprehensive classification of tidal freshwater forests that considers aspects of hydrology, salinity, and biota. Hydrology of these sites is primarily dictated by local tides and river discharge, but it

---

* The U.S. Government's right to retain a non-exclusive, royalty-free licence in and to any copyright is acknowledged.

is also influenced by local rainfall, evapotranspiration, surface runoff, and subsurface drainage. Almost 70 years ago, Beaven and Oosting (1939) qualitatively described the hydrology of a tidal freshwater forest dominated by baldcypress (*Taxodium distichum* [L.] L.C. Rich.) and swamp tupelo (*Nyssa biflora* Walt.); in general, however, investigative research of tidal freshwater systems first focused on the herbaceous components of tidal freshwater marshes (Doumlele 1976, 1981; McCormick 1977; Whigham and Simpson 1977; Odum et al. 1978). Odum et al. (1984) included the forested edge as a part of the marsh, listing a baldcypress-tupelo community type as the most landward portion of a tidal freshwater marsh, forming an ecotonal boundary between the marsh itself and wooded swamp or upland forest at approximately the level of mean high water (average high tide). The hydrologic characteristics of tidal freshwater forested wetlands are directly related to the conditions which prevail at the upper extreme (laterally or upstream) of tidal freshwater marshes.

The hydrology of tidal freshwater forested wetlands has been studied even less than other aspects of this type of ecosystem. Very few of these wetlands have had a water-level recorder installed within the swamp to directly measure surface inundation rates and groundwater levels. More often, hydrology is inferred from a continuously recording gage in a nearby channel. In many cases, wetland hydrology data consist of manual measurements of water depth taken during site visits, resulting in individual water level data points that are weeks or months apart. Channel hydrology may provide a fairly accurate estimate of adjacent wetland hydrology in cases where the terrain is flat and/or the soil is relatively permeable (Zampella et al. 2001; Burt et al. 2002); however, areas away from the channel may be slow to drain and are usually inundated for longer periods than indicated by the stream gage.

Researchers have become increasingly aware of the importance of actually recording water levels in all types of wetlands instead of relying on estimates derived from gages in waterways adjacent to the studied wetland. Tidal freshwater systems are no exception. Chapters 4-13 of this book present specific tidal freshwater forested wetlands and include descriptions of the hydrologic characteristics of each site. Some authors present actual hydrographs from continuous recorders installed within the swamp; however, we use our own previously unpublished data to produce within-forest hydrographs in this chapter.

The hydrology of tidal baldcypress swamps and other riverine tidal forested communities are described in this chapter. We also discuss the hydrology of the coastal margins of groundwater-dominated forested wetlands which may transition directly into brackish or salt marsh. First, we introduce the forcing functions of freshwater input, tides, sea level, and salinity; these factors are described in the context of how they interact at the upper extreme of tidal influence near the level of mean high water. We rely heavily on hydrologic data from the national network of tide and stream gages available from the National Ocean Service, Center for Operational Oceanographic Products and Services (NOAA 2006a) and the U.S. Geological Survey, National Water Information System (USGS 2006). Second, we describe the hydrology of specific tidal freshwater forests as reported in the literature, supplemented by references to chapters in this book. The effects of elevation, microtopography, and wind on tidal inundation patterns are discussed. Following that is a discussion of the relation of hydrology to the biological and biogeochemical functions of tidal forests. Finally, we describe how changing hydrologic conditions might affect the sustainability of tidal freshwater forests.

2.2. Forcing functions

Freshwater input (rainfall, river flow, and ground water) and the physical forcing of the tide act together to control water levels in tidal freshwater wetlands. Sea level is the moving base that controls the elevation at which these two interact. The physical interaction between salt water and fresh water controls salinity, which is a major environmental factor influencing the biology and chemistry of tidal wetlands. The forcing functions that define the characteristics of tidal freshwater forests are the same that regulate all estuarine wetlands, from saline to fresh. In fact, the international classification system of the Ramsar Convention on Wetlands couples tidal freshwater swamp forests with mangrove forests into one functional group, "intertidal forested wetlands," under the classification "marine/coastal wetlands" (Ramsar Convention Bureau 1990). This discussion, however, will focus on the upper extreme of tidal influence and how the forcing functions interact there to affect tidal freshwater forested wetlands.

2.2.1. Freshwater input

Annual precipitation in coastal areas of the southeastern United States ranges from approximately 600 mm in south Texas to 1,600 mm on the northern rim of the Gulf of Mexico and to near 1,300 mm along the Atlantic coast (Muller and Grymes 1998). In south Texas, high evaporation and low river discharge limit freshwater availability, and there is very limited occurrence of swamp forest near the coast (Shew et al. 1981). From east

Texas eastward to the rest of the Gulf of Mexico and south Atlantic coasts, freshwater runoff is plentiful. High rainfall provides freshwater flow to sustain tidal forested wetlands in Barataria Basin in the Mississippi River deltaic plain of Louisiana (Figure 2.1a) which have been cut off from river flow by levees. Rainfall, along with its effect on ground water, also dominates the hydrology of freshwater forests in the karst limestone topography of Florida and the sandy soils of wetland communities directly behind coastal beaches along the entire coastline of the southeastern United States. In these cases, the tidal signature may only be apparent in groundwater fluctuations.

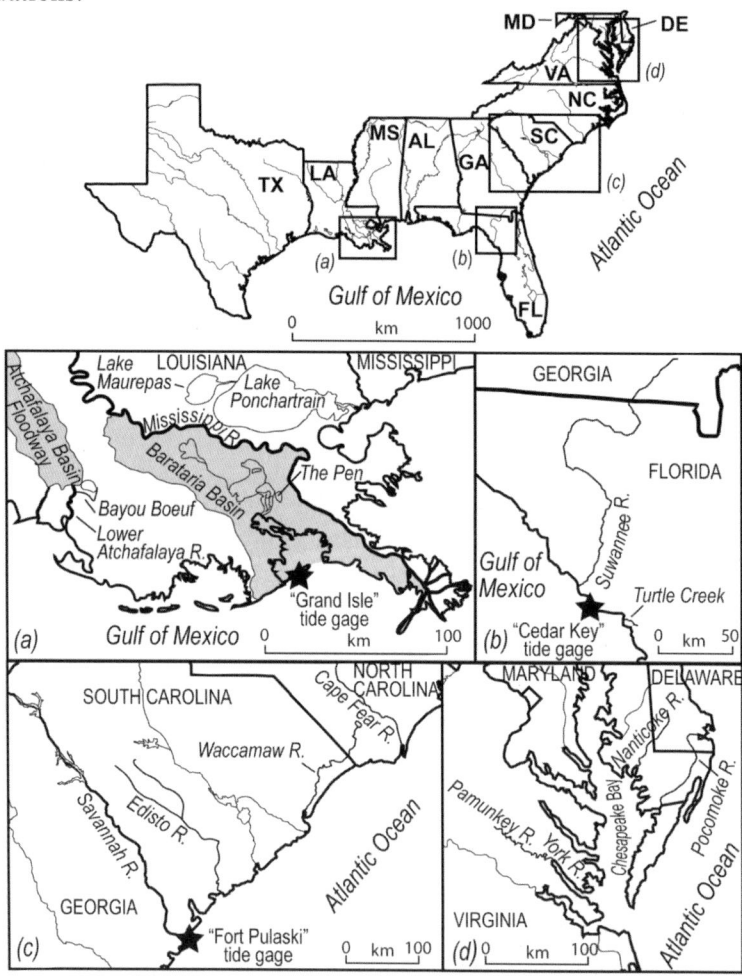

Fig. 2.1. Location of tidal freshwater forested wetland study sites referenced in the text. The Atchafalaya Basin Floodway and Barataria Basin are shaded gray. Stars refer to the location of tide gages.

River flow does not correlate to local rainfall (see Chapter 1). A river's discharge depends on rainfall and evapotranspiration over its entire drainage basin as well as on antecedent groundwater conditions. Poff et al. (1997) identified five critical components of river flow regime that regulate ecological processes in river ecosystems: the magnitude, frequency, duration, timing, and rate of change of hydrologic conditions. The interaction of these five river flow components with coastal tides regulates ecological processes in tidal freshwater wetlands. The Suwannee River in Florida and the Savannah River in South Carolina and Georgia (Figure 2.1b, c) have comparable drainage basin areas and mean annual precipitation, resulting in similar mean annual discharge rates (Table 2.1). In general, within a given climatic regime, rivers with similar drainage basin areas will have similar mean annual discharge rates. Similarity of climate also puts southeastern United States rivers in the same category with respect to seasonal pattern of river discharge; the month of October is the time of year of lowest average river discharge throughout the southeastern United States, although there is high variability from year to year for each river (USGS 2006). Human activities such as dredging, channelization, and damming of rivers all affect the interaction of coastal tides with river flow regime. The spatial distribution of freshwater forests in riverine tidal settings depends on the interaction of river discharge rate with tide range.

2.2.2. Tide

The timing and magnitude of tides are determined by a complex interaction of astronomy, bathymetry, and meteorology. Coastal water levels undergo regular daily, monthly, and annual fluctuations, all of which vary by region. Most mean tide ranges are larger along the Atlantic coast (1–2 m) than within the Gulf of Mexico (0.3–1.0 m) (Table 2.1, Figure 2.2). Barataria Basin has the lowest tide range of the three sites in Table 2.1, and although isolated from river flow, it has the highest rainfall. The tide range at the mouth of the Savannah River is more than twice that of the Suwannee River. Thus, given their similarity of climate and discharge, tide regime has more effect than does rainfall or river flow on differences in tidal swamp characteristics in these two river systems.

The Atlantic coast of the southeastern United States has semidiurnal tides, whereas the Gulf of Mexico has areas with diurnal tides and areas with mixed tides (Figures 2.2 and 2.3). Lunar cycling causes a repeating 28-day bimodal cycle of tide levels timed to the full and new moons (Figure 2.4). Prevailing winds as well as unpredictable storm winds that cause episodic surge events can affect water levels at any time. Tropical cyclones

Table 2.1. Hydrologic statistics for three areas of the southeastern United States where tidal freshwater forested wetlands occur. NA = not applicable.

System	Coastal tide station	Geographic coordinates of coastal tide station	Mean annual precipitation basinwide [mm][a]	River drainage basin area [km^2][a]	Mean annual river discharge [m^3/s][a]	Tides			
						Magnitude class	Frequency class	Mean tide range [m][b,c]	Mean diurnal range [m][b,d]
Barataria Basin[e]	Grand Isle, Louisiana	29° 15.8' N 89° 57.4' W	~1,500	NA	NA	Lower microtidal	Diurnal	0.32	0.32
Suwannee River	Cedar Key, Florida	29° 8.1' N 83° 1.9' W	1,340	24,967	294	Upper microtidal	Mixed	0.86	1.16
Savannah River	Fort Pulaski, Georgia	32° 2.0' N 80° 54.1' W	1,140	27,414	319	Lower mesotidal	Semidiurnal	2.11	2.29

[a] Data from Benke and Cushing (2005).
[b] Tide means are calculated for the 19-year period from 1983 through 2001 (data from NOAA 2006a).
[c] Mean tide range: mean high water minus mean low water.
[d] Mean diurnal range: mean higher high water minus mean lower low water for sites with two high tides and two low tides daily.
[e] Barataria Basin is a rain-fed system which is isolated by levees from riverine input but open to tidal influence from the coast.

Chapter 2 - Hydrology of Tidal Freshwater Forests   35

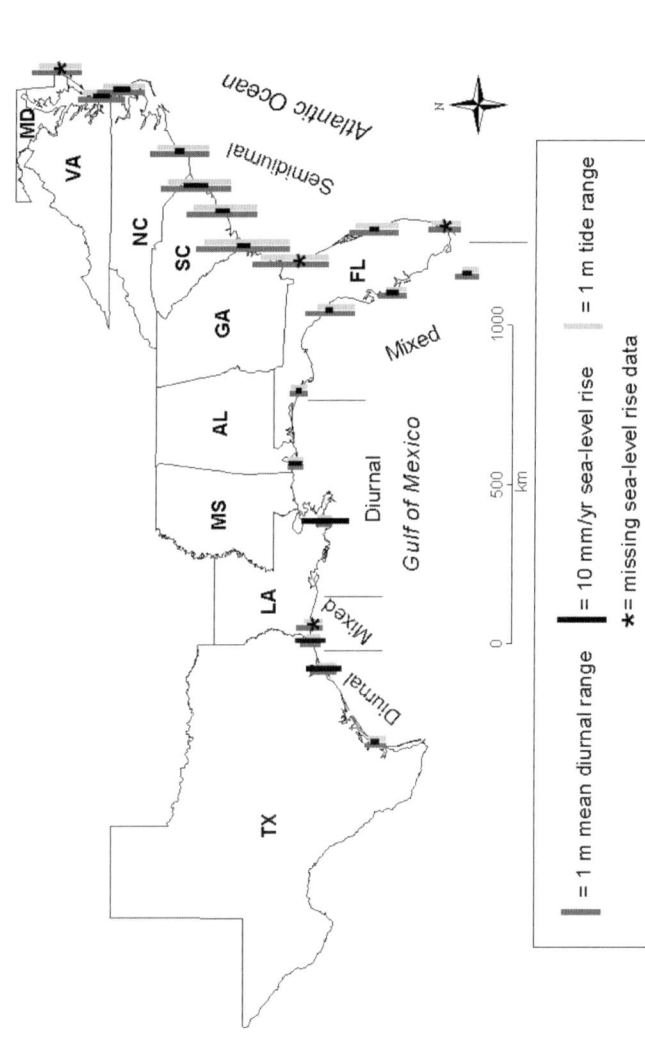

Fig. 2.2. Tide type, tide means, and annual sea-level rise for selected sites in the southeastern United States (data from NOAA 2006a). Annual sea-level rise is calculated from monthly means for stations with 25 years or more of data. Tide means are calculated for the 19-year period from 1983 through 2001. Mean diurnal range = mean higher high water minus mean lower low water (for sites with two high tides and two low tides daily). Mean tide range = mean level of all high tides minus mean level of all low tides. Semidiurnal = two high tides and two low tides daily, with approximately equal amplitudes. Diurnal = one high tide and one low tide daily. Mixed = relatively large inequality in heights of two high tides and/or two low tides daily (Hicks et al. 2000).

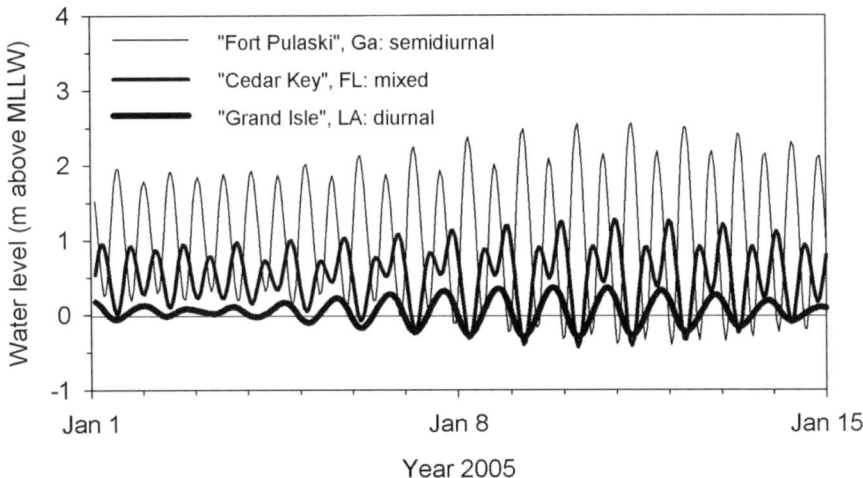

Fig. 2.3. Example of predicted tide levels at three locations that exhibit semidiurnal, mixed, and diurnal tides (data from NOAA 2006a). Predicted tides are calculated from the harmonic oscillations of coastal water levels caused by gravitational pull of the Sun and the Moon, dependent on the geomorphology of the coastline and ocean basins and disregarding local wind and river flow. MLLW = Mean Lower Low Water = mean level of the lower of two low tides daily.

may affect coastal water levels (Figure 2.4) locally or across the entire southeastern United States depending on size and pathway of each storm. In the microtidal environment of Louisiana, wind has a greater affect than do astronomical tides on coastal water levels (Baumann 1987), which is also true for some coastal areas bordering Chesapeake Bay in Maryland (NOAA 2006a, see Chapter 5).

At what point is a riverine freshwater wetland considered tidal? The National Ocean Service defines the point where the mean tide range is less than 6 cm as the head of the tide (Hicks et al. 2000). The depth of a river channel controls how far the tide travels up the channel. The tide continues upstream until the bottom elevation is near that of high tide. Upstream of the head of the tide, water level in the stream is related only to the flow of the stream and is not considered tidal. The head of the tide is not to be confused with the point of tide reversal; the point of tide reversal is the farthest point upstream from the coast at which the tidal current actually flows upstream (Hicks et al. 2000). Below the point of tide reversal, current within the main river channel reverses flow depending on whether the tide is flooding or ebbing. Between the head of the tide and the point of tide reversal, water in the channel continues to flow downstream even during a flood tide that is raising the water level. The concept of a tidal wave

Chapter 2 - Hydrology of Tidal Freshwater Forests 37

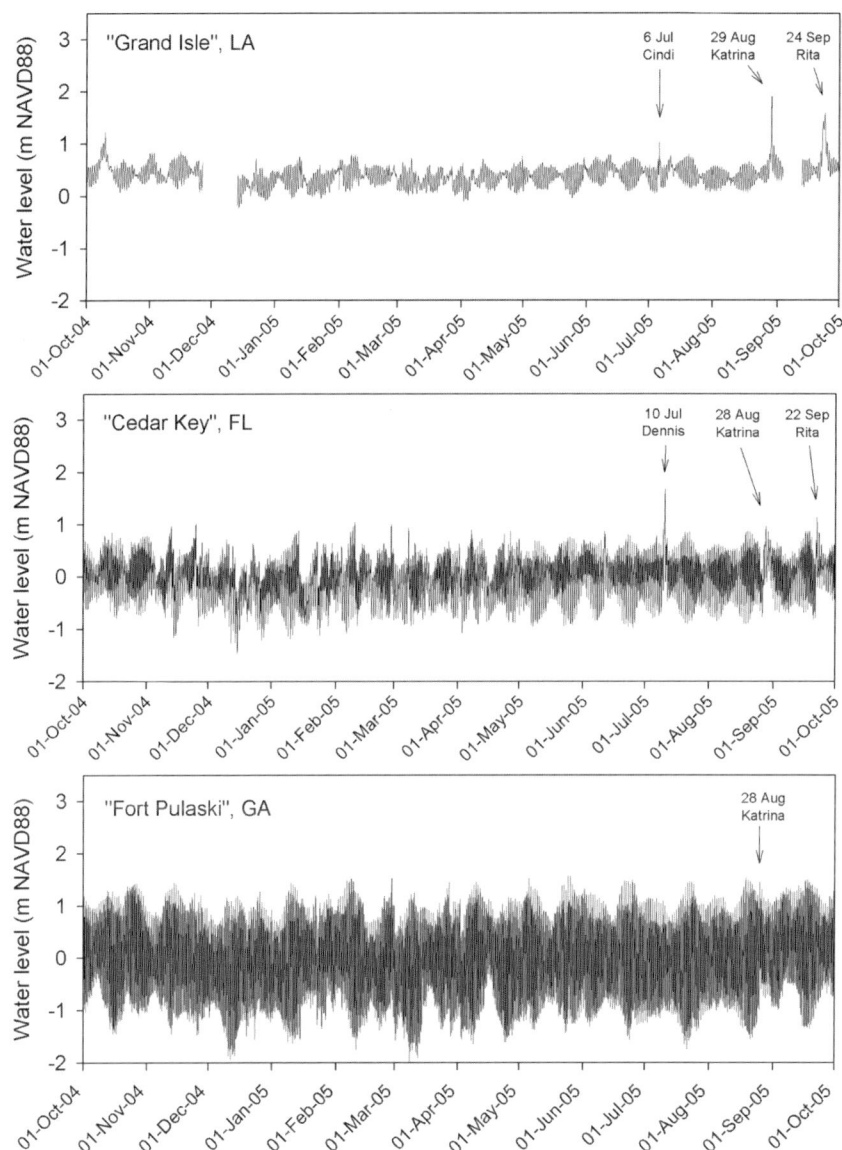

Fig. 2.4. Hourly observed tide levels for Water Year 2005 (01 Oct 2004–30 Sep 2005) (data from NOAA 2006a). The 28-day bimodal lunar cycle is evident from May through August at Grand Isle and Cedar Key. Arrows refer to high water levels caused by hurricanes. NAVD88 = North American Vertical Datum of 1988.

moving upstream against a current is not hard to grasp if one compares it to tossing a stone into a moving current: the resulting ripple radiates in all directions from the point that the stone enters the water, independent of the direction of flow of the current, although the faster the current downstream, the shorter the distance the ripple moves upstream.

The head of the tide and the point of flow reversal are not static points. They move up and down a stream channel in direct relation to river discharge and tide range. The Savannah River at river km 98 is upstream from all tidal influence (Figure 2.5). The tide range at river km 44 varies from ~0.0 to ~0.9 m during the period 2004–2006. There is a clear inverse relationship between river discharge and tide range at these two locations on the Savannah River, even though they are more than 50 km apart. Daily tide range decreases as river discharge increases. At high discharge, the force of water flowing downstream impedes the progression of the tidal wave upstream, which is characteristic of the tidal portion of all rivers (Henderson 1970; USACE 2001).

The head of the tide is farther upstream than tidal swamps occur on large rivers such as the Mississippi River. The hydrograph near the head of

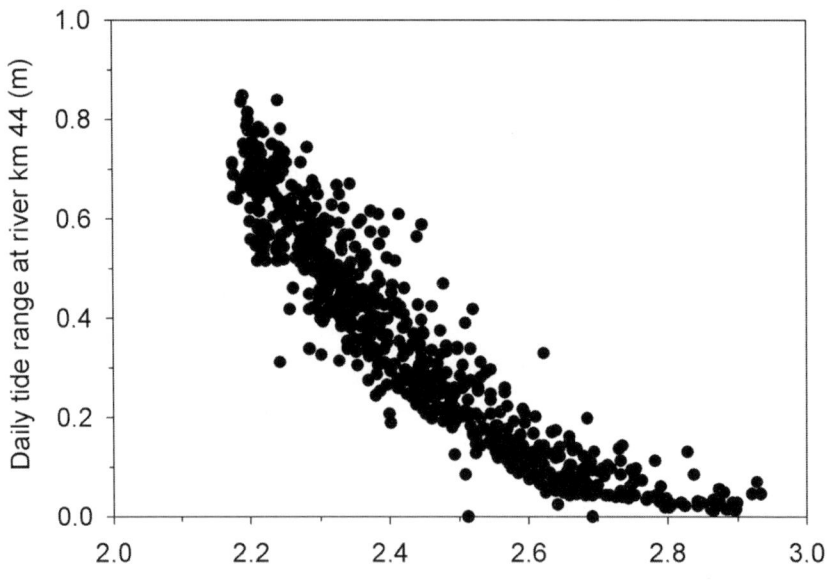

Fig. 2.5. Relationship within the Savannah River between daily tide range at river km 44 ("Above Hardeeville" gage) and logarithm of mean daily discharge at river km 98 ("Near Clyo" gage) for the period 01 Oct 2004–30 Sep 2006 (data from USGS 2006).

the tide on large rivers is characterized by a large range of water-level elevations. There is no tidal signature at the highest flows (USGS 2006). The low flow necessary for a tidal signature to become apparent exposes bare river bank below the elevation at which trees occur. Tree seedlings that become established in what could be considered the tide-influenced portion of the river bank usually die during a subsequent rise in the river level.

Tide range diminishes inland from the coast along rivers that narrow and become shallower upstream (Henderson 1970). Dredging of rivers may

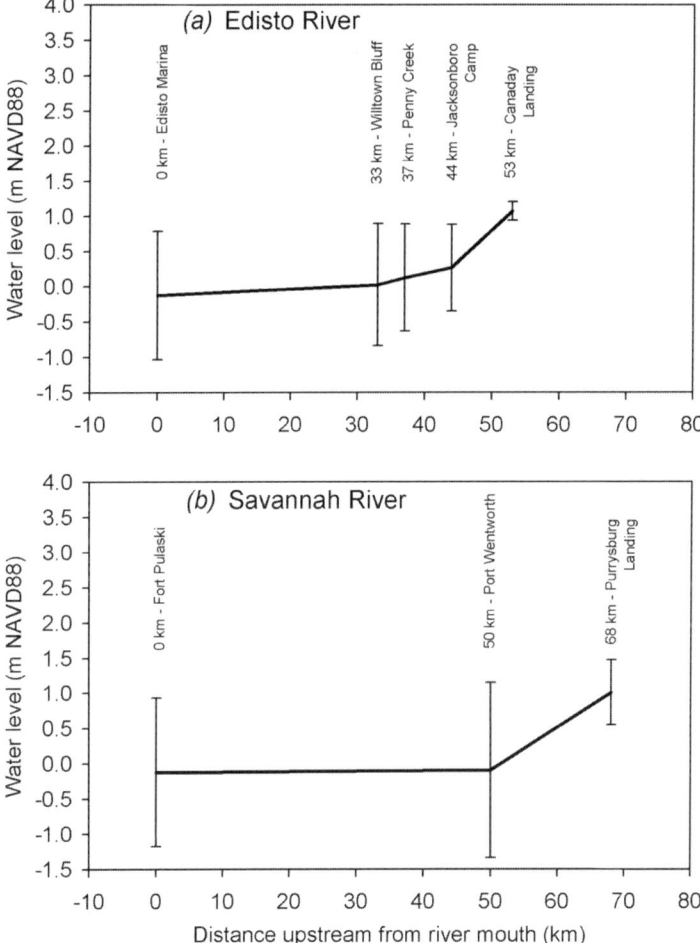

Fig. 2.6. Mean tide level and tide range (represented by vertical bars) referenced to NAVD88 (North American Vertical Datum of 1988) of the coastal reaches of two rivers in South Carolina (data from NOAA 2006a). The top of each vertical bar represents the level of mean high water.

have the opposite effect by magnifying tide range. The tide range of the Edisto River in South Carolina steadily declines with distance upstream and shows a rapid decline over the last 9 km (Figures 2.1c and 2.6a). The level of mean high water is ~80 cm NAVD88 (North American Vertical Datum of 1988) for the first 44 km and rises above 1.0 m NAVD88 at 53 km. In general, the Edisto River follows the tidal behavior of an unmodified stream. Not far down the coast, the Savannah River has a dredged shipping channel, 12.8 m deep from the coast to Savannah, Georgia, and 11 m deep from there to Port Wentworth, Georgia (USACE 2004; USACE 2006). The tide range of the Savannah River increases 37 cm from the coast at the "Fort Pulaski" gage to the end of the shipping channel in Port Wentworth; the tide range then decreases from Port Wentworth to Purrysburg Landing, South Carolina (Figure 2.6b). The level of mean high water is ~1.0 m NAVD88 for 50 km inland and rises to almost 1.5 m NAVD88 at 68 km. This is a classic example of the influence of dredging on tide range.

The Savannah River flow regime is also affected by three major reservoirs and several low-head locks and dams (USACE 2006; see Chapter 12). The most obvious affect of the dams is to dampen peak flows and raise the level of minimum flows (Smock et al. 2005), allowing the tide to dominate the hydrograph of the coastal reach of the Savannah River for a more prolonged period of time.

2.2.3. Sea level

Coastal water levels have cyclic annual variation in addition to daily and monthly cycling of the tides. All three sites depicted in (Figure 2.4) have mean water levels lowest in late winter and highest in early fall. Despite the large difference in tide range between Grand Isle, Louisiana, and the "Fort Pulaski" gage, mean monthly water levels are surprisingly similar (Figure 2.7). This same seasonal trend was reported by Baumann (1987) for the Gulf of Mexico. Seasonal trends in relative mean sea level are attributable to large-scale oceanic thermal expansion and contraction as well as to seasonal atmospheric pressure differentials (Chen et al. 2000; Yan et al. 2004). Sea level is approximately 30–40 cm higher in September and October than in January. The seasonally high sea level in the fall can be significant for flooding of tidal freshwater wetlands, especially in coastal Louisiana, with its broad and relatively flat Mississippi River deltaic plain. As mentioned in the last section, the fall is also the time of lowest average river discharge throughout the southeastern United States, which lowers the supply of fresh water to the coast (USGS 2006). Higher sea level at this

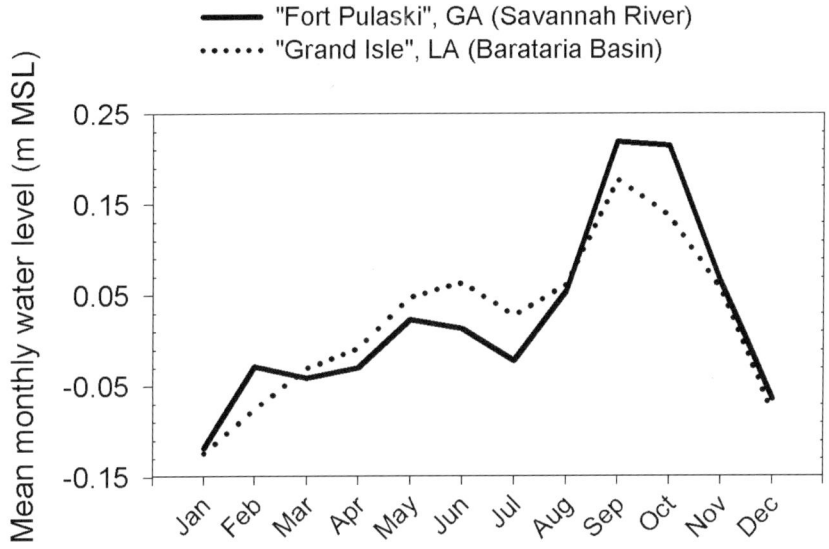

Fig. 2.7. Five-year (1999–2004) mean monthly water level relative to mean sea level (MSL) at two coastal tide gages in the southeastern United States (data from NOAA 2006a).

time of year increases the chance for saltwater intrusion into tidal freshwater wetlands.

Superimposed on the regular cycles of tide variation is long-term global eustatic sea-level rise (see Chapter 1). Eustatic sea-level rise couples with large-scale tectonic movement and/or local subsidence to manifest in apparent sea-level rise that varies widely throughout the Gulf of Mexico and southern Atlantic Ocean (Figure 2.2). The current rate of apparent sea-level rise at Grand Isle is almost 10 mm/yr (NOAA 2006a). A rise in sea level of almost 30 cm in 30 years on top of a 30 cm tide range at Grand Isle is a much greater percentage increase than at the "Fort Pulaski" gage, where apparent sea-level rise is currently less than 3 mm/yr in the presence of a tide range greater than 2 m (NOAA 2006a). Accordingly, tidal swamp forests in Louisiana are being affected by sea-level rise to a much greater degree than are those along the Savannah River (Conner and Day 1988; Conner and Brody 1989).

2.2.4. Salinity

It may seem contradictory to discuss the salinity of freshwater ecosystems; however, it makes perfect sense to state that the measured salinity of a tidal freshwater system can be 0 ppt. Indeed, freshwater swamps at the extreme upper elevation of the tide are characterized as such, dominated by water with negligible salinity year round. It is at the lower edge of tidal freshwater systems that salinity becomes a controlling, sometimes chronic, factor. By our own definition, tidal freshwater forests are characterized by freshwater species on the edge of saltwater habitat. The U.S. Fish and Wildlife Service wetland classification system (Cowardin et al. 1979) classifies tidal freshwater forested wetlands as either estuarine intertidal or palustrine with a tidal modifier. The difference between the two systems depends on the use of 0.5 ppt as the lower salinity limit of estuarine wetlands and the upper salinity limit of palustrine wetlands when measured during the period of average annual low flow. The wetland classification system (Cowardin et al. 1979) does not have tidal forests in the riverine system, even though many tidal freshwater forests are adjacent to and flooded daily by flowing rivers.

River discharge, topography, and proximity to tidal influence interact to form many different patterns of salinity in estuaries. Near the coast, deep rivers with high freshwater discharge tend to stratify with fresh water on top and a saltwater wedge on bottom. Tidal freshwater swamps nearby could experience little to no saltwater influence even though water at the bottom of the adjacent river has high salinity. Shallow estuaries with low riverine input can be well mixed from top to bottom. Tidal freshwater swamps located near the head of such an estuary may experience salinity values which rise and fall with the tide.

The size of a river channel and magnitude of its discharge affect the salinity in its lower reaches. The Waccamaw River, South Carolina, is a system with a relatively small drainage basin and low discharge (Figure 2.1c) (see Chapter 9). Surface salinity remains very low at high and moderate discharge but spikes sharply at the lowest flows (Figure 2.8a). Freshwater conditions (salinity <0.5 ppt) are common in the Waccamaw River. In contrast, surface salinity in the larger Savannah River remains at 2–5 ppt throughout normal discharge as well as during drought (Figure 2.8b). Only during extremely high flow (e.g. January–July 2003) does salinity drop to almost fresh conditions for an extended time. In the lower Savannah River, freshwater conditions are uncommon, even though base river flow (200–300 m^3/s) is almost an order of magnitude higher than in the Waccamaw River. The depth of the dredged Savannah River channel allows a significant saltwater wedge to intrude upstream along the bottom, which mixes with the freshwater above (see Chapter 12).

The absence of river input has an enormous effect on salinity in coastal Louisiana. Tidal swamps occur mostly without direct river influence except

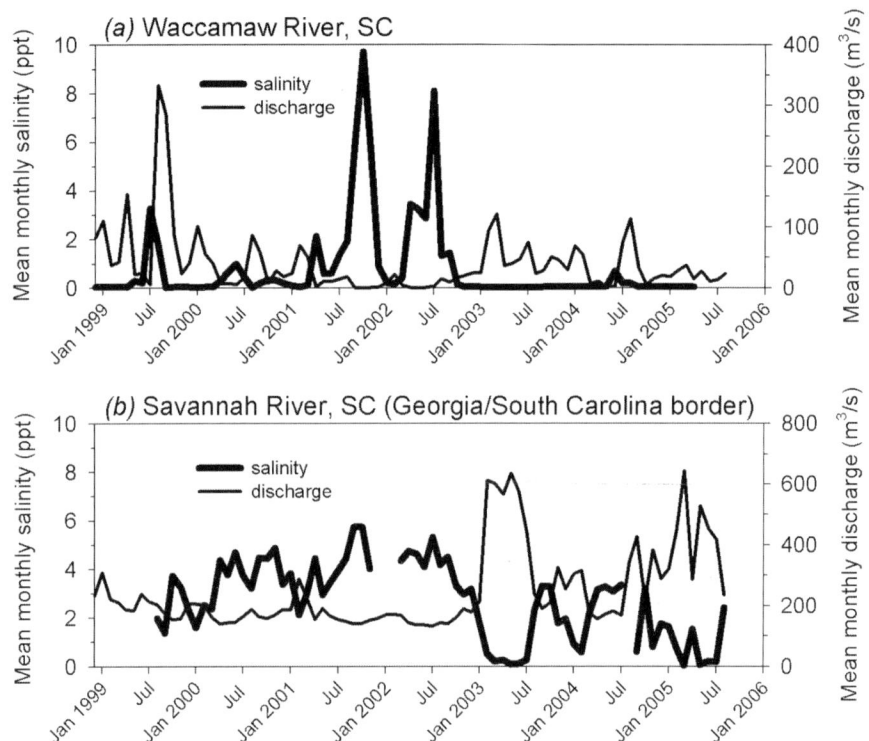

Fig. 2.8. Mean monthly surface salinity and mean monthly discharge for two rivers in South Carolina (data from USGS 2006). (a) Waccamaw River: salinity from "Hagley Landing near Pawleys Island" station (~river km 34); discharge from "Conway Marina" station (~river km 89). (b) Savannah River: salinity from "Port Wentworth" station (~river km 33); discharge from "Hardeeville" station (~river km 61).

near the Lower Atchafalaya River, where ship navigation canals allow river water to penetrate areas away from the main channel (Swarzenski 2001). Forested wetlands in the vicinity of Bayou Boeuf at Amelia, Louisiana (Figure 2.1a) are influenced by the Atchafalaya River and were unaffected by a local drought from 1999 to 2000, as indicated by continued fresh conditions during that period (Figure 2.9a); river discharge was sufficient to keep salt water from intruding into the area. In contrast, the swamp forests of Barataria Basin are isolated from river influx by extensive flood-control levees but are open to the Gulf of Mexico. Fresh water enters primarily as runoff from local precipitation. The Pen is an embayment in Barataria Basin approximately 50 km from the Gulf of Mexico (Figure 2.1a). Monthly average salinity in The Pen was high throughout the 1999–2000

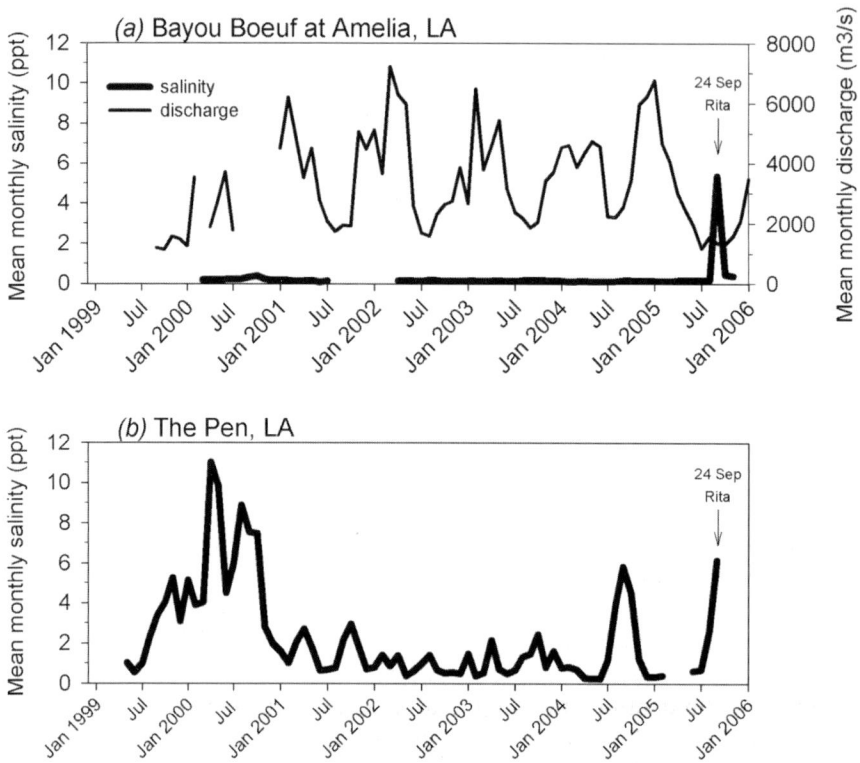

Fig. 2.9. Mean monthly surface salinity at two sites in coastal Louisiana near tidal freshwater forested wetlands. (a) Bayou Boeuf at Amelia is influenced by the Atchafalaya River (data from USGS 2006: salinity from Bayou Boeuf at "Amelia" station, discharge from Lower Atchafalaya River at "Morgan City" station). (b) The Pen, in Barataria Basin, has no river influence (salinity data from LADNR 2006, Naomi Siphon Diversion Project, station BA03C-60). Arrows refer to high mean monthly salinity in September 2006 as a result of the Hurricane Rita storm surge.

drought, reaching over 10 ppt during that period (Figure 2.9b). The passing of Hurricane Rita in the fall of 2005 raised water levels sufficiently in Bayou Boeuf and in The Pen such that monthly average salinity was raised considerably (Figure 2.9a, b).

2.3. Hydrology of tidal freshwater forested wetlands

Beaven and Oosting (1939) provided a descriptive narrative of a tidally influenced baldcypress-swamp tupelo forest on the Pocomoke River, Maryland (Figure 2.1d):

> "The portions of the swamp adjacent to the river are regularly flooded and drained by the tide ... A few hundred yards from the river drainage at low tide is less complete and there is some standing water at all times. Even where ditches have been dug, the level contour, fallen logs, cypress knees and accumulation of plant debris prevent much run-off at low tide. The land rises very gradually from the river and as it reaches an elevation above high tide it becomes less swampy with a gradual transition from swamp to upland forest."

In a typical riverside tidal baldcypress swamp, there is a season when the forest is flooded and drained daily (or twice daily). At other times of the year the river rises and floods the forest floor for long periods. There may or may not be a tidal signature superimposed on that high water, but the tide during those periods does not affect the flooding duration of the forest; only the depth of flooding is affected. When river discharge is low and/or astronomical or wind forces cause the water levels to lower, the forest floor goes dry for an extended period. During this drought the tide influences the groundwater level, which is never far below the surface.

Hydrographs of many types of tidal freshwater forested wetlands follow the pattern described in the previous paragraph. The elevation of the soil surface in tidal forests ranges above and below mean high water. Thus, hydroperiods of tidal freshwater forests range from rarely flooded to permanently flooded, with concomitant changes in forest tree species. Hydroperiod may be defined as any time unit of inundation per time unit of study period, such as hours flooded per day or days flooded per year, and can be expressed as a percentage. Careful attention must be paid to defining the timing and duration of flooding that constitute the calculation of hydroperiod. A wetland which is inundated 50% of the time on an annual basis could describe a riverine backswamp that is continuously flooded for 6 months and continuously dry for 6 months. Another wetland with a 50% annual inundation rate is a tidal wetland that is flooded for 12 hours and drained for 12 hours each day. By definition, there must be a tidal signature evident in the hydrograph of a tidal wetland, either above or below the surface, during some period in a normal annual cycle. Following are descriptions of the interaction of the tide with elevation, microtopography, and wind to create the range of hydrologic conditions present in tidal freshwater forests.

2.3.1. Elevation: the significance of mean high water

Swamp floor elevations in tidal baldcypress swamps are typically equivalent to the level of mean high water in the adjacent water body (Table 2.2, Figure 2.10a), which is true regardless of the tide range or distance from the coast. Codominant tree species in these swamps are usually swamp tupelo and/or water tupelo (*Nyssa aquatica* L.). As the elevation of the swamp floor becomes higher relative to the level of mean high water, baldcypress becomes less dominant and bottomland hardwood species become more dominant (Figure 2.10b). The Pamunkey River site in Virginia (Figure 2.1d) is a higher elevation tidal bottomland hardwood forest dominated by ash (*Fraxinus* spp.), swamp tupelo, and red maple (*Acer rubrum* L.) (Rheinhardt and Hershner 1992; see Chapter 7). This same type of tidal forest was described by Peterson and Baldwin (2004) on the Nanticoke River in Maryland (Figure 2.1d) (see Chapter 6). The Pocomoke River study area (Figure 2.1d) consists of two sites which vary in elevation and have baldcypress grading into mixed bottomland hardwoods at both sites (see Chapter 5).

Mean tide ranges at three study sites were higher than at the coast (Suwannee River at 5 km, Savannah River, and Pamunkey River) (Table 2.2). The rest of the sites had attenuated tide ranges. Magnification or attenuation as the tidal wave travels upstream is a function of river morphology, tide type and range, and distance upstream.

Other tidal freshwater forests that occur above mean high water are the coastward edges of hydric hammocks (Figure 2.10c) (Vince et al. 1989; Williams et al. 1999a; see Chapter 10 for a definition of hydric hammock), maritime forests (forests bordering coastal beaches) (Bellis 1995), and Atlantic white cedar (*Chamaecyparis thyoides* (L.) B.S.P.) communities (Laderman 1989; see Chapter 4). Hydric hammocks and maritime forests are primarily rain-fed as the source of fresh water and may be considered tidal when they are located close enough to the coast that a tidal signature is evident in a groundwater hydrograph (Bellis 1995; Williams et al. 1999b). Even though they occur above mean high water, proximity to the coast makes hydric hammocks and maritime forests vulnerable to inundation from storm surges. These wetlands may also be subject to overtopping by flood waters when located in proximity to a river (Light et al. 2002). Cowardin et al. (1979) classify hydric hammocks in general as nontidal

Table 2.2. Elevation and tide range of selected tidal freshwater forested wetland study sites in the southeastern United States. Numbers in parentheses are estimates[a].

Study site	Forest type[b]	Distance to coast [km]	Study site forest floor elevation [m][c,d]	Study site mean high water elevation [m][c,e]	Study site mean tide range [m][c,f]	Coast mean tide range [m][c,f]	References[g]
Louisiana: Barataria Basin The Pen	baldcypress swamp	(50)	(0.3) NAVD88	(0.3–0.4) NAVD88	(0.1–0.2)	0.3	This chapter
Florida: Suwannee River	baldcypress swamp	5	0.4–0.5 NGVD29	0.6 NGVD29	0.9	0.8	Light et al. 2002
Florida: Suwannee River	baldcypress swamp	43	1.0 NGVD29	1.0 NGVD29	0.3	0.8	Light et al. 2002
Georgia: Savannah River	baldcypress swamp	(35)	(1.4) NGVD29	(1.4) NGVD29	(2.3–2.5)	2.1	This chapter
North Carolina: Cape Fear River (Indian Creek site)	baldcypress swamp	(60)	0.5–0.7 NAVD88	(0.6) NAVD88	1.2	1.3	C.T. Hackney (pers. communication); Chapter 8
Maryland: Pocomoke River	baldcypress/ mixed hardwoods	(155–175)[h]	(0.4–1.4) MTL	(0.2) MTL	(0.4)	0.8	D.E. Kroes (pers. communication); Chapter 5
Virginia: Pamunkey River ("Elsing" site)	mixed hardwoods	(100)[h]	(0.9–1.0) MTL	(0.4–0.6) MTL	(0.8–1.2)	0.8	Rheinhardt and Hershner 1992
Florida: Turtle Creek	cabbage palm hammock	(2–4)	0.4–1.3 NAVD88	(0.4) NAVD88	(0.9)	0.9	Williams et al. 1999a

[a] Estimated distances were measured on a map. Estimated elevations and tide ranges are from the nearest one or two gages available at NOAA (2006a).
[b] Baldcypress swamp: baldcypress is dominant or co-dominant. Mixed hardwoods: baldcypress is unimportant or absent. Cabbage palm hammock: cabbage palm is dominant or co-dominant.
[c] *NGVD29* National Geodetic Vertical Datum of 1929. *NAVD88* North American Vertical Datum of 1988. *MTL* Mean tide level (a datum which sets zero at halfway between mean high water and mean low water). NGVD29 and NAVD88 are fixed geodetic references, while MTL is calculated from tide gage measurements and is unique to each gage (Hicks et al. 2000).
[d] Swamp-floor elevation may include ranges from hollows to hummocks for one to many sites.
[e] "Tide means are calculated for the 19-year period from 1983 through 2001.
[f] Mean tide range = mean high water minus mean low water.
[g] References are sources of information for forest type, site location, elevation, and some tide data; all other tide information is from NOAA (2006a). "This chapter" refers to data that are presented in this chapter for the first time.
[h] The coast refers to the mouth of Chesapeake Bay (Chesapeake Bay Bridge-Tunnel tide station) for Virginia and Maryland sites.

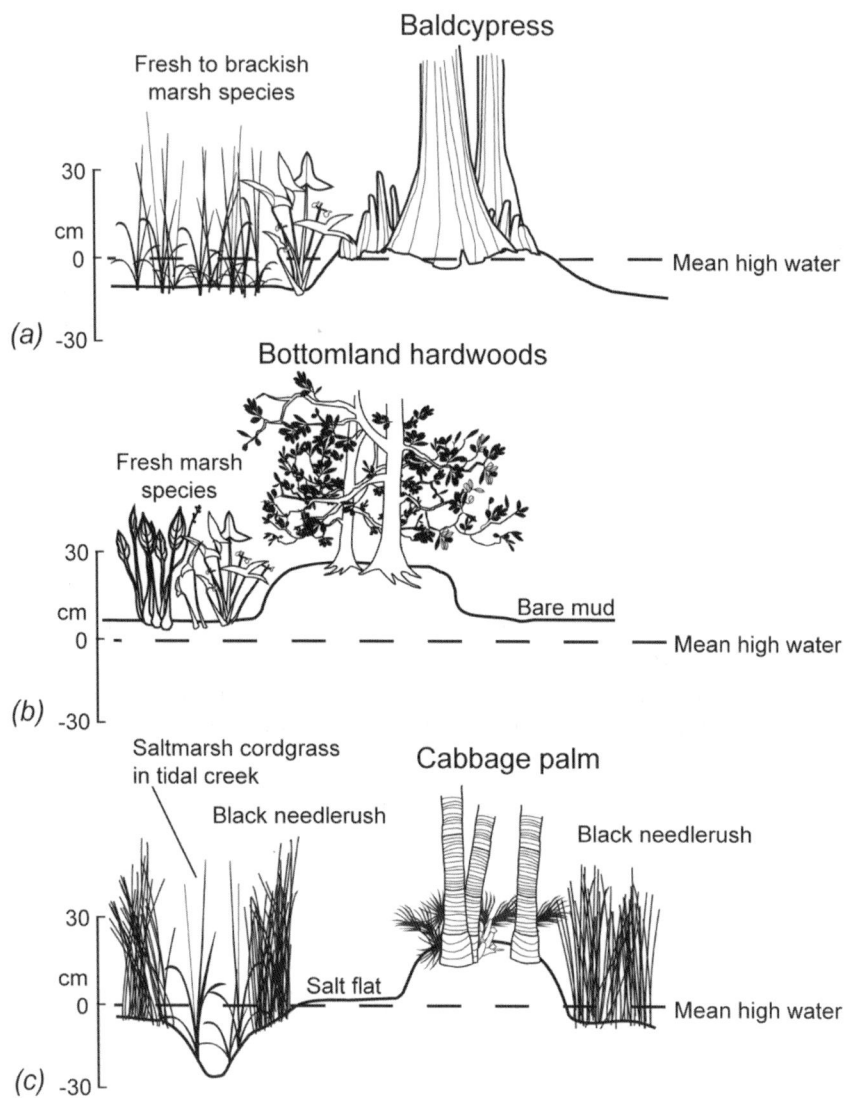

Fig. 2.10. Typical profiles of tidal freshwater forested wetlands showing hummock and hollow topography (high and low elevations, respectively) and position with respect to mean high water. (a) Baldcypress swamp (drawn from descriptions by Light et al. 2002 and personal observation; see Chapters 9 and 11). (b) Bottomland hardwood swamp (after Rheinhardt and Hershner 1992; see Chapters 5, 6, and 7). (c) Cabbage palm hydric hammock (drawn from descriptions by Vince et al. 1989; Williams et al. 1999a; Doyle et al. 2003; see Chapter 10).

palustrine forested wetlands with seasonally flooded water regimes, but sometimes the coastward edges carry the modifier "seasonal-tidal." Characteristic of hydric hammocks and maritime forests are hydrographs dominated by nonflowing water, or "still-water wetlands" (Wharton et al. 1977) and relatively salt-tolerant yet nonhalophytic trees such as cabbage palm (*Sabal palmetto* (Walt.) Lodd. ex J.A. & J.H. Schultes), live oak (*Quercus virginiana* P. Mill.), and southern redcedar (*Juniperus virginiana* var. *silicicola* (Small) J. Silba). Tidally influenced hydric hammocks are often found directly upslope of salt marsh, in particular irregularly flooded salt marsh dominated by black needlerush (*Juncus roemerianus* Scheele) grading into tidal creeks dominated by saltmarsh cordgrass (*Spartina alterniflora* Loisel.) (Stout 1984; Vince et al. 1989; Williams et al. 1999a). Salt flats that are barren or covered by saltmarsh succulent plant species sometimes occur at the base of cabbage palm island hammocks right at the level of mean high water, presumably because of the lack of tidal flushing and high evaporation rates which combine to create hyper-saline conditions (Figure 2.10c) (Stout 1984; Doyle et al. 2003). Atlantic white cedar, on the other hand is salt intolerant and is usually found upslope of freshwater marshes or swamps, often at the toe of slopes where ground water is plentiful (Beaven and Oosting 1939; Laderman 1989; see Chapter 4). Laderman (1989) reported that Atlantic white cedar almost always occurs above the level of high tide; however, a hydrograph of an Atlantic white cedar community appears in Chapter 4 which exhibits a tidal signature.

Permanently to semipermanently flooded forests of the Mississippi River deltaic plain in Louisiana typically occur below the level of mean high water. While many areas of coastal forest in Louisiana have an irregular microtidal signature, they are so low in elevation that they are very vulnerable to saltwater intrusion, particularly during surge events. The obvious threat of periodic hurricanes is compounded by minor events such as sustained wind from the south, which pushes water up the estuaries and can occur at any time of the year. Most of Louisiana's swamp forests are classified by Cowardin et al. (1979) as nontidal, semipermanently flooded palustrine forests, but with only slight differences in hydrology they are considered semipermanent-tidal or permanent-tidal. Day JW Jr et al. (2006) describe a tidal freshwater forested wetland near Bayou Boeuf (Figure 2.1a) that is continuously flooded 20–40 cm deep throughout an entire year of study with less than a few centimeters of daily tidal fluctuation. There are permanently flooded baldcypress forests in the Atchafalaya Basin Floodway (Figure 2.1a) that have distinct 10–15 cm daily tidal fluctuations during low flow of the Atchafalaya River (Day RH, unpublished data). Since the substrate in these swamps is never exposed except in an extreme drought, perhaps they would better be classified as subtidal. Higher elevation swamps within the Atchafalaya Basin Floodway are only flooded during higher discharge of the Atchafalaya River when no tidal signature is apparent.

2.3.2. Microtopography: hummocks, hollows, and hydroperiod

As in all wetland ecosystems, microtopography plays a role in controlling hydroperiod in tidal swamps and associated species assemblages according to their flood tolerances. Rheinhardt (1991, 1992) placed water-level recorders within five tidally influenced bottomland hardwood forests along the Pamunkey River, a tributary of the York River which flows into lower Chesapeake Bay (Figure 2.1d) (see also Rheinhardt and Hershner 1992 and Chapter 7). The Pamunkey River study site listed in Table 2.2 ("Elsing") has an elevation of 0.85–1.00 m above mean tide level (1.35–1.50 m above mean low water) in an area where the tide range varies from 0.8 to 1.2 m. All but one of the sites were characterized by hummocks and hollows (topographic highs and lows, respectively), common microtopographical characteristics of tidal freshwater swamps. The hummocks support the woody vegetation of the forest and were approximately 15 cm in elevation above the hollows, which were almost always saturated (Figure 2.10b). The hollows are bare mud or contain herbaceous vegetation similar to that found in tidal freshwater marshes. Water level was recorded continuously during three growing seasons, and the tops of the hummocks were almost never inundated. Only the most downriver site experienced flooding above the hummock surface during the growing season, and that was less than 0.1% of the time. During the growing season, the surface of the hollows was inundated 5%–20% of the time, reaching up to 15 cm of water depth, saturating but not overtopping hummock surfaces. The soil of the hollows 0–15 cm below the surface was flooded 20%–100% of the time. The surface of the hollows was often inundated twice daily during spring tide conditions.

Hummock and hollow topography has also been described in tidal forested wetlands of Maryland's Eastern Shore (Harrison et al. 2004; Peterson and Baldwin 2004; see Chapters 5 and 6), on the Cape Fear River, in North Carolina (Figure 2.1c) (see Chapter 8), and as hummocks and sloughs (up to 1.2 m difference in elevation) along the Apalachicola River in Florida (Leitman et al. 1984). Duberstein (2004) described hummocks and hollows in tidal freshwater forests of the Savannah River floodplain; he hypothesized that hydroperiods of Savannah River tidal forests were longer than those measured by Rheinhardt (1992) on the Pamunkey River based on higher organic matter content of the hummocks of the Savannah River sites (see Chapter 12). Water level in a tidal forest of the Savannah River (Figure 2.11) shows a pattern similar to a hydrograph published by Rheinhardt (1992) (reproduced in Chapter 7) and also to hydrographs in Chapters 6, 8, and 9. The higher high tides inundate the hummocks for the first 7 days on the graph, while the lower high tides only flood the hollows. Following that is a period during which no tide inundates the surface of the hummocks or hollows. During this period, the tides impede subsurface drainage, and the soil of the hollows is kept saturated to within 20 cm of the surface.

Light et al. (2002) reported that hummocks are present at the bases of trees at nearly all forested wetland transects studied on the Suwannee River, but are more common in lower tidal areas, where they are large and well defined (see Chapter 11). Light et al. (2002) did not have water-level recorders within forested sites, but they calculated inundation rates based on surveyed elevations of forested sites and river stage. The mud floors (hollows) in tidal swamps remained nearly 100% saturated because the elevation of the mud floors was close to the elevation of median daily high tide stage. Some hummock tops had better drained soils that were not continuously saturated.

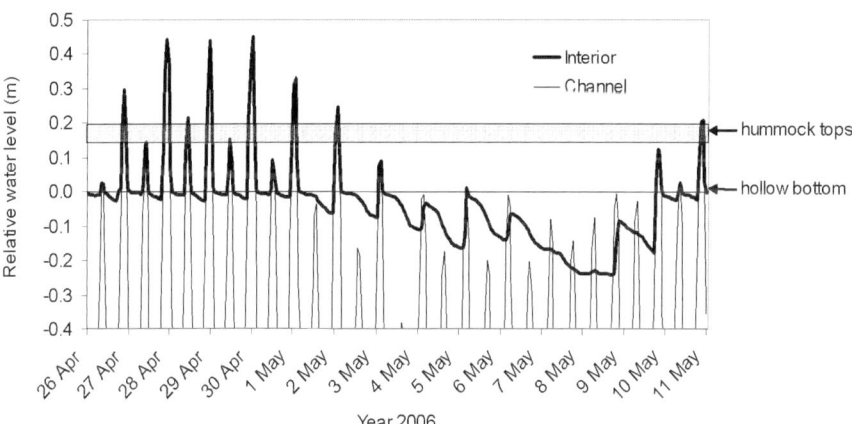

Fig. 2.11. Hourly water levels from stage recorders installed within the interior of a tidal forested wetland and in an adjacent tidal channel at ~river km 35 on the Savannah River, Georgia (referenced as "this chapter" in Table 2.2). The interior gage is installed within a hollow bottom; the soil surface of the bottom is 0.0 m on the relative water level scale of the graph. The tops of the surrounding hummocks are 15–20 cm above the level of the hollow bottom.

Even hammocks have hummocks. According to Light et al. (2002), referring to transects in the lower Suwannee River floodplain (and using the term "mud floor" instead of "hollow"), "... some lower tidal forests have a distinct hummock-mud floor topography that supports hammock species on the hummocks and swamp species on the mud floor." Hydric hammocks in saltier environments downstream (or coastward) of the tree line may take the form of islands (hummocks) surrounded by salt marsh (hollows) (Figure 2.10c). Light et al. (2002) referred to these as maritime hammocks. Williams et al. (1999a) published hydrographs (partially replotted in Chapter 10) of groundwater fluctuations underneath coastal hydric hammocks near Turtle Creek, Florida, which exhibit tidal influence (Figure 2.1b). Ground water level was usually 0.5–1.0 m below the surface of the hammocks, rose occasionally to within 5–10 cm below the surface of one of the hammocks, but never inundated the tops of the hammocks for an entire year of continuous recordings.

The existence of hummocks and hollows is related to the elevation of tidal freshwater forests relative to mean high water. The very upper end of the tide cycle supplies just enough water to inundate the hollows to an extent that they do not support the vegetation that exists on the top of the hummocks. The hollows function as the upper extreme of tidal creeks, providing the benefits of regular flooding and draining without the erosive force of stronger tidal forces downstream. The hummock tops provide a well-watered/well-drained environment with more aeration than forested wetlands with continuously saturated soils. Rheinhardt and Hershner (1992) hypothesized that hummocks are formed and maintained by the accumulation of fallen trees, where tree species may germinate and grow, forming root balls which persist after the fallen trees have decomposed. Leitman et al. (1984) mentions a lack of hummocky terrain at sites along the Apalachicola River that are slightly higher in elevation than tidally influenced forests or that are cut off from tidal influence by natural and artificial levees. Perhaps at the upper extreme of tidal influence, the slow movement of water back and forth with the tide may help loosen the soil matrix in soils not bound by the root balls of trees, providing a gentle erosional force to facilitate the development of hollows. However they are formed, higher inundation rates prevent germination of tree species in the hollows.

2.3.3. Wind and water level

The wind may play as much a role as does river flow in controlling water levels in many coastal areas of the southeastern United States. Inundation of tidal freshwater wetlands on the lower Pocomoke River (a tributary to Chesapeake Bay) (Figure 2.1d) is controlled more by wind direction and velocity than by river storm flow events (see Chapter 5). The mouth of Chesapeake Bay has a mean tide range of 0.78 m (Table 2.2), which is attenuated over a distance of ~150 km to ~0.4 m in the lower Pocomoke River. Also, the lower Pocomoke River has a much wider channel cross section than does the river farther upstream; therefore rainstorm flow peaks are attenuated. Finally, the orientation of Chesapeake Bay is generally north-south, and the Pocomoke River is located on the eastern side of the bay; therefore prevalent winds from the west and south have the greatest effect on water levels in the tidal swamps of the lower Pocomoke River.

Wind plays an even greater role in upper Barataria Basin, where tidal swamps are isolated by levees from any major riverine influence. A water-level recorder was installed within a tidal baldcypress forest near The Pen, an embayment of the Barataria Bay system (Figure 2.1a). The entrance from the Gulf of Mexico is to the south. The hydrograph shows a tidal signature but is dominated by wind-driven water levels (Figure 2.12a, b). The average daily wind speed and direction coincide with dominant peaks and valleys of the hydrograph (Figure 2.12a, b). Winds from the south (arrows pointing up on the graph) (Figure 2.12b) push the water levels higher, and winds from the north push water out of the system, lowering water level. This is a shallow, well-mixed system, so salinity rises and falls from <1 ppt to >10 ppt as the water mass moves back and forth with the tide and changes in wind direction (Figure 2.12c). The nearest gage from which predicted tides are available is at Grand Isle, on the Louisiana coast about 50 km to the south (Figure 2.12d). The observed Grand Isle data show the same pattern as The Pen gage, and the difference between observed and predicted water level correlates well with the wind vectors from the Louis Armstrong New Orleans International Airport, even though New Orleans is 85 km to the north of Grand Isle and 35 km north of the study site. The general trend of the predicted mean tide level (without wind influence) is higher in October and slowly falls to a low in December. The timing of average daily winds was such that in October the south wind pushed the highest tides even higher and in December the north wind pushed the lowest tides even lower. The Pen was inundated for most of October and November by water with salinity 2–10 ppt. Salinity in The Pen dropped to near 0 ppt in December, but water level within the swamp also dropped. The swamp floor was dry in December, so there was no freshwater flushing effect, and groundwater salinity remained high within the swamp (personal observation).

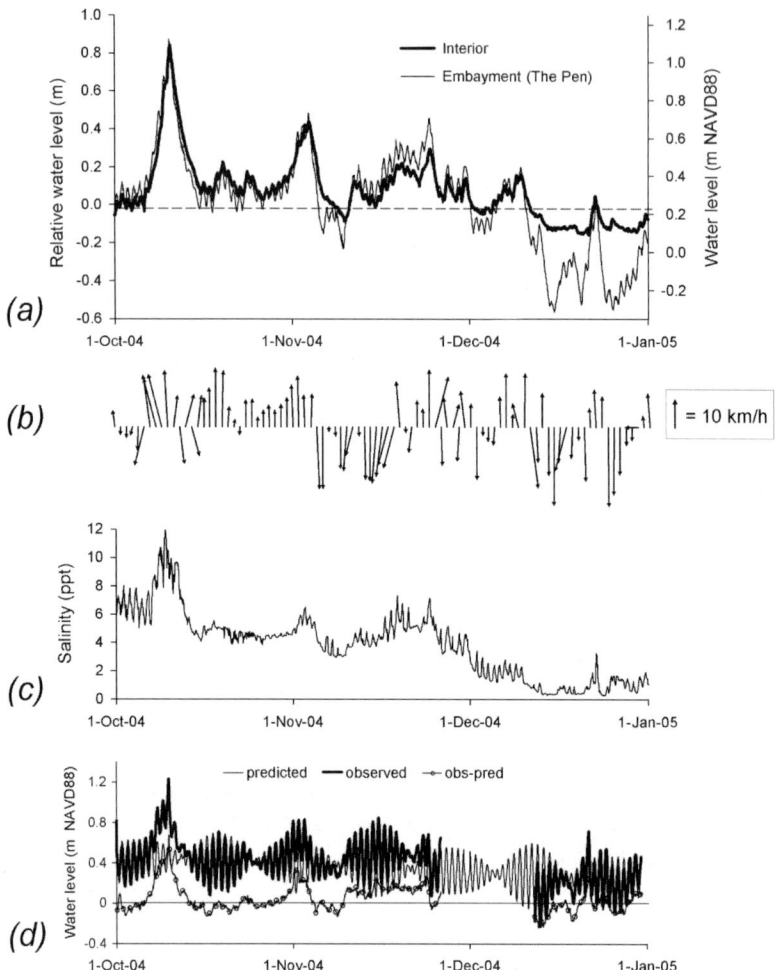

Fig. 2.12. (a) Hourly water levels from a stage recorder within the interior of a tidal freshwater forested wetland and from a tide gage in a nearby embayment (The Pen) (Naomi Siphon Diversion Project, station BA03C-60) (LADNR 2006) in Barataria Basin, Louisiana for a 3-month period (referenced as "this chapter" in Table 2.2). The interior gage is installed within a hollow bottom; the soil surface of the bottom is 0.0 m on the relative water level scale of the graph. (b) Daily wind vectors at the Louis Armstrong New Orleans International Airport (NOAA 2006b). These are resultant speed and direction, defined as the vector sum divided by the number of observations. Arrows pointing up are winds from the south. (c) Hourly salinity records from The Pen (LADNR 2006). (d) Hourly water levels of predicted and observed tides and their difference (observed minus predicted) at Grand Isle, Louisiana (data from USGS 2006). *NAVD88* North American Vertical Datum of 1988.

2.4. Hydrology, biology, and biogeochemistry

There is generally a decline in plant species diversity from freshwater to saltwater wetlands (forested or herbaceous), and in both there is a change in species composition with annual inundation rate. Beneath the canopy of tidal freshwater forested wetlands is a community of herbaceous plants which may change with the season and from year to year depending on degree of flooding and salinity of the floodwater; the tree species, however, have to tolerate the changing conditions to survive to maturity. Chronic exposure to inundation by water with salinity of approximately 2 ppt leads to conversion of tidal forested wetland to oligohaline or brackish marsh on the Cape Fear River (see Chapter 8). Light et al. (2002) reported that soil porewater salinity was highest below the root zone in regularly inundated tidal freshwater forests along the Suwannee River, while Williams et al. (1999a) recorded salinities up to ~23 ppt in the groundwater of rarely flooded coastal hydric hammocks. In salt marsh ecosystems, increased productivity has been noted for the creek tidal zone related to daily aeration of the subsurface by tidal action (King et al. 1982; Portnoy and Valiela 1997; Mendelssohn and McKee 1988). Daily flooding and draining can also theoretically increase productivity in tidal freshwater forests. It has been shown experimentally that seedlings of baldcypress and black willow (*Salix nigra* Marsh.) have increased growth rates with hydroperiods that simulate tidal frequency (Conner 1995; Day RH et al. 2006); however, this could only occur in an ideal situation where there is enough freshwater flow that no salt water is introduced with the tidal action. Brown and Lugo (1982) compared freshwater and saltwater forested wetlands in the greater Caribbean Sea basin and the southeastern United States but did not include tidal freshwater forests. They found that the number of tree species, complexity index, and aboveground biomass were higher in freshwater than in saltwater forested wetlands.

Physical and biological aspects of tidal freshwater and salt marshes were compared by Odum (1988), and even though trees were considered part of the marsh landscape, most of the data described herbaceous vegetation ecosystems. One of the greatest differences was in the composition of the anaerobic microbial communities. Generally, fermentors and methanogens dominate in tidal freshwater, and sulfur reducers dominate in the salt marsh. Daily cycling between anaerobic and aerobic conditions in the wetland soil environment, even if limited to near the swamp surface, can influence accumulation of porewater sulfides, cycling of trace metals, decomposition of organic matter, and trace gas emissions (Odum 1988; also see Chapter 8). Prolonged aeration during dry periods may result in pyrite oxidation, lowered pH, and release of metals, possibly in concentrations that are toxic to the vegetation (Portnoy and Giblin 1997). Verhoeven et al. (2001) compared tidal versus nontidal and forested versus herbaceous freshwater riverine wetlands in the United States and Europe. Soil organic matter, total nitrogen, total phosphorus, and bulk density were higher in tidal freshwater wetland soils than in nontidal wetland soils. One explanation given was the possibility that nutrient-rich riverine sediments were regularly deposited in the tidal freshwater wetlands, presumably as river flow is slowed by the upstream advance of the tide. This explanation is most plausible in the upper reaches of tide effect where the possibility of tidal scour and export is minimal. The sediment and nutrient trapping function of forested wetlands is important along alluvial reaches, and may peak in the upper reach of the tide, only to diminish downstream where tidal influence is greater (see Chapter 5). Verhoeven et al. (2001) also found that forested wetlands had higher soil organic matter and total nitrogen but lower bulk density and total phosphorus than did freshwater herbaceous wetlands, possibly because of the low rates of decay of woody plant tissues compared to the high rates of decomposition of herbaceous species. Again, this assumes *in situ* deposition of autochthonous material in the absence of appreciable tidal scour. More detail on the effects of hydrology on soil biogeochemistry is presented in Chapter 3.

2.5. Forest sustainability

Hydraulic forces move riverine freshwater and tidal salinity effects many kilometers upstream and downstream on daily, monthly, annual and multiyear cycles. The life spans of trees are decades to centuries, during which time the environmental conditions of hydroperiod and salinity regime are prone to change. How can a forest exist in such a variable environment? The trees in the forest have to deal with two thresholds, flood tolerance and salt tolerance. Tolerances vary among tree species, within a species (e.g., salt-tolerant genotypes), and throughout the life cycle of individuals (seed, seedling, sapling, adult) (Hook 1984; Allen et al. 1996; Krauss et al. 1998; see Chapter 14). Both flood and salt tolerance may vary between seedling establishment, sapling survival, and adult growth and productivity within an individual tree (Conner and Askew 1992; Conner 1994; Krauss et al. 1999). Within a given time period in a tidal freshwater swamp, a certain environmental condition allows a tree seed to germinate. The tree either dies or survives depending on the sustained conditions which may or may not change in the individual's lifetime. Global climate patterns affect the magnitude and seasonality of the fresh water input to the coast, the most critical component to the sustenance of tidal freshwater wetlands (see Chapter 1). Certainly in the southeastern United States the conditions under which some baldcypress forests were established were much fresher and less waterlogged than now. Over time, with relative sea-level rise and saltwater intrusion, many baldcypress trees have died, yet salt-tolerant individuals have survived where conditions are no longer favorable for regeneration to occur (Conner and Day 1988; Conner and Brody 1989; see Chapters 8 and 13). The coastal edges of cabbage palm hammocks are suffering a similar fate (Williams et al. 1999a, b). The upper extreme of riverine tidal freshwater forested wetlands are healthy and thriving (see Chapters 6 and 7); however, ghost forests (forests of mostly dead standing trees surrounded by emergent marsh or open water) are a common sight on the coastward edges, as attested to by the photos in Chapters 8, 9, 10, and 14.

2.6. Summary

Tidal freshwater forested wetlands occur anywhere that tides influence the hydrology of freshwater forests. If a tidal forest is healthy, it is usually located at or above the level of mean high water. Hydrology of tidal freshwater forests is primarily dictated by local tides and river discharge. Tide ranges vary by coastal reach across the southeastern United States from microtidal to mesotidal. Actual tide heights are driven by both astronomical and meteorological influences as well as by river channel and estuarine basin geomorphology. The predictable astronomical tides vary daily and seasonally and are well known for most coastal rivers and outlets. Meteorological forces of prevailing winds, frontal passages, and tropical storms can greatly amplify tidal highs and lows and often dominate over astronomical influences in microtidal settings. The upper end of tidal fluctuation in channels with relatively uniform bottom slope occurs at the point where the bottom elevation is approximately at high tide. Salinity and flooding are boundary controls on the downstream extent of tidal freshwater forests. Salinity, as well as duration and timing of flooding, controls tree species diversity, primary productivity, and soil biochemistry. Daily flooding and draining may increase forest productivity when salinity influence is minimized. What are the optimum conditions? More research is needed to understand the specific response of the productivity of tidal freshwater forested wetlands to the many patterns of inundation caused by the interaction of freshwater input and tides.

Hummocks and hollows typify the microtopography of tidal freshwater forested wetlands. The hollows function as the headwaters of tidal creeks within tidal freshwater swamps. Hollows are characteristically saturated 100% of the time and are dominated by herbaceous vegetation. The tops of the hummocks are flooded much less often and support the tree species. The hummocks provide a well-drained environment with more aeration than in forested wetlands with continuously saturated soils.

Conditions are continuously changing in many tidal freshwater forests. Eustatic sea level is rising. The land in many areas of the Southeastern United States is sinking by geologic process. Humans have changed many rivers by building dams, dredging harbors, or both. Tidal freshwater forested wetlands are both complex and threatened. The seemingly incompatible terms "freshwater" and "tidal" relegate these forested wetlands to a relatively narrow position in the landscape at the upstream edge of the estuarine mixing zone. There must be some degree of stability over a time period which includes the life span of trees, but the inexorable rise in coastal water levels at present is causing the coastal fringe of these systems to become increasingly characterized by dead or dying trees.

2.7. Acknowledgments

We would like to thank Thomas W. Doyle, Dennis F. Whigham, and Bobby D. Keeland for constructive advice in the early stages of the draft manuscript, Richard Keim and Michael G. Waldon for reviewing the manuscript, Ken W. Krauss for supplying data for hydrographs in Louisiana and Georgia, Diane K. Baker for illustrations, Andrew S. From for help with ArcView, and Beth Vairin and Victoria Jenkins for editing. Financial support was provided by the U.S. Geological Survey Global Climate Change Program.

References

Allen JA, Pezeshki SR, Chambers JL (1996) Interaction of flooding and salinity stress on baldcypress (*Taxodium distichum*). Tree Phys 16:307–313

Baumann R (1987) Physical variables. In: Conner WH, Day JW Jr (eds.) The ecology of Barataria Basin, Louisiana: An estuarine profile. Biological Report 85 (7.13). U.S. Fish and Wildlife Service, Washington, pp 8–17

Beaven GF, Oosting HJ (1939) Pocomoke swamp: a study of a cypress swamp on the eastern shore of Maryland. Bull Torrey Bot Club 66:367–389

Bellis VJ (1995) Ecology of maritime forests of the southern Atlantic Coast: a community profile. Biological Report 30. National Biological Service, Washington

Benke AC, Cushing CF (eds) (2005) Rivers of North America. Elsevier, Amsterdam

Brown S, Lugo AE (1982) A comparison of structural and functional characteristics of saltwater and freshwater forested wetlands. In: Gopal B, Turner RE, Wetzel RG, Whigham DF (eds) Wetlands ecology and management. National Institute of Ecology and International Scientific Publications, Jaipur, India, pp 109–130

Burt TP, Pinay G, Matheson FE, Haycock NE, Butturini A, Clement JC, Danielescu S, Dowrick DJ, Hefting MM, Hillbricht-Ilkowska A, Maitre V (2002) Water table fluctuations in the riparian zone: comparative results from a pan-European experiment. J Hydrol 265:129–148

Chen JL, Shum CK, Wilson CR, Chambers DP, Tapley BD (2000) Seasonal sea level change from TOPEX/Poseidon observation and thermal contribution. J Geodesy 73:638–647

Conner WH (1994) The effect of salinity and waterlogging on growth and survival of baldcypress and Chinese tallow seedlings. J Coast Res 10:1045–1049

Conner WH (1995) Baldcypress seedlings for planting in flooded sites. In: Edwards MB (ed) Proceedings of the eighth biennial southern silvicultural research conference. General Technical Report SRS-1. U.S. Department of Agriculture, Forest Service, Southern Research Station, Asheville, pp 430–433

Conner WH, Askew GR (1992) Response of baldcypress and loblolly pine seedlings to short-term saltwater flooding. Wetlands 12:230–233

Conner WH, Brody M (1989) Rising water levels and the future of southeastern Louisiana swamp forests. Estuaries 12:318–323

Conner WH, Day JW Jr (1988) Rising water levels in coastal Louisiana: Implications for two coastal forested wetland areas in Louisiana. J Coast Res 4:589–596

Cowardin LM, Carter V, Golet FC, Larse ET (1979) Classification of wetlands and deepwater habitats of the United States. FWS/OBS-79/31. U.S. Fish and Wildlife Service, Biological Services Program, Washington

Day JW Jr, Westphal A, Pratt R, Hyfield E, Rybczyk J, Kemp GP, Marx B (2006) Effects of long-term municipal discharge on the nutrient dynamics, productivity, and benthic community structure of a tidal freshwater forested wetland in Louisiana. Ecol Eng 27:242–257

Day RH, Doyle TW, Draugelis-Dale RO (2006) Interactive effects of substrate, hydroperiod, and nutrients on seedling growth of *Salix nigra* and *Taxodium distichum*. Environ Exp Bot 55:163–174

Doumlele DG (1976) Primary production and plant community structure in a tidal freshwater marsh. M.A. thesis, The College of William and Mary

Doumlele DG (1981) Primary production and seasonal aspects of emergent plants in a tidal freshwater marsh. Estuaries 4:139–142

Doyle TW, Day RH, Biagas JM (2003) Predicting coastal retreat in the Florida Big Bend region of the Gulf Coast under climate change induced sea-level rise. In: Ning ZH, Turner RE, Doyle T, Abdollah K (eds) Integrated assessment of the climate change impacts on the Gulf Coast region. Foundation Document. Louisiana State University Press, Baton Rouge, pp 201–209

Duberstein J (2004) Freshwater tidal forest communities sampled in the lower Savannah River floodplain. M.S. thesis, University of Florida

Harrison JW, Stargo P III, Aguirre MC (2004) Forested tidal wetland communities of Maryland's Eastern Shore: Identification, assessment, and monitoring. Maryland Department of Natural Resources, Natural Heritage Program, Annapolis

Henderson FM (1970) Open channel flow. Macmillan, New York

Hicks SD, Sillcox RL, Nichols CR, Via B, McCray EC (2000) Tide and current glossary. National Oceanic and Atmospheric Administration, National Ocean Service, Silver Springs

Hook DD (1984) Waterlogging tolerance of lowland tree species of the south. South J Appl For 8:136–149

King GM, Klug MJ, Wiegert RG, Chalmers AG (1982) Relationship of soil water movement and sulfide concentration to *Spartina alterniflora* production in a Georgia salt marsh. Science 218:61–63

Krauss KW, Chambers JL, Allen JA (1998) Salinity effects and differential germination of several half-sib families of baldcypress from different seed sources. New For 15:53–68

Krauss KW, Chambers JL, Allen JA, Luse BP, and DeBosier AS (1999) Root and shoot responses of *Taxodium distichum* seedlings subjected to saline flooding. Environ Exp Bot 41:15–23

Laderman AD (1989) The ecology of the Atlantic white cedar wetlands: a community profile. Biological Report 85(7.21). U.S. Fish and Wildlife Service, Washington

LADNR (2006) Louisiana Department of Natural Resources, Strategic online natural resources information system. Available at http://sonris-www.dnr.state.la.us/www_root/sonris_portal_1.htm, accessed 14 November 2006

Leitman HM, Sohm JE, Franklin MA (1984) Wetland hydrology and tree distribution of the Apalachicola River floodplain, Florida. Water-Supply Paper 2196. U.S. Geological Survey, Washington

Light HM, Darst MR, Lewis LJ, Howell DA (2002) Hydrology, vegetation, and soils of riverine and tidal floodplain forests of the lower Suwannee River, Florida, and potential impacts of flow reductions. Professional Paper 1656A. U.S. Geological Survey, Tallahassee

McCormick J (1977) Productivity of freshwater tidal marsh vegetation (Abstract). Bull NJ Acad Sci 22:41

Mendelssohn IA, McKee KL (1988) *Spartina alterniflora* die-back in Louisiana: time course investigation of soil waterlogging effects. J Ecol 76:509–521

Muller RA, Grymes JM III (1998) Regional climates. In: Messina MG, Conner WH (eds) Southern forested wetlands ecology and management. Lewis Publishers, Boca Raton, pp 87–101

NOAA (2006a) National Oceanic and Atmospheric Administration, National Ocean Service, Center for Operational Oceanographic Products and Services, Tides and currents. Available at http://tidesandcurrents.noaa.gov/

NOAA (2006b) National Oceanic and Atmospheric Administration, National Weather Service Forecast Office, New Orleans climate information. Available at http://www.srh.noaa.gov/lix/html/msy/climate.htm

Odum WE (1988) Comparative ecology of tidal freshwater and salt marshes. Ann Rev Ecol Systematics 19:147–176

Odum WE, Dunn ML, Smith TJ III (1978) Habitat value of tidal freshwater wetlands. In: Greeson PE, Clark JR, Clark JE (eds) Wetland functions and values: the state of our understanding. American Water Resources Association, Minneapolis, pp 248–255

Odum WE, Smith TJ III, Hoover JK, McIvor CC (1984) The ecology of tidal freshwater marshes of the United States east coast: A community profile. FWS/OBS-83/17. U.S. Fish and Wildlife Service, Biological Services Program, Washington

Peterson JE, Baldwin AH (2004) Variation in wetland seed banks across a tidal freshwater landscape. Am J Bot 91:1251–1259

Poff NL, Allan JD, Bain MB, Karr JR, Prestegaard KL, Richter BD, Sparks RE, Stromberg JC (1997) The natural flow regime: a paradigm for river conservation and restoration. Bioscience 47:769–784

Portnoy JW, Giblin AE (1997) Biogeochemical effects of seawater restoration to diked salt marshes. Ecol Appl 7:1054–1063

Portnoy JW, Valiela I (1997) Short-term effects of salinity reduction and drainage on salt marsh biogeochemical cycling and *Spartina* (cordgrass) production. Estuaries 20:569–578

Ramsar Convention Bureau (1990) The Ramsar convention on wetlands: Proceedings of the fourth meeting of the conference of the contracting parties, Montreux, Switzerland, 27 June–4 July 1990, Vol 1. Ramsar Convention Bureau, Gland, Switzerland

Rheinhardt RD (1991) Vegetation ecology of tidal freshwater swamps in lower Chesapeake Bay, USA. Ph.D. thesis, The College of William and Mary, School of Marine Science

Rheinhardt RD (1992) A multivariate analysis of vegetation patterns in tidal freshwater swamps of lower Chesapeake Bay, USA. Bull Torrey Bot Club 119:192–207

Rheinhardt RD, Hershner C (1992) The relationship of below-ground hydrology to canopy composition in five tidal freshwater swamps. Wetlands 12:208–216

Shew DM, Baumann RH, Fritts TH, Dunn LS (1981) Texas Barrier Islands Region ecological characterization: Environmental synthesis papers. FWS/OBS-81/32. U.S. Fish and Wildlife Service, Biological Services Program, Washington

Smock LA, Wright AB, Benke AC (2005) Atlantic Coast rivers of the southeastern United States. In: Benke AC, Cushing CF (eds) Rivers of North America. Elsevier, Amsterdam, pp 73–124

Stout JP (1984) The ecology of irregularly flooded salt marshes of the northeastern Gulf of Mexico: a community profile. Biological Report 85(7.1). U.S. Fish and Wildlife Service, Washington

Swarzenski CM (2001) Surface-water hydrology of the Gulf Intracoastal Waterway in south-central Louisiana, 1996–99. Professional Paper 1672. U.S. Geological Survey, Lafayette

USACE (2001) HEC-RAS, river analysis system hydraulic reference manual. U.S. Army Corps of Engineers, Hydraulic Engineering Center, Davis

USACE (2004) Savannah estuary and wetlands study. Section 905 (B) Report. U.S. Army Corps of Engineers, Savannah District

USACE (2006) U.S. Army Corps of Engineers, Savannah District, Savannah harbor expansion project management plan. Available at http://www.sas.usace.army.mil/shexpan/description.htm

USGS (2006) U.S. Geologic Survey, Water Resources Division, National Water Information System. Available at http://waterdata.usgs.gov/nwis/

Verhoeven JTA, Whigham DF, van Logtestijn R, O'Neill J (2001) A comparative study of nitrogen and phosphorus cycling in tidal and nontidal riverine wetlands. Wetlands 21:210–222

Vince SW, Humphrey SR, Simons RW (1989) The ecology of hydric hammocks: A community profile. Biological Report 85(7.26). U.S. Fish and Wildlife Service, Washington

Wharton CH, Odum HT, Ewel KC, Duever MJ, Lugo A, Boyt R, Bartholomew J, DeBellevue E, Brown S, Brown M, Duever LC (1977) Forested wetlands of Florida – their management and use. Center for Wetlands, University of Florida, Gainesville

Whigham D, Simpson R (1977) Growth, mortality and biomass partitioning in freshwater tidal populations of wild rice (*Zizania aquatica* var. *aquatica*). Bull Torrey Bot Club 104:347–351

Williams K, Ewel KC, Stumpf RP, Putz FE, Workman TW (1999a) Sea-level rise and coastal forest retreat on the west coast of Florida. Ecology 80:2045–2063

Williams K, Pinzon ZS, Stumpf RP, Raabe EA (1999b) Sea-level rise and coastal forests on the Gulf of Mexico. Open-file Report 99-441. U.S. Geological Survey, St. Petersburg

Yan Z, Tsimplis MN, Woolf D (2004) Analysis of the relationship between the North American oscillation and sea-level changes in northwest Europe. Int J Climatol 24:743–758

Zampella RA, Dow CL, Bunnell JF (2001) Using reference sites and simple linear regression to estimate long-term water levels in coastal plain forests. J Am Water Res Assoc 37:1189–1201

Chapter 3 - Soils and Biogeochemistry of Tidal Freshwater Forested Wetlands

Christopher J. Anderson and B. Graeme Lockaby

School of Forestry and Wildlife Sciences, 3301 Forestry and Wildlife Sciences Building, Auburn, AL 36849

3.1. Introduction

Tidal freshwater forested wetlands are located along the upper tidal reach of coastal rivers. These forests have a highly complex biogeochemistry in large part because they are influenced by coastal tides but are far enough up river that they are not regularly inundated by saltwater. In addition to freshwater tides, these forests also receive water at different frequencies from river floods, saltwater surges, and groundwater sources. The quantity and timing of each source varies seasonally and annually, often resulting in dramatic shifts in soil physiochemical conditions. In the southeastern United States, soil complexity of tidal freshwater forested wetlands is amplified by a variety of site-specific factors such as microtopography, local climate (e.g., wind driven tides), elevation, proximity to river mouth, vegetation type and cover, and physiographic origin. All of these factors result in soil conditions that are highly variable over space and time, and difficult to generalize.

Compared to other wetlands, there have been very few studies on freshwater tidal forested wetlands and among those, very few have evaluated soil conditions in detail. The scarcity of tidal forested research is likely attributed to their narrow range of occurrence. Perhaps the most commonly cited source for general community descriptions in the southeast United States is Wharton et al. (1982) as part of their comprehensive review of bottomland hardwood swamps. More detailed analyses and/or experimental research on edaphic conditions have been conducted along Gulf coast rivers such as the Apalachicola River (Coultas 1984) and Suwannee River (Light et al. 2002) in Florida. Studies on the Atlantic Coast include the

Pamunkey River, a tributary of the lower Chesapeake Bay in Virginia (Rheinhardt 1991, 1992); the White Oak River (Kelley 1993; Kelley et al. 1990; Megonigal 1996; Megonigal and Schlesinger 2002) and Cape Fear River (Hackney et al. 2002, 2005) in North Carolina; and the Savannah River in South Carolina (Duberstein 2004). Other information comes from numerous county soil surveys produced by the U.S. Department of Agriculture-Natural Resources Conservation Service (USDA-NRCS) (fka, USDA-Soil Conservation Service). Older editions of the USDA-NRCS soil surveys tended to aggregate coastal wetlands and may have contributed to the impression that these were fairly homogeneous soils. However, as newer USDA-NRCS surveys have become available, more intensive classification has occurred (J. DeWit, personal communication; Light et al. 2002), and it has become apparent that floodplain soils are among the most heterogeneous of forested ecosystems.

Because prolonged periods of inundation are common, tidal freshwater wetland soils are noted for being highly anaerobic and containing high amounts of organic matter. Consequently, these soils have the potential to retain nutrients and other material from the inflowing waters through sedimentation, denitrification, plant uptake/detritus storage, sorption and microbial immobilization. Many freshwater tidal wetlands in the southeastern United States accumulate fine sediments in the form of silts and clays which, because of their high sorption affinity, can contribute to the retention of phosphorus and other nutrients (Mitsch and Gosselink 2000). High organic matter contents in tidal freshwater wetland soils also contribute to high cation exchange capacity and the retention of heavy metals and other potential pollutants (Simpson et al. 1983). Where soils are highly reduced, carbon mineralization is limited and dominated by methanogenesis, sulfate reduction, or both in areas where sulfur availability (associated with oceanic water) fluctuates over time (Hackney et al. 2005). As seen through the examples in this chapter, there is considerable temporal and spatial variability in soil processes among and within these tidal swamps.

3.2. Geomorphology

3.2.1. Geological history

The geomorphology of coastlines along both the southeast Atlantic and the Gulf of Mexico has been primarily shaped by multiple changes in sea level that occurred during the late Pleistocene and early Holocene. Following sea level declines in the late Pleistocene (approximately 18,000 years ago), shorelines along the Gulf and southeast Atlantic reached southern extremities that were approximately 100–250 km and 100 km, respectively, seaward of present positions (Hunt 1967). The concurrent increase in channel gradient stimulated the incision of fluvial valleys near coastal areas and, consequently, rivers and floodplains extended to the margins of present-day continental shelves (Smith 1988). Evidence of buried fluvial channels in the Gulf has been found at points 30 m below present sea levels.

Successive glacial and inter-glacial periods during this time caused intense frost and precipitation that led to the rapid transport of surface material to coastal areas. As water levels rose during the post-Wisconsin period of the Holocene, tidal wetlands rapidly expanded into the inundated river systems (Odum et al. 1984). Consequently, as sea levels rose, fluvial valleys were drowned along both coastlines. Along the Gulf coast, the process of sea level inundation within incised fluvial valleys may have been more pronounced to the east of Mobile Bay than to the west (Smith 1988). It might be expected that alluviation would become more pronounced above the zone of oceanic inundation and result in wider alluvial deposition zones there as sea levels rose. However, relatively steep gradients of the Gulf coastal plain may have prevented broad alluvial floodplains from forming with some exceptions such as that of the Mobile-Tensas system.

Along the southeast Atlantic coast most tidal wetlands were formed recently and under the same general process. Unlike some tidal wetlands along the Gulf of Mexico that are declining in size (Louisiana's deltaic wetlands), others along the Atlantic coast are thought to be still expanding. It has been shown that human agricultural activity over the past 300 years has extended the spatial extent of tidal wetlands along the Atlantic coast by increasing the amount of sediment deposition in coastal areas (Frommer 1980). Data collected by Orson et al. (1992) showed that sediment accumulation rates in a Delaware River estuary marsh were 0.04 cm yr^{-1} prior to European colonization. After two centuries of intensive land management (including diked wetlands), tidal marshes became established again and began accumulating sediment at 0.97 cm yr^{-1}. In other estuaries, recent sediment accumulation has likely been tempered by better soil conservation practices (Craft and Casey 2000), the construction of dams (Bednarek 2001), and the channelization of riverbeds that results in the transport of sediments beyond the estuary (Day et al. 2005).

Local relief is lower along the southeast Atlantic coastal plain and the Gulf coast west of Mississippi (<30 m) than along the eastern Gulf (30-90 m) (Walker and Coleman 1987). Differences in relief have probably influenced the width of fluvial valleys in these areas, and according to Walker and Coleman (1987) there are differences between the two coasts in terms of vegetation and landforms. The southeast Atlantic coast can be characterized by salt marshes and tidal creeks whereas the Gulf coast is typified by narrow salt marshes and broad brackish marshes. The southeast Atlantic marshes do not extend between Jacksonville and St. Augustine, Florida, and are replaced there by tidal mangrove swamps. There is no alluviation associated with that coastal zone (Sherman 2005). Walker and Coleman further note that the marsh communities along the southeast Atlantic are more stable than their Gulf coast counterparts due to more rapid subsidence along the Gulf.

One of the most unique ecosystems along the southeast Atlantic coast is that of the St. John's River which runs from south to north on the eastern side of Florida. There is no alluviation associated with the St. John's since it has a very low gradient from its inception to its mouth (Smock et al. 2005). In addition, 59% of the St. John's length is tidally influenced, in contrast to 4% and 12% for the Apalachicola and Suwannee respectively (McPherson and Hammett 1991). Unfortunately, detailed soil information is not available for the St. John's River floodplains.

As alluvial depositional areas such as deltas extend into deeper water, rates of alluviation are slowed and subsidence of the deltaic plain may occur. As a result, stream courses may wander within the plain to areas where gradients are steeper and flow is less impeded (Shirley and Ragsdale 1966). For instance in the Lower Mississippi River Delta, this is the reason that the river course would shift to the southwest without anthropogenic intervention. Rivers of the southeast Atlantic coastal plain may be migrating in a southerly direction as evidenced by high bluffs and wide floodplains on the southern and northern sides of rivers, respectively (Brinson 1990).

3.2.2. Formation

Relative to other tidal wetlands, freshwater forests generally occur at the very fringe of tidal influence. Throughout the coastal plain, the most extensive forests seem to occur along larger rivers that are influenced by a wide tidal range (Duberstein 2004). The salinity of tidewaters is also important, and Rheinhardt (1992) reported that for tidal freshwater forested wetlands along the Pamunkey River, salinity rarely rises above 0.05 ppt except at the extreme downriver extent of the swamp. Extending upriver from tidal salt marshes, freshwater marshes often occur (sometimes extensively) between the salt marshes and tidal forests. Freshwater tidal forested wetlands such as those at the Suwannee River (Light et al. 2002) and Pamunkey River (Rheinhardt 1992) have been found to occur as a mosaic of forest and marsh and the cover of understory vegetation can vary significantly based on canopy closure.

One commonly reported feature of tidal freshwater forested wetlands is the mosaic of hummocks and hollows that occurs along the forest floor. The hummocks are elevated sections or "islands" above the forest floor that consist of a network of interwoven tree roots or moss-covered remnants of tree trunks or large branches (Peterson and Baldwin 2004). The reported heights of these hummocks relative to the surface of the hollows can range between 15-20 cm, and often correspond to the mean water level of the swamp (Rheinhardt and Hershner 1992). Trees in freshwater tidal forested wetlands are frequently limited to those contributing to hummock formations with herbaceous or no vegetation in the hollows. There is also evidence that hydrology related to tidal range and flood duration may influence the relative proportions of hummocks and hollows. Rheinhardt (1992) reported that an ash/blackgum (*Fraxinus/Nyssa*) community located along a lower reach of the Pamunkey River was 65.5% hollow compared to only 25.0% for a maple/sweetgum (*Acer/Liquidambar*) community ~5 km upriver (where flood durations were shorter). We are aware of no study that has extensively compared edaphic conditions between hummocks and hollows. It is expected that hollows would have more prolonged anaerobic periods and consequently contain higher levels of organic matter; however, Rheinhardt (1992) found that concentrations between hummocks and hollows were comparable in tidal swamps along the Pamunkey River. Nevertheless, given the differences in elevation and hydrology, substantial differences in the soil biogeochemistry of hummocks and hollows are likely.

Given the importance of tidal range to these wetlands, it should be emphasized that the process of sea-level rise is continuing, with rates estimated between 0.1–0.2 cm yr^{-1} worldwide (Gornitz et al. 1982; Warrick and Oerlemans 1990). Rising sea levels may eventually influence all tidal wetlands but some coastal forests may be particularly susceptible (Chapter 1). Rates of sea-level rise along the Gulf of Mexico may be higher than average as indicated by estimates of 1.0 cm yr^{-1} along the Mississippi River Delta (Reed 2002) and 0.23 cm yr^{-1} near Pensacola, Florida (Penland et al. 1987). The impact of increased water levels along coastal river forests may be amplified by altered sediment loads and hydrology from the construction of 20th century dams, agriculture, and development (Walker and Coleman 1987). Consequently, the geomorphic relationship among coastal forest systems, humans, and marine environments remains highly dynamic.

3.3. Soil taxonomic classification

Across the southeast, USDA-NRCS county soil surveys indicate that the following soil orders may be found near river outlets: entisols, inceptisols, histosols, alfisols, mollisols, and ultisols. The presence of some of these orders on a floodplain may seem to conflict with expectations in terms of geomorphic context. The soil orders that might be anticipated to occur on floodplains include those associated with active depositional positions (e.g., entisols and inceptisols). Histosols would also be expected in areas where surface saturation persists for long periods and, as a result, decomposition is inhibited. Entisols and inceptisols exhibit less profile development (i.e., fewer horizons) than some other orders and are said to be young in a geological sense since they are morphologically homogenous in a vertical direction. Conversely, older soils with well developed profiles (e.g., ultisols, alfisols, and mollisols) are usually associated with very stable geomorphic positions such as uplands, as opposed to active alluviation zones. Consequently, the presence of the latter orders on floodplains contributes greatly to soil diversity there. In general, USDA-NRCS surveys indicate that entisols and inceptisols are more common on floodplains along the lower reaches of Gulf coast rivers while ultisols, alfisols, and mollisols, along with entisols and inceptisols, are mapped on the Southeast Atlantic coast. While it is tempting to seek geomorphological reasons for the apparent contrast between the Gulf and Atlantic coasts, this probably indicates nothing more than variation in the level of detail provided in soil mapping among various locations.

As noted, soil maps of lower reaches of rivers emerging along the Atlantic coast often exhibit greater diversity than those of the Gulf coast (e.g., the Mobile and Apalachicola Rivers). Although soil complexes along coastal reaches of the Satilla River (Rigdon and Green 1980) are similar to those described for the Mobile and Apalachicola Rivers, the lower reaches of the Altamaha River (Rigdon and Green 1980) are mapped as aqualfs and aquults (wet soils with a high degree of profile development) and aquents (wet soils with little profile development). Along the Edisto River (Stuck 1982), aquents and aquepts are mapped along the lower reaches of the river; however, udults and aquults occur closer to the coast. The soils of the forested reaches along the Edisto, nearest the coast, are mapped as wet mollisols. Sulfaquents dominate as the forest transitions to marsh nearer the coast. Similarly, although aquents are mapped on the lowest forested reaches of the Greater Pee Dee River (Stuckey 1982), aquults and udults dominate its floodplains a few km upriver from the coast. Soils studied in the tidal forests of the White Oak River estuary (Megonigal and

Schlesinger 2002) were classified as thermic typic medisaprists (Barnhill 1981, 1992) and tidal forests along the Pamunkey River (Rheinhardt 1992) were mapped as ferric mediasaprists of the Mattan series (Hodges et al. 1988).

Along the Gulf coast, soils found along the lower reaches of rivers such as the Mobile (McBride and Burgess 1964; Hickman and Owens 1980) and Apalachicola (Sasser et al. 1994) are predominantly mapped as aquents, aquepts, and saprists. These represent wet mineral (and, in the case of saprists, organic) soils with little horizonation. Further from river channels, udults and aquults are found. As vegetation changes from forest to marsh, soils near the river mouth are typically of the Bohicket series (typic sulfaquent), a wet entisol with high organic matter content.

Published research conducted in freshwater tidal wetlands has generally focused on the forest interior. However, studies that have been more inclusive of the entire tidal forest range have confirmed a wide range of soil characteristics (Doumlele et al. 1984; Light et al. 2002). Light et al. (2002) evaluated 96 soil profiles along an elevation and distance-to-river-mouth gradient for coastal forests along the Suwannee River (including non-tidal floodplains). After analyzing a subset of 67 profiles, they found soils were highly variable and represented 7 orders and 18 taxonomic sub-groups. Visually, soils ranged from upland in appearance to deep mucks. In the upper tidal reaches of the Suwannee River (22–37 km from the river mouth), forest soils were more diverse and included entisols, mollisols, inceptisols and histosols. This portion of the forest was considered to be only partially influenced by tides and the occurrence of highly organic histosols only occurred at the lowest elevations. In the lower tidal swamp reach of the river (<22 km from the river mouth) where tides and salinity are more influential, histosols became much more prevalent and as the elevation of sampled soils decreased there was a general increase in the overall depth of the surficial muck layer (in some cases >2 m). Most forest soils along the lower tidal reach of the Suwannee River were classified as typic or terric haplosaprists (Light et al. 2002). Studying soils in the tidal freshwater forests of the Apalachicola River, Coultas (1984) classified three out of five soil pedons (collected to depths >125 cm) as typic sulfihemists based on existing soil concentrations of sulfur > 0.75% (although surface soil concentrations were typically much lower).

In the southeastern United States, higher proportions of the watersheds of blackwater rivers lie in the coastal plain compared to those of redwater systems which are primarily associated with the piedmont. Blackwater rivers are also characterized by higher organic loads derived from extensive swamps and floodplain forests that dominate the coastal plain landscape

(Schilling and Lockaby 2005). In contrast, redwater rivers often have high grades which impart a greater capacity for carrying sediment (and, consequently, nutrient) loads (Lockaby et al. in press). As a result of the geomorphic differences between the system types, it might be expected that a comparison of redwater vs. blackwater floodplain soils would yield morphological distinctions. However, based on available mapping information, there is little evidence of a taxonomic differential between soils near the outlets of blackwater vs. redwater rivers in the southeast United States. Nevertheless, there are clear distinctions between floodplains of these two river types in terms of biogeochemistry (Schilling and Lockaby 2006).

3.4. Biogeochemistry

3.4.1. Soil organic matter

Soils of freshwater tidal wetlands tend to be highly organic (Table 3.1). In the southeast, organic matter in tidal wetlands is usually highly decomposed muck (saprist soils) as opposed to some coastal wetlands in northern regions where organic matter can often accumulate as fibrist peat (Odum et al. 1984). In terms of concentration, Wharton et al. (1982) listed tidal freshwater swamps among floodplain communities in the southeast as having the highest concentrations of soil organic matter. The high level of organic matter is the result of suppressed decomposition under anaerobic soil conditions and moderate to high autochthonous production within the forest (Wharton et al. 1982). As with most wetlands, the accumulation of organic matter in tidal freshwater swamps is closely linked to hydrology and several authors have reported deep mucky soils occurring in the lowest elevations of the forest (Doumlele 1984; Light et al. 2002). Comparing different forest communities along the Pamunkey River, Rheinhardt (1992) found that an ash-blackgum community had significantly higher organic matter (40.5%) than a maple-sweetgum community (27.8%) located further upriver, and attributed this difference to longer flood durations in the ash-blackgum community.

The physiographic and geomorphic setting of the river also plays a role in soil organic matter content. Estimates from the literature indicate that freshwater tidal forests in blackwater river swamps have higher organic matter concentrations (average 46%) than redwater river swamps (average 31%) (Table 3.1, although see Rheinhardt [1992] and Duberstein [2004]). Reported soil organic matter concentrations often reflect the most highly inundated sections of the forest, however it should be noted that there is

Table 3.1. Soil characteristics and dominant tree communities for freshwater tidal forests in the southeastern United States. Multiple river listings are provided for different sampling sites.

River, Location	River-type	Soil depth (cm)	pH	Organic Matter (%)	Cond. (mmhos/cm)	Textural components			Tree community	Source
						clay	silt	sand		
Apalachicola, FL	redwater	8 - 30	5.3	15		46	36	18	Tupelo gum, cypress	Wharton et al. 1982
Apalachicola, FL	redwater	0 - 20	5.6	33	0.1				Mixed hardwood, cypress	Coultas 1984
Apalachicola, FL	redwater	0 - 30	5.5	28	0.9	85	14	1	Mixed hardwood, cypress	Coultas 1984
Apalachicola, FL	redwater	2 - 40	4.9	9	0.2	77	23	0	Mixed hardwood, cypress	Coultas 1984
Apalachicola, FL	redwater	2 - 32	5.8	14	0.1	78	22	0	Mixed hardwood, cypress	Coultas 1984
Apalachicola, FL	redwater	0 - 15	5.7	25	0.2	73	22	5	Mixed hardwood, cypress	Coultas 1984
Choctawhatchee, FL	redwater	8 - 30	5.5	36					Swamp tupelo, bay	Wharton et al. 1982
Escambia, FL	redwater	8 - 30	5.4	15		28	17	55	Sweet bay, cypress, white cedar	Wharton et al. 1982
Pamunkey, VA	redwater	0 - 15	4.8	41					Ash-blackgum	Rheinhardt 1992
Pamunkey, VA	redwater	0 - 15	4.9	25					Maple-sweetgum	Rheinhardt 1992
Patuxent, MD	redwater	0 - 10		44					Shrub	Verhoeven et al. 2001
Savannah, GA	redwater	0 - 13	5.6	77	6.7				Swamp tupelo, tag alder, shrub	Duberstein 2004
Savannah, GA	redwater	0 - 13	5.7	46	3.0				Water oak, swamp bay, tupelo	Duberstein 2004
Aucilla, FL	blackwater	8 - 30	6.3	19		14	13	73	Mixed	Wharton et al. 1982
Sopchoppy, FL	blackwater	8 - 30	5.7	77					Sweet bay, swamp tupelo	Wharton et al. 1982
St. Marks, FL	blackwater	8 - 30	6.2	45		6	21	73	Swamp tupelo, sweet bay	Wharton et al. 1982
Suwannee, FL	blackwater	8 - 30	6.0	40		17	23	60	Swamp tupelo, cypress-ash	Wharton et al. 1982
Suwannee, FL	blackwater	8 - 30	6.4	41		20	30	50	Swamp tupelo, cypress-ash	Wharton et al. 1982
Yellow, FL	blackwater	8 - 30	5.2	53					Sweet bay, white cedar	Wharton et al. 1982

significant spatial variability associated with forest geomorphology. Ridges and other high elevations within the forest (less often reported) will contain more mineral soil conditions and can make generalizations difficult to interpret.

3.4.2. Methanogenesis

Studies examining the redox potential of soils in freshwater tidal wetlands have reported levels capable of supporting methanogenesis (Megonigal and Schlesinger 2002; Hackney et al. 2002, 2005). The emission of methane from wetlands in general is a substantial component of the atmospheric methane budget, and Whalen (2005) has reported that emission rates of 100 mg CH_4 m^{-2} d^{-1} are common for wetland ecosystems. Because of the carbon reserves in freshwater tidal swamps, these wetlands have the potential to produce substantial quantities of methane. A series of studies on methane emissions were conducted at the White Oak River estuary along the east coast of North Carolina. Kelley et al. (1995) found large seasonal changes in methane production in freshwater tidal swamps that did not occur in permanently inundated creek sites nearby. Methane production was mediated by bacterial oxidation at low tide and at the oxygenated sediment surface. While very little CH_4 was produced during the winter season, large fluxes (up to 1000 mg CH_4 m^{-2} d^{-1}) were measured during the growing season. Nearby, non-vegetated sites that were permanently submerged had only moderate increases in CH_4 flux during the summer. Overall methane flux was higher along the vegetated river banks (tidal areas) than in the permanently inundated sites.

Within tidal freshwater swamps, there is evidence that CH_4 emissions have high spatial and temporal variability. Temporal variability of emissions can be attributed to changes in water level associated with tide events (daily fluctuation) and seasonal variability inherent to temperate climate conditions. Kelley et al. (1995) found that the highest daily CH_4 emission rates occurred when water levels coincided with the soil surface (Figure 3.1). They surmised that when tidewaters were below the sediment surface, the aerobic layer at the sediment-water interface increased in area and greater CH_4 oxidation occurred. Alternately, when tidewaters flowed into the swamp and standing water was present, CH_4 emissions declined because of a diffusion barrier caused by the standing water. Furthermore, inflowing tidewaters may provide an aerobic environment for methane oxidation resulting in a reduction of net emissions. Incoming water can also impede methane emission by inundating plant surfaces that normally convey belowground gases to the atmosphere. Many macrophytes provide a

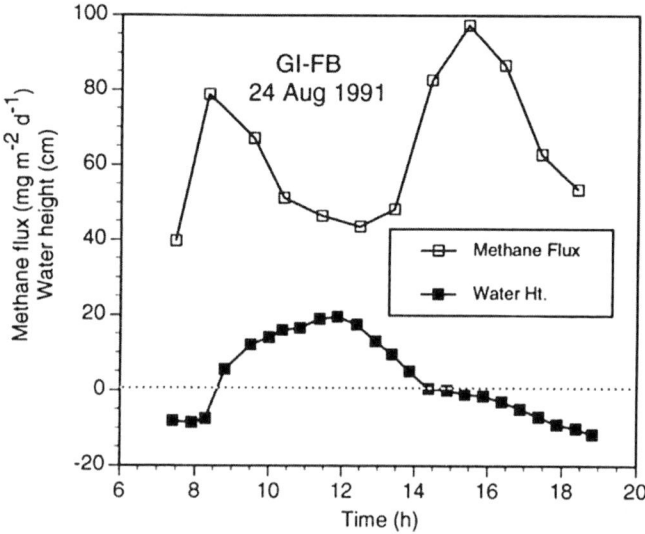

Fig. 3.1. Measured fluxes using static chamber techniques and water level over a 12-h tidal cycle at a tidal swamp in the White Oak River Estuary. Dotted line represents the sediment surface (water height = 0 cm) (Kelley et al. 1995 with permission; copyright (1995) by the American Society of Limnology and Oceanography, Inc.).

conduit for gas exchange between the root zone and the atmosphere through aerenchyma tissue. When plant surfaces become inundated, water can again act as a diffusion barrier for methane emission (Chanton et al. 1992).

Interestingly, it appears that the tidal hydrology of freshwater tidal swamps may suppress CH_4 emissions compared to other forested wetlands. Methane flux rates reported by both Kelley et al. (1995) and Megonigal and Schlesinger (2002) were low compared to ranges reported for swamps in Virginia (83-155 mg CH_4 m^{-2} d^{-1}, Wilson et al. 1989) and Louisiana (146-912 mg CH_4 m^{-2} d^{-1}, Alford et al. 1997) during the growing season. Very low rates were reported by Megonigal and Schlesinger (2002) who estimated a peak monthly mean flux of only 17 mg CH_4 m^{-2} d^{-1} at an upstream tidal forest along the White Oak River although there was considerable spatial variability within sites (one plot had measurements >200 mg CH_4 m^{-2} d^{-1}).

Annual variability of methane production has been reported by Hackney et al. (2005) for tidal swamps and marshes along the Cape Fear River. Using a series of monitoring stations, they have monitored shifts between methanogenesis and sulfate reduction as the primary mode of C mineralization since 2000 (Hackney et al. 2002, 2005). Oceanic saltwater contains higher levels of sulfate so that when tidal ranges expand, microbes that utilize sulfates as an electron acceptor can become prevalent and outcompete methanogens (Mitsch and Gosselink 2000). Examining conditions four times a year, they have found irregular shifts between the two biochemical processes corresponding with upriver influence on the tidal range of oceanic saltwater. When tidal shifts have occurred (as a result of reduced upriver flows during droughts) and the range of saltwater influence has expanded to previously freshwater reaches, they have found that the dominant microbial community readily shifts from methanogenic to sulfate reducing. Other hydrologic differences attributed to tidal range and microtopography can have substantial effects on soil biogeochemistry. Megonigal and Schlesinger (2002) found that for soils in a site further upriver (where soils were exposed from flooding 60% of the year compared to 40% at a site downriver), the hydrologic difference elicited a 27% higher CH_4 oxidation capacity in upper site soils, a vertically deeper occurrence of peak CH_4 production in the soil profile, and an herbaceous stratum that was less flood tolerant.

There is also evidence that the presence and/or quantity of macrophytes in a tidal freshwater swamp may regulate CH_4 emissions. Although flooding may impede some gas emissions via plant tissue to the atmosphere (Chanton et al. 1992), it is likely that these plants still represent an important conduit for CH_4 transmission. Kelley et al. (1995) found discrepancies between two techniques used to estimate CH_4 flux (diffusion models using dissolved methane concentrations v. static chambers) and suggested that the higher rates detected by chambers was attributed to plant-mediated gas transport. Likewise, Megonigal and Schlesinger (2002) explained that the sparse herbaceous cover in their sites (due to dense canopy cover) may have attributed to the low CH_4 emissions detected in their study.

3.4.3. Salinity

Salinity is another critical factor controlling the range and distribution of tidal freshwater swamps. Most of these swamps occur where water salinities are normally below 0.5 ppt however they are exposed to occasional tidal surges that can induce saline conditions in normally freshwater areas (Simpson et al. 1983; Peterson and Baldwin 2004). These surges are caused by storms/hurricanes, tide pushing winds, and upstream droughts that reduce freshwater inputs and increase the tidal extent of saltwater into the forest. Because the frequency and intensity of tidal surges varies over time, it is likely that soil salinity in freshwater wetlands also varies year-to-year. While all tidal swamps are exposed to periodic surges of saltwater, long-term decreases in upstream freshwater flow have the potential to cause more permanent changes to the ecology of the forest. Along the Cape Fear and Suwannee River, researchers have investigated the potential effects of human river management on existing tidal ranges, upriver soil salinities, and the ecology of existing freshwater tidal wetlands (Hackney et al. [2002] and Light et al. [2002], respectively).

In tidal freshwater forests along the Suwannee River (Light et al. 2002) and Apalachicola River (Coultas 1984), the highest salinities within the soil profiles were generally observed below the surface layer and root zone. Along the Apalachicola, tidal forest soils had surface conductivities <1.0 mmhos cm^{-1} (Table 3.1) while subsurface conductivity ranged between 2.9 and 7.7 mmhos cm^{-1}. It is likely that the regular inflow of freshwater effectively flushes occasional salt deposition from the surface. Light et al. (2002) detected a general trend of increasing salinity in Suwannee River tidal forests as transects got closer to the river mouth. In the lower reaches of the tidal forest (those closest to the Gulf) surface soil salinities were often >4 mmhos cm^{-1} and, unlike soils in the upper tidal reaches, the subsurface soils tended to be as saline (or more) than at the surface reflecting the greater exposure to tidal saltwater. It was also noted that swamps in the upper tidal reach that were isolated from regular surface flows but within reach of storm surges, there was not an accumulation of salts in the soils as would be expected. This was attributed to the prevalence of groundwater flushing from a shallow limestone aquifer that is close to the surface.

Soils are considered to be saline if soil conductivity levels exceed 4 mmhos cm^{-1}; however, impacts of salinity on plant growth and species occurrence occur at lower levels. In the Suwannee River floodplains, water tupelo (*Nyssa aquatica* L.) apparently does not occur in lower reaches due to high salinity levels there (> 4 mmhos cm^{-1}). The impacts of high salinity on tree growth and species occurrence are associated with moisture stress imposed by hydrophilic salts near roots. In the cypress/tupelo forests of the lower Mobile Delta, across plots with similar tree ages and past management histories, heights of dominant trees (a well established index of site quality) were considerably lower at soil conductivity levels above approximately 1.75 mmhos cm^{-1} (Figure 3.2). Mobile Delta tree heights averaged 22.3 m on sites with soil conductivities below 1.75 mmhos cm^{-1} and 17.4 m on soils above that conductivity level. Similarly, basal areas averaged 18.7 and 37.1 m^2 ha^{-1} respectively above and below 1.75 mmhos cm^{-1} (Figure 3.3). Low heights and basal areas in the Mobile Delta were associated with average exchangeable Na levels of 817 mg kg^{-1} (upper meter)

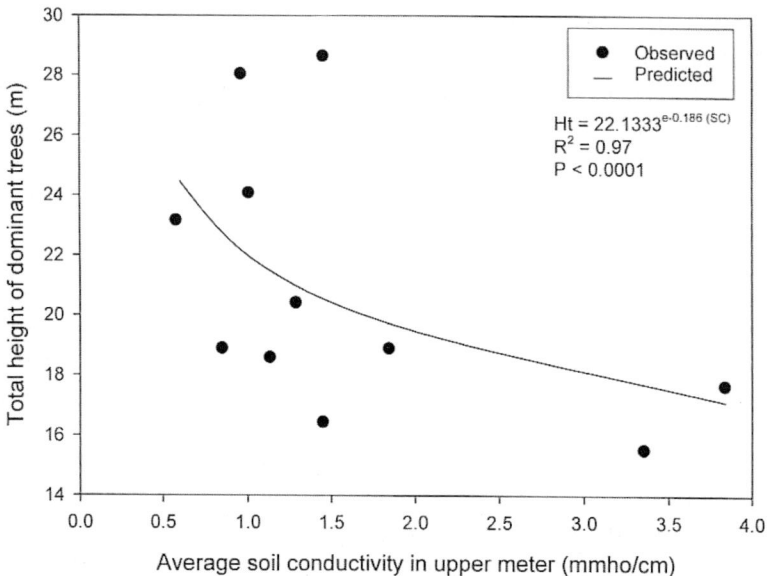

Fig. 3.2. Comparison of total height of dominant trees and soil conductivity in cypress-tupelo stands of the Mobile River Delta (Jim DeWit – unpublished data).

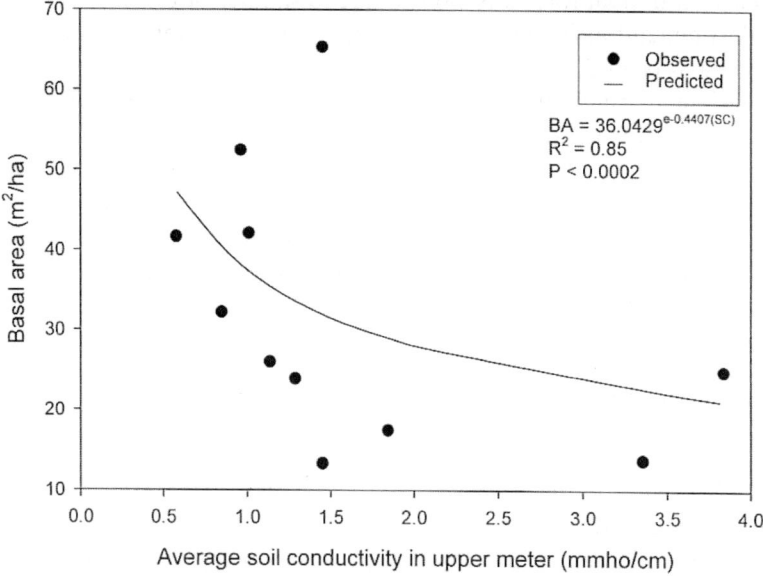

Fig. 3.3. Comparison of total basal area and soil conductivity in cypress-tupelo stands of the Mobile River Delta (Jim DeWit – unpublished data).

while greater heights and basal areas occurred on soils averaging 238 mg kg^{-1} Na. The low basal areas associated with higher conductivity levels in the Mobile Delta are in agreement with the low standing crops of aboveground biomass previously noted in fringe wetland forests such as mangroves (Lugo 1990). In the latter forests, a high root to shoot ratio was often observed.

As would be expected, soil conductivity levels and forest productivity within the Mobile Delta generally rose and fell respectively in a southerly direction toward Mobile Bay (Jim DeWit, unpublished data). However, there is also significant variation locally in the southern portion of the Mobile Delta. DeWit (personal communication) indicates that soils located on natural berms adjacent to the tidal creeks tend to be higher in Na and conductivity and lower in productivity compared to soils in low areas behind the berms. The soils behind the berms often support highly productive tupelo gum stands and can be classed as thapto-histic entisols (entisols that have formed over a buried organic soil). Also, soils adjacent to streams that form within the lower delta are apparently more subject to tidal influences and tend toward higher salinity compared to counterparts adjacent to streams with headwaters further north. Forest productivity is lower on soils adjacent to lower-delta streams vs. soils near streams with watersheds that extend north of the tidal zone.

There is no indication from the limited amount of data available that decomposition in tidal forests is slowed due to salinity or other factors. Actually, decomposition in these systems may be more rapid than elsewhere (see discussion in 3.4.4), an indication of the effectiveness of halophiles and other microbial communities that are adapted to saline environments. Similarly, in the data gathered by DeWit in the Mobile River Delta, no relationship was apparent between soil conductivity and organic matter concentrations in the upper 1 m of soil. This suggests that salinity does not affect soil organic matter accumulation there and that the presence of histosols and near-histosols in coastal forests may be due primarily to wetness as it is in freshwater wetlands.

3.4.4. Soil chemistry and nutrient cycling

Only a handful of studies have reported nutrient concentrations for soils in tidal freshwater swamps in the Southeast (Table 3.2). Compared to other floodplain soils, the concentrations reported in tidal freshwater swamps are much higher. Wharton et al. (1982) explained that one reason for high levels of Ca and Mg may be that several of the blackwater river forests examined have groundwater inputs from spring-fed streams that pass through a

Table 3.2. Soil nutrient concentrations for freshwater tidal forests along different rivers in the southeastern United States. Multiple river listings are provided for different sampling sites.

River, Location	River-type	Soil depth (cm)	S (%)	N (%)	Macronutrients P (ppm)	Ca (ppm)	K (ppm)	Mg (ppm)	Na (ppm)	Source
Apalachicola, FL	redwater	8 - 30			8.0	1676	92	180	53	Wharton et al. 1982
Apalachicola, FL	redwater	0 - 20	3.22	0.76						Coultas 1984
Apalachicola, FL	redwater	0 - 30		0.78						Coultas 1984
Apalachicola, FL	redwater	2 - 40	0.04	0.07						Coultas 1984
Apalachicola, FL	redwater	2 - 32	0.27	0.47						Coultas 1984
Apalachicola, FL	redwater	0 - 15	0.15	0.74						Coultas 1984
Choctawhatchee, FL	redwater	8 - 30			18.0	1213	164	616	664	Wharton et al. 1982
Escambia, FL	redwater	8 - 30			15.2	776	116	456	224	Wharton et al. 1982
Pamunkey, VA	redwater	0 - 15			10.8	806	68			Rheinhardt 1992
Pamunkey, VA	redwater	0 - 15			8.0	578	67			Rheinhardt 1992
Patuxent, MD	redwater	0 - 10		1.21						Verhoeven et al. 2001
Savannah, GA	redwater	0 - 13			62.4	6804	284	1569	889	Duberstein 2004
Savannah, GA	redwater	0 - 13			60.3	3901	239	514	276	Duberstein 2004
Aucilla, FL	blackwater	8 - 30			12.0	2852	35	400	80	Wharton et al. 1982
Sopchoppy, FL	blackwater	8 - 30			29.0	1324	36	628	183	Wharton et al. 1982
White Oak, NC	blackwater	0 - 27		1.20 – 1.74[a]						Kelley 1993
White Oak, NC	blackwater	0 - 27		1.36 – 1.92[a]						Kelley 1993
St. Marks, FL	blackwater	8 - 30			41.2	2462	97	556	145	Wharton et al. 1982
Suwannee, FL	blackwater	8 - 30			55.0	5784	323	870	1011	Wharton et al. 1982
Suwannee, FL	blackwater	8 - 30			46.0	5488	78	499	65	Wharton et al. 1982
Yellow, FL	blackwater	8 - 30			14.4	1404	144	588	196	Wharton et al. 1982

[a] %N range based on soils analyzed at 3-cm depth increment.

limestone substratum. High levels of these same nutrients were also reported for several of the alluvial rivers listed in (Table 3.2). The high concentrations of many listed nutrients may be a reflection of high soil organic matter with nutrients sorbed to or bound within soil organic matter. More research on the form of soil nutrients in these forest soils is needed.

Wharton et al. (1982) and Schilling and Lockaby (2005) have suggested that soils associated with redwater systems are more fertile in terms of base cations than those of blackwater counterparts. These potential distinctions may be due to more active alluviation in the redwater systems. Stanturf and Schoenholtz (1998) also note that alluviation of coastal reaches of redwater river floodplains is usually associated with clayey, Piedmont sediments whereas any alluviation along blackwater rivers is derived from coastal sediments that originate locally. Analyzing tidal swamp soils along the Apalachicola River (a redwater river), Coultas (1984) found the concentration of fine sediment (silts and clay) at or near 100% for most of the pedons examined and Wharton et al. (1982) reported concentrations as high as 82%. In contrast, the concentration of sand in blackwater tidal swamps sampled by Wharton et al. (1982) ranged from 50 to 73%.

Soils in tidal freshwater swamps tend to be acidic with a reported pH range between 4.8 and 6.4 (Table 3.1). This is slightly lower than what is typically reported for alluvial non-tidal swamps (pH of 6 to 7) and comparable to the range reported for cypress domes and other perched-basin swamps with high concentrations of organic matter (Mitsch and Gosselink 2000). In swamps with high organic matter content, soil pH generally reflects the high amount organic acids generated. Our limited literature review (Table 3.1) found that the pH range for redwater swamps (4.8–5.8) is slightly lower than blackwater swamps (5.2–6.4), which is somewhat surprising given the higher organic matter concentrations in the latter. However, pH in blackwater swamps may be moderated somewhat by spring-fed groundwater that commonly emerges from a limerock substratum along the Gulf coast of Florida (Clewell 1991).

Research is also needed to identify important nutrient cycling processes in tidal freshwater swamps. Because prolonged flooding and highly reduced soils are common, denitrification is likely the most important mechanism for N removal from these wetlands. Lockaby and Walbridge (1998) found that denitrification rates reported for southern forested wetlands ranged from 0.5 to 350 kg ha^{-1} y^{-1}. We would expect rates for tidal freshwater swamps to be on the high end of this scale based on the high input of N from rivers.

Based on research related to tidal fluctuations, forest litter accumulation and decomposition in these swamps are likely controlled by tidal fluctuations and movement within the forest. Twilley (1982) described decomposition in selected mangrove systems as rapid with rates being directly related to C:N ratios. Decay rates (k) for mangrove forests in Florida averaged 2.87 (i.e., a turnover time of 0.34 years) (Lugo 1990). Lugo noted that forest floor mass was highly variable in these systems due to fluctuations in water movement. Often, high accumulations could be found at the tidal fringe although very little forest floor mass might accrue in other areas due to the flushing action of the tide. In the latter areas, rapid turnover rates may be primarily due to waterborne export rather than decomposition (Lugo 1990). Similarly, Twilley (1998) states that turnover rates in mangrove systems rise as tidal inundations increase and are directly related to the ecological setting (i.e., rates rank riverine > fringe > basin) and that litter mass loss rates may be greatly accelerated by the presence of crabs. Decomposition rates reported in tidal forests along the South Carolina coast (Ozalp 2005) were rapid (e.g., average k = 1.8, turnover time = 0.55 years). The k rates of Twilley, Lugo, and Ozalp may be compared to an average k from riverine forests of 1.01 (Lockaby and Walbridge 1998).

Plant uptake has been cited as a major pathway for nutrient cycling in tidal freshwater marshes (Simpson et al. 1983) and forested floodplains in the Southeast (Lockaby and Walbridge 1998). In freshwater tidal wetlands, nitrogen and phosphorus are imported into these wetlands largely as inorganic ions and delivered at various degrees by tidal waters, river flooding, groundwater seepage, and precipitation. During the growing season, these nutrients are rapidly assimilated into plant material. After the growing season and plant senescence, much of the nutrients associated with plant detritus are exported from the marsh through tidal flushing. Simpson et al. (1983) reported up to 80 percent of the nitrogen and phosphorus associated with plant litter was lost within 2 months of senescence in a Delaware River tidal freshwater marsh. It is uncertain how much detritus export occurs in tidal swamps but there are circumstances to suggest that the dynamics may be different than marshes. Because of their generally closer proximity to the coast, tidal energies may be higher through marsh systems than the upper tidal reaches occupied by forests. Also, the structure (trees) and microtopography of hummocks and hollows may further dampen tidal energies that would potentially export detritus (although initially, senescent marsh stems would dampen tidal energies as well).

Like N, there is no known account regarding the import-export dynamics of P in tidal freshwater swamps. In acid soils such as those in tidal freshwater swamps, the adsorption of phosphates to aluminum and iron has been identified as a key P retention mechanism (Lockaby and Walbridge 1998). In alluvial rivers where fine sediments occur, we expect that the deposition of P adsorbed to sediment is another key retention process. Although at tidal swamps along the Pamunkey River, Rheinhardt and Hershner (1992) observed that most sediment accumulation seemed to occur along the river edge at low-lying levees.

Studies on nutrient cycling efficiencies in tidal forests are limited to mangrove systems. In his 1990 review, Lugo used mangrove (*Rhizophora* spp.) data to compare nutrient cycling efficiencies (ratio of litterfall mass to nutrients in litterfall, Vitousek 1984) among various mangrove systems. Comparisons indicated that efficiencies of nitrogen cycling were high relative to other forest systems, those of calcium were similar to those of upland forests, and those of phosphorus were highly variable.

3.5. Research needs

It is apparent that relatively little information exists on the soils and biogeochemistry of tidal freshwater forests compared to other wetland types. While saltwater and freshwater tidal marshes have been studied to a much greater extent, we know very little about the regulation of biogeochemical and energy exchange between the sea and coastal forests. The few soil studies in coastal swamps have shown that these areas are highly diverse with variable conditions among and within swamps. A limited number of studies have focused on evaluating potential forest impacts from human induced changes in hydrology (Hackney et al. 2002; Light et al. 2002). While the latter is a very critical topic, information is lacking on fundamental biogeochemical processes such as decomposition, mineralization, and nutrient requirements for these systems. Consequently, a major imbalance is evident when the societal importance of these ecotones (in particular, considering the history of catastrophic events along southeastern coastlines) is compared to our knowledge of how they function.

References

Alford DP, Delaune RD, Lindau CW (1997) Methane flux from Mississippi River deltaic plain wetlands. Biogeochemistry 37:227-236

Barnhill WL (1981) Soil survey of Jones County, NC. U.S. Department of Agriculture, Soil Conservation Service, Washington

Barnhill WL (1992) Soil survey of Onslow County, NC. U.S. Department of Agriculture, Soil Conservation Service, Washington

Bednarek AT (2001) Undaming rivers: a review of the ecological impacts of dam removal. Environ Manage 27:803-814

Brinson MM (1990) Riverine forests. In: Lugo AE, Brinson MM, Brown S (eds) Ecosystems of the world 15: Forested wetlands. Elsevier, New York, pp 87-141

Chanton JP, Whiting GJ, Showers WJ, Crill PM (1992) Methane flux from *Peltandra virginica*: stable isotope tracing and chamber effects. Global Biogeochemical Cycles 6:15-31

Clewell AF (1991) Florida rivers: the physical environment. In: Livingston RJ (ed) The rivers of Florida. Springer-Verlag, New York, pp 17-30

Coultas CL (1984) Soils of swamps in the Apalachicola, Florida estuary. Florida Scientist 47:98-107

Craft CB, Casey WP (2000) Sediment and nutrient accumulation in floodplain and depressional freshwater wetlands of Georgia, USA. Wetlands 20:323-332

Day JW Jr, Barras J, Clairain E, Johnston J, Justic D, Kemp GP, Ko J-Y, Lane R, Mitsch WJ, Steyer G, Templet P, Yanez-Arancibia A (2005) Implications of global climatic change and energy cost and availability for the restoration of the Mississippi delta. Ecol Engr 24:253-265

DeWit J (2006) Personal communication. Director of Land Management, Molpus Timberland Management

Doumlele DG, Fowler K, Silberhorn GM (1984) Vegetative community structure of tidal freshwater swamp in Virginia. Wetlands 4:129-145

Duberstein J (2004) Freshwater tidal communities sampled in the lower Savannah River floodplain. M.S. thesis, University of Florida

Frommer NL (1980) Morphological changes in some Chesapeake Bay tidal marshes resulting from accelerated soil erosion. Z Geomorph NF 34:242-254

Gornitz V, Lebedeff S, Hansen J (1982) Global sea level trend in the past century. Science 215:1611-1614

Hackney CT, Posey M, Leonard LL, Alphin T, Avery GB (2002) Monitoring the effects of a potential increased tidal range in the Cape Fear River ecosystem due to deepening Wilmington Harbor, North Carolina. Year 1:August 1, 2000-July 31, 2001. U.S. Army Corps of Engineers, Wilmington District (Contract No. DACW 54-00-R-0008), Wilmington

Hackney CT, Posey M, Leonard LL, Alphin T, Avery GB, Brooks G, DuMond DM (2005) Monitoring the effects of a potential increased tidal range in the Cape Fear River ecosystem due to deepening Wilmington Harbor, North Carolina. Year 1:June1, 2003-May 31, 2004.

U.S. Army Corps of Engineers, Wilmington District (Contract No. DACW 54-00-R-0008), Wilmington

Hickman GL, Owens C (1980) Soil survey of Mobile County, Alabama. U.S. Department of Agriculture, Soil Conservation Service, Washington

Hodges RL, Sabo PB, Straw RW (1988) Soil survey of New Kent County, Virginia. U.S. Department of Agriculture, Soil Conservation Service, Washington

Hunt CB (1967) Physiography of the United States. W.H. Freeman and Co, San Francisco

Kelley CA (1993) Physical controls on methane production and flux from organic-rich wetland environments, Ph.D. thesis, University of North Carolina

Kelley CA, Martens CS, Chanton JP (1990) Variations in sedimentary carbon remineralization rates in the White Oak River estuary, North Carolina. Limnol Oceanogr 35:372-383

Kelley CA, Martens CS, Ussler III W (1995) Methane dynamics across a tidally influenced riverbank margin. Limnol Oceanogr 40:1112-1129

Light HM, Darst MR, Lewis LJ, Howell DA (2002) Hydrology, vegetation, and soils of riverine and tidal forests of the Lower Suwannee River, Florida, and potential impacts of flow reductions. Professional Paper 1656A. U.S. Geological Survey, Tallahassee

Lockaby BG, Walbridge MR (1998) Biogeochemistry. In: Messina MG, Conner WH (eds) Southern forested wetlands: Ecology and management. Lewis Publishers, Boca Raton, pp 149-172

Lockaby BG, Conner WH, Mitchell JD (In press) Floodplains. In: The Encyclopedia of ecology. Elsevier, London

Lugo AE (1990) Fringe wetlands. In: Lugo AE, Brinson MM, Brown S (eds) Ecosystems of the world 15: Forested wetlands. Elsevier, New York, pp 143-169

McBride EH, Burgess LH (1964) Soil survey of Baldwin County, AL. U.S. Department of Agriculture, Soil Conservation Service, Washington

McPherson BF, Hammett KM (1991) Tidal rivers of Florida. In: Livingston RJ (ed) The rivers of Florida: Ecological Studies 83. Springer-Verlag, New York, pp 31-46

Megonigal JP (1996) Methane production and oxidation in a future climate. Ph.D. thesis, Duke University

Megonigal JP, Schlesinger WH (2002) Methane-limited methanotrophy in tidal freshwater swamps. Global Biogeochemical Cycles 16(4):1088

Mitsch WJ, Gosselink JG (2000) Wetlands, 3rd edn. John Wiley & Sons, Inc., New York

Odum WE, Smith TJ III, Hoover JK, McIvor CC (1984) The ecology of tidal freshwater marshes of the United States east coast: A community profile. FWS/OBS-83/17. U.S. Fish and Wildlife Service, Washington

Orson RA, Simpson RL, Good RE (1992) A mechanism for the accumulation and retention of heavy metals in tidal freshwater marshes of the upper Delaware River estuary. Estuar Coast Shelf Sci 34:171-186

Ozalp M (2005) Above-ground productivity, litter decomposition, and nutrient dynamics of a coastal floodplain forest, Bull Island, South Carolina. Ph.D. thesis, Clemson University

Penland S, Ramsey K, McBride RA (1987) Relative sea level rise and subsidence measurements in the Gulf of Mexico based on national ocean survey tide gauge records. Proc. of the 10th National Conference of The Coastal Society, New Orleans

Peterson JE, Baldwin AH (2004) Variation in wetland seed banks across a tidal freshwater landscape. Am J Bot 91:1251-125

Reed DJ (2002) Sea level rise and coastal sustainability: geological and ecological factors in the Mississippi River Delta plain. Geomorphology 48:233-243

Rheinhardt RD (1991) Vegetation ecology of tidal swamps of the lower Chesapeake Bay, USA. Ph.D. thesis, Virginia Institute of Marine Sciences, College of William and Mary

Rheinhardt RD (1992) A multivariate analysis of vegetation patterns in tidal freshwater swamps of lower Chesapeake Bay, USA. Bull Torrey Bot Club 119:192-207

Rheinhardt RD, Hershner C (1992) The relationship of below-ground hydrology to canopy composition in five tidal freshwater swamps. Wetlands 12:208-216

Rigdon TA, Green AJ (1980) Soil survey of Camden and Glynn Counties, GA. U.S. Department of Agriculture, Soil Conservation Service, Washington

Sasser LD, Monroe KL, Schuster JN (1994) Soil survey of Franklin County, FL. U.S. Department of Agriculture, Soil Conservation Service, Washington

Schilling EB, Lockaby BG (2005) Microsite influences on productivity and nutrient circulation within two southeastern floodplain forests. Soil Sci Soc Am J 69(4):1185-1195

Schilling EB, Lockaby BG (2006) Relationships between productivity and nutrient circulation within two contrasting southeastern U.S. floodplain forests. Wetlands 26(1):181-192

Sherman DJ (2005) North America, coastal geomorphology. In: Schwartz ML (ed) Encyclopedia of coastal science. Springer, The Netherlands, pp 721-728

Shirley ML, Ragsdale JA (1966) Deltas. Houston Geologic Society. Houston

Simpson RL, Good RE, Leck MA, Whigham DF (1983) The ecology of freshwater tidal wetlands. BioScience 33:255-259

Smith WE (1988) Geomorphology of the Mobile Delta. Bulletin 132. Geologic Survey of Alabama, Tuscaloosa

Smock LA, Wright AB, Benke AC (2005) Atlantic coast rivers of the Southeastern United States. In: Benke AC, Cushing CE (eds) Rivers of North America. Elsevier, New York, pp 73-122

Stanturf JA, Schoenholtz SH (1998) Soils and landforms. In: Messina MG, Conner WH (eds) Southern forested wetlands: Ecology and management. Lewis Publishers, Boca Raton, pp 123-147

Stuck WM (1982) Soil survey of Colleton County, SC. U.S. Department of Agriculture, Soil Conservation Service, Washington

Stuckey BN (1982) Soil survey of Georgetown County, SC. U.S. Department of Agriculture, Soil Conservation Service, Washington

Twilley RR (1982) Litter dynamics and organic carbon exchange in black mangrove basin forests in a Southwest Florida estuary. Ph.D. thesis, University of Florida

Twilley RR (1998) Mangroves. In: Messina MG, Conner WH (eds) Southern forested wetlands: Ecology and management. Lewis Publishers, Boca Raton, pp 445-473

Verhoeven JTA, Whigham DF, van Logtestijn R, O'Neill J (2001) A comparative study of nitrogen and phosphorus cycling in tidal and nontidal river wetlands. Wetlands 21:210-222

Vitousek PM (1984) Litterfall, nutrient cycling, and nutrient limitation in tropical forests. Ecology 65:285-298

Walker HJ, Coleman JM (1987) Atlantic and Gulf Coast province. In: Geomorphic Systems of North America. Centennial Special Volume 2. Geologic Survey of America, Inc., Boulder

Warrick RA, Oerlemans J (1990) Sea level rise. In: Houghton JT, Jenkins JT, Ephaurms JJ (eds) Climatic change: The IPPC scientific assessment. Cambridge University Press, Cambridge, pp 257-281

Whalen SC (2005) Biogeochemistry of methane exchange between natural wetlands and the atmosphere. Environ Engr 22:73-94

Wharton CH, Kitchens WM, Pendleton EC, Sipe TW (1982) The ecology of bottomland hardwood swamps of the southeast: A community profile. FWS/OBS-81/37. U.S. Fish and Wildlife Service, Biological Services Program, Washington

Wilson JO, Crill PM, Bartlett KB, Sebacher DI, Hariss RC, Sass RL (1989) Seasonal variation of methane emissions from a temperate swamp. Biogeochemistry 8:55-71

Chapter 4 - Plant Community Composition of a Tidally Influenced, Remnant Atlantic White Cedar Stand in Mississippi*

Bobby D. Keeland and John W. McCoy

U.S. Geological Survey, National Wetlands Research Center, 700 Cajundome Blvd., Lafayette, LA 70506

4.1. Introduction

Atlantic white cedar (*Chamaecyparis thyoides* [L.] B.S.P.) is a common tree species along the Atlantic coast from Maine to Florida but is relatively uncommon along the Gulf of Mexico Coast (Little 1950a; Laderman 1987, 1989). Korstian and Brush (1931) observed large populations of Atlantic white cedar in Alabama, Florida, and Mississippi, but these stands were not thoroughly evaluated and subsequent logging may have degraded or eliminated some. Stands in Alabama have been severely depleted (Ward 1987). Only one stand in Mississippi (Eleuterius and Jones 1972) has been quantitatively studied and results published in the literature. Ward and Clewell (1989) presented the most comprehensive description of known Atlantic white cedar locations along the Gulf of Mexico coast, but the majority of stands they discussed included only a general location and little community composition data. In addition to the stand along Bluff Creek (described by Eleuterius and Jones 1972), Ward and Clewell (1989, page 11) reported that in Mississippi, Atlantic white cedar occurs along "...tributaries of the Escatawpa River and Pascagoula River, Jackson County; and branches of the Catahoula River, Pearl River County." McCoy and Keeland (2005) provided a more comprehensive description of Atlantic white cedar locations in Mississippi. They listed this species in five counties with the main concentration near the Alabama border in Jackson County. This sparse occurrence of Atlantic white cedar in Mississippi may reflect how few large stands remain.

---

* The U.S. Government's right to retain a non-exclusive, royalty-free licence in and to any copyright is acknowledged.

Atlantic white cedar is an obligate wetland species (Reed 1988), but prolonged flooding results in high mortality of seedlings (Brown and Atkinson 1999) and slow death of mature stems (Laderman 1989). Although hydrology is one of the most important factors controlling species composition of Atlantic white cedar wetlands, little data are available on water regimes in these stands (Laderman 1989). Golet and Lowry (1987) provided the first long-term research on Atlantic white cedar swamp hydrology. Atkinson et al. (2003) stated that studies reporting on water table dynamics in Atlantic white cedar swamps were mostly still lacking. Laderman (1989) reported that Atlantic white cedar is almost always found growing above the high tide, and where there is a tidal influence, the flux is very small and infrequent. Several studies, however, have provided evidence that seems to support the occurrence of Atlantic white cedar in tidal zones. Moore and Carter (1987) described several Atlantic white cedar stands in North Carolina (Alligator River, Bull's Neck Swamp, Scuppernong River) where saltwater flooding was thought to be at least partially responsible for Atlantic white cedar mortality. Storm tides were specifically discussed in relation to the Scuppernong River. The U.S. Army Corps of Engineers (1982) also described a tidal influence on an Atlantic white cedar stand on the Alligator River in Dare County, North Carolina. In this stand, the tidal influence was caused primarily by wind rather than lunar tides. No Atlantic white cedar stands affected by tides have been previously reported in Mississippi.

The objective of this chapter is to describe a preliminary study on the hydrology and plant species composition of an Atlantic white cedar stand located within the tidal zone of coastal Mississippi. This chapter will deviate somewhat from the format of subsequent case study chapters because so little is known about remnant, tidal Atlantic white cedar stands in general, and our selected stand in particular. Our research program on this topic is in its infancy, much work remains to be done.

4.2. Site description

4.2.1. Location

The Grand Bay study site is located on Grand Bay National Wildlife Refuge, about 16 km east of Pascagoula, Mississippi, and 33 km southwest of Mobile, Alabama. The site is about 9 km overland from the Gulf of Mexico but is situated 31.8 km upstream from the Gulf on the Pascagoula and Escatawpa Rivers. The site is bounded to the east by a sharp slope leading up to a pine savannah and the Mississippi Welcome Center along Interstate 10 (I-10). The slope represents the edge of the floodplain. From the edge of the pine stand the soil surface drops about 3-4 m over a distance ranging from 5-20 m. North of the site is the Escatawpa River, and a canal extending south from the river forms the western border. Interstate 10 is located along the southern boundary (Figure 4.1).

4.2.2. Soils

Soil types are of a sandy nature, grading from Maurepas muck on the northwest corner of the site, along the river, to an Axis mucky sandy clay loam in the swamp and lower slope (NRCS 2006). The pine savannah is a Wadley loamy sand. All of these soil series are strongly acidic. The Maurepas and Axis soils are very poorly drained with rapid to moderate permeability. The Maurepas soils are associated with large backswamps, whereas the Axis soils formed in thick loamy marine sediments associated with coastal marshes.

Spoils from the canal have impounded water in the swamp, resulting in a perched wetland that is directly connected to the river during high flows. Openings have eroded through the spoil banks and allowed at least partial draining of the wetland during low flows on the Escatawpa River. The Maurepas and Axis soil types are also found downstream along the river in addition to natural levee soils of the Nugent and Jena series. These two levee soils are both deep and well drained to excessively drained with surfaces ranging from less than a meter to several meters above the river.

4.2.3. Climate

Climate at the site is hot and humid during the summer with mild winters. High temperatures average 32.1°C in July while low temperatures average 3.9°C in January. Killing frosts are rare in the area. The 30-year average precipitation for the southeast Mississippi region is 170 cm. Precipitation over the last 10 years has averaged slightly higher, at 178.9 cm, even though rainfall in 1999 and 2000 were below average (134.1 cm and 134.4 cm, respectively).

Hurricanes are common along this part of the Mississippi coast, with an average return frequency of approximately 0.16 annually for Category 1 storms based upon the 1851 to 2004 record (Chapter 1). In addition to

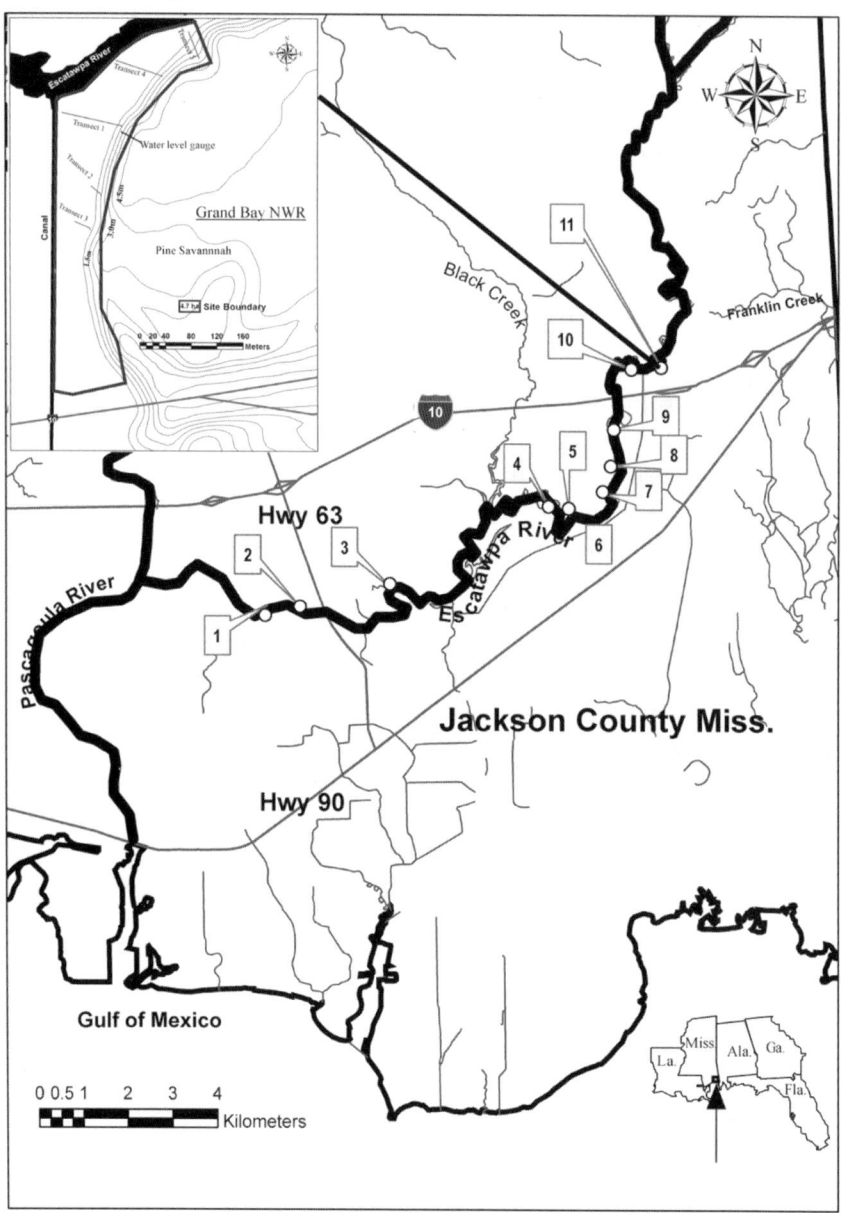

Fig. 4.1. The Lower Escatawpa River, showing river stations where salinity measurements were taken adjacent to individual or small groups of Atlantic white cedar, except for the three lowest stations (1, 2, 3). Inset shows the main study area on Grand Bay National Wildlife Refuge.

wind damage, periodic surge with saline waters can be problematic for Atlantic white cedar stands.

4.3. Hydrology

4.3.1. Hydrologic methods

Water levels within the site were recorded every four hours from May 3, 2004 through January 4, 2006 with an electronic water level recorder (Model 138, Infinities USA, Inc. Port Orange, Florida, USA). The water-level well was situated at the toe of the slope, adjacent to the swamp, and consisted of a perforated PVC pipe, installed to approximately 1 m below the soil surface, backfilled with coarse sand, and capped with bentonite according to the methods of Sprecher (2000). Limitations of the pressure transducer used in the water-level recorder resulted in greater accuracy during mid to low flows, but truncated measurements during high flows. In addition, stage data at 15-minute intervals for the Escatawpa River at the I-10 bridge near the downstream end of the study site were obtained from the U.S. Geological Survey, Water Resources Division (Gage 0248018020). Surface and in-channel water salinities were measured with a conductivity meter (Model 30, YSI Inc., Yellow Springs, Ohio, USA), and attained from the aforementioned USGS gage.

4.3.2. Water levels

Topographic relief within the study site resulted in natural variations in water levels unrelated to stage on the Escatawpa River. Although portions of the swamp were continuously flooded during the study, the entire swamp was inundated during most of the 2005 growing season. Ground level for water level measurements was arbitrarily chosen as the sharp transition between slope and swamp. Water levels averaged 7.4 cm below the soil surface between May 3 and October 31, 2004. During the same period of 2005, water levels averaged 8.4 cm above the soil surface. The slope and the levee were almost always dry except during very high flows on the river.

In addition to the effects of topography, water levels on the site were tied to stage in the river, tidal variations, and the integrity of the spoil bank along the canal. Between the beginning of water level measurements and a flood in mid-September 2004, water levels at the study site followed stage in the river, with gradually decreasing water levels during low stage and rapidly increasing water levels during higher stage (Figure 4.2). Tidal fluctuations were at times prominently displayed (Figure 4.3). Following the mid-September flood in 2004, hydrology within the swamp appeared perched with water levels static during low stage in the river. Shortly after Hurricane Rita (about September 30, 2005) water levels once again decreased gradually between high-water events (Figure 4.2).

Salinity within the water of the Grand Bay study site and the adjacent portion of the Escatawpa River was generally below 0.1 parts per thousand (ppt). During the storm surge associated with Hurricane Katrina, water

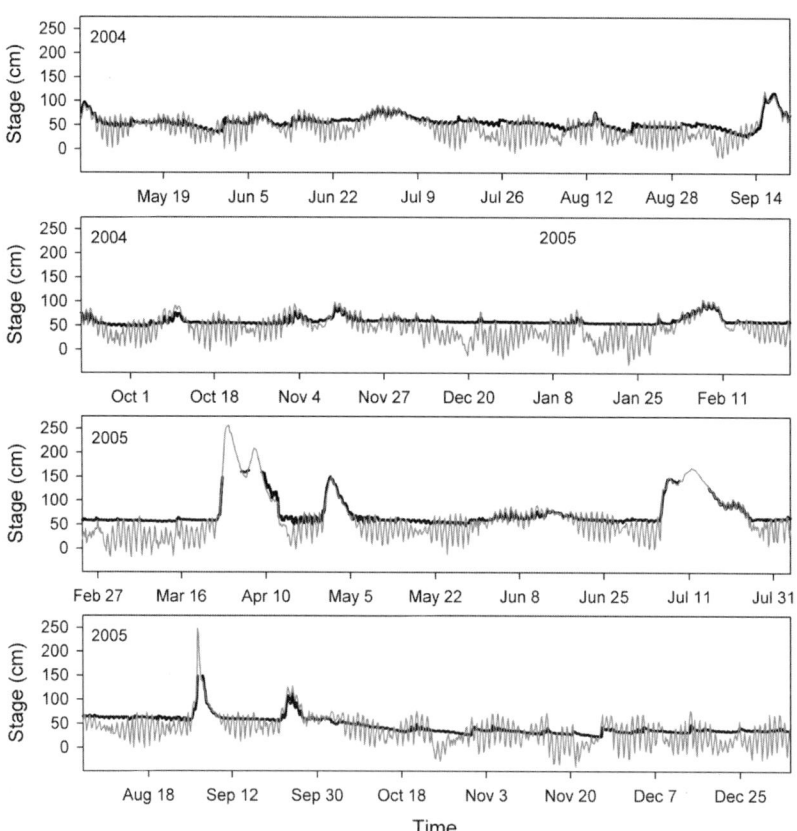

Fig. 4.2. Water levels at the Grand Bay Atlantic white cedar site (thick line) and at the Interstate 10 bridge (thin line). A stage of 64 cm at the Grand Bay site represents ground level.

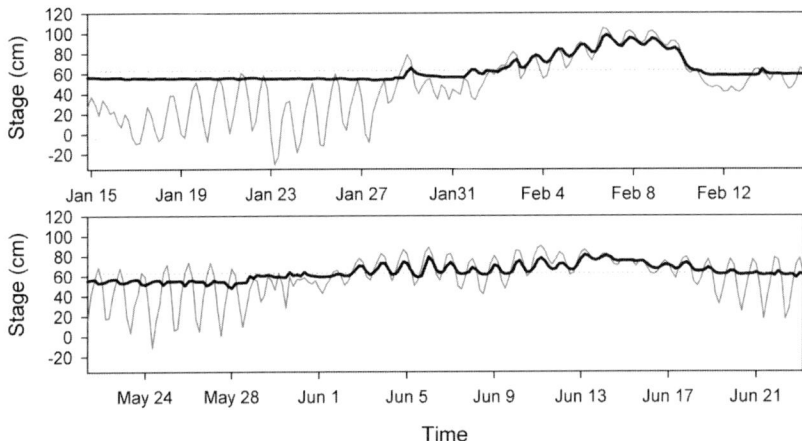

Fig. 4.3. Water levels at the Grand Bay Atlantic white cedar site (thick line) and at the Interstate 10 bridge (thin line) for two 4-week time periods in winter and late spring, 2005. The dotted line represents ground level at the toe of the slope on the Grand Bay National Wildlife Refuge study site (see text).

levels rose at least 90 cm at the study site (2.5 m at the I-10 bridge). Salinity at the I-10 bridge increased to 1.56 ppt at the peak of the storm surge. Water levels remained elevated for at least a week. These conditions are similar to more inland tidal freshwater forested wetlands along the Atlantic Coast (see Chapters 5-8).

4.4. Vegetation

4.4.1. Vegetation transects

Forest vegetation on the site was sampled along five belt transects. Each transect was 10 m wide and of variable length, extending perpendicular from the slope toward the Escatawpa River. Within each transect, all trees greater than 2.5 cm in diameter (at 140 cm height; here after referred to as dbh) were identified to species and measured. All stems greater than or equal to 15 cm dbh were considered overstory and smaller stems were considered to make up the midstory of the canopy. Shrub plots of 10 m^2 each were spaced at 10 m intervals along each transect to sample stems less than 2.5 cm dbh. The herbaceous vegetation was sampled for species composition and percent cover by using 1 m² subplots placed at 10-m intervals along each transect.

4.4.2. Vegetation composition

Vegetation on the site varied by forest strata and by location. Each vegetative layer (i.e., including the canopy [consisting of the overstory and the midstory] and the shrub/ground cover layers) and each habitat type (i.e., including slope, swamp, and levee: Section 4.4.3), are discussed below. The shrub and ground cover layers are discussed together because of the sparse nature of these layers.

4.4.2.1. Overstory (dbh ≥ 15.0 cm)

Over all, the forest overstory at the Grand Bay site was fairly dense, with an average of 635 stems/ha and 42.29 m² of basal area/ha (Table 4.1). The dominant species included water tupelo (*Nyssa aquatica* L.), swamp tupelo (*Nyssa biflora* Walt.), loblolly pine (*Pinus taeda* L.), Atlantic white cedar, sweetbay (*Magnolia virginiana* L.), and baldcypress (*Taxodium distichum* [L.] L.C. Rich.). These species represent a combined importance value of 238.9 (of 300 total, see Table 4.1). All but Atlantic white cedar are common canopy dominants along the northern Gulf Coast and typically grow to 30 m or more in height. The trees with the largest diameters consisted of swamp tupelo (55.5 cm), loblolly pine (50.7 cm), water tupelo (48.8 cm), and Atlantic white cedar (46.5 cm). Mean diameters for each of these species were much less than the maximum (Table 4.1). Eight additional species, including water oak (*Quercus nigra* L.), red maple (*Acer rubrum* L.), swamp titi (*Cyrilla racemiflora* L.), southern magnolia (*M. grandiflora* L.), buckwheat tree (*Cliftonia monophylla* [Lam.] Britt. Ex.), sweetgum (*Liquidambar styraciflua* L.), ash (*Fraxinus* sp.) and Darlington's oak (*Q. hemisphaerica* Bartr. Ex Willd.), made up the remainder of the overstory (Table 4.1).

4.4.2.2. Midstory (dbh between 2.5 and 15 cm)

Twenty five tree species were encountered within the midstory, with swamp titi as the overwhelming dominant (909 stems/ha; IV 106.2; Table 4.2). Co-dominants consisted mostly of small individuals of canopy species, such as red maple, water tupelo, sweetbay, and Atlantic white cedar. Tree species in the midstory that were not found in the overstory included swamp bay (*Persea palustris* [Raf.] Sarg.), laurel oak (*Q. laurifolia*

Table 4.1. Diameter, density, basal area, and importance values of the dominant overstory trees growing in a remnant Mississippi Atlantic white cedar stand [1].

Common Name	DBH[2] (cm)	Density (stems/ha)	Basal area (m²/ha)	Relative Frequency	Relative Density	Relative Dominance	IV[3]
Water tupelo	27.9	164.6	10.08	16.1%	25.9%	23.8%	65.9
Swamp tupelo	29.7	116.6	8.08	11.3%	18.4%	19.1%	48.8
Loblolly pine	32.9	92.6	7.87	6.5%	14.6%	18.6%	39.7
Atlantic white cedar	31.5	54.9	4.27	9.7%	8.6%	10.1%	28.4
Sweetbay	22.1	54.9	2.12	14.5%	8.6%	5.0%	28.2
Baldcypress	34.5	48.0	4.49	9.7%	7.6%	10.6%	27.9
Water oak	31.8	30.9	2.46	8.1%	4.9%	5.8%	18.7
Red maple	23.1	24.0	1.01	8.1%	3.8%	2.4%	14.2
Swamp titi	21.1	13.7	0.48	4.8%	2.2%	1.1%	8.1
Southern magnolia	21.4	17.2	0.62	3.2%	2.7%	1.5%	7.4
Buckwheat tree	16.7	6.9	0.15	3.2%	1.1%	0.4%	4.7
Sweetgum	37.3	3.4	0.38	1.6%	0.5%	0.9%	3.0
Ash	24.5	3.4	0.16	1.6%	0.5%	0.4%	2.5
Darlington's oak	20.8	3.4	0.12	1.6%	0.5%	0.3%	2.4
Totals		634.5	42.29	100.0%	99.9%	100.0%	299.9

[1] Includes all stems greater than or equal to 15.0 cm diameter at breast height (at 140 cm above the ground).
[2] DBH - quadratic mean diameter.
[3] IV – refers to importance value, a summation of relative frequency, relative density, and relative dominance. IV sums to 300 except where rounding errors occur.

Table 4.2. Diameter, density, basal area, and importance values of midstory trees growing in a remnant Atlantic white cedar stand in Mississippi[1].

Common Name	DBH[2] (cm)	Density (stems/ha)	BA (m²/ha)	Relative Freq.[3]	Relative Density	Relative Dom.[3]	IV[4]
Swamp titi	5.9	909.0	2.48	18.1%	52.3%	35.8%	106.2
Red maple	9.1	164.6	0.86	12.9%	9.5%	12.4%	34.8
Water tupelo	9.9	109.8	0.85	6.9%	6.3%	12.2%	25.4
Sweetbay	8.6	116.6	0.68	7.8%	6.7%	9.8%	24.3
American holly	5.6	92.6	0.23	6.9%	5.3%	3.3%	15.5
Atlantic white cedar	8.6	34.3	0.20	5.2%	2.0%	2.9%	10.1
Buckwheat tree	8.0	41.2	0.21	4.3%	2.4%	3.0%	9.7
Swampbay	8.0	27.4	0.14	5.2%	1.6%	2.0%	8.8
Swamp tupelo	10.1	24.0	0.19	3.4%	1.4%	2.8%	7.6
Darlington's oak	9.1	20.6	0.13	4.3%	1.2%	1.9%	7.4
Laurel oak	10.5	17.2	0.15	3.4%	1.0%	2.2%	6.6
Blueberry	5.6	34.3	0.08	2.6%	2.0%	1.2%	5.8
Southern magnolia	8.3	27.4	0.15	1.7%	1.6%	2.2%	5.5
Baldcypress	11.4	10.3	0.11	2.6%	0.6%	1.5%	4.7
Persimmon	9.5	17.2	0.12	1.7%	1.0%	1.8%	4.5
Loblolly pine	13.9	6.9	0.10	1.7%	0.4%	1.5%	3.6
Live oak	12.6	6.9	0.09	1.7%	0.4%	1.2%	3.3
Water oak	8.7	6.9	0.04	1.7%	0.4%	0.6%	2.7
Deerberry	4.8	20.6	0.04	0.9%	1.2%	0.5%	2.6
Ash	4.6	10.3	0.02	1.7%	0.6%	0.2%	2.5
Fringe tree	5.7	17.2	0.04	0.9%	1.0%	0.6%	2.5
Holly	3.2	6.9	0.01	1.7%	0.4%	0.1%	2.2
Wax myrtle	4.2	6.9	0.01	0.9%	0.4%	0.1%	1.4
Elliott's blueberry	3.4	6.9	0.01	0.9%	0.4%	0.1%	1.4
Red bud	6.2	3.4	0.01	0.9%	0.2%	0.1%	1.2
		1739.4	6.95	100.0%	100.3%	100.0%	300.3

[1] Includes all stems greater than 2.5 cm but less than 15.0 cm diameter at breast height.
[2] DBH - quadratic mean diameter.
[3] Freq. = Frequency, Dom. = Dominance.
[4] IV – refers to importance value, a summation of relative frequency, relative density, and relative dominance. IV sums to 300 except where rounding errors occur.

Michx.), common persimmon (*Diospyros virginiana* L.), and live oak (*Q. virginiana* P. Mill.). Nine shrub/small tree species including, American holly (*Ilex opaca* Ait.), buckwheat tree, blueberry (*Vaccinium* sp.), deerberry (*V. stamineum* L.), fringe tree (*Chionanthus virginicus* L.), holly (*Ilex* sp.), wax myrtle (*Morella cerifera* [L.] Small), Elliott's blueberry (*V. elliottii* Chapman) and eastern redbud (*Cercis canadensis* L.) were also found in the midstory (Table 4.2).

4.4.2.3. Shrubs (dbh < 2.5 cm) and ground layer vegetation

The ground layer was almost completely absent throughout most of the site. Many of the subplots contained large areas of bare ground or litter.

Flooding within the swamp no doubt inhibited development of a herbaceous layer, but low light under a full canopy was probably the proximate cause of a limited ground layer on the levee. The only area with a well-developed ground layer was the slope, where light levels were much higher, possibly a result of controlled burning on the adjacent pine savannah. Of the species that occurred in the shrub and herbaceous vegetation subplots, the majority consisted of shrub species such as coastal sweetpepperbush (*Clethra alnifolia* L.), hawthorn (*Crataegus* L.), St. Andrew's-cross (*Hypericum hypericoides* [L.] Crantz), dahoon (*Ilex cassine* L.), large gallberry (*I. coriacea* [Pursh] Chapman), inkberry (*I. glabra* [L.] Gray), Georgia holly (*I. longipes* Chapman ex Trel.), yaupon (*I. vomitoria* Ait.), maleberry (*Lyonia ligustrina* [L.] DC.), flameleaf sumac (*Rhus copallinum* L.), saw palmetto (*Serenoa repens* [Bartr.] Small), farkleberry (*Vaccinium arboreum* Marsh.), and grape (*Vitis* L.), in addition to those shrubs listed in Table 4.2.

Seedlings of several overstory trees were also counted in the subplots, including southern red oak (*Quercus falcata* Michx.) and species listed in Tables 4.1 and 4.2. Only one cedar was encountered in the subplots. Additional species found in the subplots but not listed in Table 4.2 included several vines, such as crossvine (*Bignonia capreolata* L.), saw greenbrier (*Smilax bona-nox* L.), laurel greenbrier (*S. laurifolia* L.), common greenbrier (*S. rotundifolia* L.), climbing dogbane (*Trachelospermum difforme* [Walt.] Gray, and only a few herbaceous species, such as sedges (*Carex* L.), Carolina spiderlily (*Hymenocallis caroliniana* [L.] Herbert), Arizona bog violet (*Viola affinis* Le Conte), meadow spike-moss (*Selaginella apoda* [L.] Spring), and Virginia chainfern (*Woodwardia virginica* [L.] Sm.). Several additional species were noted in the study area but were not counted in any of the subplots. These species included American witchhazel (*Hamamelis virginiana* L.), poison ivy (*Toxicodendron radicans* [L.] Kuntze), fetterbush lyonia (*Lyonia lucida* [Lam.] K. Koch), mountain azalea (*Rhododendron canescens* [Michx.] Sweet), smallflower blueberry (*Vaccinium virgatum* Ait.), netted chainfern (*Woodwardia areolata* [L.] T. Moore), and cinnamon fern (*Osmunda cinnamomea* L.).

4.4.3. Variation in vegetation due to topographic position

4.4.3.1. Slope

Vegetation along the slope consists of swamp titi, loblolly pine, swamp tupelo, Atlantic white cedar, and red maple, with small amounts of live oak, water oak, sweetbay, swampbay, laurel oak, sweetgum, buckwheat tree, and baldcypress. A very few individuals of Chinese tallow tree (*Triadica sebifera* [L.] Small) were observed on the slope, but all individuals of this highly invasive tree species were less than 1 m tall. Baldcypress and swamp tupelo trees encountered in this zone were found at the base of the slope. Although many of the Atlantic white cedars in this zone were also found at the base of the slope, some Atlantic white cedar trees were growing up the slope and out onto the pine stand. Most of the Atlantic white cedars growing on the pine stand, however, had been killed by controlled burns. The Atlantic white cedar trees growing up on the edge of the pine stand ranged up to 50 years in age.

Approximately 20 percent of the Atlantic white cedar trees encountered across the study site were dead, with the majority being fire-killed trees on the slope. Most of the species listed in the shrub and ground layer vegetation section above (Section 4.4.2.3.) were found on or near the top of the slope. Several cinnamon ferns were observed near the middle to lower sections of the slope.

4.4.3.2. Swamp

More than one third of the site consists of baldcypress/water tupelo swamp. At the base of the slope, the transition into the swamp was sharp whereas the transition from swamp to levee was more gradual. The center of the swamp was semipermanently flooded, and the soil surface of peripheral areas was occasionally exposed. All of the water tupelo and the majority of swamp tupelo were found in this habitat. Other species growing in the swamp included baldcypress, green ash, red maple, swamp bay, and a few scattered Atlantic white cedars. The Atlantic white cedar trees growing in the swamp were found on slightly elevated soils; however, the typical hummock and pool (i.e., "hollow") topography associated with Atlantic white cedar swamps, and tidal freshwater swamps in general, was mostly absent. Most of the Atlantic white cedar saplings encountered in the study were found in the swamp, but many of these saplings appeared severely stressed. Buckwheat trees and swamp titi were found along the transition from slope to swamp and from swamp to levee.

4.4.3.3. Levee

Nearer the river, the land surface rose onto a natural levee. The levee was highest at the northeast corner of the study site where it merged with the slope (Figure 4.1). During flood events on the river, this upstream end of the study site occasionally received fresh deposits of sand. Downstream, to the southwest along the Escatawpa River, the height of the levee decreased. An older levee angled northwest toward the river, providing additional topographic complexity. This complexity of the levee zone supported the greatest diversity of tree species of the area. Every species found on the site, with the exception of loblolly pine, water tupelo, and persimmon was found in the levee zone. The blueberries, deerberry, and Darlington's oak were found only on the levee. Spoil banks along the canal (see Figure 4.1) increase the height of the levee at the point where the canal intersects the river. This small area was much higher than the surrounding land and supported a large number of Atlantic white cedar trees. South of the river the spoil bank supported few Atlantic white cedars. The violets and netted chainferns were found on the levee, often near the transition to swamp.

The levee zone extends downstream from the main study site and supports a periodically broken, narrow band of Atlantic white cedar off site to river kilometer 24.1 (Figure 4.4). Vegetation along these levee systems was very similar to the main study site, but several additional species were encountered. Tree species included longleaf pine (*Pinus palustris* P. Mill.), southern redcedar (*Juniperus virginiana* var. *silicicola* [Small] J. Silba), Carolina ash (*Fraxinus caroliniana* P. Mill.), camphor tree (*Cinnamomum camphora* [L.] J. Presl.), and sand live oak (*Quercus geminata* Small). Additional shrub species included eastern baccharis (*Baccharis halimifolia* L.), buttonbush (*Cephalanthus occidentalis* L.), bedstraw St. Johnswort (*Hypericum galioides* Lam.), spicebush (*Lindera* Thunb.), and elderberry (*Sambucus nigra* ssp. *canadensis* [L.] R. Bolli). Herbaceous species included seaside goldenrod (*Solidago sempervirens* L.), threeawn (*Aristida* L.), longleaf spikegrass (*Chasmanthium sessiliflorum* [Poir.] Yates), creeping burhead (*Echinodorus cordifolius* [L.] Griesb), scarlet calamint (*Clinopodium coccineum* [Nutt. ex Hook.] Kuntze), fragrant flatsedge (*Cyperus odoratus* L.), pickerelweed (*Pontederia cordata* L.), rosette grass (*Dichanthelium* [A.S. Hitchc. & Chase] Gould), torpedo grass (*Panicum repens* L.), Jamaica swamp sawgrass (*Cladium mariscus* [L.] Pohl ssp. *jamaicense* [Crantz] Kűkenth.), sugarcane plumegrass (*Saccharum giganteum* [Walt.] Pers.), and royal fern (*Osmunda regalis* L.). Southern red cedar, baldcypress, loblolly pine, saw palmetto, elderberry, eastern baccharis, and yaupon were all found just above high tide in areas where salinity of the river was as high as 4.8 to 5.7 ppt. The plant species composition of the downstream levees was determined during a one-day boat trip. A more complete survey would probably show additional species.

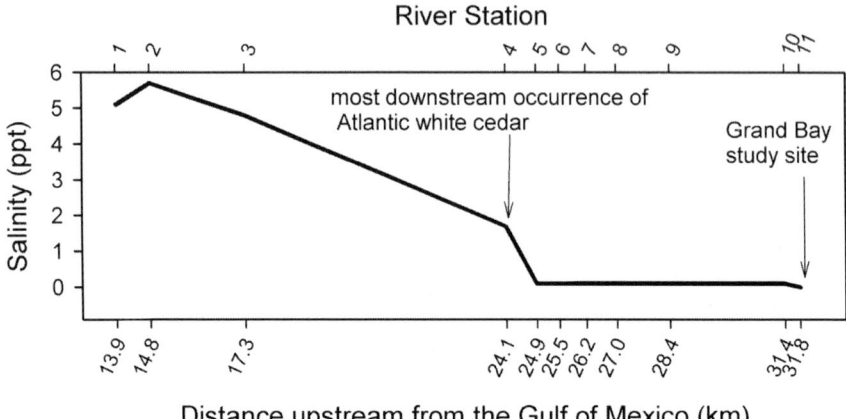

Fig. 4.4. Salinity at various stations on the lower Escatawpa River. Station locations are shown in Figure 4.1. Distances shown are from the Gulf of Mexico along the Pascagoula and Escatawpa Rivers.

4.5. Discussion

4.5.1. Interactions among hydrology, salinity, and fire

Although the Escatawpa River is subject to relatively large fluctuations in stage compared with other areas that support Atlantic white cedar stands, water levels at the Grand Bay site rarely fell to more than 25 cm below the soil surface and were not observed any lower than -40 cm. High flows are usually of short duration. During periods of low flow in the river, tidal fluctuations probably help to increase soil moisture, even though flooding does not occur above the soil surface. Soils on the Grand Bay site always appeared moist, and the pine savannah was always observed to be wet, with standing water in many places. Seepage of this water down the slope supplied the vegetation with an abundant supply of moisture, yet aboveground flooding did not occur at all on the upper portion of the slope. These conditions appear to be optimal for the survival and growth of Atlantic white cedar. This observation is in agreement with Clewell and Ward (1987) who indicated that Atlantic white cedar occupies areas where soils are perennially moist or wet from constant seepage of groundwater but are only briefly, if at all, inundated. Soil moisture levels on the higher portions of the levee are unknown, yet Atlantic white cedar appeared to be growing quite well. Eleuterius and Jones (1972) reported Atlantic white cedar on bluffs up to 18 m above Bluff Creek. Unfortunately, little research has been conducted on the hydrology of south Mississippi wetland soils in general, and tidal swamp soils along the Gulf Coast in particular.

The occasional blocked drainage from the swamp portion of the Grand Bay study site could have implications for survival and growth of Atlantic white cedar and other tree species. If the drainage is blocked during a storm that pushed sea water upstream, the swamp can retain some saline water to the detriment of the forest (see Williams 1993 and Conner 1993). Any seedlings in the swamp would therefore be subjected to long-term flooding and the potential of brackish water. Each of these factors can adversely affects seedling survival and growth. This situation probably occurred to some extent following Hurricane Katrina in late August and September, 2006 (Figure 4.2). The tidal surge at the I-10 bridge had a salinity of 1.56 ppt. It is unknown how much of this surge passed over the study site, but due to the proximity of the bridge to the southern edge of the site we must assume that the site was impacted as water from the surge mixed with water already in the swamp. Following the surge, some of this water was retained until the swamp began to partially drain following high water associated with Hurricane Rita. Subsequent high stage in the Escatawpa River would help flush any residual salinity out of the swamp, but depending on the amount of time between high flows some damage to seedlings could have occurred.

When gaps in the spoil bank are open, such as observed during the spring and summer of 2004 and following Hurricane Rita in September 2006, the swamp partially drains during low stage on the river. Low water levels in the swamp would allow soil exposure and the potential for seedling establishment. Dense shade beneath the swamp canopy, however, could preclude seedling survival.

Atlantic white cedar trees are not typically tolerant of salinity, and Moore and Carter (1987) discussed several studies where Atlantic white cedar has been killed by flooding with brackish water. Eleuterius and Jones (1972) found that Atlantic white cedar extended along the lower Pascagoula River until it became sharply reduced as the limit of saltwater penetration was approached. We observed the same occurrence of Atlantic white cedar along the Escatawpa River. Some trees were found at the front of saltwater penetration and densities increased with distance upstream. The Atlantic white cedar trees at the Grand Bay study site are usually not affected by salinity, but brackish water can be pushed upstream and onto Atlantic white cedar stands during large storms, such as from the recent surge of Hurricane Katrina (2005). Salinity as high as 1.56 ppt is known to impact Atlantic white cedar seedlings (Sedia and Zimmermann 2006) and may decrease the vigor of mature trees. The effects of a short-term salinity pulse on Atlantic white cedar are unknown, but it must be detrimental to at least some degree. The good vigor of most of the Atlantic white cedar trees along the Escatawpa River, however, suggests that the pulse of brackish water from Hurricane Katrina did not have serious long-term impacts. The presence of some dead and several stressed Atlantic white cedar trees along the lower part of the river raises questions about the ability of Atlantic white cedar to withstand salt surges of greater frequency or duration.

Numerous authors have reported Atlantic white cedar to be a fire-maintained species (Little 1950a, Laderman 1989), but Eleuterius and Jones (1972) reported no evidence of fire in the stand near Vancleave, Mississippi. Laderman (1989) attributed the lower frequency of destructive fires along the Gulf Coast to the incised topography and the constantly moist soils and leaf litter. Clewell and Ward (1987) suggested that this rarity of fire favored the development of a mixed forest, rather than the large monocultures often observed in more northerly locations. The forest at Grand Bay National Wildlife Refuge showed extensive evidence of fire. Although fire has been most intense in the pine savannah, followed by the slope, numerous trees of several species along the levee zone also had fire scars or charred bark. Only trees in the swamp seemed to have been missed by fire. Almost every Atlantic white cedar tree with any amount of charred bark was dead, even though many still had an intact (though dead) canopy. Apparently very little heat from a fire is required to kill Atlantic white cedars. It is not clear whether fire is common on the study site naturally or if fire is only a factor during the periodic controlled burns. Regardless of the nature and frequency of fire on the pine savannah, fire does not seem to be detrimentally affecting the main portion of the Atlantic white cedar stand at Grand Bay or downstream.

Atlantic white cedar generally supports few fungi (Fowells 1965); however, Little and Garrett (1990) list witches' broom (*Gymnosporangium ellisii* [Berk.] Farlow) as a rust fungus that infects and may contribute to mortality of Atlantic white cedar. Witches' broom was common throughout the Grand Bay Atlantic white cedar stand, but most trees had little to no infestation, and no trees were observed to be heavily parasitized, as was reported for an Atlantic white cedar stand in Vancleave, Mississippi (Eleuterius and Jones 1972). Although many dead Atlantic white cedar trees encountered in our study had been infected with witches' broom, the levels of infestation did not seem sufficient to have been the primary cause of death. Observations of witches' broom in other Atlantic white cedar stands in Mississippi and Alabama suggest that it is widespread in the area but generally causes minimal damage to the trees.

4.5.2. Forest structure in Mississippi tidal Atlantic white cedar stands

The tree and shrub composition at the Grand Bay study site was made up of species typically found throughout the southeastern United States. In total, 18 tree species and 10 shrub species were encountered in the study transects. One additional tree and 12 shrub species were observed on the site but not within the transects, and downstream of the main study site, five additional tree and five additional shrub species were encountered. All together, 24 tree species and 27 shrub species were noted along a 7.7 km reach of the Escatawpa River. These numbers are higher than the results of Eleuterius and Jones (1972) who reported a total of 12 tree and 21 shrub species along a 11.25 km length of Bluff Creek, near the town of Vancleave, Mississippi, and much higher than the 22 woody species reported for stands in Florida. Compared with many of the Atlantic white cedar stands along the Atlantic Coast, where the number of woody species can be as low as four (Loomis et al. 2003), six (Bernard 1963; Stockwell 1999 [only tree species reported]), eight (Zampella et al. 1999), or even 18 (Loomis et al. 2003), this Mississippi stand is extremely diverse.

Atlantic white cedar stems encountered on our study ranged from 2.9 to 46.5 cm dbh. The size class distribution of Atlantic white cedar stems in the canopy indicates some regeneration (Figure 4.5), but very few stems less than 6 cm dbh were observed. Only one Atlantic white cedar seedling was found in the subplots. These data indicate a general lack of Atlantic white cedar regeneration at the Grand Bay study site. Such limited regeneration in stands with high stem density and a full canopy is not surprising as several studies have indicated that Atlantic white cedar is shade intolerant (Buell and Cain 1943; Ash et al. 1983; Laderman 1989; Motzkin et al. 1993). Eleuterius and Jones (1972), however, noted large numbers of seedlings and saplings, some growing vigorously in the shade of a slash pine-dominated upland forest. McCoy and Keeland (2005) also noted extensive cedar regeneration at some Mississippi locations. An increase in the number of Atlantic white cedar seedlings observed at the Grand Bay site after Hurricane Katrina suggests that partial removal of the forest canopy and the resultant increase in light at ground level allows higher survival. As Atlantic white cedar is known to regenerate poorly under large amounts of logging slash (Korstian and Brush 1931; Little 1950b), and Hurricane Ka-

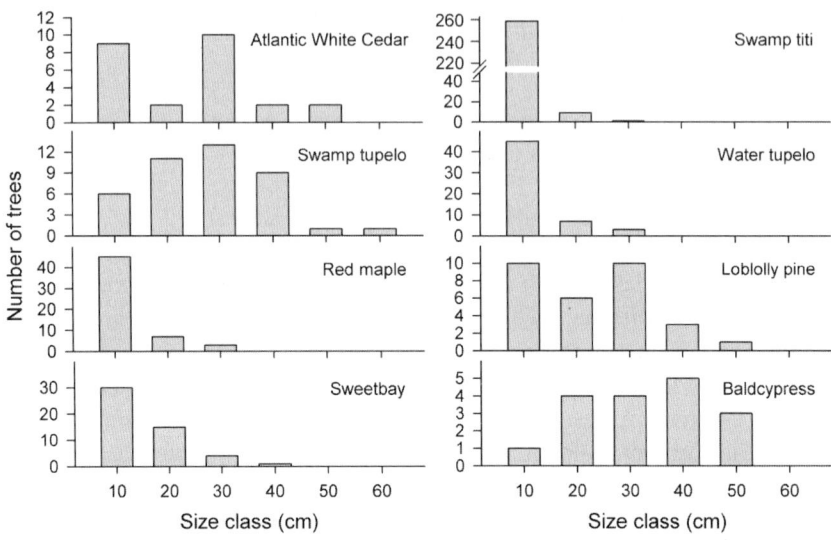

Fig. 4.5. Size class distribution of the eight most important canopy tree species. Note the differences in y-axis range for each plot.

trina left a lot of debris, the expected survival of these additional seedlings is questionable.

With an IV of only 28.4, Atlantic white cedar was not the dominant species on the Grand Bay site (Table 4.1). Water tupelo (IV=65.9), swamp tupelo (IV=48.8), and loblolly pine (IV=39.7) each exhibited greater importance than Atlantic white cedar. This regionally low IV is in great contrast to the many monotypic Atlantic white cedar stands along the Atlantic Coast (Laderman 1987, 1989), but is in agreement with Clewell and Ward (1987) and Ward and Clewell (1989) who stated that Atlantic white cedar along the Gulf Coast are often sparsely distributed among hardwoods. Eleuterius and Jones (1972) reported that the site near Vancleave, Mississippi, was a mixed forest, but that "[Atlantic white] cedar clearly dominated all areas sampled." In our study, Atlantic white cedar would occasionally be found in small groupings of several trees, but in general, hardwoods dominated the site and Atlantic white cedar trees were merely scattered throughout.

Of the 25 species other than Atlantic white cedar that were observed in the forest canopy, only five were of relatively high importance in the overstory (Table 4.1). The dominance of the two tupelo species is in agreement with several studies (Korstian and Brush 1931; Eleuterius and Jones 1972; Laderman 1989), but the relatively low importance of red maple (IV=14.2) is unusual. Red maple is one of the most common species that co-occurs with Atlantic white cedar (Levy 1987; Laderman 1989). Loomis et al. (2003) and Shacochis et al. (2003) listed red maple as a prominent species, usually of secondary importance to Atlantic white cedar, in most of the stands they studied in North Carolina and Virginia.

The high relative importance of loblolly pine contrasts with the findings of Eleuterius and Jones (1972) who reported slash pine as dominant in the nearby Atlantic white cedar stand at Vancleave. Loblolly pine apparently did not occur at the Vancleave site. Ward and Clewell (1989) listed both loblolly and slash pine as dominants in Atlantic white cedar stands of Florida, but these species did not appear to occur together in the same stands. Both sweetbay (IV=28.2) and baldcypress (IV=27.9) are common in Atlantic white cedar swamps throughout the Southeast (Korstian 1924; Eleuterius and Jones 1972; Dill et al. 1987; Ward and Clewell 1989). Stem density distribution plots of seven dominant or co-dominant canopy species (Figure 4.5) indicate some increase in swamp titi, red maple, water tupelo, sweetbay, and loblolly pine. Both swamp tupelo and baldcypress appear to have limited recruitment into the canopy and increasing size of the existing stems.

Within the midstory, 16 of 25 species were smaller individuals of the tree species making up the overstory. Of these, swamp titi was the overwhelming dominant (IV=106.2) with more stems per hectare than all other species combined. This species is often dominant in the midstory of Atlantic white cedar stands along the Gulf Coast (Eleuterius and Jones 1972; Ward and Clewell 1989) and the southern Atlantic Coast (Moore and Carter 1987) and is dominant in the overstory of other Atlantic white cedar swamps in Jackson County, Mississippi (observation by the authors, no data presented). Most shrub species in the midstory were of low overall importance (Table 4.2).

Few individual plants and large amounts of open space were found on the ground layer subplots. Woody vines and tree/shrub seedlings accounted for 99% of the vegetative cover. Only eight herbaceous species were found in the subplots with an additional two species observed outside of the subplots. Ground cover in the swamp and near the river was low presumably because of the dense canopy cover and flooding. The slope was oriented away from the sun in the morning and the trees of the swamp shade the slope later in the afternoon. The upper portion of the slope, where sunlight was more available, supported the greatest amount of ground cover. The nonwoody species included sparse amounts of sedges, grasses, cinnamon fern, violets, and spider lilies. Twelve additional herbaceous species were found associated with Atlantic white cedar downstream of the main study plot.

4.6. Research needs for tidally influenced Atlantic white cedar swamps

This chapter is a preliminary observation about Atlantic white cedar ecology in southern Mississippi. More intense investigations are warranted. McCoy and Keeland (2005) identified the location of several Atlantic white cedar populations and provided a very brief description of the vegetative composition at each location. A quantitative analysis of the plants and animals that populate each of these sites is needed. Additional research topics for future investigation could include (1) analysis of Atlantic white cedar regeneration ecology along the Gulf Coast, (2) a dendroecological analysis of stand development patterns and the response of the various Atlantic white cedar populations to past climate changes, (3) the interaction of landscape position, soils, and hydrology on Atlantic white cedar occurrence and reproduction, (4) biogeochemical interactions within Gulf Coast Atlantic white cedar stands, (5) mapping the aerial extent of coastal Atlantic white cedar stands along with tidal overlays, and (6) identifying the salt tolerance of Atlantic white cedar trees growing under tidal flood regimes. All of these studies would provide much needed information for managers.

4.7. Acknowledgments

The authors appreciate field assistance provided by Ron Weiland (Mississippi Museum of Natural Science), Jim Kelly (Eco-Logic Restoration Services), Susan Walls (USGS, National Wetlands Research Center), Jason Sullivan and Dana Drake (IAP World Services), and reviews by Ken Krauss and William Conner. Funding for the study was provided by the U.S. Geological Survey. Any use of trade, product, or firm names is for descriptive purposes only and does not imply endorsement by the U.S. Government.

References

Ash AN, McDonald CB, Kane ES, Pories CA (1983) Natural and modified pocosins: Literature synthesis and management options. FWS/OBS-83-04. U.S. Fish and Wildlife Service, Division of Biological Services, Washington

Atkinson RB, DeBerry JW, Loomis DT, Crawford ER, Belcher RT (2003) Water tables in Atlantic white cedar swamps: Implications for restoration. In: Atkinson RB, RT Belcher, DA Brown, JE Perry (eds) Atlantic white cedar restoration ecology and management, proceedings of a symposium. Christopher Newport University, Newport News, pp 137-150

Bernard JM (1963) Lowland forest of the Cape May formation in southern New Jersey. Bull NJ Acad Sci 8:1-12

Brown DA, Atkinson RB (1999) Assessing the survivability and growth of Atlantic white cedar [*Chamaecyparis thyoides* (L.) B.S.P.] in the Great Dismal Swamp National Wildlife Refuge. In: Shear T, Summerville KO (eds) Atlantic white cedar: Ecology and management symposium. General Technical Report SRS-27. U.S. Department of Agriculture, Forest Service, Asheville, pp 1-7

Buell MF, Cain RL (1943) The successional role of southern white cedar, *Chamaecyparis thyoides*, in southeastern North Carolina. Ecology 24:85-93

Clewell AF, Ward DB (1987) White cedar in Florida and along the northern Gulf Coast. In: Laderman A (ed) Atlantic white cedar wetlands. Westview Press, Boulder, pp 69-82

Conner WH (1993) Artificial regeneration of baldcypress in three South Carolina forested wetland areas after Hurricane Hugo. In: Brissette JC (ed) Proceedings of the seventh biennial southern Silvicultural research conference. General Technical Report SO-93. U.S. Department of Agriculture, Forest Service, pp 185-188

Dill NH, Tucker AO, Seyfried NE, Naczi RFC (1987) Atlantic white cedar on the Delmarva Peninsula. In: Laderman AD (ed) Atlantic white cedar wetlands. Westview Press, Boulder, pp 41-55

Eleuterius LN, Jones SB (1972) A phytosociological study of white-cedar in Mississippi. Castanea 37:67-74

Fowells HA (1965) Silvics of forest trees of the United States. Handbook No 271. U.S. Department of Agriculture, Washington, pp 151-156

Golet FC, Lowry DJ (1987) Water regimes and tree growth in Rhode Island Atlantic white cedar swamps. In: Laderman AD (ed) Atlantic white cedar wetlands. Westview Press, Boulder, pp 91-110

Korstian CF (1924) Natural regeneration of southern white cedar. Ecology 5:188-191

Korstian CF, Brush WD (1931) Southern white cedar. Technical Bulletin 251. U.S. Department of Agriculture, Washington

Laderman AD (1987) Atlantic white cedar wetlands. Westview Press, Boulder

Laderman AD (1989) The ecology of the Atlantic white cedar wetlands: A community profile. Biological Report 85(7.21). U.S. Fish and Wildlife Service, Washington

Levy GF (1987) Atlantic white cedar in the Great Dismal Swamp and the Carolinas. In: Laderman AD (ed) Atlantic white cedar wetlands. Westview Press, Boulder, pp 567-568

Little S (1950a) Ecology and silviculture of white cedar and associated hardwoods in southern New Jersey. Bulletin 56. Yale University School of Forestry, New Haven

Little S (1950b) Observations on the minor vegetation of the Pine Barren swamps in Southern New Jersey. Bull Torrey Bot Club 78(2):153-160

Little S, Garrett PW (1990) Atlantic white-cedar. In: Burns RM, Honkala BH (tech. coords.), Silvics of North America. Vol. 1: Conifers. Agriculture Handbook 654. U.S. Department of Agriculture, Forest Service, Washington

Loomis DT, DeBerry JW, Belcher RT, Shacochis KM, Atkinson RB (2003) Floristic diversity of eight Atlantic white cedar sites in southeastern Virginia and northeastern North Carolina. In: Atkinson RB, Belcher RT, Brown DA, Perry JE (eds) Atlantic white cedar restoration ecology and management, proceedings of a symposium. Christopher Newport University, Newport News, pp 91–100

McCoy JW, Keeland BD (2005) Locations of Atlantic white cedar in the coastal zone of Mississippi. In: Burke MK, Sheridan P (eds) Atlantic white cedar: Ecology, restoration, and management, proceedings of the Arlington Echo Symposium. General Technical Report SRS-91. U.S. Department of Agriculture, Southern Research Station, Asheville, pp 44-53

Moore JH, Carter JH III (1987) Habitats of white cedar in North Carolina. In: Laderman AD (ed) Atlantic white cedar wetlands. Westview Press, Boulder, pp 177-188

Motzkin G, Patterson WA III, Drake NER (1993) Fire history and vegetation dynamics of a *Chamaecyparis thyoides* wetland on Cape Cod, Massachusetts. J Ecol 81:391-402

NRCS (2006) Soil survey for Jackson, County, Mississippi. U.S. Department of Agriculture, Natural Resources Conservation Service, Washington

Reed Jr PB (1988) National list of plant species that occur in wetlands: National summary. Biological Report 88(24). U.S. Fish and Wildlife Service, Washington

Sedia E, Zimmermann GL (2006) Effects of salinity and flooding on Atlantic white-cedar seedlings. Abstract of paper presented at The Ecology and Management of Atlantic White-Cedar Conference, Atlantic City

Shacochis KM, DeBerry JW, Loomis DT, Belcher RT, Atkinson RB (2003) Vegetation importance values and prevalence index values of Atlantic white cedar stands in Great Dismal Swamp and Alligator River National Wildlife Refuges. In: Atkinson RB, Belcher RT, Brown DA, Perry JE (eds) Atlantic white cedar restoration ecology and management, proceedings of a symposium. Christopher Newport University, Newport News, pp 227-233

Sprecher SW (2000) Installing monitoring wells/piezometers in wetlands. Technical Notes Collection (ERDC TN-WRAP-00-02). www.wes.army.mil/el/wrap. U.S. Army Engineer Research and Development Center, Vicksburg

Stockwell KD (1999) Structure and history of the Atlantic white cedar stands at Appleton Bog, Knox County, Maine, USA. Nat Areas J 19:47-56

U.S. Army Corps of Engineers (1982) Draft environmental impact statement, Prulean Farms, Inc., Dare County. Wilmington District, Regulatory Functions Branch, Wilmington

Ward DB (1987) Commercial utilization of Atlantic white cedar (*Chamaecyparis thyoides*, Cupressaceae). Econ Bot 43(3):386-415

Ward DB, Clewell AF (1989) Atlantic white cedar (*Chamaecyparis thyoides*) in the southern states. Florida Sci 52(1):8-47

Williams TM (1993) Salt water movement within the water table aquifer following Hurricane Hugo. In: Brissette JC (ed) Proceedings of the seventh biennial southern silvicultural research conference. General Technical Report SO-93. U.S. Department of Agriculture, Forest Service, pp 177-183

Zampella RA, Laidig KJ, Lathrop RG, Bognar JA (1999) Size-class structure and hardwood recruitment in Atlantic white cedar swamps of the New Jersey pinelands. J Torrey Bot Soc 126:268-275

Chapter 5 - Sediment, Nutrient, and Vegetation Trends Along the Tidal, Forested Pocomoke River, Maryland*

Daniel E. Kroes[1,2], Cliff R. Hupp[2], and Gregory B. Noe[2]

[1]*U.S. Geological Survey, 3535 S Sherwood Forest Blvd., Baton Rouge, LA 70816*
[2]*U.S. Geological Survey, 430 National Center, Reston, VA 20192*

5.1. Introduction

The Pocomoke River Swamp was once considered to be an almost impenetrable wilderness, with conditions strongly resembling the Dismal Swamp of North Carolina and Virginia (Beaven and Oosting 1939). The original width of the forested wetland, as evidenced by black organic soils, extended as much as two to three times beyond the active-floodplain edge (pre-channelization) along the upper reaches of the river (Beaven and Oosting 1939). Currently, the Pocomoke River Swamp is extant only by the river and its active (flooded annually) floodplain that ranges from 0.35 km in width along upper reaches to 3.6 km along lower tidal reaches.

Blackwater systems, such as those found on the Coastal Plain, are characterized by very low-stream gradients, wide floodplains, organic stained waters that may appear black, very low total suspended solids, and long hydroperiods (Wharton et al. 1982; Hupp 2000). The historically (pre-channelized) blackwater Pocomoke River, Maryland is a minor tributary to the Chesapeake Bay (Figure 5.1), draining portions of the Delmarva Peninsula with climatic conditions typical of the coastal Mid-Atlantic Region. Under historical, natural conditions, the non-tidal Pocomoke River sequestered organic material for a period of time sufficient to develop 1-2 m of peat deposits at non-tidal upstream sites. Floodplain sediments in the past

---

* The U.S. Government's right to retain a non-exclusive, royalty-free licence in and to any copyright is acknowledged.

W.H. Conner, T.W. Doyle and K.W. Krauss (eds.), Ecology of Tidal Freshwater Forested Wetlands of the Southeastern United States, 113–137.
© 2007 *Springer.*

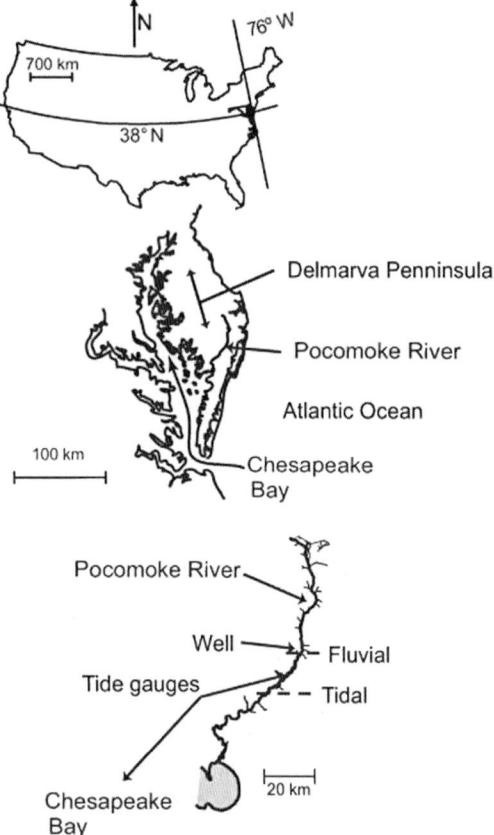

Fig. 5.1. Study sites are located along the Pocomoke River, a minor tributary to the Chesapeake Bay.

ranged from sapric-histosols (organic soils in which the original plant parts are not recognizable) in the headwaters and upper reaches (Beaven and Oosting 1939) to sandy and silty loams in the lower non-tidal reaches (Perkins and Bacon 1928).

The majority of the Pocomoke River drainage basin was ditched (channel created where no channel previously existed) and tributaries were channelized (natural channel modified to facilitate efficient drainage) prior to 1938 (Beaven and Oosting 1939; Ross et al. 2004). The main stem channelization of the Pocomoke began in November 1940 and was dedicated on September 25, 1946 (M. P. Sigrist, USDA, personal communication, 2002). Ditching has increased the drainage density of the basin by nearly 300%. Channelization and incision has made floodplain inundation rare and has drained the groundwater from large portions of the upper basin floodplain (Kroes and Hupp in review). The Pocomoke River is currently 40% channelized, 50% tidal and embayed, and the remaining 10% has a natural channel with a minor tidal range; only a few small tributaries remain unchannelized.

5.2. Geological history and historical land use

The surface of the southern Delmarva Peninsula is a flat to gently rolling central ridge bordered on the west and east by low plains sloping toward the Chesapeake Bay and the Atlantic Ocean. These surfaces are believed to be remnants of the original depositional surfaces, which formed during emplacement of the marginal-marine and fluvial-estuarine deposits (Mixon 1985; Bricker et al. 2003). The surficial and subsurface deposits (2-60 m) include unconsolidated sand, gravel, silt, clay and peat of Quaternary age (<1.5 million years before present [mybp]). These sediments overlie hundreds of meters of Tertiary (2-65 mybp) greensands and clay-silt. The Tertiary deposits show major paleovalleys running southeastward across the northern, central, and southern parts of the Penninsula (Bricker et al. 2003). The largest paleovalley, crossing the Penninsula near Eastville, Virginia (80 km south of the Pocomoke), was eroded to a depth of 50 m below present sea level and is believed to mark a main drainway of the ancestral Susquehanna-Potomac river system (Mixon 1985).

The Delmarva Peninsula was settled in the early 1600s. Subsistence hoe and shovel-based agriculture was practiced until ploughs were adopted in the late 1700s to early 1800s. The switch from hand tool to plough-based agriculture occurred between the generations of Charles Carroll of Carollton (signer of the Declaration of Independence) and his son Charles Carroll Jr., at that time the major landowners of colonial Maryland. This increase in farm efficiency resulted in an expansion of agriculture. As a result, farm journals of the Carroll estate record intense ditching efforts after 1800. In 1867, the Pocomoke River Improvement Company was founded in order to drain the Cypress Swamp (headwaters of the Pocomoke River) (Scharf 1888). Today, agricultural fields constitute 37% of the drainage basin (Maryland Department of Planning 2002). In some areas, ditches currently constitute 3 of every 4 stream miles on the Delmarva Penninsula (Kroes and Hupp in review).

Forest cover trends in this area are consistent with much of the eastern United States. Forest cover prior to settlement has been estimated at 80 to 90%. The lowest percentage of forest occurred during the Civil War (1861-1865) with as little as 20% cover. Since that time forest cover has increased; currently forests constitute approximately 45% of land cover within the Pocomoke River watershed (Maryland Department of Planning 2002). Almost all of the forests have been harvested at some time; some are currently managed for timber production.

Poultry rearing has been an important economic contributor to the area since around 1900. The first chicken feed plant was constructed in the area in 1928. Since that time, poultry rearing operations have increased to supply approximately 9% of the chicken production in the United States (USDA, NASS 2002 data).

5.3. Climatic drivers and hydrological characterization

The Delmarva Peninsula receives 1.14 m of total precipitation and 0.29 m of snow in an average year. The average temperature is 12.6°C and the growing season is about 198 days (NOAA, NCDC 2006). The Peninsula is located in an area where hurricanes play a very minor role in its ecology. Since 1854, no hurricane strength storms have made landfall, but in 1999, Hurricane Floyd skirted the coast. While hurricanes are rare, storms can exert a strong influence on the ecology of the Peninsula. On average, a major storm (tropical storm, tropical depression, extratropical storm) crosses the Peninsula every three years (NOAA 2006). Ninety percent of these major storms travel from southwest to northeast. Because of this typical storm trajectory, the associated winds and storm surge may cause the change from forested to marsh environments as saline Chesapeake Bay water is forced up the tidal rivers along the west side of the Peninsula killing salt intolerant trees (Newell et al. 2004). In effect, pushing salt stressed trees over the brink of death. Normal storms generally do not have the wind fetch, duration, and strength necessary to create a large enough surge to push Bay water far up the river.

Low-lying marshes and islands along the Chesapeake Bay and Atlantic Ocean, like Blackwater National Wildlife Refuge, are threatened by sea-level rise of 3 mm/yr, subsidence of variable rates and scales, erosion due to biological factors such as mute swans and nutria (currently eradicated on the Peninsula) and increasing development along shorelines (Larsen et al. 2004). Increases in soil and groundwater salinity due to rising sea level cause many forested wetlands on the Peninsula to retreat inland; these forests are commonly replaced by marsh vegetation. The floodplain along the lower Pocomoke River is one of these areas in transition from non-tidal forests to tidal forests to tidal marshes.

Our study examined sediment, nutrients, and vegetation at two different hydrologic regimes along the unchannelized Pocomoke River: 1) a tidal and wind-driven floodplain (Blades, hereafter Tidal) and 2) a fluvial dominated floodplain with minor tidal influences (Porters, hereafter Fluvial). At the Tidal site, some portions of the floodplain are inundated by overbank water levels almost daily, with complete site inundation during storms (Figure 5.2a). High-water levels at the tidal site are primarily driven by winds blowing from the south or southwest, typical of storms crossing the Chesapeake Bay during the summer. The closest gauge site is located 11 km upstream of the tidal site (NOAA tide station #8571359). At this station, wind and stream flow may have equal influence on water levels. Streamflow has incrementally less influence downstream as the channel increases in cross-sectional area (without significant tributary input) and the floodplain increases in width.

Upstream, at the Fluvial site, the floodplain is primarily inundated during the winter by groundwater derived streamflow (baseflow), with total site inundation occurring during relatively minor (2-3 cm) rainfall events. Inundation during the growing season occurs by storm flows originating from severe thunderstorms or tropical depressions during the summer. This fluvial site shows a minor (0.1-0.3 m) tide and wind signature when stream discharges are primarily contained within the channel (Figure 5.2B).

5.4. Geospatial description

Upstream from the mouth of the river, the floodplain is covered by salt tolerant marsh vegetation such as cordgrass (*Spartina* spp.) and needlegrass rush (*Juncus roemerianus* Scheele) until approximately river kilometer 13 (Figure 5.3). In this region, baldcypress (*Taxodium distichum* [L.] L.C. Rich.) and wax myrtle (*Morella cerifera* L.) establish on higher areas of the floodplain such as the levees. The density of baldcypress increases on the levees further upstream. In the vicinity of river kilometer 17, red maple (*Acer rubrum* L.) and loblolly pine (*Pinus taeda* L.) begin to occupy the levee. Full forest cover of the floodplain occurs around 25 kilometers from the mouth. From this area upstream, the floodplain is dominated by baldcypress, water tupelo (*Nyssa aquatica* L.), red maple, sweetgum (*Liquidambar styraciflua* L.), loblolly pine, American hornbeam (*Carpinus caroliniana* Walt.), and green ash (*Fraxinus pennsylvanica* Marsh.) (Figure 5.4). At approximately river kilometer 55, the embayed portion of the Pocomoke ends; from this point upstream species of oak become common along with the aforementioned species (Figure 5.5).

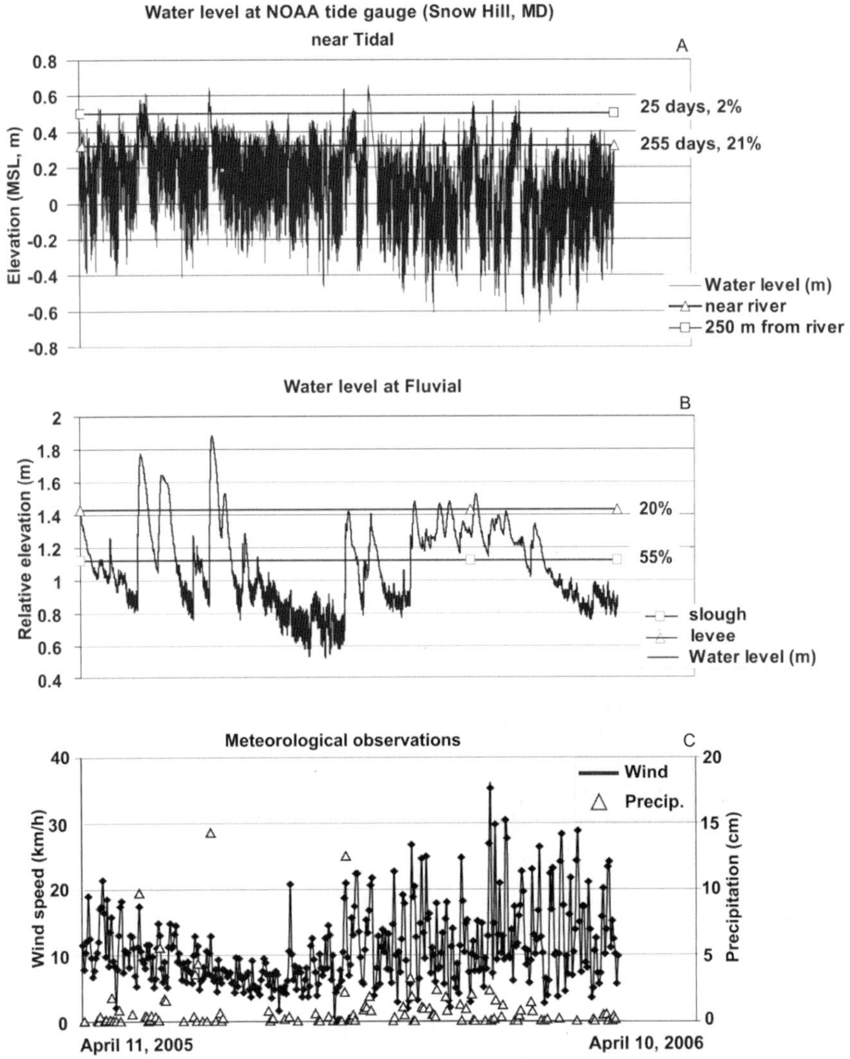

Fig. 5.2. Hydrographs for the Tidal (A) and Fluvial site (B). The hydrograph for the Tidal site is from the closest tide gauge located 11 km upstream of the site. Durations of inundation are indicated for floodplain surfaces. Wind speed and precipitation (C) are compared with these hydrographs.

Chapter 5 - Vegetation Trends Along Pocomoke River, Maryland   119

Fig. 5.3. Floodplain vegetation at river kilometer 6. The floodplain is dominated by marsh vegetation at this location.

Fig. 5.4. Bank vegetation at river kilometer 40 in the vicinity of the Tidal site. The floodplain banks are dominated by baldcypress.

Fig. 5.5. Floodplain vegetation at the Fluvial site (58 km). The floodplain is dominated by American hornbeam, red maple, and ash.

5.5. Ecological characterization

The ecological continuum from marsh to forest described above appears to be driven by salinity and hydroperiod and is typical of embayed rivers entering the Chesapeake Bay. Two sites were compared along the Pocomoke River, separated by hydrologic regime and 19 river km, to exemplify the differences between tidal and fluvial floodplains. One major tributary enters the river between the two sites.

The Tidal site has a large, clearly defined channel with a cross-sectional area of 360 m^2 (Figure 5.6A). The floodplain topography is dominated by hummocks (raised mounds held together by root masses) and hollows (areas typified by unconsolidated sediment between hummocks). The vertical distance between top of hummock to dense sediment at the bottom of the hollow typically is approximately 0.6 m. The hollows at this site have a nearly constant water level within 0.1-0.2 m of the hummock tops that appears to be maintained by tidal water, rainfall, groundwater discharge, and poor drainage. This tidal site has a diurnal tidal range of 0.6-0.7 m that inundates portions of the floodplain for short durations. During the period of study (1997-2006), water levels never exceeded 0.85 m above mean sea level. The average surface of the floodplain is lowest near the channel and increases about 0.18 m in elevation over a distance of 250 m from the channel (Figure 5.7A). These systems are notoriously difficult to traverse (Wharton et al. 1982; Doumlele et al. 1985) and have received, until recently, little ecological (Rheinhardt 1992) or hydrogeomorphic (Light et al. 2002) study.

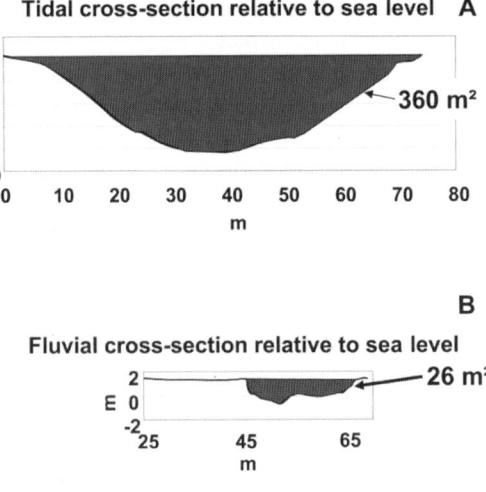

Fig. 5.6. Channel cross-sections at Tidal (A) and Fluvial (B) study sites.

The Fluvial site is typical of a fourth-order southeastern blackwater river (Hupp 2000). This site has a clearly defined channel with a cross-sectional area of 26 m² (Figure 5.6B). There are clearly defined flowpaths (sloughs) across the floodplain. These sloughs hold water substantially longer than the surrounding floodplain surface (Ross et al. 2004) at this site for approximately 55% (Figure 5.7A) of a normal year, and generally for extended periods (weeks - months) during the winter months (Kroes and Hupp unpublished well data). Sloughs are important conduits for transmission of water and sediment and increase the riparian connectivity of areas otherwise relatively distant from the channel (Hupp and Noe 2006). This fluvial site has an elevation range of 0.3 m with the highest elevation near the river, on the natural levee, and the lowest point mid-floodplain in the main slough. Pad elevations show the general elevational trend (Figure 5.7B).

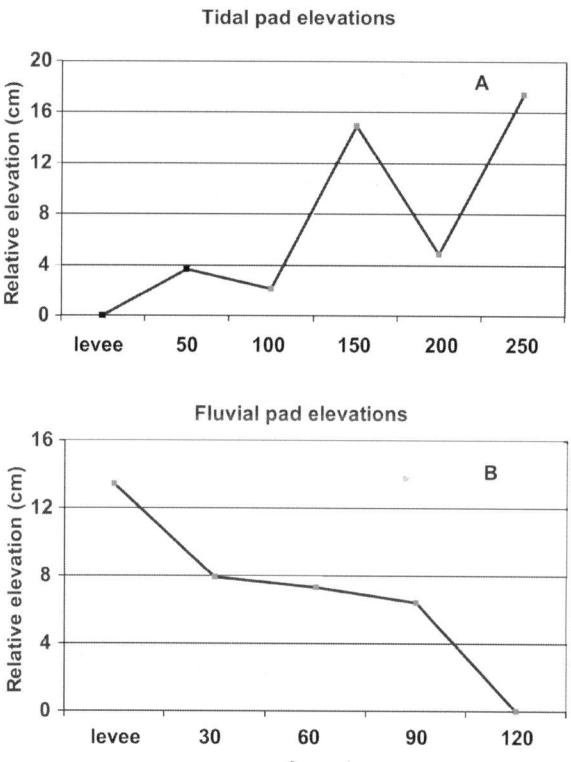

Fig. 5.7. Relative elevations of clay pads at the Tidal site (A) and Fluvial site (B). The Tidal site shows an increase in elevation as distance from the channel increases. At the Fluvial site (B) elevations on the floodplain are highest near the channel and decrease with distance.

5.6. Methods

These sites were investigated for characterization and contrasting of the hydrogeomorphic forms and processes (sedimentation) and vegetation. In 1997, 50 cm x 50 cm white feldspar marker horizons (clay pads, Hupp et al. 1993) were placed along transects at several locations along the Pocomoke River, including the two described here, in order to determine sedimentation rates and analyze recently deposited sediment samples (Kroes and Hupp in review). At Tidal, two transects of six pads each were placed perpendicular to the channel with a spacing of 50 m between pads. At Fluvial, three transects of five pads each were placed with a pad spacing of 30 m. The rates reported here were determined from cumulative deposition during the period 1997-2005. Nutrient and loss-on-ignition (LOI) data reported here were determined from the period 1997-2003.

Hydroperiod data for the sites were determined by two methods: a NOAA tide gauge (Tidal), and a combination surface/groundwater well (Fluvial). The tide gauge was installed in April 2005 and is located 11 km upstream of the Tidal site (6 minute intervals). Another tide gauge (NOAA #8633532) located 35 km south west of the mouth of the Pocomoke River, in the Chesapeake Bay, records similar tidal amplitude with approximately a 4 hour offset (Bay to Snow Hill) between high and low tides. Surveyed river stage at the Tidal site indicated similar tidal amplitude with a timing offset of ±30 minutes from the Snow hill gauge. A ground/surface water well was installed at the Fluvial site in August 2001 (15 minute intervals). The borehole was set to a depth of 1.2 m and measures water to a stage of 2 m above the floodplain surface. Sloughs at the Fluvial site have high connectivity to the river, i.e. river stage is equal to slough stage, and little to no ponding is present at the site. Surveyed standing water elevations in the sloughs and river stage are accurately represented by the well at this location.

Deposition was measured by cutting a plug from the clay pad with the sediments above it and measuring the accumulation. Sediments were collected from three 20-cm^2 cores from each clay pad in 2004. Each core was carefully extracted from the coring tubes and sediment above the marker horizon was collected and composited. These samples were analyzed in 2004 for organic content LOI, total carbon (C), total nitrogen (N), and total phosphorus (P). Samples were dried to a constant mass at 60° C and then weighed. Dried sediment samples were then ground with a mortar and pestle to pass through a 0.5 mm sieve. Coarse organic matter was pre-ground with a Wiley Mill. The organic and mineral content of the sediments was determined by loss-on-ignition at 400°C for 16 hr in a muffle furnace

(Nelson and Sommers 1996). Total C and total N concentrations were determined with a Carlo-Erba CHN elemental analyzer. Preliminary analyses indicated that inorganic C was a negligible proportion of total C (Noe, unpublished data). Total P concentrations were measured in digested sediments by ICP-OES analysis. Sediments were digested at high temperature and pressure by repeated microwave-assisted digestion following sequential addition of HNO_3, HCl and HF, and then HBO_3 acids.

Woody vegetation was analyzed from 400 m² plots at both sites near the river (0 m), mid-floodplain (50 m), and backswamp (100 m) (Hupp and Schening 1997). At Tidal, the same plot location protocol was used with the addition of another plot at 250 m from the river. Species composition, importance value basal area, and density were determined from trees with a diameter at breast height (DBH) greater than 2.5 cm (Doumlele et al. 1985; Rheinhardt 1992; Megonigal et al. 1997). Importance value was calculated by combining a species values of relative density, relative basal area, and, relative frequency. Selected trees were cored with an increment borer and aged to calculate growth rates (Tidal n=26, Fluvial n=55).

5.7. Results and discussion

5.7.1. Sediment

The data from the sediment analyses show a distinct contrast between the fluvial and tidal sites. Deposition rates at both sites are typical for forested wetlands on the Atlantic Coastal Plain (Hupp 2000). The range of sediment deposition rates were also comparable between these two sites ranging from 2-6 mm/yr (Figure 5.8), with a mean rate of 3.1 mm/yr at Tidal and 4.0 mm/yr at Fluvial. Soil bulk density at Tidal averaged 0.1 g/cm³ and 0.33 g/cm³ at Fluvial. LOI was significantly different between sites (ANOVA df=21 p=0.001); Tidal averaged 70% LOI with a range of 20-90%. LOI increases with distance away from the channel indicating decreased mineral input from river water (Figure 5.8A) The high LOIs indicate that this tidal site traps mineral sediment primarily within 50 m of the river's edge. High deposition rates and LOI indicate that there is either high primary productivity (autochthonous) or that large volumes of organic debris (allochthonous) are being trapped deep within the swamp, in conjunction with a very long hydroperiod.

Fluvial averaged 23% LOI with a range of 15-25%. LOI was lowest at 50 m from the river. Frequent low level flooding during the winter and fall wash leaf litter toward the backswamp or downstream, preventing organic

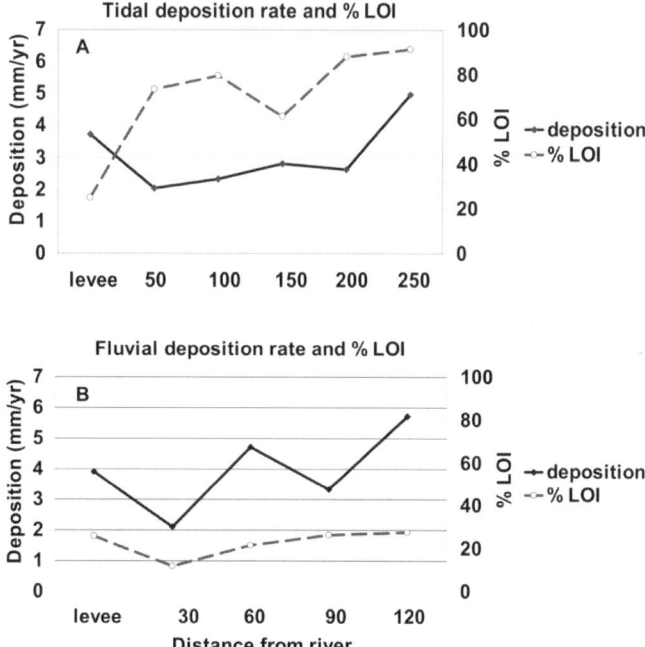

Fig. 5.8. Deposition rates and percent LOI (Loss on ignition) for Tidal (A) and Fluvial (B) sites. At Tidal (A) % LOI is lowest near the river and increases rapidly with distance from the channel. At Fluvial (B) LOI is lowest at intermediate distances.

material accumulation (Figure 5.8B). Low-level flooding, common during late fall, winter, and early spring (Figure 5.2), erodes the levee and high intermediate floodplain and deposits sediment in the backswamp (Kroes and Hupp in review). Deeper flooding during late spring and summer results in coarse sediment deposition on the levee and intermediate floodplain, as well as fine sediment deposition across the floodplain.

These data indicate that the sites store different sediment through differing processes. The Tidal site stores primarily organic material through constant inundation or saturation, similar to backwater flooding, while the Fluvial site stores primarily alluvial mineral sediment as a result of flows across the floodplain.

5.7.2. Nutrients

The degree to which floodplains act as nutrient sinks or sources depends highly on nutrient load in the flooding waters, hydroperiod, geomorphology, and other factors (Noe and Hupp in review). The majority of phosphorus in the Pocomoke River is associated with clays and other mineral sediments, whereas nitrogen is associated with organic matter (Noe and Hupp 2005). Nutrient concentrations in recently deposited sediments showed significant (t-test, df=19, %C, p=0.004, %N, p=0.07, P (mg/g), p<0.001) dissimilarity between the Tidal and Fluvial sites. Phosphorus concentrations at Tidal were 50% lower than at Fluvial. The rate of phosphorus accumulation ($g/m^2/yr$) at Fluvial was 6 times greater than at Tidal.

At the Tidal site, there was a decrease in phosphorus concentrations and an increase in C:P ratios at distances greater than 100 m from the channel (Figure 5.9A), indicating a change in sediment source. Phosphorus concentrations decreased more than N, as shown by N:P (Figure 5.10A), suggesting an inverse gradient of mineral and organic matter deposition from the channel to the backswamp. The lowest concentrations of nutrients were found at the highest elevation pads, those 150-250 m away from the channel. Throughout the Tidal site, high C:P ratios (390-1350) in the sediment indicate poor organic matter quality for microbial activity (Brinson 1977); this source becomes extremely poor at 200-250 m from the river (Figure 5.10A). Higher phosphorous concentrations in the sediment near the river at Tidal suggest alluvial sources, or alternatively, sorption of phosphorous to sediments in areas that are regularly inundated by phosphorous-rich river water (near the river). Due to the relatively young age of the tested sediments (5 yr), and the similar rate of deposition throughout the Tidal site, diagenic processes are unlikely to have caused the observed phosphorous gradient, suggesting that sediments are primarily from non-alluvial sources in the backswamp of Tidal. If the deposited sediments in the backswamp of Tidal are autochthonous in origin, then the very high C:P ratios in this sediment suggest a negative feedback loop whereby low rates of organic matter mineralization due to high sediment C:P ratios limit phosphorous availability to plants, which in turn produce phosphorous-poor litter.

The Fluvial site maintains a high level of connectivity with the river. The pads located in the middle of the site (farthest away from the channel and slough) had the lowest concentrations of nutrients (Figure 5.9B). The highest concentrations were found on the lowest pads located along the slough with a hydroperiod of approximately 200 days/yr (Figure 5.2). Relatively constant nutrient ratios with low C:N and C:P ratios across this site suggest that these pads receive a relatively high level of nutrients from the same alluvial source (Figure 5.10B). These data suggest that the tidal Pocomoke stores little alluvial nutrients, with primary nutrient storage occurring along non-tidal reaches (40% of the river, of which 80% is channelized).

5.7.3. Hydroperiod

The differences in nutrient storage and ratios exhibited between the Tidal and Fluvial sites can be explained by hydroperiod and depth of inundation. In order for significant sediments and nutrients from the watershed to be deposited on a floodplain there must be flow across the floodplain during a period of time when sediment concentrations are high (alluvial streams during rising flood stages). There is an order of magnitude difference between channel cross sections and the Tidal floodplain is three times

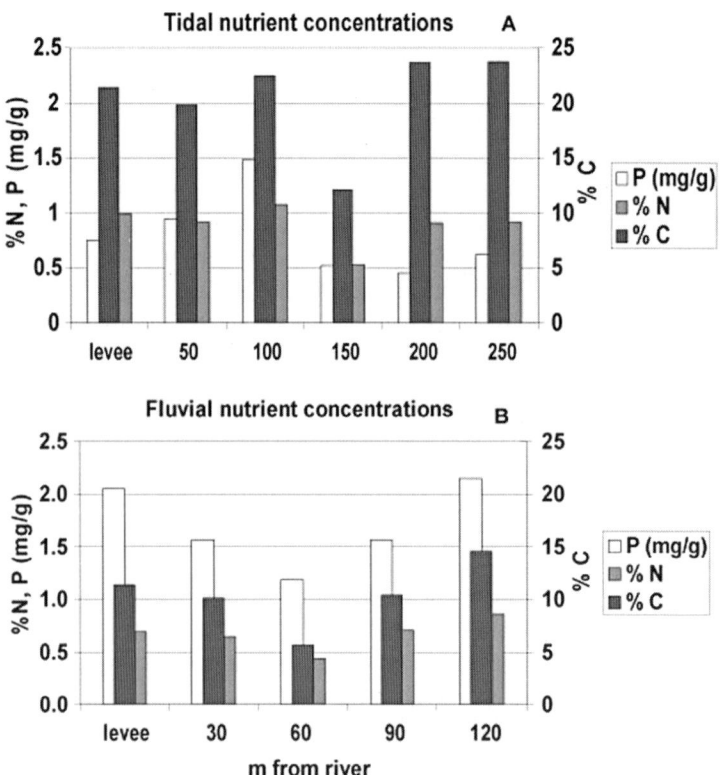

Fig. 5.9. Nutrient concentrations at Tidal (A) and Fluvial (B) sites. Tidal has higher carbon and lower nitrogen and phosphorous concentrations than Fluvial. Phosphorus concentrations at Tidal are lowest far from the channel. At Fluvial P and N concentrations are highest near water flow paths (channel and slough).

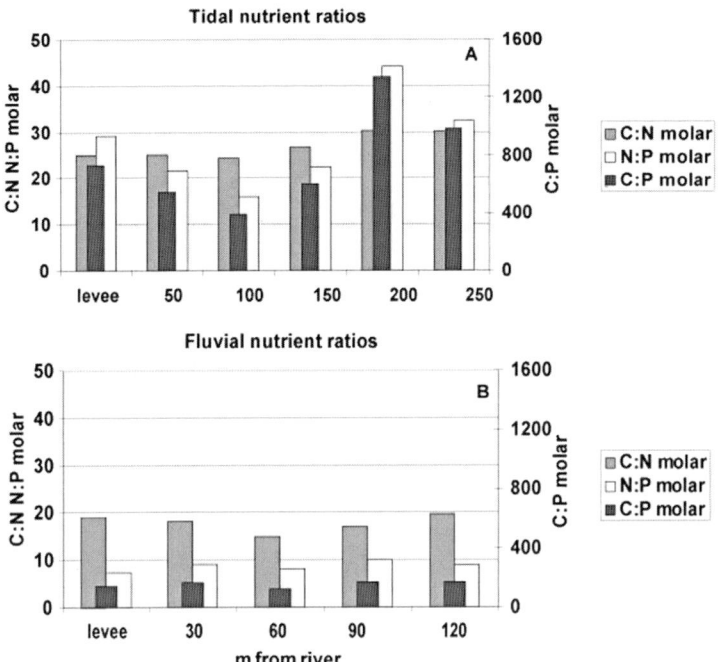

Fig. 5.10. Nutrient ratios at the Tidal (A) and Fluvial (B) sites. Higher nutrient ratios, like those at Tidal, indicate a limiting environment relative to carbon supplies.

the width of the Fluvial floodplain making storm flow inundation of the Tidal floodplain much less probable (Figure 5.11). At Tidal, the lowest pads located near the channel were inundated 255 days/yr but only 21% of the time while the highest pads were inundated by river water 25 days/yr and less than 2% of the time (Figure 5.2). Near the NOAA tide station (11km upstream), wind is equal to or greater than the effect of storm flow on the river stage (Figure 5.12). The channel at the tide gauge has a cross-sectional area of 288 m^2 (Davis 2005) in comparison to 360 m^2 at our Tidal site. At the Tidal site, inundation is controlled more by wind direction and velocity. On the Pocomoke River, there is generally a 1-day lag between storm and peak river discharge at the Tidal Site. Sediment peak discharge most commonly occurs before peak water discharge (Leopold et al. 1964). If wind conditions facilitate inundation of the entire floodplain during a storm, then the peak sediment load (drainage basin originating) comes after the tidal floodplain is already inundated with sediment poor tidal waters. This "preloaded" condition prevents sediment rich waters from diffus-

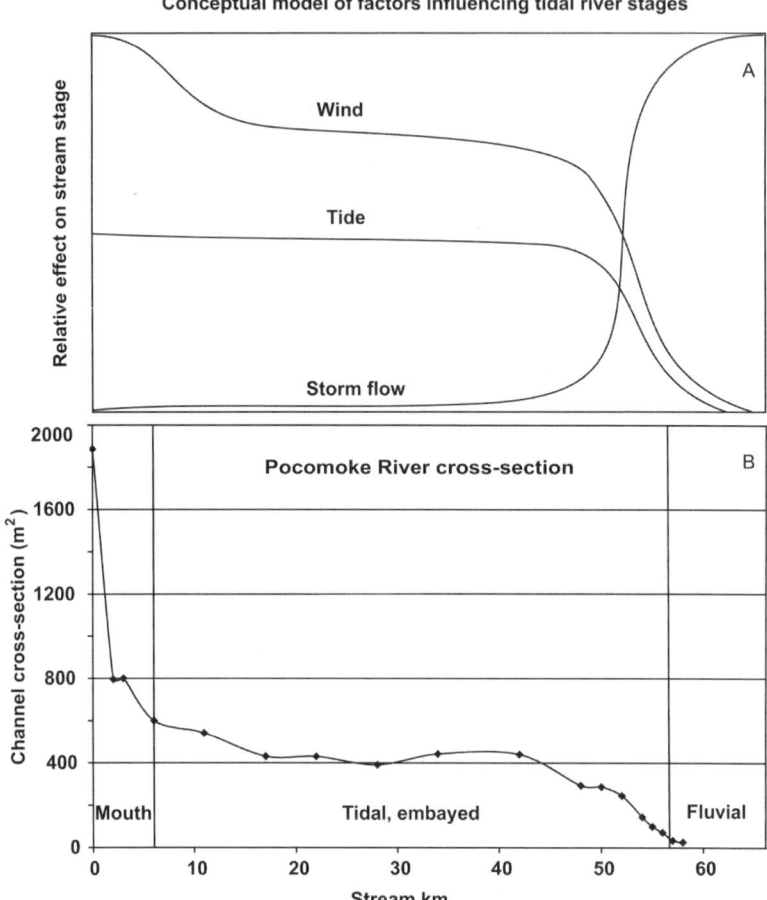

Fig. 5.11. A conceptual model of the factors influencing water levels in the lower Pocomoke River (A), in comparison with measured cross-sectional data measured by Davis (2005) and Kroes and Hupp (in review).

ing across the floodplain (Figure 5.12). For example, steady winds blowing from the west and south have the greatest effect on water levels for the tidal portion of the Pocomoke (Figure 5.12: Period 1 had an average wind direction and speed of 203° at 19 km/h; Period 2: 262° and 16 km/h; Period 3: 203° and 22 km/h). Despite high stream stage at Fluvial (Figure 5.12B) at the beginning of Period 2, water levels were normal at the tide gauge (Figure 5.12A) with low wind speeds (Figure 5.12C). Suspended sediment loads coming from the watershed are further reduced by a rapid water velocity decrease and sediment settling caused by the damming effect of Chesapeake Bay water blown upstream. Lastly, tide dominated sys-

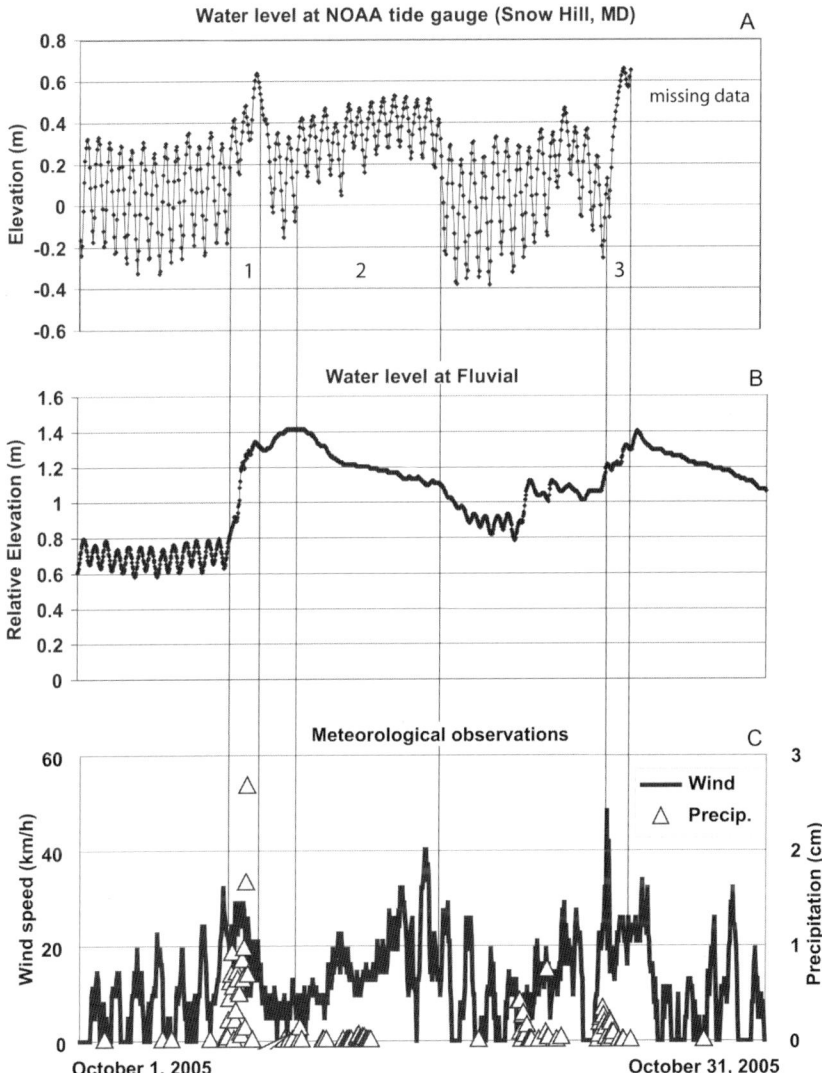

Fig. 5.12. Hydrographs and meterological data with sub-hourly data. Periods of interest are indicated with vertical black lines and are matched by time and date.

tems on the Coastal Plain are inherently less affected by alluvial flooding than non-tidal systems because of their large floodplain capacities (typically underfit; Hupp 2000) and their proximity to sea level.

Tide data near the Tidal site were recorded from September through December 2000 (Davis 2005) and since April 2005 (NOAA tide station #8571359); additionally, single-stage suspended sediment samplers were established at both sites in 1997. During the tidal record, data indicate that water levels did not exceed 0.3 m of depth on any portion of the Tidal site. Samplers set to collect at 0.65 m above floodplain surface never collected water during their 9 year implementation at the Tidal site. Low depth of flooding combined with high surface roughness (high Mannings coefficient of the hummock and hollow topography), and the presence of low sediment water in place on the floodplain may preclude alluvial sediment import to the Tidal swamp. Water samples collected from this site corroborate this interpretation (Ross et al. 2004). These patterns are similar to the perirheic zone concept developed for non-tidal floodplains (Mertes 1997).

5.7.4. Vegetation

The differences between the Tidal and Fluvial sites were more pronounced in woody vegetation composition. At the Tidal site, the levee was dominated by baldcypress, red maple, and green ash (Figure 5.13A). Green ash maintained a high importance value until around 100 m from the channel where water tupelo increased in importance. At 250 m from the channel, there was a complete species composition shift where water tupelo and sweetgum became the dominant species, with ironwood and loblolly pine present.

The levee at the Fluvial site was dominated by American hornbeam, red maple, sweetgum, and water tupelo. The intermediate zone is dominated by American hornbeam, red maple, swamp chestnut oak (*Quercus michauxii* Nutt.), and willow oak (*Quercus phellos* L.). The backswamp, 100 m from the channel, was dominated by ash, overcup oak (*Quercus lyrata* Walt.), and baldcypress (Figure 5.13B). The reversal of baldcypress presence pattern was clearly evident particularly along the stream banks of the lower tidal reaches (Figure 5.4).

The Tidal site had an average density of 2650 stems/ha. Basal areas at this tidal site ranged from 39-71 m^2/ha. The Fluvial site had an average density of 820 stems/ha with basal areas ranging from 19–35 m^2/ha (Figure 5.14). These values are within normal ranges for the mid-Atlantic tidal (Doumlele et al. 1985; Rheinhardt 1992; Megonigal et al. 1997) and fluvial floodplains (Spencer et al. 2001) A reduction in basal area was documented at 250 m from the river at the Tidal site, possibly as a result of decreasing nutrient availability (Figure 5.9A). This pattern of increased basal areas and densities with tidal water flux has been observed at other study rivers in the Chesapeake Bay area. Along the Pamunkey River there was a 15% increase in basal area and a 40% increase in density from fluvial to

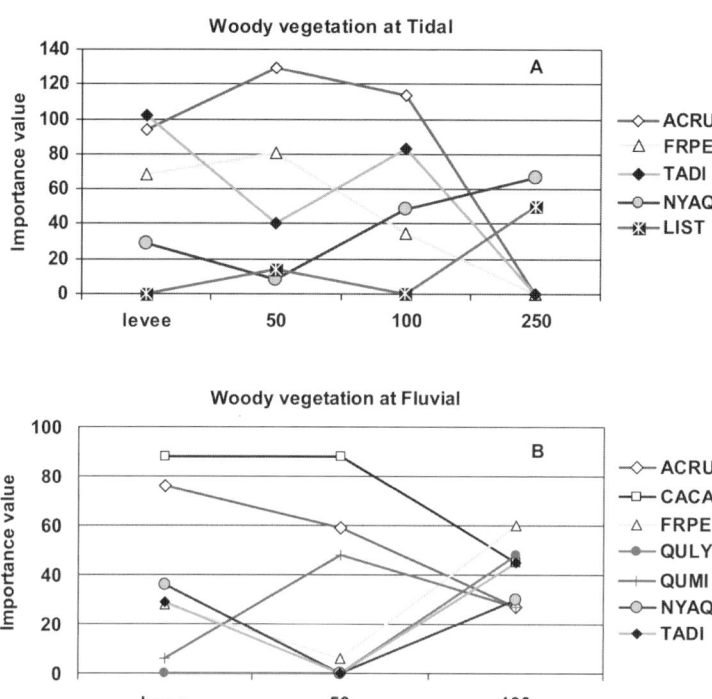

Fig. 5.13. Tree species importance value at Tidal (A) and Fluvial (B) sites. At Tidal, baldcypress (TADI), red maple (ACRU), and green ash (FRPE) were dominant near the river, and water tupelo (NYAQ) and sweetgum (LIST) dominated far away from the channel. Fluvial was dominated by American hornbeam (CACA), and red maple near the channel, with swamp chestnut oak (QUMI) increasing in importance mid-levee. Green ash, laurel oak (QULY), American hornbeam, and baldcypress dominated the backswamp. Species abbreviation is standard notation with first 2 letters indicating genus and the last two indicating species (i.e., bald cypress, *Taxodium distichum,* is TADI).

tidal. The Mattaponi River showed a 28% increase in basal area and a 40% increase in density (Hupp and Schening 1997).

The canopy trees along the Pocomoke River were mature. Individual tree ages were up to 238 years old with a median age of 83 (n=55) at the Fluvial site and 121 at the Tidal site with a median age of 55 (n=26) (Hupp and Schening 1997). At the Tidal site, tree canopies appeared to be thin and generally in poor condition, possibly explaining the overall high density values as greater light availability to the sub-canopy vegetation layers may allow for greater survival of saplings. Despite the appearance of poor canopy condition, trees at the Tidal site exhibited an average dbh growth

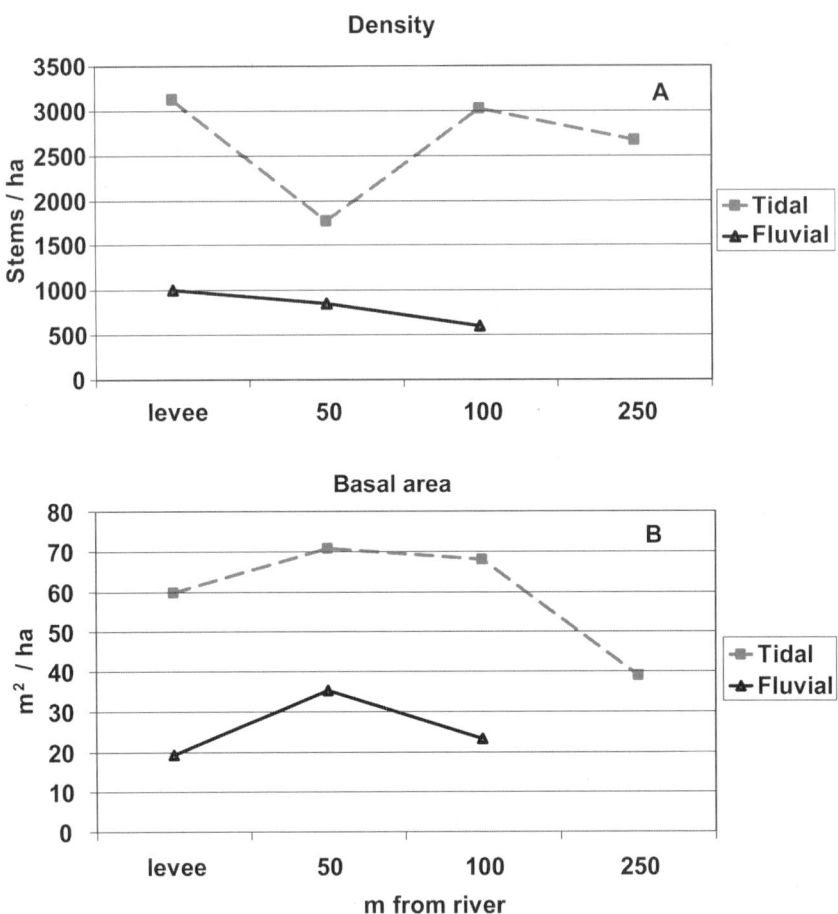

Fig. 5.14. Tree density (A) and basal area (B) at the Tidal and Fluvial sites. Tidal exhibited higher density and basal area than Fluvial.

rate of 0.39 cm/yr whereas at the Fluvial site dbh growth rates were 0.17 cm/yr. Trees of similar age and species exhibit 30-60% faster growth rates at the Tidal site. The extended duration of inundation at the Fluvial site may suppress growth, whereas tidal flushing at the Tidal site may prevent the build up of compounds inhibiting growth (Falcão and Vale 1995). However, the sediments at the Tidal site are perennially saturated and at 250 m from the river are not flushed regularly by tidal action, whereas at the Fluvial site there was a clear tidal signature of 0.1-0.2 m in the groundwater of the floodplain. These vegetation data in combination with nutrient and hydrologic data suggest that between the Tidal and Fluvial sites, hydrologic regime limits productivity more than nutrient availability.

5.8. Conclusion

In comparison to fluvial floodplains, tidal floodplains function differently. Our sites along the Pocomoke are inundated a similar number of days. However, the timing, duration, and depth of inundation vary greatly. Large portions of the Fluvial site were inundated for most of the winter and weeks during the summer. In the absence of steady wind, the tidal site was inundated frequently for short periods of time (hours). During our study, the Fluvial site was inundated frequently by up to 2 meters of water; the Tidal site was not flooded by more than 0.65 m since 1997.

As a result of differing hydrologic regimes, these areas store different sediments from different sources. Fluvial areas are inundated by storm flows originating from rainfall on the watershed and thus store alluvial sediments and nutrients. Depending on the watershed and site location sediments at fluvial sites may range from primarily mineral to organic. These sediments are predominantly from upstream sources with some autochthonous organic material. Tidal sites, being inundated by wind driven waters and to a minor degree stormflow, store a mix of sediments from organic production, downstream marsh and bank erosion, and fine alluvial sediments. At our Tidal site, these sediments were primarily organic. Nutrient data indicate that farthest from the river, the sediments may be primarily of autochthonous origin. In short, the alluvial sediment and nutrient trapping function of riparian areas (Brinson 1993; Hupp et al. 1993) remains substantial and important along alluvial reaches, however, this function may diminish along tidal reaches. Tidal reaches play an important role in the storage of organic carbon.

5.9. Research needs

5.9.1. Current research

Current research on the Pocomoke River consists of ongoing projects of the NAWQA program (USGS National Water-Quality Assessment) and other water quality assessment programs (Maryland Department of Natural Resources). With the exception of the water quality studies along the tidal reaches, to the authors' knowledge, there are no ongoing projects. There are other ongoing projects along the non-tidal river including Sediment Elevation Table (SET) studies to determine subsidence rates along channelized reaches and radioisotopic fingerprinting of suspended sediments to document source areas (both by the USGS). We have maintained a surface/groundwater well at Porter's Crossing (Fluvial) since August 2001 (partial record).

5.9.2. Future research

Future research should be directed toward quantifying the effect of tidal action in the groundwater at sites similar to the Fluvial site described above. The flushing of the groundwater in conjunction with high nutrient concentrations in the sediment may decrease the nutrient storage function of these systems to an unknown degree. Root physiology may differ between tidal and fluvial floodplains. Additionally, research should be devoted to understanding what makes tidal freshwater forests so productive and dense.

5.10. Acknowledgments

Thanks to all of our field personnel who made this study possible and to all landowners and managers who allowed us to use their lands for our studies. This research was funded by the USGS Chesapeake Bay Priority Ecosystems Initiative and the National Research Program.

References

Beaven GF, Oosting HJ (1939) Pocomoke Swamp: A study of a cypress swamp on the eastern shore of Maryland. Bull Torrey Bot Club 66:376-389

Bricker O, Newell W, Simon N (2003) Bog iron formation in the Nassawango watershed, Maryland. Open File Report 03-346. U.S. Geological Survey, http://pubs.usgs.gov/of/2003/of03-346

Brinson MM (1977) Decomposition and nutrient exchange of litter in an alluvial swamp forest. Ecology 58:601–609

Brinson MM (1993) Changes in the functioning of wetlands along environmental gradients. Wetlands 13:65-74

Davis EV (2005) Circulation and transport processes for the Pocomoke River, a tributary to a partially mixed estuary. M.S. thesis, University of Maryland

Doumlele DG, Fowler K, Silberhorn GM (1985) Vegetative community structure of a tidal freshwater swamp in Virginia. Wetlands 4:129-145

Falcão M, Vale C (1995) Tidal flushing of ammonium from intertidal sediments of Ria Formosa, Portugal. Aq Ecol 29:239-244

Hupp CR (2000) Hydrology, geomorphology, and vegetation of Coastal Plain rivers in the southeastern United States. Hydrol Proc 14:2991-3010

Hupp CR, Noe GB (2006) Sediment and nutrient accumulation within lowland bottomland ecosystems: An example from the Atchafalaya River Basin, Louisiana. In: Proceedings hydrology and management of forested wetlands. American Society of Agricultural and Biological Engineers, pp 175-187

Hupp CR, Schening M (1997) Patterns of sedimentation and woody vegetation along black-and brown-water riverine forested wetlands. Assoc Southeastern Biol Bull 44:140.

Hupp CR, Woodside MD, Yanosky TM (1993) Sediment and trace element trapping in a forested wetland, Chickahominy River, Virginia. Wetlands 13:95-104

Kroes DE, Hupp CR (in review) Floodplain sedimentation and subsidence along channelized and unchannelized reaches of the Pocomoke River, Maryland. Hydrol Proc

Larsen C, Clark I, Guntenspergen G, Cahoon D, Caruso V, Hupp CR, Yanosky T (2004) The Blackwater NWR inundation model. Rising sea level on a low-lying coast: Land use planning for wetlands. Open File Report 04-1302. U.S. Geological Survey, http://pubs.usgs.gov/of/2004/1302

Leopold LB, Wolman MG, Miller JP (1964) Fluvial processes in geomorphology. W.H. Freeman and Co., San Francisco

Light HM, Darst MR, Lewis LJ, Howell DA (2002) Hydrology, vegetation and soils of riverine and tidal floodplain forest of the lower Suwannee River, Florida and potential impacts of flow reductions. Professional Paper 1656-A. U.S. Geological Survey, Tallahassee

Maryland Department of Planning (2002) Land use land cover GIS data http://www.mdp.state.md.us/zip_downloads_accept.htm

Megonigal JP, Conner WH, Kroeger S, Sharitz RR (1997) Aboveground production in southeastern floodplain forests: A test of the subsidy-stress hypothesis. Ecology 78(2):370–384.

Mertes LAK (1997) Documentation and significance of the perirheic zone on inundated floodplains. Water Resour Res 33:1749–1762

Mixon RB (1985) Stratigraphic and geomorphic framework of uppermost Cenozoic deposits in the southern Delmarva Peninsula, Virginia and Maryland. Professional Paper 1067-G. U.S. Geological Survey, p G1-G53

Nelson DW, Sommers LE (1996) Total carbon, organic carbon, and organic matter. In: Sparks DL (ed) Methods of soil analysis, Part 3: Chemical Methods-SSSA book series 5, Soil Science Society of America, Madison, pp 1002-1005

Newell W, Clark E, Bricker O (2004) Distribution of Holocene sediment in Chesapeake Bay as interpreted from submarine geomorphology of submerged landforms, selected core holes, bridge borings, and seismic profiles. Open File Report 04-1235. U.S. Geological Survey, http://pubs.usgs.gov/of/2004/1235

NOAA (National Oceanic Atmospheric Administration) (2006) Historical hurricane tracks. http://hurricane.csc.noaa.gov/hurricanes/viewer.html

NOAA (National Oceanic Atmospheric Administration), NCDC (National Climatic Data Center) (2006) US climate normals. http://hurricane.ncdc.noaa.gov/cgi-bin/climatenormals/climate normals.pl

Noe GB, Hupp CR (2005) Carbon, nitrogen, and phosphorus accumulation in floodplains of Atlantic Coastal Plain rivers, USA. Ecol Appl 15:1178-1190

Noe GB, Hupp CR (in review) Seasonal variation in nutrient retention during inundation of a short-hydroperiod floodplain. River Res Appl

Perkins SO, Bacon SR (1928) Soil survey Worcester County, Maryland, U.S. Government Printing Office, Washington

Rheinhardt RD (1992) A multivariate analysis of vegetation patterns in tidal freshwater swamps of lower Chesapeake Bay, USA. Bull Torrey Bot Club 119:193-208

Ross KM, Hupp CR, Howard AD (2004) Sedimentation in floodplains of selected tributaries of the Chesapeake Bay. In: Bennett SJ, Simon A (eds) Riparian vegetation and fluvial geomorphology. American Geophysical Union, Water Sci Appl 8:187-208

Scharf JT (1888) History of Delaware. LJ Richards & Co, Philadelphia

Spencer DR, Perry JE, Silberhorn GM (2001) Early secondary succession in bottomland hardwood forests of southeastern Virginia. Environ Manage 27:559–570

USDA, NASS (U.S. Department of Agriculture, National Agriculture Statistics Service) (2002) Census of agriculture volume 1 county level data. http://www.nass.usda.gov/census/census02/volume1/md/st24_2_013_013.pdf

Wharton CH, Kitchens WM, Pendleton EC, Sipe TW (1982) The ecology of bottomland hardwood swamps of the southeast: A community profile. FWS/OBS-81/37. U.S. Fish and Wildlife Service, Washington

Chapter 6 - Vegetation and Seed Bank Studies of Salt-Pulsed Swamps of the Nanticoke River, Chesapeake Bay

Andrew H. Baldwin

Department of Environmental Science and Technology, University of Maryland, College Park, MD 20742

6.1. Introduction

Despite their classification as tidal freshwater swamps, research conducted at the Nanticoke River in Maryland and Delaware suggests that ecologically significant salinity gradients develop during droughts. While average salinity in these swamps may meet the criterion for "fresh water" of <0.5 parts per thousand (ppt) (Cowardin et al. 1979), significant variation in swamp vegetation and seed banks are observed as distance upstream increases. During a drought year, a salinity gradient ranging from 0 to 7 ppt developed across the upper estuary, suggesting that vegetation patterns were related to periodic saltwater intrusion events.

The Nanticoke River is located on the Delmarva Peninsula in Maryland and Delaware and is a major tributary of the Chesapeake Bay (Figure 6.1). The Nanticoke watershed encompasses about 86,200 ha in Maryland and 127,500 ha in Delaware. The Nature Conservancy (TNC) designated the watershed as both a Bioreserve and a Last Great Place in 1991 (TNC 1998). Almost 200 rare, threatened, or endangered species of plants (20 globally rare) and 70 rare, threatened, or endangered species of animals (5 globally rare) occur within several significant community types, including freshwater intertidal wetlands.

The freshwater intertidal wetlands of the Nanticoke River watershed include about 2,000 ha of freshwater tidal marsh, 2,800 ha of freshwater tidal swamp forest, and 360 ha of freshwater tidal shrub swamp (McCormick and Somes 1982). The marsh wetlands consist primarily of annual and perennial emergent herbaceous vegetation typical of tidal freshwater

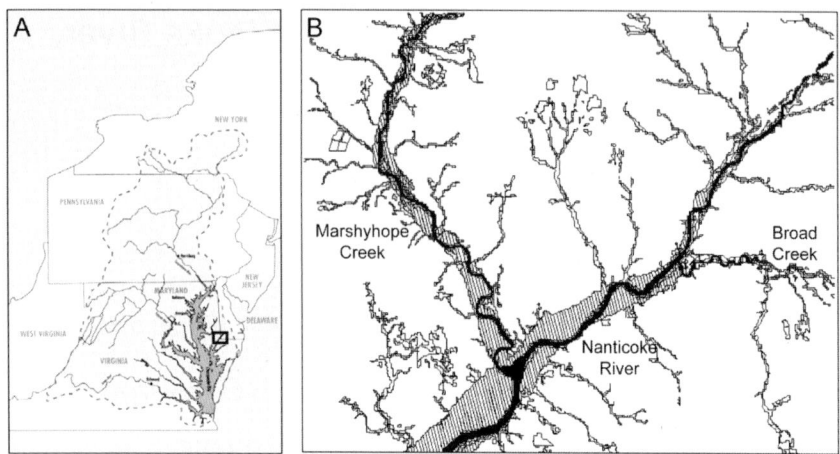

Fig. 6.1. Location of Nanticoke River study area. A: the study area location is outlined in a box, showing its location on the Delmarva Peninsula within the boundaries of the Chesapeake Bay watershed. B: the portions of major tributaries (Nanticoke River main stem, Marshyhope Creek, and Broad Creek) discussed in this chapter (in solid black) and associated low-salinity tidal wetlands (in patterned shading).

marshes of the Atlantic Coast (Simpson et al. 1983; Odum et al. 1984; Tiner 1993). The swamp wetlands have a hummock/hollow microtopography and species-rich vegetation similar to the ash-black tupelo (*Fraxinus pennsylvanica-Nyssa sylvatica*) tidal freshwater swamps described by Rheinhardt in Virginia (1992 and in Chapter 7). Hummocks in many of the Nanticoke swamps support a diverse group of forbs, graminoids, and ferns, while the hollows may have composition similar to the dominant species of freshwater tidal marsh if sufficient light is available. Typically, marshes fringe the river in a relative narrow strip (e.g., < 50 m), with extensive swamps occurring inland and at a slightly lower elevation.

This chapter summarizes major research results from studies in tidal swamps in the Nanticoke River, as well as presents general historical, hydrological, and geospatial information. For convenience, the "tidal freshwater" zone refers to the upper portions of the estuary within which most of the tidal swamps occur, recognizing that salinity concentrations can vary across this zone and may considerably exceed 0.5 ppt.

6.2. Historical land use change and geologic history

The Nanticoke River lies on the western edge of the Delmarva Peninsula (comprised of parts of Delaware, Maryland, and Virginia) (Figure 6.1). This is a low relief region forming the eastern boundary of the Chesapeake Bay that is part of the Coastal Plain Province of the southeastern United States (Schmidt 1993). Most of the coastal plain consists of unconsolidated sediments, and the lower reaches of the Nanticoke River and other rivers flowing into the Chesapeake Bay are underlain by sediments deposited during the Quaternary Period, below which lie older Tertiary deposits (Schmidt 1993). Coastal wetlands of the Atlantic Coastal Plain, as well as those of the Gulf Coast, formed during a period of slow sea-level rise during the late Holocene Epoch, the period of time since the last ice age ended about 10,000 years ago (Michener et al. 1997). The Chesapeake Bay itself is a drowned river valley, carved by the Susquehanna River and its tributaries when sea level was more than 100 m lower than today, eventually becoming flooded by rising seas (White 1989; Tiner and Burke 1995).

Before John Smith began exploring the Chesapeake Bay in 1608, Native Americans had altered vegetation by creating clearings for agriculture in the primary forest and burning forested areas around villages, but most of the region remained forested (e.g., 95% in Maryland) (Schneider 1996). European settlement, timber harvest, and agriculture (including intensive tobacco farming in the Nanticoke watershed) beginning in the early 17th century transformed the landscape. By 1840, 40-50% of the bay region was cleared of forest, resulting in increases in sedimentation rates, eutrophication, anoxia, and turbidity in the bay (Cooper 1995).

Wetlands were also impacted by European settlement. About 93,000 ha of wetlands (45% of total land area) occurred in the 2070-km^2 Nanticoke River watershed before European settlement, but these had been reduced to 57,000 ha, or 62% of their original extent, by 1998 (Tiner 2005). Estuarine wetland losses, however, only exhibited a 4% loss (9,570 ha to 9,160 ha). Between 1982 and 1989, there was a reduction in tidal freshwater swamps (estuarine forested, E2FO) in Maryland of 310 ha, or 4% of the total (Tiner and Burke 1995). The major causes of wetland loss were conversion to agricultural lands and, for estuarine wetlands, relative sea-level rise (Tiner 2005).

Currently, about 42,000 people live in the Nanticoke watershed, and developed land comprises about 2.4% of the watershed (Chesapeake Bay Program Watershed Profile, www.chesapeakebay.net). Approximately 43% of the watershed is devoted to agriculture and 28% is forested. Runoff from farm fields containing nutrients and sediment has been identified as the primary threat to the conservation targets in the watershed, and freshwater intertidal wetlands were judged to be under the greatest stress (TNC 1998).

6.3. Climatic drivers and hydrological characterization

6.3.1. Climatic conditions

Data recorded in Vienna, Maryland (about 5 km south of the study area in the Nanticoke River watershed) between 1991 and 2001 indicate that mean temperatures range from -2° to 9°C in January up to 20° to 30°C in July (National Climatic Data Center, lwf.ncdc.noaa.gov). Mean annual precipitation between 1993 and 2001 was 1057 mm.

6.3.2. Tidal regime

Tidal measurements for the Nanticoke River are available from a National Oceanographic and Atmospheric Administration (NOAA) water level recording station (#8571773) in Vienna, MD, 5 km downstream from our tidal swamp study area. Water level measurements at this station are available since April 28, 2005, while only water level predictions are available before this time. At this station, the diurnal tidal range (difference between mean higher high water and mean lower low water) in the Nanticoke River is 0.7 m. Due to variables such as wind and precipitation, measured water level is generally different from predicted water level (Figure 6.2).

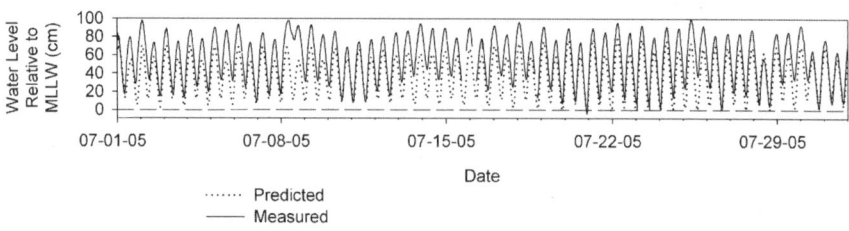

Fig. 6.2. Predicted and measured water level of the Nanticoke River, NOAA Station 8571773, Vienna, MD, for July 2005. Plotted values are elevation above mean lower low water.

In addition to NOAA data, water level recorders (Ecotone WL-80, Remote Data Systems, Whiteville, North Carolina) were set up at various study plots throughout the tidal freshwater reaches of the Nanticoke River watershed. Data from select tidal freshwater swamp locations in the watershed for July 2001 illustrate the variability in hydroperiod in tidal swamps within the watershed (Figure 6.3). The predicted water level at the Vienna station (panel A; actual measurements are not available for this period) shows a typical 2-week variation in amplitude between spring and neap tides. The higher-amplitude spring tides caused increases in standing water at all four swamp sites, extending at least 25 km upstream from Vienna and across three different tributaries (Nanticoke River main stem, Marshyhope Creek, and Broad Creek). However, the depth and frequency of flooding differed considerably between the sites. Water level at the two sites farthest upstream (panels D and E) ranged from <5 cm below the surface to about 25 cm above the surface for higher-amplitude tidal cycles. In contrast, sites located toward the lower reaches of the tidal freshwater zone (panels B and C) experienced shallower surface flooding and greater draining across tidal cycles. The most downstream site pictured (panel B), did not experience any surface flooding over a 5-day period in mid-July, during which groundwater levels dipped to 15 cm below the surface. These findings indicate that, despite considerable variation in channel water level (up to 80 cm, panel A), tidal swamps are not likely to be flooded with standing water during every tidal cycle. Additionally, the results suggest that swamps farthest downstream in the tidal freshwater zone experience less tidal inundation than upstream swamps.

6.3.3. Trends in relative sea level

Relative sea level in the region (Cambridge, Maryland station) is increasing at a rate of 3.52 mm/year (35 cm/century) with a standard error of 0.24 mm/yr based on monthly mean sea-level data from 1943 to 1999 (tidesandcurrents.noaa.gov). This rate of relative sea-level rise is among the highest on the U.S. Atlantic coast (Stevenson and Kearney 1996), resulting in loss of brackish marshes in the Chesapeake Bay (Stevenson et al. 1985; Kearney et al. 1988; Kearney and Stevenson 1991; Downs et al. 1994). Kearney et al. (1988) found that between 1938 and 1985 in the Nanticoke River, tidal freshwater marshes were relatively stable and were accreting sufficiently to keep pace with relative sea-level rise, while deterioration rates increased and accretion rates decreased proceeding downstream along the estuary. In the more downstream reaches, marsh loss has been the result of a number of contributing and interrelated factors, the

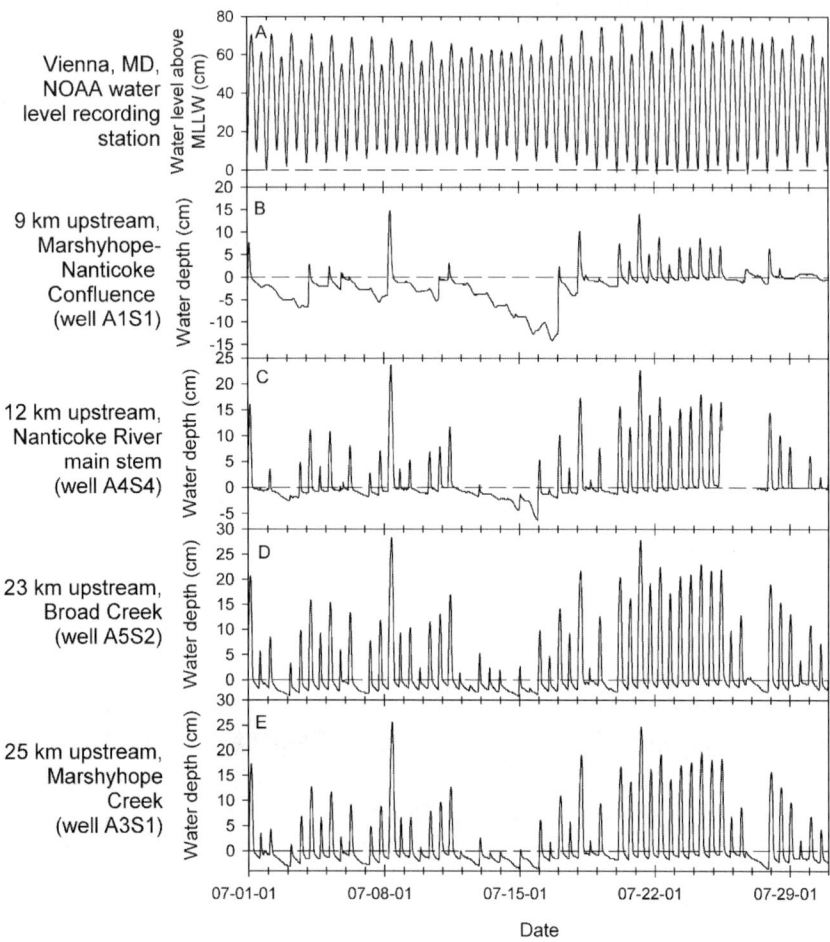

Fig. 6.3. Tidal characteristics of the Nanticoke River and associated tidal swamp sites. A: Predicted water level (measured water level not available) at a NOAA recording station in the Nanticoke River channel 5 km downstream from our swamp study area (see text). B-E: Surface water and groundwater level at four swamp study sites at various distances upstream from the NOAA station on three tributaries of the Nanticoke watershed.

most important being submergence (i.e., the long-term increase in water levels due to land subsidence) and sea-level rise that have increased flooding stress on marsh vegetation (Stevenson and Kearney 1996). Little information is available concerning loss of tidal freshwater swamps due to sea level-rise in the Nanticoke River or elsewhere in the Chesapeake Bay.

6.4. Geospatial description

In the Nanticoke River watershed, tidal swamps are often separated from the river channel by a band of marsh (Figure 6.4). This marsh consists of a band of "low marsh" directly adjacent to the channel, and a "high marsh" farther away from the channel. The high marsh often occurs on a natural levee formed by sediment deposition during flood events. Due to lower input of riverine sediment, the tidal swamps may occur at a slightly lower elevation than the high marsh even though they visually appear to be at a higher elevation, as in Figure 6.4.

As mentioned previously, the tidal swamps of the Nanticoke watershed can contain a diverse group of herbaceous plants. This typically occurs where the tree canopy is more open, which allows greater penetration of sunlight (Figure 6.5). Contributing to the herbaceous plant diversity is a hummock-hollow microtopography resulting from fallen trees and uplifted roots (Figure 6.6). The hummocks also provide sites for germination and establishment of woody plants.

Fig. 6.4. Vegetation zonation typical of the tidal freshwater reaches of the Nanticoke River. At the lowest elevation is "low marsh" dominated by pond-lily. At a slightly higher elevation is "high marsh" dominated by a diverse mixture of perennial and annual species. Farthest from the river is tidal swamp, visible to the right of the photograph.

Fig. 6.5. Diverse understory and relatively open canopy of a tidal freshwater swamp on the Nanticoke River.

Fig. 6.6. Variation in understory vegetation due to hummock and hollow microtopopgraphy. The ferns and forb plants visible toward the lower left side of the photo are growing on a hummock created by a fallen log, while the broad-leaved plants with arrow-shaped leaves are green arrow arum growing at a lower-elevation hollow.

6.5. Ecological characterization of Nanticoke tidal swamps

Research conducted in the Nanticoke River swamps has primarily focused on describing vegetation and seed banks and relating variation in vegetation and seed banks to environmental variables. Accordingly, the general characteristics of standing vegetation in tidal swamps of the Nanticoke and how vegetation varies across the estuarine gradient are first summarized. The results of landscape-scale seed bank studies are then reported, followed by a summary of data on environmental variables and their relationship to vegetation patterns.

6.5.1. Vegetation community composition

6.5.1.1. Species composition

The tidal swamps of the Nanticoke River are floristically diverse, with over 40 species of trees and shrubs and 100 herbaceous species observed in 24 swamp study plots. The dominant tree species (>2.5 cm diameter at breast height, DBH) across the tidal freshwater reaches of the watershed are green ash (*Fraxinus pennsylvanica* Marsh.), red maple (*Acer rubrum* L.), and swamp tupelo (*Nyssa biflora* Walt.) (Figure 6.7), although there is considerable variation between locations in relative abundance of species. Subdominant trees include sweetbay (*Magnolia virginiana* L.), American hornbeam (*Carpinus caroliniana* Walt.), Atlantic white cedar (*Chamaecyparis thyoides* [L.] B.S.P.) as well as species with a shrub-like growth form and DBH >2.5 cm such as swamp azalea (*Rhododendron viscosum* [L.] Torr.), common winterberry (*Ilex verticillata* L.), and arrowwood (*Viburnum dentatum* L.). Understory species classified as shrubs (<2.5 cm DBH, >1m tall) included species that also occurred as small trees such as arrowwood and swamp azalea, as well as coastal sweetpepperbush (*Clethra alnifolia* L.), northern spicebush (*Lindera benzoin* [L.] Blume), and highbush blueberry (*Vaccinium corymbosum* L.).

Herbaceous plants can be divided fairly distinctly into those primarily occurring in hollows (microtopographic lows) or on hummocks (microtopographic highs). The average elevation of the hummocks is 15 cm higher than the hollows, creating large differences in hydroperiod over horizontal distances of a meter or less and considerable interspersion between species of hummocks and hollows. The dominant species in hollows include halberdleaf tearthumb (*Polygonum arifolium* L.), jewelweed (*Impatiens capensis* Meerb.), and green arrow arum (*Peltandra virginica* [L.]

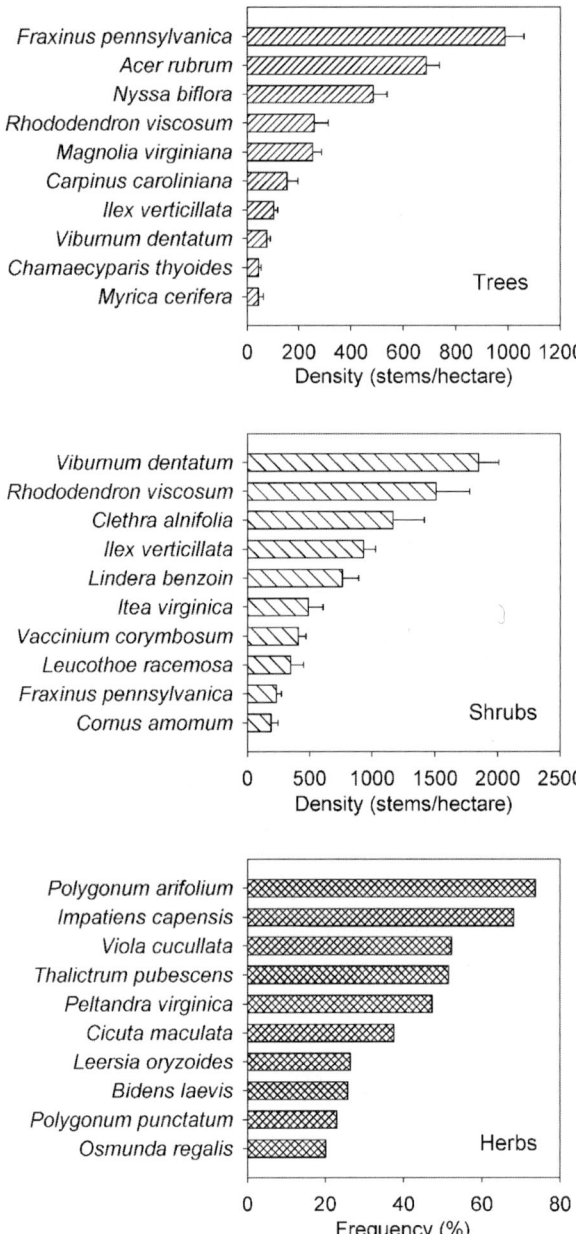

Fig. 6.7. Abundance of the 10 most abundant species of trees, shrubs, and herbs occurring in tidal swamps of the Nanticoke River.

Schott), while those of hummocks include marsh blue violet (*Viola cucullata* Ait.), king of the meadow (*Thalictrum pubescens* Pursh), and common water hemlock (*Cicuta maculata* L.).

Taken together, the assemblage of numerous shrub, tree, and hollow and hummock herbaceous plants create a forest structure that is spatially complex in both vertical stratification and horizontal interspersion of microhabitats. These freshwater tidal swamp ecosystems arguably contain a higher degree of floristic biodiversity than other types of temperate coastal wetlands, in some cases as high as or higher than tidal freshwater marshes.

6.5.1.2. Variation in forest structure across the estuarine gradient

Most of the tidal swamps of the Nanticoke watershed occur within the "tidal freshwater" zone, as discussed previously, yet there are significant trends related to distance upstream across this zone (Figure 6.8). For example, tree basal area, density of shrub stems and woody seedlings, and the Leaf Area Index (LAI) of woody plant communities all increase linearly and significantly progressing upstream (panels A-D, Figure 6.8). As might be expected due to greater density, size, and foliage area, light availability below the woody canopy (measured as Photosynthetically Active Radiation, PAR) was found to decrease significantly progressing upstream (panel E, Figure 6.8). Presumably because of lower light availability, the cover of herbaceous plants also decreased significantly moving upstream (panel F, Figure 6.8).

While there is certainly variability in the data portrayed in Figure 6.8, the influence of the estuarine gradient on structuring tidal swamp plant communities cannot be dismissed. It is also interesting that this variation was noted across three different tributaries, suggesting that distance upstream along a given tributary is a useful predictor of vegetation structure. The evidence of upstream linear variation in swamp community structure is also surprising, even though the gradient could be considered "weak" relative to more saline coastal wetland types (>0.5 ppt). However, it is possible that short-term salinity pulses during droughts may be more important in structuring the swamps than the average, low salinity level they experience over the long term.

6.5.2. Landscape variation in seed banks

As for standing vegetation, variation in seed banks also occurs across the tidal freshwater reaches of the Nanticoke (Figure 6.9; Peterson and Baldwin 2004a). Soil samples that were subjected to the seedling emer-

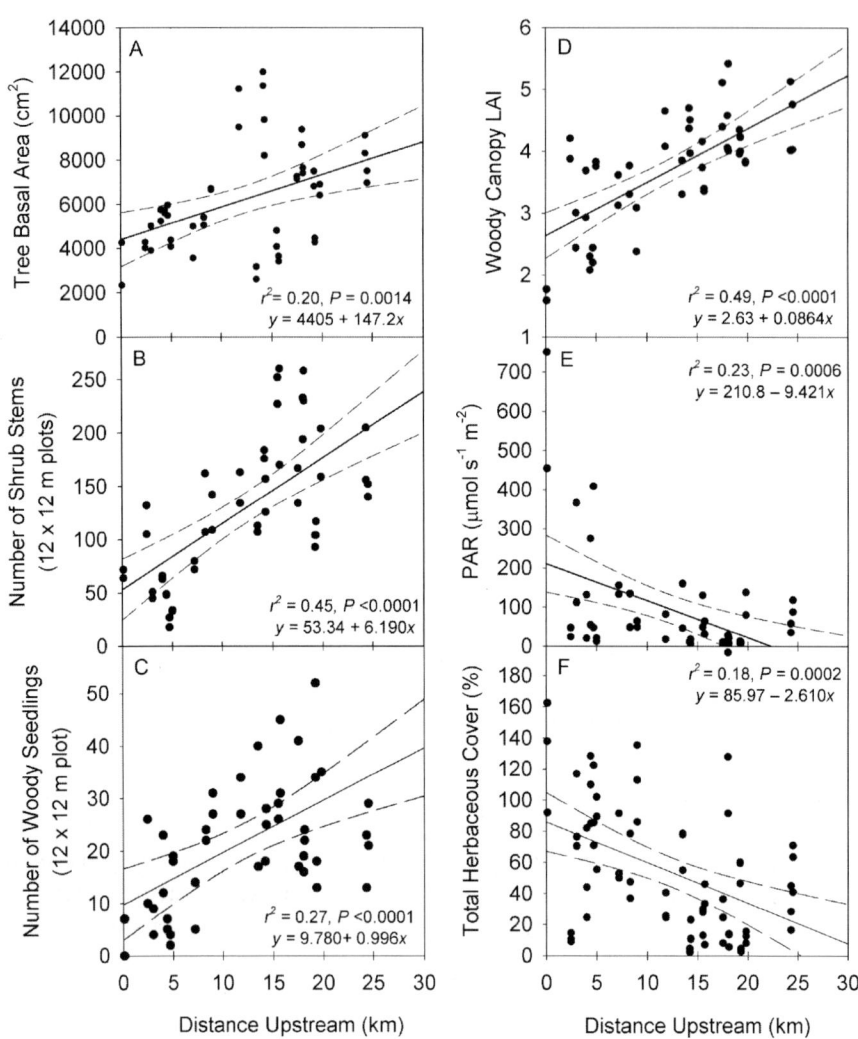

Fig. 6.8. Relationship between distance upstream from the most downstream study plot and woody and herbaceous vegetation parameters, leaf area index (LAI), and Photosynthetically Active Radiation (PAR). The Vienna monitoring station mentioned in Figure 6.3 and the text is 5 km downstream from the most downstream study plot. Solid lines are best-fit, dashed lines are 95% confidence intervals.

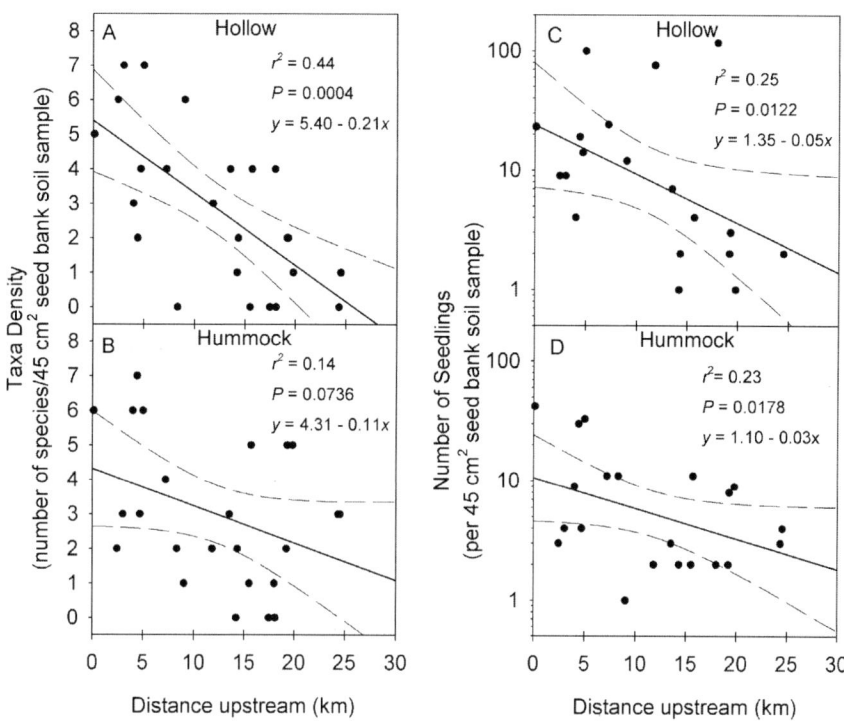

Fig. 6.9. Relationship between distance upstream from the most downstream study plot and taxa density and number of seedlings emerging from seed bank samples. A and C: taxa density and seedling number for swamp hollows. B and D: taxa density and seedling number for swamp hummocks. Adapted from data originally published by Peterson and Baldwin (2004a).

gence method of seed bank analysis revealed a significant linear reduction in taxa density (panels A and B, Figure 6.9) and significant logarithmic reduction in emerging seedling density (panels C and D, Figure 6.9) as distance upstream increased. This pattern occurred in both hollow and hummock microhabitats. The decrease in diversity and density of seeds in the seed bank may be due to a decrease in seed production related to the upstream decrease in herbaceous plant cover mentioned previously.

6.5.3. Environmental variables and their relationship to vegetation

6.5.3.1. Salinity

Together with hydroperiod (Section 6.3), salinity is one of the dominant variables structuring vegetation of coastal wetlands (Mitsch and Gosselink 2000). This also appears to be true in tidal swamps despite their low salinity relative to other coastal wetland types (e.g., brackish and salt marshes). At Sharptown, Maryland, located roughly in the middle of the upstream-downstream range of tidal swamps in the Nanticoke River watershed, the long-term salinity average is less than 0.5 ppt (Figure 6.10), the upper limit for classification as fresh water (Cowardin et al. 1979). However, the maximum salinities recorded between 1986 and 2004 increased an order of magnitude above the long-term average during the months of July though October, when the salinity profile moves upstream in Chesapeake Bay rivers, particularly during drought years. The maximum value reported at the Sharptown station between 1986 and 2004 was 4.84 ppt, which was also the maximum salinity value recorded at the station in August of 2002, indicating that salinity in tidal swamps of the Nanticoke was as high or higher in August 2002 than at any other time since recording began.

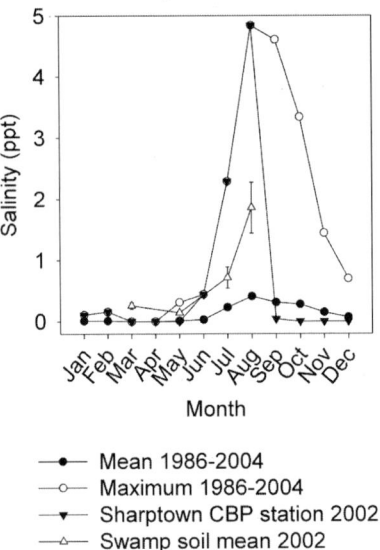

Fig. 6.10. Summary of salinity measurements from Maryland Department of Natural Resources monitoring station ET6.1 on the Nanticoke River in Sharptown, Maryland. Also shown are measurements of soil porewater salinity at swamp study sites in 2002, plotted as mean ± SE.

The salinity in swamp soil porewater averaged across 25 river-kilometers was 1.9 ppt in August 2002 (Figure 6.10). However, the salinity concentrations recorded in August 2002 ranged from 7.1 ppt at the most downstream swamp site to 0.1 ppt at the most upstream sites (Figure 6.11). The salinity gradient was less steep in July 2002, when the maximum salinity recorded was 3.2 ppt, and also in March and May 2002, when all salinity concentrations across were below 0.8 ppt. These results indicate that infrequent increases in salinity occurring during particularly dry years can result in interstitial salinities at levels typical of those of brackish or mesohaline marshes (>5 ppt, Cowardin et al. 1979). Because of the general sensitivity of woody plant species of temperate wetlands to salinity of 2 ppt or higher (Pezeshki and Chambers 1986; Pezeshki et al. 1989; Conner et al. 1997; Kozlowski 1997), it may be that these rare pulse events are much more important in determining the structure of tidal swamps than the long-term average salinity level.

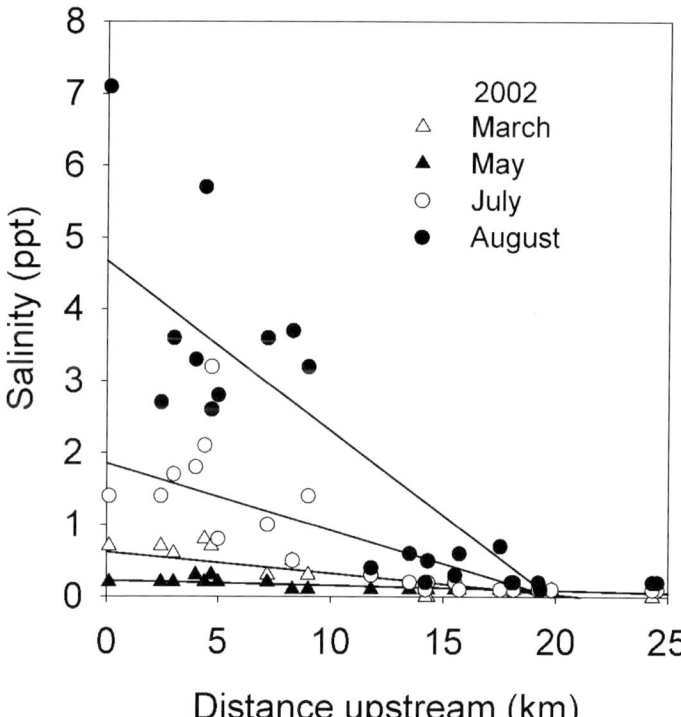

Fig. 6.11. Relationship between distance upstream (from the most downstream study site) and swamp soil porewater salinity.

6.5.3.2 Other edaphic variables

Given the periodic intrusion of sea water into tidal swamps of the Nanticoke watershed, it may be assumed that concentrations of some soil constituents also vary along the upper reaches of the estuary. The fact that sulfate (SO_4^{2-}), potassium (K^+), and boron (B) occur at concentrations several orders of magnitude higher in sea water relative to freshwater (Mitsch and Gosselink 2000) helps explain why total S, K, and B concentration in soil decreased as distance upstream increased (Figure 6.12). In contrast, concentrations of the heavy metals zinc (Zn) and copper (Cu), which occur in surface runoff, increased in concentration proceeding upstream (Figure 6.12). Soil concentrations of calcium (Ca) increased slightly and magnesium (Mg) decreased proceeding upstream (data not shown), but the Ca:Mg ratio increased strongly proceeding upstream. The Ca:Mg ratio is inversely related to chloride and sulfide concentrations in some coastal wetlands (Chris Swarzenski, U.S. Geological Survey, personal communication), and thus may be a sensitive index of saltwater intrusion.

Organic matter content varied between 20-80%, but did not vary in relation to distance along the estuary. Similarly, there was not a strong relationship between the N:P ratio and distance along the estuary, but data suggest that the majority of the swamp sites studied are P-limited. This conclusion is made assuming that the N:P ratios of vegetation are similar to those of soils, and that vegetation N:P ratios >16 indicate P-limitation, as suggested by Koerselman and Meuleman (1996). P-limitation in the Nanticoke River swamps might be explained by a settling of sediments (with adsorbed P) tending to occur primarily in the marshes fringing the river, with fewer sediments reaching the landward swamps.

Sediment accretion rates measured using feldspar marker horizons during a one-year period (March 2001- March 2002) increased proceeding upstream in tidal swamps (Figure 6.13). However, there was no significant relationship between accretion and distance upstream before this period (data not shown). Accretion rates ranged from zero (which in some areas appeared to be erosional) to over 2 cm/yr. The increase in sedimentation rates proceeding upstream may be due to the processes that result in a maximum of suspended matter in estuarine waters immediately upstream of the brackish zone, i.e., salinity-induced flocculation and reduced flow velocity at the salt wedge (Meade 1972; Officer 1981).

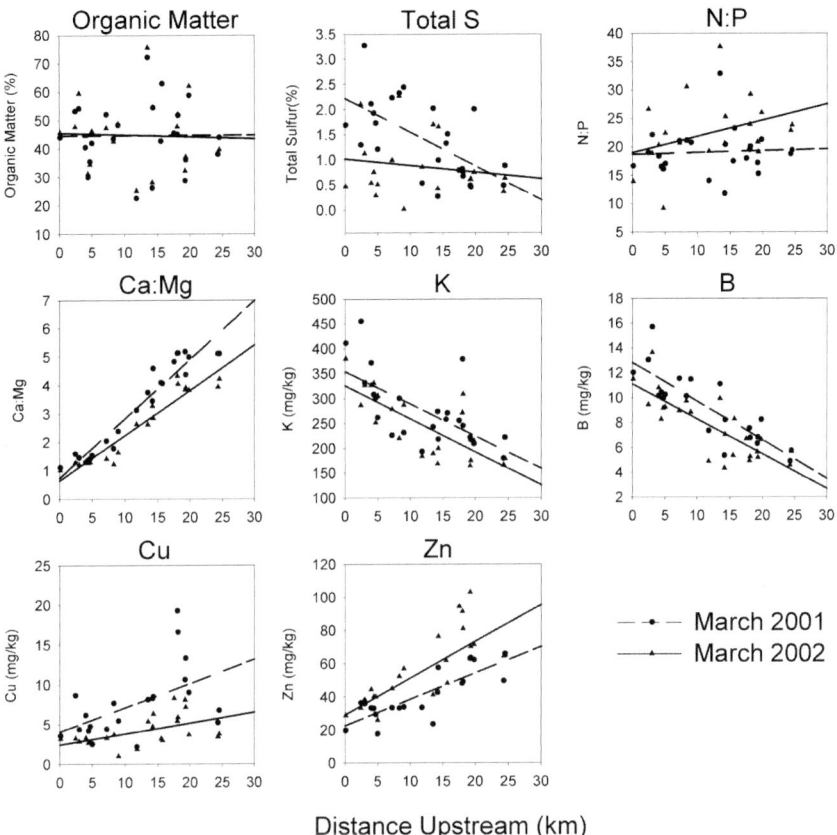

Fig. 6.12. Relationship between distance upstream across the tidal freshwater zone of the Nanticoke River watershed and various soil constitutents and element ratios.

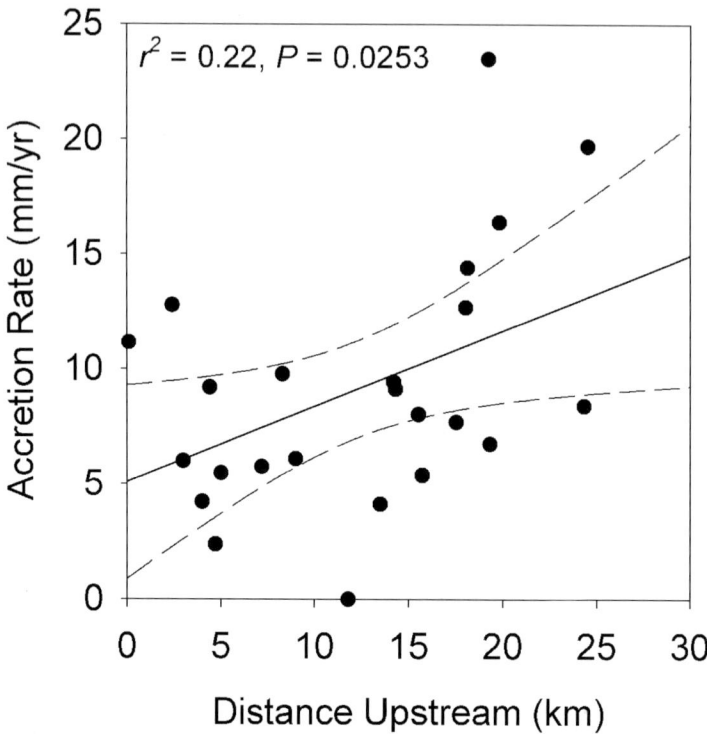

Fig. 6.13. Variation in accretion rate in swamps across the tidal freshwater zone of the Nanticoke River watershed. Accretion was measured using feldspar marker horizons between March 2001 and 2002. No significant relationship between accretion rate and distance upstream was detected for 2000-2001 or 2002-2003. Dashed lines are 95% confidence intervals.

6.6. Research needs

Studies conducted on tidal swamp forests of the Nanticoke River indicate that there are important gradients occurring within the tidal freshwater zone that affect the vegetation structure and seed banks. These findings suggest that saltwater intrusion events during drought years play a particularly important role in structuring these forests. However, little is known about how the frequency and magnitude of saltwater intrusion events affect the structure of these forests. Furthermore, under conditions of rising sea level, saltwater intrusion events will likely become more frequent and severe, and the potential responses of tidal swamps to increased salinity pulsing are unknown. Rising sea level, combined with subsidence, may also increase the frequency and duration of inundation and soil waterlogging in tidal swamp forests, assuming they are unable to accrete and increase in elevation at a rate sufficient to keep pace with sea-level rise. The potential interactions between higher salinity concentrations and pulses and changes in hydroperiod deserve greater research attention, given the high biodiversity of these ecosystems and their likely sensitivity to small increases in salinity.

6.7. Synthesis and conclusions

Information about the Nanticoke River from governmental monitoring stations and studies summarized here demonstrates the importance of periodic increases in salinity during dry years. These intrusions of saline water may reduce growth rates of trees and shrubs in the lower reaches of their range, facilitating the establishment of diverse herbaceous understory plant communities. These herbaceous communities may include salt tolerant species (e.g., green arrow arum occurs in wetlands that regularly experience 7 ppt salinity or higher; Anderson et al. 1968). Salt-intolerant species might avoid saltwater flooding by growing on hummocks or else grow rapidly in hollows between droughts. Conversely, at locations farther upstream, the lower frequency, magnitude, and duration of saltwater intrusion events may allow for more rapid growth of canopy trees, reducing light availability and the abundance of understory vegetation. Therefore, while the Cowardin et al. (1979) salinity modifier cutoff of 0.5 ppt for fresh water has unquestioned value for classification, it is important to recognize the temporal variability of salinity in these tidal swamps. Because of the evidence that salinity is important in structuring vegetation in these ecosystems, it may be argued that these ecosystems should be referred to as "low-salinity" or "salt-pulsed" rather than "freshwater" tidal swamps.

While salinity appears to be a driving factor controlling the structure of tidal swamps of the Nanticoke watershed, the role of hydroperiod cannot be dismissed outright. Data presented here (Figure 6.3) from a limited number of stations suggest that the depth, frequency, and duration of inundation may be lower at the downstream end of the tidal swamp range than farther upstream. Because of the inhibitory effects of inundation on seed germination and plant growth (McKee and Mendelssohn 1989; Baldwin et al. 1996; Peterson and Baldwin 2004b), the greater abundance of herbaceous plant communities might be explained in part by less inundation in the downstream tidal swamps. However, more beneficial hydrologic conditions for plant growth might also be predicted to enhance tree and shrub growth as well, which does not appear to be the case.

A number of environmental variables other than salinity and hydroperiod also varied linearly proceeding upstream across the tidal swamp landscape. Results suggest that the Ca:Mg ratio in soils is related to the degree of periodic intrusion of saline water and thus a valuable tool for monitoring small changes in relative sea level in tidal swamps and possibly other coastal wetlands.

Taken together, these findings emphasize the linear nature of tidal swamp ecosystems. The patterns reported here are from 24 swamp sites across three tributaries of the Nanticoke River. Regression of various parameters against distance upstream in any tributary is a way of compressing all tributaries into a single line. The result that linear relationships were observed between many environmental and vegetation parameters and distance clearly demonstrates that strong gradients exist across the tidal swamp landscape. This leads to what is perhaps the most important conclusion of this chapter: that tidal swamps do not exist in a uniform, benign set of "freshwater" environmental conditions, but rather are influenced by small variations in salinity and perhaps hydroperiod and other variables. Thus, these diverse, productive wetlands may be among the most susceptible ecosystems to small changes in water level and salinity resulting from global climate change.

References

Anderson RR, Brown RG, Rappleye RD (1968) Water quality and plant distribution along the upper Patuxent River, Maryland. Chesapeake Sci 9:145-156

Baldwin AH, McKee KL, Mendelssohn IA (1996) The influence of vegetation, salinity, and inundation on seed banks of oligohaline coastal marshes. Am J Bot 83:470-479

Conner WH, McLeod KW, McCarron JK (1997) Flooding and salinity effects on growth and survival of four common forested wetland species. Wet Ecol Manage 5:99-109

Cooper SR (1995) Chesapeake Bay watershed historical land use: Impact on water quality and diatom communities. Ecol Appl 5:703-723

Cowardin LM, Carter V, Golet FC, LaRoe ET (1979) Classification of wetlands and deepwater habitats of the United States. FWS/OBS-79/31. U.S. Fish and Wildlife Service, Washington

Downs LL, Nicholls RJ, Leatherman SP, Hautzenroder J (1994) Historic evolution of a marsh island: Bloodsworth Island, Maryland. J Coastal Res 10:1031-1044

Kearney MS, Stevenson JC (1991) Island land loss and marsh vertical accretion rate evidence for historical sea-level changes in Chesapeake Bay. J Coastal Res 7:403-415

Kearney MS, Grace RE, Stevenson JC (1988) Marsh loss in Nanticoke Estuary, Chesapeake Bay. Geog Rev 78:206-220

Koerselman W, Meuleman AFM (1996) The vegetation N:P ratio: a new tool to detect the nature of nutrient limitation. J Appl Ecol 33:1441-1450

Kozlowski TT (1997) Responses of woody plants to flooding and salinity. Tree Physiol Monograph 1:1-29

McCormick J, Somes Jr HA (1982) The coastal wetlands of Maryland. Maryland Department of Natural Resources, Coastal Zone Management, Jack McCormick and Associates, Inc., Chevy Chase

McKee KL, Mendelssohn IA (1989) Response of a freshwater marsh plant community to increased salinity and increased water level. Aquat Bot 34:301-316

Meade RH (1972) Transport and deposition of sediments in estuaries. Geol Soc Am Mem 133:91-120

Michener WK, Blood E, Bildstein KL, Brinson MM, Gardner LR (1997) Climate change, hurricanes and tropical storms, and rising sea level in coastal wetlands. Ecol Appl 7:770-801

Mitsch WJ, Gosselink JG (2000) Wetlands, 3rd edn. John Wiley and Sons, New York

Odum WE, Smith III TJ, Hoover JK, McIvor CC (1984) The ecology of tidal freshwater marshes of the United States east coast: A community profile. FWS/OBS-83/17U.S. Fish and Wildlife Service, Washington

Officer CB (1981) Physical dynamics of estuarine suspended sediments. Mar Geol 40:1-14

Peterson JE, Baldwin AH (2004a) Variation in wetland seed banks across a tidal freshwater landscape. Am J Bot 91:1251-1259

Peterson JE, Baldwin AH (2004b) Seedling emergence from seed banks of tidal freshwater wetlands: Response to inundation and sedimentation. Aquat Bot 78:243-254

Pezeshki SR, Chambers JL (1986) Effect of soil salinity on stomatal conductance and photosynthesis of green ash (*Fraxinus pennsylvanica*). Can J For Res 16:569 573

Pezeshki SR, DeLaune RD, Patrick WH Jr (1989) Assessment of saltwater intrusion impact on gas exchange behavior of Louisiana Gulf Coast wetland species. Wet Ecol Manage 1:21-30

Rheinhardt R (1992) A multivariate analysis of vegetation patterns in tidal freshwater swamps of lower Chesapeake Bay, U.S.A. Bull Torrey Bot Club 119:192-207

Schmidt MF (1993) Maryland's geology. Tidewater Publishers, Centreville

Schneider DW (1996) Effects of European settlement and land use on regional patterns of similarity among Chesapeake forests. Bull Torrey Bot Club 123:233-239

Simpson RL, Good RE, Leck MA, Whigham DF (1983) The ecology of freshwater tidal wetlands. BioScience 33:255-259

Stevenson JC, Kearney MS (1996) Shoreline dynamics on the windward and leeward shores of a large temperate estuary. In: Nordstrom KF, Roman CT (eds) Estuarine shores: Evolution, environments and human alterations. John Wiley and Sons Ltd, New York, pp 233-259

Stevenson JC, Kearney MS, Pendleton EC (1985) Sedimentation and erosion in a Chesapeake Bay brackish marsh system. Mar Geol 67:213-235

The Nature Conservancy (1998) Nanticoke River Bioreserve Strategic Plan. Chevy Chase, MD and District of Columbia Field Office and Delaware Field Office

Tiner RW (1993) Field guide to coastal wetland plants of the southeastern United States. The University of Massachusetts Press, Amherst

Tiner RW (2005) Assessing cumulative loss of wetland functions in the Nanticoke River watershed using enhanced National Wetlands Inventory data. Wetlands 25:405-419

Tiner RW, Burke DG (1995) Wetlands of Maryland. U.S. Fish and Wildlife Service, Ecological Services, Region 5, Hadley, and Maryland Department of Natural Resources, Annapolis

White CP (1989) Chesapeake Bay--Nature of the Estuary. Tidewater Publishers, Centerville

Chapter 7 - Tidal Freshwater Swamps of a Lower Chesapeake Bay Subestuary

Richard D. Rheinhardt

Department of Biology, East Carolina University, Greenville, NC 27858

7.1. Introduction

Chesapeake Bay, the largest estuary in North America, incorporates all the physical conditions suitable for the development of tidal freshwater swamps: subdued topography, extensive freshwater input from numerous rivers, and tides that extend far inland from the coast. One would expect that Chesapeake Bay would support an extensive coverage of tidal swamps and a wide variety of types. However, wetland inventory data do not confirm a high coverage of freshwater tidal swamps throughout Chesapeake Bay tidal rivers. This may be due to a lack of reliable estimates of the Bay-wide extent and distribution of tidal swamps. This scarcity of inventory data may in turn be due to an historic lack of commercial interest in extracting resources from them. However, a better understanding of the location and extent of tidal swamps in Chesapeake Bay could help stimulate additional research. More information on their structure and function would be beneficial because (1) commercially and recreationally important anadromous fish species spawn in the tidal freshwater reaches (Bilkovic et al. 2002a), and larval anadromous fish utilize creeks within tidal swamps during early development (Bilkovic et al. 2002b), (2) migratory waterfowl and resident herons use tidal swamps for feeding and nesting, (3) tidal swamps harbor a large number of herbaceous plant species, some of them rare, and (4) tidal swamps may provide a natural laboratory for studying consequences of sea-level rise.

Although data on the structure and function of tidal swamps are sparse for Chesapeake Bay as a whole, somewhat more is known about swamps along the Pamunkey River subestuary, where extensive tidal swamps occur. Not only does the Pamunkey River possess a large amount of tidal swamps (2,236 ha), it also supports swamps over a long reach of river (42 km). Consequently, a wide variety of tidal conditions exist under which tidal swamps can occur, making the Pamunkey an ideal place to study their ecology. The objective of this chapter is to summarize information on biotic and abiotic factors related to the ecology of tidal swamps in the Pamunkey River subestuary. These relationships will be used to explain the processes that form and maintain tidal swamps in the lower Chesapeake Bay and how human alterations could detrimentally affect them.

7.1.1. Location

The Pamunkey River originates in the Piedmont Physiographic Province of Virginia. The Pamunkey is a redwater river in its nontidal portion, which means that it carries sediments derived from erosion of Piedmont uplands. At the point where it becomes tidal, east of Route 360, it has drained more than 3,500 km^2 of the Piedmont (Figure 7.1). At its downriver extent, the Pamunkey joins with the Mattaponi River at West Point, Virginia to form the beginning of the York River, one of the four main sub-estuaries of Chesapeake Bay. At this confluence, maximum salinity is

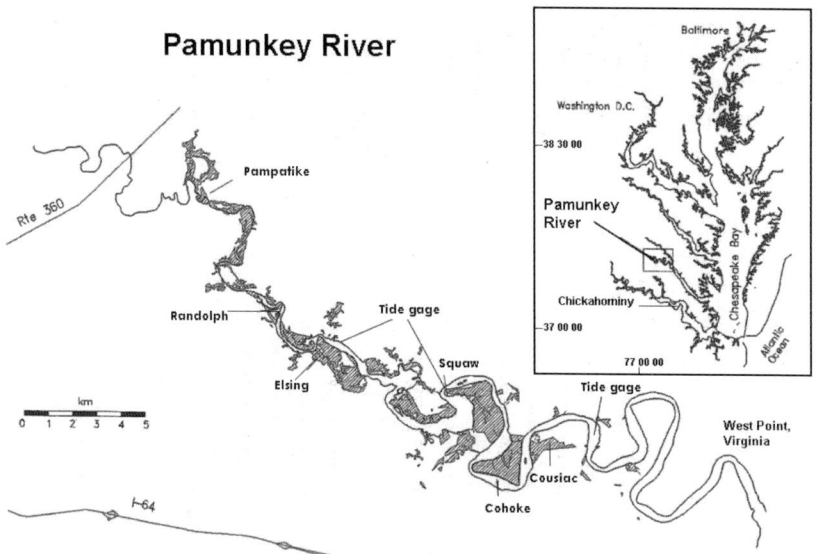

Fig. 7.1. Location of Pamunkey River tidal freshwater swamps. Tidal swamps are hatched. From Cousiac to Pampatike is 41.5 km by river. The five swamps with water level monitoring equipment are labeled. Adapted from Rheinhardt (1992b).

about 10 ppt during drought periods, but by about 30 km in river distance upstream (at Cousiac Swamp), where tidal swamps begin, salinity declines to approximately 0.5 ppt (fresh enough for trees to tolerate). Interestingly, there is a relatively abrupt transition from freshwater marsh to freshwater swamp at Cousiac Swamp.

Cousiac Swamp lies at the back edge of a tidal freshwater marsh (Cousiac Marsh) on its eastern edge, and trees do not reach the edge of the river until beyond the next bend in the river (Figure 7.1). Trees in this swamp are clearly stressed, particularly at the transition to marsh. Presumably, this transition zone defines the most upriver extent of drought-driven salinity intrusions in the Pamunkey River. Although long-term salinity measurements are unavailable, salinity intrusions are probably rare much further upriver from Cousiac Swamp, although a salinity wedge may extend more upriver in the deeper water of the river channel. Tidal swamps occur along a 42-km freshwater tidal reach between the Route 360 bridge and Cousiac. Some of these swamps are larger than 300 ha in size and most are larger than 50 ha.

7.1.2. Climate and tidal regime

Southern Chesapeake Bay lies in a temperate, coastal climate, with mild winters and hot, humid summers. Normal air temperatures range from 2.4°C in January to 23.2°C in July (NOAA 1989, 1990). The frost-free season is 230 days long (early April to early November). Normal annual precipitation is 113 cm and is evenly distributed throughout the year.

Tidal reaches in the Pamunkey River follow a mixed semi-diurnal periodicity (i.e., two high and low tides per day with one high tide higher than the other). The highest high tides and lowest low tides occur twice per month (i.e., spring tides) while the lowest high tides and highest low tides occur halfway between spring tide periods (i.e., neap tides). Maximum tidal amplitude in the Pamunkey River is about 1.2 m, about the same as at the mouth of Chesapeake Bay. However, tidal amplitudes vary along the length of the river (Rheinhardt 1991). Some swamps associated with the upriver sections of tidal excursion experience tides only irregularly, perhaps during low river stages, and fluctuations are below ground. However, most Pamunkey tidal swamps are influenced by semi-diurnal tidal fluctuations with tides flooding aboveground.

7.2. Geologic history

Chesapeake Bay and its subestuaries are drowned river valleys formed over the past 10,000 years during a period of rising sea level (Colman and Mixon 1988). In fact, the river valleys of the four main subestuaries of Chesapeake Bay (Potomac, Rappahannock, York, and James) are "drowned" westward all the way to the Fall Line (the geologic boundary between the Piedmont and Coastal Plain Physiographic Provinces). Freshwater input is most pronounced at the heads of these subestuaries where drainage from Piedmont tributaries reaches sea level just below the Fall Line. Natural channel constrictions at headward reaches of the subestuaries also funnel tides as they move upriver, leading to increases in amplitude upriver. Amplification of tides is important because tidal swamps appear to be more extensive where tidal amplitude is high (Duberstein 2004). Large freshwater input combined with tidal constrictions partially explain why tidal swamps are so extensive along the upper section of the Pamunkey.

Tidal freshwater swamps along the Pamunkey have developed in concert with rising sea level. Consequently, swamp soils consist of a combination of thick, organic-rich sediments (histosols derived from *in situ* production of organic matter) and recent alluvium from erosion of Piedmont soils. Soils in these tidal swamps are classified as Mattan (Hodges et al. 1988), which consist of a surface stratum comprised of a mixture of herbaceous and woody plant remnants, including partially decomposed logs. The peaty soils are 40-128 cm thick. Deeper sediments consist of Pleistocene riverine terrace deposits (Mixon et al. 1989). The Pamunkey River is currently experiencing a relative rate of sea-level rise of approximately 4-6.5 mm yr^{-1} (Neubauer et al. 2002), far more rapid than the 1.4 mm yr^{-1} historic average (Bailey: http://pubs.usgs.gov/fs/fs102-98/).

7.3. Exploration and historic land use

The Pamunkeys, a tribe of the Powhatan Confederacy, are the earliest known aboriginal inhabitants along the Pamunkey River. Their population was estimated to be about 1,000 at the time of first English contact by Captain John Smith in 1608 (Pollard 1894). Tribal villages were established on bluffs along the river, but Pamunkeys probably utilized the lower lying tidal swamps in much the same way they do today: to hunt game and fish, and to harvest wild rice and medicinal plants.

When Europeans first colonized the region, they developed an economy based on producing and exporting tobacco. Many plantations for this purpose were established along the lower reaches of Chesapeake Bay subestuaries, especially in the flat, outer coastal plain. However, the dissected topography of the inner coastal plain, particularly in near the Pamunkey River, was not conducive to tobacco farming and so the area has remained mostly forested (Rheinhardt, personal observation). In contrast to tidewater areas in South Carolina and Georgia, conversion of tidal swamps for rice cultivation never took place along the Pamunkey or other tidal wetlands in southeastern Virginia.

Exploitation of upland forests began soon after arrival of European colonists in the mid-1600s. Although tidal creeks provided a readily available conduit for transportation of extracted timber, it would have been difficult to cut and transport timber from the mucky interior of tidal swamps using 18th century technology. Even when large-scale, mechanized timber extraction became possible in the late-1800s, baldcypress (*Taxodium distichum* [L.] L.C. Rich) would probably have been the only commercially viable species worth harvesting. Unfortunately, records of timber extraction from tidal swamps have not been successfully located, so it has not been possible to reconstruct historic forest compositions.

7.4. Ecological characterization

The tidal freshwater portion of the Pamunkey River occurs along a 42-km stretch of river. Geomorphic constrictions in the river and meanders magnify tidal amplitude in places, but tidal influences are generally less pronounced toward the most upriver reaches of the river. Hydrologic variations along the river would be expected to reflect variations in wetland species composition along the river and may explain variations in microtopography among swamps. The following sections summarize data from studies of tidal swamps along the Pamunkey with a focus on this variation of physical factors along the river, particularly as they relate to plant species distribution patterns.

Doumlele et al. (1985) sampled the vegetation of a tidal swamp located toward the downriver end of the Pamunkey's freshwater section. They found that 96% of the canopy in that stand was composed of only four flood-tolerant tree species: green ash (*Fraxinus pennsylvanica* Marsh.), black tupelo (*Nyssa sylvatica* Marsh.), red maple (*Acer rubrum* L.), and American hornbeam (*Carpinus caroliniana* Walt.). The trees were so stressed that they were small in stature (7-10 m tall) and many had dead branches. Doumlele et al. (1985) surmised that canopy species richness was low because only the most flood-tolerant species could tolerate semidiurnal flooding. Trees were also restricted to hummocks (patches of higher elevation). As a result, the canopy was somewhat open, allowing much light to penetrate to the forest floor. There were also treeless patches where only freshwater tidal marsh species grew.

In a swamp near the Doumlele et al. (1985) site, Fowler (1987) found that primary production was comparable to other highly productive swamp communities in the Southeast. Annual aboveground biomass production was 12.3 Mg ha^{-1}, with approximately 60% attributed to woody production and 40% to herbaceous production. About 80% of total woody production was attributed to swamp tupelo (*Nyssa biflora* Walt.) and green ash, whereas green arrow arum (*Peltandra virginica* (L.) Schott) and Asian spiderwort (*Murdannia keisak* [Hassk.] Hand.-Maz.) accounted for about 50% of total herbaceous production. This result is similar to a productivity study of cypress-tupelo (*Taxodium-Nyssa*) dominated tidal freshwater swamps in South Carolina, which showed biomass productivity of 7.9-11.4 Mg ha^{-1} yr^{-1} (Ratard 2004).

Although the Doumlele et al. (1985) and Fowler (1987) studies provided basic information on two downriver tidal swamps, there was a need to obtain information from a wider selection of swamps, both up and downriver, to determine the degree to which composition and physiognomy vary along the range of tidal influence. The remainder of this chapter focuses on results of a study by the author designed to quantify some of this variation and determine how species composition patterns are related to underlying environmental gradients. Much of the background for this chapter can be found in Rheinhardt (1991) and related publications (Rheinhardt 1992a, b; Rheinhardt and Hershner 1992).

7.4.1. Methods

To relate species composition to environmental factors, water level recorders with data loggers were placed in six tidal swamps along the Pamunkey, from one of the most downriver swamps to one located 41.5 km upriver. Water level fluctuations of all wells were measured every 6 minutes (0.1 hr) and compared with data from a series of tide gages monitored by the Virginia Institute of Marine Science. Water level recorders in swamps varied in the length of time they were monitored, but all of them recorded data through at least several monthly tidal cycles.

Vegetation was sampled in the 6 gaged sites and 17 additional stands. All strata were sampled and recorded by species: basal area and density of canopy trees (>2.5-cm dbh), densities of saplings and shrubs (>1.5-m tall and <2.5 cm dbh), % coverage and frequency of herbaceous plants (estimated from 1-m² plots), and density of vines (>1.5-m tall). From these data, importance values (IV) were determined for each species. For canopy species, IV= mean of relative dominance (basal area) and relative density; subcanopy and vine IV= relative density; and herb IV= mean of % cover and relative frequency.

In each stand, a 2-l soil sample was collected to 15 cm depth and analyzed for pH, total N (ppm), P (ppm), K (ppm), Ca (ppm), Mg (ppm), Zn (ppm), Al (ppm), Cu (ppm), Fe (ppm), soluble salts, and organic matter content (%). Percent coverage of hollow microsites and distance upriver were also recorded for each stand. Vegetation data were ordinated using Detrended Correspondence Analysis (DCA), a multivariate statistical technique used to compare composition among samples (McCune and Grace 2002). All environmental factors were statistically compared with the resulting DCA ordination diagram using PC-ORD (Vers. 4) software (McCune and Mefford 1999).

7.4.2. Hydrologic regime and substrate

Almost all tidal swamps have a low berm fringing the main stem of the river. The berms tend to be low-lying (0.3 m). In some places, tidal freshwater marshes occur between the river proper and forest instead of a berm (Figure 7.2). Meandering tidal creeks punctuate berms at intervals, allowing tidal flow into and out of the interior of swamps (Figure 7.3). In most swamps, the topographic surface is a mosaic of hummocks and hollows (Figure 7.4). Woody species and their roots are restricted to hummocks, which are composed of gnarled, tightly compacted tree roots with hypertrophied lenticels and very little soil.

Hummocks vary in size from 1-10 m², large enough to support 1-3 large trees and a few shrubs. In contrast, hollows are composed of highly organic, soupy muck, in which only flood-tolerant herbaceous species grow. Standing water usually occurs in hollows, but rarely deep enough to cover hummocks. In effect, woody species are perched at the upper range of tidal flooding. It seems that hummocks are formed by fallen trees that become nurse logs for tree seedlings. Thus, large downed wood is needed for the formation of hummocks, which in turn, is critical for the vertical accretion of tidal forests.

Flooding of hollows usually occurs in response to the higher of the two daily tides, except during spring tides, when both tides might cause flooding in hollows (Figure 7.5), and during neap and apogeal periods, when

Fig. 7.2. Tidal freshwater swamp with a fringing tidal marsh between river and swamp.

Fig. 7.3. Aerial photograph of Elsing Swamp, looking northeast. A tidal creek flowing into the tidal swamp is in the lower right of photograph. Tide gage data maintained by Virginia Institute of Marine Science.

Fig. 7.4. Idealized diagram of hummock and hollow microtopography. MLW= mean low water. Adapted from Rheinhardt and Hershner (1992).

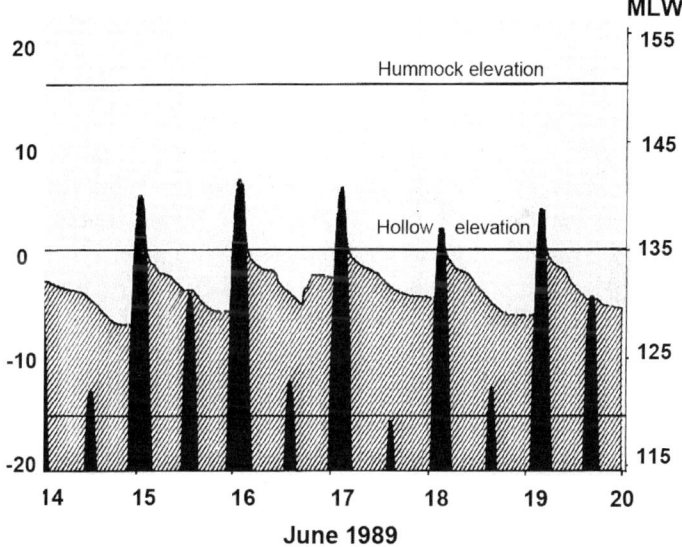

Fig. 7.5. Water level and tidal gage fluctuations over a 6-day period in an Ash/Tupelo stand. Lag time between swamp and gage of 0.7 h has been subtracted. The hatched line represents the water level in the monitored gage, the solid area is the top portion of the tidal curve in the channel superimposed on the gage data, which matched exactly. Vertical numbers on left are elevations (cm) above and below surface of hollow. Right side is elevation (cm) above mean low water (MLW). Adapted from Rheinhardt (1992b).

neither tide floods hollows. Although hydrographs of tidal swamps occur in synchrony with tidal fluctuations in the river, fluctuations are delayed relative to those in the river, with the amount of delay perhaps being a function of distance from the river. In all cases, water level drops in concert with the falling tide until the surface of the hollow is reached. At that point, water level drops slowly, with only a 5-10 cm drop before the next lower high tide delays a further decline. The elevation of tidally forced fluctuations relative to ground level varies among swamps, and as is shown below, is related to the composition of the plant community.

As with any natural gradient, there appears to be a continuum from semi-diurnally flooding to irregular flooding by tides. In the most upriver portion of the river, toward the limit of tidal excursion, tidal flooding occurs only during spring tides (twice monthly), during freshets, at high tides during storm surges associated with tropical depressions, or in some combination of these conditions. Where tidal flooding is irregular, there is no hummock/hollow microtopographic pattern apparent. Instead, the swamp surface resembles a typical river floodplain and the substrate is relatively firm.

7.4.3. Vegetation patterns in tidal swamps

Twenty-two tree species were recorded in the 23 tidal stands sampled along a 42 km length of the river (Table 7.1). However, the canopy was typically co-dominated by only 4-5 hydrophytic tree species in semi-diurnally flooded swamps: ash species (primarily pumpkin ash (*Fraxinus profunda* [Bush] Bush) and green ash), swamp tupelo, and red maple. However, two stands were also codominated by baldcypress. Together, these five tree species comprised 87-100% of total canopy IV. Based on leading dominants, this community was defined as an Ash/Tupelo community type. Stands with appreciable baldcypress were classified as a Baldcypress variant of the Ash/Tupelo type.

Ash, swamp tupelo, red maple, and baldcypress are probably the most flood-tolerant wetland tree species occurring north of the James River (the southern-most subestuary of Chesapeake Bay), thus it is not surprising that they dominate the most regularly flooded tidal swamps. However, it was surprising that baldcypress was not an important component in all the Ash/Tupelo sites, considering that baldcypress commonly codominates tidal swamps further south (Ozalp 2003; Duberstein 2004; Ratard 2004), including many tidal stands along the Chickahominy River, Virginia (personal observation). It is unlikely that baldcypress was selectively logged

Table 7.1. Vegetation summary of types. Canopy IV (Importance Value) was mean of relative basal area and relative density of tree species >2.5 cm dbh, IV of Subcanopy was relative densities of saplings and shrubs (>1.5-m tall and <2.5 cm dbh), Vine IV was relative density of vines (>1.5-m tall), and Herbaceous IV was mean of % cover and relative frequency from 1-m^2 plots. The Ash/Tupelo type and the Cypress variant are flooded once to twice daily. The Maple/Sweetgum type was flooded regularly by tides, but not every day. The Ash/Sweetgum/Elm type is probably tidal only within the root zone, and then perhaps only sporadically, such as during storm surges and freshets.

	Ash/Tupelo	Baldcypress Variant	Maple/ Sweetgum	Ash/ Sweetgum/Elm
Number of stands	13	2	6	2
Canopy				
ash	40.6	19.7	13.2	23.0
swamp tupelo	34.8	22.6	19.7	2.4
red maple	21.0	30.9	36.6	14.8
sweetgum	0.5	3.0	21.6	17.1
baldcypress	0.9	21.3	--	--
elm	--	--	--	14.9
American beech	--	--	0.1	8.4
bitternut hickory	--	--	--	6.6
swamp chestnut oak	0.3	--	0.5	5.3
Total IV of other species	2.8	2.5	8.2	7.5
No. of additional species	7	2	8	6
species richness (mean)	4.5	5.0	6.0	11.0
Subcanopy				
northern spicebush	20.6	20.4	0.9	--
common winterberry	19.8	7.6	2.9	3.4
American hornbeam	11.3	6.8	22.0	48.5
American holly	7.1	13.7	1.7	--
highbush blueberry	2.1	13.9	0.1	--
sweetbay	1.6	9.0	5.8	10./
Canadian serviceberry	0.8	--	11.2	--
coastal sweetpepperbush	0.7	4.4	9.0	21.0
American elder	0.1	--	8.6	14.9
Total IV of other species	35.9	24.3	37.8	2.1
No. of additional species	12	8	9	1
species richness (mean)	8.6	10.5	8.2	5.0
Vines				
common greenbriar	51.4	42.9	0.2	14.6
poison ivy	16.6	10.3	0.9	12.5
groundnut	11.1	6.0	76.2	56.1
wild yam	6.6	19.3	1.2	--
crossvine	5.2	11.6	6.5	--
dodder	0.1	--	3.6	13.0
Total IV of other species	9.0	9.9	11.3	3.7
No. of additional species	3	2	3	1
species richness (mean)	6.0	5.0	5.3	3.5

Table 7.1. Continued

	Ash/Tupelo	Baldcypress Variant	Maple/ Sweetgum	Ash/ Sweetgum/Elm
Number of stands	13	2	6	2
Herbs				
halbredleaf tearthumb	19.4	14.9	--	23.1
bromelike sedge	9.0	13.2	6.2	19.0
lizard's tail	7.6	7.3	1.7	--
green arrow arum	7.4	2.8	--	--
Asian spiderwort	4.7	1.3	--	--
broadleaf uniola	4.6	--	0.2	--
poison ivy	3.9	0.4	--	--
aster	3.5	2.2	10.9	--
jewelweed	3.5	1.4	0.8	--
stout woodreed	2.9	3.2	2.9	--
common water hemlock	2.7	1.3	5.2	9.5
golden ragwort	2.1	2.3	1.1	14.0
Virginia dayflower	1.9	--	15.3	6.3
deertongue	1.4	--	1.2	9.8
small-spike false nettle	1.3	4.7	1.7	1.9
partridgeberry	1.3	4.1	--	--
Jack in the pulpit	0.6	0.8	2.5	--
fowl manna grass	0.1	3.1	--	--
smooth tickclover	0.1	--	3.6	--
graceful sedge	0.1	--	3.6	1.3
netted chainfern	--	8.0	--	--
crowned beggarticks	--	1.4	3.9	--
white turtlehead	--	0.6	3.4	--
asplenium ladyfern	--	--	8.0	--
Total IV of other species	21.9	27.1	27.8	15.1
No. of additional species	18	23	24	6
species richness (mean)	19.6	28.5	14.0	10.5

from the swamps because there were no stumps in any of the stands and baldcypress stumps decay very slowly.

In less regularly flooded tidal swamps, red maple and sweetgum (*Liquidambar styraciflua* L.) were the leading dominants, together comprising 50-72% of total IV. These were defined as a Maple/Sweetgum community type. On average, only six tree species occur in Maple/Sweetgum stands and soils are not as organic. In contrast, stands located more toward the terminus of tidal excursion had 11 canopy tree species on average, including many less flood-tolerant species such as American elm (*Ulmus americana* L.), slippery elm (*U. rubra* Muhl.), swamp chestnut oak (*Quercus michauxii* Nutt.), and American beech (*Fagus grandifolia* Ehrh.). These stands were defined as an Ash/Sweetgum/Elm type, but given that only two stands were sampled, the true variation in composition of such sites is unknown. Increasing mean canopy species richness from 4.5-6 species in regularly flooded Ash/Tupelo type to 6 species in the less regularly flooded Maple/Sweetgum type to 11 species in irregularly flooded

Ash/Sweetgum/Elm type reflects the effects of flooding stress among the types (Table 7.1). In other words, increased flooding stress reduced canopy species richness.

The three types of tidal swamp, as differentiated by canopy dominants, also differed in understory composition. The Ash/Tupelo swamps were dominated by northern spicebush (*Lindera benzoin* [L.] Blume) and common winterberry (*Ilex verticillata* L.) in the subcanopy, and halberdleaf tearthumb (*Polygonum arifolium* L.), bromelike sedge (*Carex bromoides* Schkuhr ex Willd.), lizard's tail (*Saururus cernuus* L.), and green arrow arum in the herb layer. Maple/Sweetgum swamps tended to be dominated by American hornbeam and Canadian serviceberry (*Amelanchier canadensis* [L.] Medik.) in the subcanopy, and by Virginia dayflower (*Commelina virginica* L.) and asters (*Aster* spp.) in the herb layer. The irregularly flooded Ash/Sweetgum/Elm type was dominated by American hornbeam and coastal sweetpepperbush (*Clethra alnifolia* L.) in the subcanopy, and by halberdleaf tearthumb and golden ragwort (*Packera aurea* [L.] A.&D. Love) in the herb layer (Table 7.1).

In contrast to the canopy stratum, understory species richness was higher on sites with more regular flooding (Table 7.1). Both subcanopy and herbaceous species richness were highest in the Ash/Tupelo and the Baldcypress variant and lowest in the irregularly flooded swamps. Species richness is associated with a more open canopy and more developed hummock/hollow microtopographic pattern. This combination of factors provides a variety of small-scale microhabitats, allowing species with varying tolerances to flooding and light to coexist within a small area. For example, because hollows can range from being well shaded to well lit, both shade tolerant and shade intolerant obligate wetland herbs grow in hollows (e.g., tidal freshwater marsh species such as halberdleaf tearthumb, green arrow arum, and annual wild rice [*Zizania aquatica* L.], as well as common bottomland hardwood forbs such as lizard's tail, Jack in the pulpit [*Arisaema triphylum* (L.) Schott], and royal fern [*Osmunda regalis* L.]). Likewise, both shade tolerant and shade intolerant facultative wetland and upland species grow on hummocks, such as bromelike sedge and broadleaf uniola (*Chasmanthium latifolium* [Michx.] Yates), respectively. Species richness of the herb layer is high due to this mixture of microsite conditions.

7.4.4. Ordination results

The DCA ordination shows the relative similarity of composition in the sampled stands in relation to the most statistically important environmental factors (Figure 7.6a). The distance between stands represents relative differences in species compositions, based on data from all strata. The relative lengths and directions of vectors (joint plot arrows) represent the relative strengths and directions of measured environmental gradients. Stands located toward the left of the ordination have the highest percent of their forest floors composed of hollow microtopography and highest percent organic matter content in surface soils, indicating that these stands are wetter. Stands with more hollow area and organic matter also tend to be located more downriver. The wetter, more downriver sites, concentrated toward the left side of the ordination, are the 13 Ash/Tupelo stands and the 2 Baldcypress variants of Ash/Tupelo. The joint plot arrow for ferrous iron (Fe^{+2}) is highest in stands located in the lower left corner of the ordination, where most of the Ash/Tupelo stands are concentrated. This makes sense in that the most reduced subsurface conditions likely occur in soils of regularly flooded (and always saturated) stands.

Stands with relatively more hummock cover, including those lacking hollows altogether, are located toward the right side of the ordination diagram. Stands toward the middle and upper left are the Maple/Sweetgum type, while stands to the far right on the ordination are Ash/Sweetgum/Elm stands. Species with the highest IVs influence where these stands are located. The leading dominants of Ash/Tupelo stands, ash and swamp tupelo, are most important in stands in the lower left of the ordination. Stands in which the IV is >30 for these two species are delineated in Figure 7.6b. Halbredleaf tearthumb is also important (IV>20) in many of these same stands, but not all. Red maple has an IV >15 in four of the six Maple/Sweetgum stands in the middle of the ordination and sweetgum has an IV>13 in all the Maple/Sweetgum stands and in the Ash/Sweetgum/Elm stand on the far right of the ordination. Distribution of these species in the ordination makes sense in that wettest stands and most flood-tolerant species are located on the left side of the ordination.

It is apparent from this ordination that species composition is related to a wetness gradient, suggesting that composition probably responds to flooding caused by tidal forcing. If so, in what way are tidal fluctuations related to relative wetness among community types? Water level monitoring data can help answer this question. Figure 7.7 shows the results of water level monitoring in five sites, three Ash/Tupelo stands and two Maple/Sweetgum stands. The duration of flooding varied among swamp types as did the range of tidal amplitude. Hollows were flooded 13-20% of the time, which also saturated the shallow root zones of hummocks. In contrast, the tops of hummocks were only flooded 1-2% of the time. Although tidal amplitude varied considerably among the five sites, neither amplitude

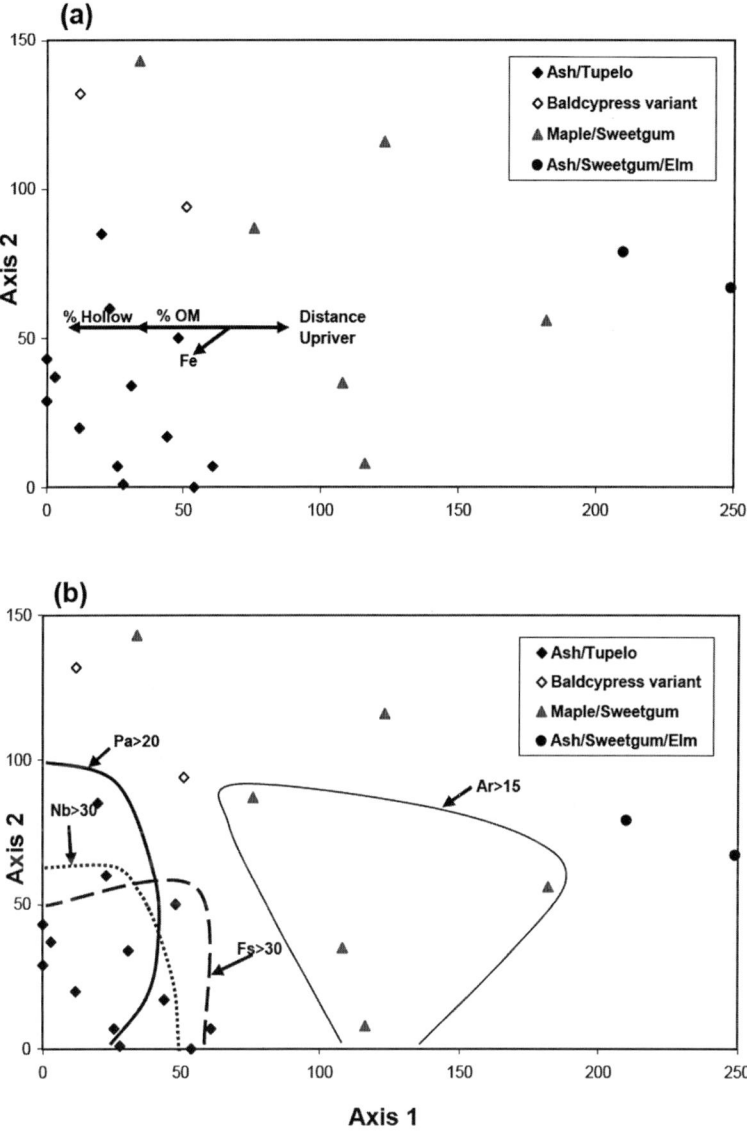

Fig. 7.6. (a) Detrended Correspondence Analysis (DCA) ordination based on importance values of vegetation in all strata. The lengths of joint plot arrows represent the relative strengths of the environmental relationships. Abbreviations: Holl=hollow, OM=organic matter, Fe=ferrous iron (ppm). Axis 1 explains 38% of the variation among stands, Axis 2 explains 15%. Joint plots all have $r^2>0.25$. (b) The same ordination as part a, but with contour lines enclosing stands where identified species exceed the indicated IVs. Pa=*Polygonum arifolium*, Nb=*Nyssa biflora*, Fs=*Fraxinus* spp., Ar=*Acer rubrum*.

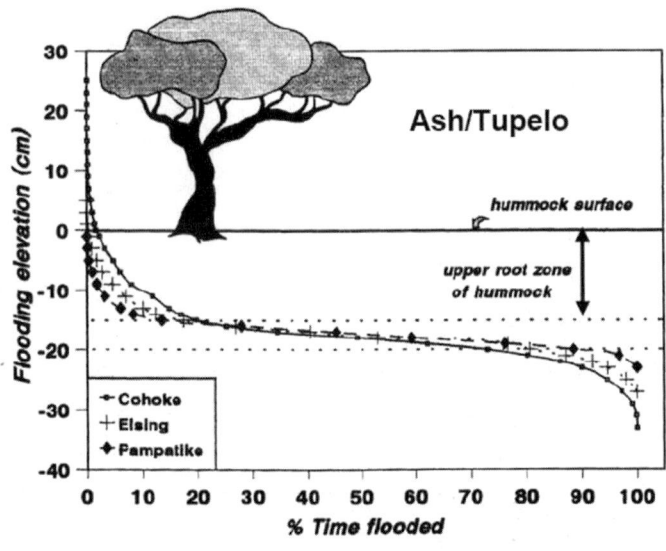

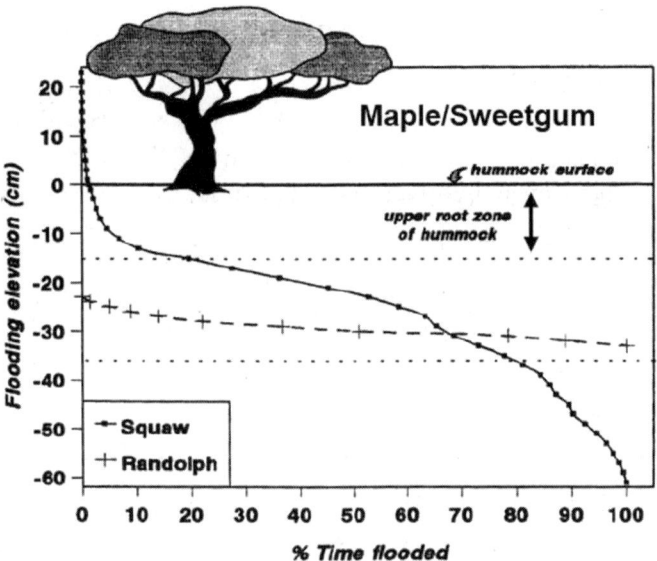

Fig. 7.7. Inundation curves for tidal swamps, percent time flooded vs. flooding elevation, relative to hummock surface. Hollow surface is about 15 cm below hummock surface, but Randolph lacks hollows. Dotted lines delimit region where rate of change is greatest for percent time flooded per unit elevation. The range in water level fluctuations varies among swamps: Cohoke (58 cm), Elsing (32 cm), Pampatike (22 cm), Squaw (86 cm), Randolph (10 cm) (Rheinhardt and Hershner 1992; used with permission).

nor flooding duration of hollows was related to composition. Rather, composition was related to mean saturation depth, i.e., the depth at which the soil is saturated for 50% or more of the time. This depth was a function of the elevation of hummocks in a given swamp in relation to tidal range. Although Maple/Sweetgum swamps tended to be located further upriver, not all were upriver, though none were located in the lower half of the river.

The three Ash/Tupelo swamps were saturated at 17-18 cm depth >50% of the time, while the two Maple/Sweetgum swamps were saturated >50% at 22 and 30 cm depths. Data are not available for the irregularly flooded Ash/Sweetgum/Elm stands. However, hydrographic data show that differences of 4-12 cm in elevation corresponded to quantitative differences in structure and species composition among tidal swamps. Thus, even a small rise in mean sea level might be expected to convert Maple/Sweetgum swamps to Ash/Tupelo swamps. In the most downriver locations, the intrusion of more saline water as sea level rises is responsible for a change from Ash/Tupelo swamp to marsh. This conversion is in evidence today in that tree dieback has occurred in the most-downriver swamp (Cousiac) and wetlands along the Pamunkey show a downriver to upriver pattern (which follows a salinity gradient) from tidal oligohaline marsh to freshwater marsh to freshwater swamp. However, short coastal plain tributaries that feed into the Pamunkey in mid- to upriver sections also appear to have large tracks of freshwater marsh rather than swamp (personal observation). This may be related to a lower sediment supply in the trunk tributaries. If this is true, then, even midriver Ash/Tupelo swamps might convert to marsh if sea-level rise is too rapid for the amount of sediment supplied. Considering all the possible changes in structure related to sea-level rise, the relative rate may be the most important factor and could affect how quickly and over what area these changes might occur.

7.5. Research needs

For Chesapeake Bay, the first research priority should be a reliable inventory of tidal swamps, including location and extent, along with an evaluation of potential threats, such as inappropriate shoreline development. Tidal swamps may be somewhat difficult to visually differentiate from other forested wetlands using remote imagery, but perhaps this could be accomplished with hyperspectral imaging or something similar.

The Pamunkey River tidal swamps are currently experiencing a relative rate of sea-level rise of approximately 4-6.5 mm yr^{-1}, far more rapid than the 1.4 mm yr^{-1} historic average. In a geologic time frame, tidal swamps have probably always migrated upriver during periods of rising sea level; tidal marshes replace tidal swamps as salinity intrudes upriver, tidal swamps replace nontidal riverine swamps in more upriver locations, and nontidal swamps replace low-lying uplands. For this migration to be successful, tidal swamps must keep pace with sea-level rise by vertically accreting sediment and organic matter at a sufficient rate and by migrating landward. Damming tributaries and mainstem rivers trap sediment needed for accretion. Where trunk tributaries support freshwater marsh rather than swamp, it might be fruitful to compare the composition of these marshes with that of marshes in the main stem of the Pamunkey, further downriver. One would suspect that the downriver marshes might possess relatively more oligohaline species than the strictly freshwater ones. If there are differences, then sediment profiles might provide insight into the relative importance of sediment for marshes and tidal swamps to maintain elevation in response to sea-level rise.

Timber harvests could likewise hinder the maintenance of hummocks. Data are needed to determine the relative importance of sediment and *in situ* organic matter production in maintaining tidal swamps and how the different community types differ in accretion rates. Macrofossil data could be obtained to gain insight into vegetation changes through time. It is possible that the first wave of forest clearing (and subsequent erosion) in the Piedmont may have expanded tidal swamps, and perhaps now we are seeing a sediment deficit that has the effect of reducing their capacity to keep up with sea level rise.

All the above research would provide insight into whether there is a threshold in which sediment deposition and productivity cannot keep pace with the rate of local sea-level rise. Answers to such questions might foster better predictions of potential changes in tidal wetland area and community type in response to various estimated rates of sea-level rise.

7.6. Conclusions

Tidal amplitudes vary along the length of the river, as does the elevation of tidal swamps in relation to tidal fluctuations. However, species composition of tidal freshwater swamps in the Pamunkey River seems to be controlled by variations in belowground (root zone) saturation which is controlled by tidal fluctuations. Thus, relative wetness among tidal swamps along the Pamunkey is related to the depth at which the root zone is saturated >50% of the time. This mean depth of saturation varied among five sites that had been instrumented with water level monitors, with depths ranging from 17-18 cm depth in Ash/Tupelo swamps to 22-30 cm depth in Maple/Sweetgum swamps. These results suggest that mean water table depth in the root zone is a more biologically appropriate measure of wetness in tidal swamps than duration of flooding or the total range in flooding height.

Multivariate ordinations based on vegetative composition of 23 stands showed that several other environmental indicators were likewise related to species composition, including percent coverage of hollows, percent soil organic matter, and Fe^{+2} concentration. All three parameters were higher in the wetter Ash/Tupelo swamps than in the other swamp types. Degree of wetness was not only related to species composition, but it was also related to species richness. However, the canopy and understory strata showed different patterns in reaction to flooding stress. Species richness was lower for the canopy under wetter conditions wherein only the most flood-tolerant trees grew. Woody subcanopy and herbaceous strata showed increased species richness with wetness. Wetter conditions stress trees, leading to an underdeveloped (open) canopy. In fact, in the wettest sites trees can only survive on hummocks created from fallen trees (nurse logs). However, high microtopographic complexity (due to the hummock and hollow pattern) and a poorly developed canopy together provided a wide variety of small-scale microhabitats for herbaceous and subcanopy species. These microhabitats allow species with varying tolerances to flooding and light to coexist within small areas. This is why overall species richness was higher in the wetter, more stressed tidal swamps. In swamps where conditions are less hydrologically stressful, the canopy is better developed (canopy closure more pronounced) and microtopographic complexity is lower. These factors are reflected in a lower variation in both light and hydrologic regimes and lower species richness overall.

As sea level rises and salinity intrudes further upriver, tidal swamps convert to marsh. However, further upriver where tidal waters are still fresh, tidal swamp must keep pace *in situ* with sea-level rise. Further work is needed to determine if there is a threshold at which productivity in swamps cannot keep pace with the rate of local sea-level rise. Answers to such questions might foster better predictions of changes in tidal wetland area and community type in response to various projected accelerated rates of sea-level rise.

7.7. Acknowledgments

Much of the original work cited in this chapter was only possible with the assistance, encouragement and comradery of the staff, scientists, and students of the School of Marine Science, College of William and Mary. I am particularly grateful for support and encouragement of Carl Hershner and Stewart Ware in data collection and analysis. Carl provided support, mentorship, and inspiration during the original data collection and continues to assist in retrieval and conversion of archived data. This chapter also benefited from the inputs of two anonymous reviewers.

References

Bilkovic D, Hershner C, Olney J (2002a) Macroscale assessment of American shad spawning and nursery habitat in the Mattaponi and Pamunkey Rivers, Virginia. North Am J Fisheries Manage 22:1176-1192

Bilkovic D, Olney J, Hershner C (2002b) Spawning of American shad (*Alosa sapidissima*) and striped bass (*Morone saxatilis*) in the Mattaponi and Pamunkey Rivers, Virginia. Fisheries Bull 100:632-640

Colman SM, Mixon RB (1988) The record of major Quaternary sea-level changes in a large coastal plain estuary, Chesapeake Bay, Eastern United States. Palaeogeography, Palaeoclimatology, and Palaeoecology 68:99-116

Doumlele DG, Fowler K, Silberhorn GM (1985) Vegetative community structure of a tidal freshwater swamp in Virginia. Wetlands 4:129-145

Duberstein J (2004) Freshwater tidal forest communities sampled in the lower Savannah River floodplain. M.S. thesis, University of Florida

Fowler K (1987) Primary production and temporal variation in the macrophytic community of a tidal freshwater swamp. M.A. thesis, Virginia Institute of Marine Science, School of Marine Science, College of William and Mary

Hodges RL, Sabo PB, Straw WR (1988) Soil survey of New Kent County, Virginia. U.S. Department of Agriculture, Soil Conservation Service. U.S. Government Printing Office, Washington

McCune B, Grace JB (2002) Analysis of ecological communities. MjM Software Design, Gleneden Beach

McCune B, Mefford MJ (1999) PC-ORD. Multivariate analysis of ecological data, Version 4.27. MjM Software Design, Gleneden Beach

Mixon RB, Berquist Jr CR, Newell WC, Johnson GH (1989) Geologic map and generalized cross sections of the Coastal Plain and adjacent parts of the Piedmont, Virginia. U.S. Department of the Interior, U.S. Geological Survey, Reston

Neubauer SC, Anderson IC, Constantine JA, Kuehl SA (2002) Sediment deposition and accretion in a Mid-Atlantic (U.S.A.) tidal freshwater marsh. Estuar Coast Shelf Sci 54:713–727

NOAA (1989) 1989 climatological data for Virginia. No. 99, Supplements No. 1-12. U.S. Department of Commerce, National Oceanic and Atmospheric Administration National Climatic Data Center, Asheville

NOAA (1990) 1989 climatological data annual summary for Virginia. No. 99, Supplement #13. U.S. Department of Commerce, National Oceanic and Atmospheric Administration, National Climatic Data Center, Asheville

Ozalp M (2003) Water quality, aboveground productivity, and nutrient dynamics during low flow periods in tidal floodplain forests of South Carolina. Ph.D. thesis, Clemson University

Pollard G (1894) The Pamunkey Indians of Virginia. Bureau of Ethnology, Smithsonian Institution, Government Printing Office, Washington, http://www.ls.net/~newriver/va/pamunkey.htm

Ratard MA (2004) Factors affecting growth and regeneration of baldcypress in a South Carolina tidal freshwater swamp. Ph.D. thesis, Clemson University

Rheinhardt RD (1991) Vegetation ecology of tidal swamps of the lower Chesapeake Bay, USA. Ph.D. thesis, Virginia Institute of Marine Science, School of Marine Science, College of William and Mary

Rheinhardt RD (1992a) Tidal freshwater swamps of lower Chesapeake Bay. Wetlands Program Technical Report 92-4. Virginia Institute of Marine Science

Rheinhardt RD (1992b) A multivariate analysis of vegetation patterns in tidal freshwater swamps of lower Chesapeake Bay, USA. Bull Torrey Bot Club 119:192-207

Rheinhardt RD, Hershner C (1992) The relationship of below-ground hydrology to canopy composition in five tidal freshwater swamps. Wetlands 12:208-216

Chapter 8 - Biological, Chemical, and Physical Characteristics of Tidal Freshwater Swamp Forests of the Lower Cape Fear River/Estuary, North Carolina

Courtney T. Hackney[1], G. Brooks Avery[2], Lynn A. Leonard[3], Martin Posey[1], and Troy Alphin[1]

[1]*Department of Biology and Marine Biology, University of North Carolina at Wilmington, Wilmington, NC 28403*
[2]*Department of Chemistry and Biochemistry, University of North Carolina at Wilmington, Wilmington, NC 28403*
[3]*Department of Earth Sciences, University of North Carolina at Wilmington, Wilmington, NC 28403*

8.1. Introduction

8.1.1. Definition

Tidal freshwater swamps of the southeastern United States, especially those found along the North Carolina Coast, are poorly characterized. They are abundant in southeastern North Carolina's coastal plain, where tide ranges about 1.35 m and affects river stages far inland. Tidal freshwater swamps differ from better-known bottomland hardwood swamps and other river-associated wetlands located further upstream or at higher elevations because soils are nearly always saturated in tidal freshwater swamps. More typical river-associated wetland soils are flooded during episodic events (e.g., spring floods), but dry to some depth for extended periods during summer when evapotranspiration is high or when no major flood events occur. In tidal freshwater swamps, flooding occurs with enough frequency to maintain saturated soils at or near the swamp soil surface. The continually flooded soils of tidal freshwater swamps create a very different environment for both plants and animals compared to typical riverine-wetlands, where soils dry to some depth for extended periods, especially during the growing season.

Closer to the coast, tidal fluctuations push saline water into the upper reaches of the estuaries and coastal rivers. Unlike the tropical mangrove systems farther south, there are no tree species along the Atlantic Coast that tolerate both the temperate climate and salt water. In these saline habitats, brackish marshes vegetated by various species of salt-tolerant grasses, sedges, and rushes dominate. Temperate, tidal, freshwater swamps are defined here as non-tropical wetlands dominated by woody vegetation and flooded by tides with sufficient frequency that the soil surface is almost always saturated. Tidal freshwater swamps flood with about the same frequency and duration as tidal marshes in the same area, but differ from these marshes with respect to the salinity of flood waters. The boundary between tidal brackish marsh and tidal freshwater swamp is extremely dynamic and varies from season to season and year to year. There are approximately 85,750 ha of tidal freshwater swamp in the Cape Fear River watershed, where it is typically classified as riverine swamp (Street et al. 2005). There are likely many times this acreage along the Atlantic Coast south of the Cape Fear River, where low gradient rivers with low sediment loads (Hupp 2000), coupled with strong coastal tides provide the ideal setting for tidal freshwater swamps (Isphording and Fitzpatrick 1992).

8.1.2. History

Early European settlers along the southeastern North Carolina coast converted all tidal freshwater swamps along the lower Cape Fear River into rice plantations (Clifton 1973), where dikes were used to control tidal flooding. These plantations were largely abandoned for agriculture at the beginning of the 20th Century (Clifton 1973), but remnant dikes document the upstream historic extent of tidal freshwater swamps within the Cape Fear River drainage (Hackney and Yelverton 1990). Over the last 100 years, the lower Cape Fear River has been modified by activities including widening and deepening of channels and extensions to the intracoastal waterway. In addition to sea-level rise, these activities have significantly altered the intrusion of both tides and saline water upstream to such an extent that areas once occupied by tidal freshwater swamps have been converted into brackish and salt marsh. Upstream wetlands that occupy floodplains and were once bottomland hardwood/softwood swamps and non-tidal riverine wetlands are now tidal freshwater swamps. Tides propagate upstream to the headwaters of the river's tributaries except where dams stop their advance (Street et al. 2005).

8.1.3. Geologic setting

The Cape Fear River (CFR) basin is the largest watershed entirely within North Carolina (Figure 8.1) and is comprised of more than 10,000 km of streams and rivers. The CFR basin includes two major physiographic regions, the Piedmont and Coastal Plain, which are traditionally separated on geologic maps by the Fall Line. The CFR main stem originates at the confluence of the Deep and Haw Rivers on the outer Piedmont in Chatham County, North Carolina. Piedmont streams in the upper CFR

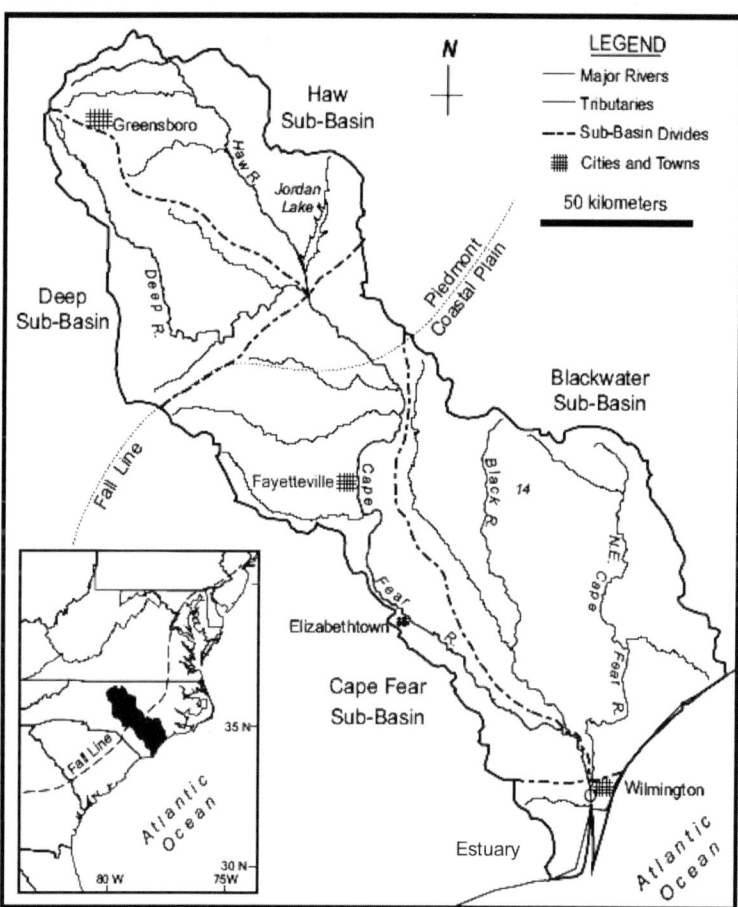

Fig. 8.1. Location map of the Cape Fear River drainage basin. The tidal wetlands described in this chapter are located in the Cape Fear and Blackwater sub-basins and the estuary. Specific site locations referred to in this chapter are shown in Figure 8.2.

basin flow through narrow incised valleys and are impounded by many small dams. In addition, the lower Haw River includes the large Jordan Lake reservoir that provides municipal water to the Raleigh-Durham metro area. The CFR then crosses the Coastal Plain, flows to the estuary, and meets the Atlantic Ocean about 40 km south of Wilmington, North Carolina. Release of water from the Jordan Lake reservoir controls the extent of saltwater incursions in the upper Cape Fear Estuary. Large metropolitan areas within the basin include Greensboro (population 223,891 in 2000), Fayetteville (121,015), and Wilmington (75,838). The latter is home to the Port of Wilmington and situated approximately 42 km from the river mouth. The main shipping channel is about 150 m wide and 13 m deep.

In the CFR basin, turbidity and sedimentation are recognized as significant threats to water quality and fish populations (Simmons 1993; NCDENR 2000). Simmons (1993) reported that Piedmont tributaries to the CFR have mean annual suspended sediment yields of about 50-150 Mg km^{-2} due to a combination of intense precipitation, the presence of erodible clay-rich soils, and flood-flow velocities in the range of 2-3 m s^{-1}. In the North Carolina Piedmont, the bedrock geology consists of a mosaic of mid-to-high grade metamorphic rocks, intrusive igneous rocks, and metavolcanic rocks. These rocks are traditionally assigned to the Milton belt, Carolina slate belt, Eastern slate belt, Raleigh belt, and the Durham basin. The Milton and Carolina slate belts, which are characterized by gneiss, schist, bedded argillites, and metavolcanic rocks (Butler and Secor 1991), underlie most of the Deep and Haw River sub basins. These rivers meet within the Durham basin, which is a Triassic sedimentary rift basin composed of mudstones, sandstones, shales, and conglomerates of the Chatham Group (Olsen et al. 1991). From the confluence of the Deep and Haw Rivers to the Fall Line, the upper CFR is underlain by high-grade metamorphic rocks of the Raleigh belt and metavolcanic rocks of the Eastern slate belt (Stoddard et al. 1991). The Piedmont soils reflect the bedrock geology and exposure to strong chemical weathering in a humid subtropical climate. Most of the region is mapped as ultisols with a lesser area of alfisols (Steila 2000).

The Coastal Plain streams in the lower CFR basin are underlain by eastward dipping sediments of Cretaceous age or younger and surface sediments dominated by quartz sand of fluvial, marine, or estuarine origin that has been extensively weathered (Soller and Mills 1991). Dominant geological units in the region include the Black Creek Formation (lignitic, mica-rich sand and clay) and Peedee Formation (marine sand and clay). Plio-Pleistocene and Holocene age unconsolidated marine sands and aeolian sands are common especially in the lower CFR valley. The streams in the Coastal Plain exhibit lower gradients and sediment loads than the Piedmont streams above the Fall Line. In this province, streams are traditionally differentiated into redwater types and blackwater types depending on their sediment load. The Piedmont-draining redwater streams on the North Carolina Coastal Plain have annual suspended sediment yields of 15-30 Mg km^{-2}, and suspended sediment concentrations in the range of 20-40 mg l^{-1}. The blackwater streams, which originate on the Coastal Plain, rarely experience suspended sediment concentrations in excess of 20 mg l^{-1} (Simmons 1993), but contain large amounts of dissolved organic matter derived from adjacent wetlands (Hubbard et al. 1990). The Coastal Plain section of the CFR is a redwater stream with a very gentle gradient, high sinuosity, and a wide floodplain that is dominated by swamp forest. Three navigation dams also are present the CFR in the reach between Wilmington and Fayetteville. The suspended sediment load in the lower CFR is much less than above the Fall Line and presumably significant storage of sediment is occurring on the Coastal Plain floodplains as signified by modern overbank deposition rates in the range of 2-9 mm yr^{-1} (Phillips 1997; Hupp 2000; Renfro 2004). Sediments trapped by these three dams limit the availability of suspended material that could be entrained within tidal swamps. These sediments are important if tidal freshwater swamps are to maintain their surface elevation in the face of rising sea level.

8.1.4. Swamp/marsh soils

It is clear that soils within tidal freshwater swamps and marshes of the lower CFR basin are starved for inorganic sediments as they are all highly organic in nature. Some sites have been shown to contain as much as 79% organic matter (determined by combustion), but generally the organic content is lower, 53-72% (Wicks 2002). There has been rapid accumulation of sediments, mostly organic, within these swamps during at least the past 50 years based on Cs^{137} data (Renfro 2004). Rates of accumulation between 0.30 and 0.95 cm yr^{-1} were found for swamps and marshes in the Cape Fear and Northeast Cape Fear Rivers, respectively (Renfro 2004). Rates of accumulation exceed the local relative sea-level rise rate of 0.22 cm yr^{-1}; however, these rates are insufficient to increase wetland surface elevation beyond the influence of tides. In spite of the low sediment concentrations particularly within waters of the blackwater streams of the northeast CFR, these wetlands are maintaining their position relative to sea-level rise through the rapid accumulation of decaying swamp/marsh vegetation.

8.1.5. Current research

In 1996, a U.S. Army Corps of Engineers' Environmental Impact Statement (USACE 1996) conducted in response to a proposal to widen and deepen the Cape Fear River and Harbor concluded that there was the potential for increased flooding of swamps and marshes upstream after the project was completed. A monitoring project was approved in 1997 (CZR, Inc. 1998) to determine the impact of the project on the plants, animals, soils, and hydrology of swamps and marshes of the lower CFR drainage. In 1996, very little data existed on tidal freshwater swamps. Twelve monitoring stations were established (Figure 8.2) and used to characterize water levels and salinity along the river and in swamps and marshes (Table 8.1) adjacent to the river. At nine of these stations, belt transects were established that extended from the river's edge to the wetland-upland boundary. Various biological, chemical, and physical characteristics of the wetlands have been evaluated within and adjacent to these transects.

8.1.6. General methodology

Data collection was initiated in 2000 and occurred continuously through May 2006. Annual reports, based on these data, are available at http://www.saw.usace.army.mil/wilmington-harbor/main.htm. Information contained in this chapter represents a small segment of data collected during the monitoring program. In addition, data collected as part of other research projects and needed to characterize these systems are summarized. The primary goal of the U.S Army Corps of Engineers funded monitoring program is to determine if channel deepening resulted in: 1) detectable changes in tidal flooding, degree, or duration, 2) changes in salinity of flood water, 3) alterations in populations of epibenthic aquatic fauna (fish and decapods), 4) benthic infaunal species changes with respect to composition and abundance, 5) alterations in the biogeochemistry of associated wetland soils, and 6) changes in salt-sensitive vascular plant species within tidal freshwater swamps of the lower CFR. The answer to the last question awaits completion of the dredging project and post monitoring analysis. Here, we use these data to characterize tidal freshwater swamps in the lower CFR. The general methodology employed is described below.

River water level was collected by continuous recording water level instruments surveyed to the NAVD88 datum and located at the stations depicted in Figure 8.2. A UNIDATA shaft-encoded water level recorder housed in an aluminum stilling well recorded water level at 1-second intervals and a UNIDATA Starlogger logs the average, maximum, and mini-

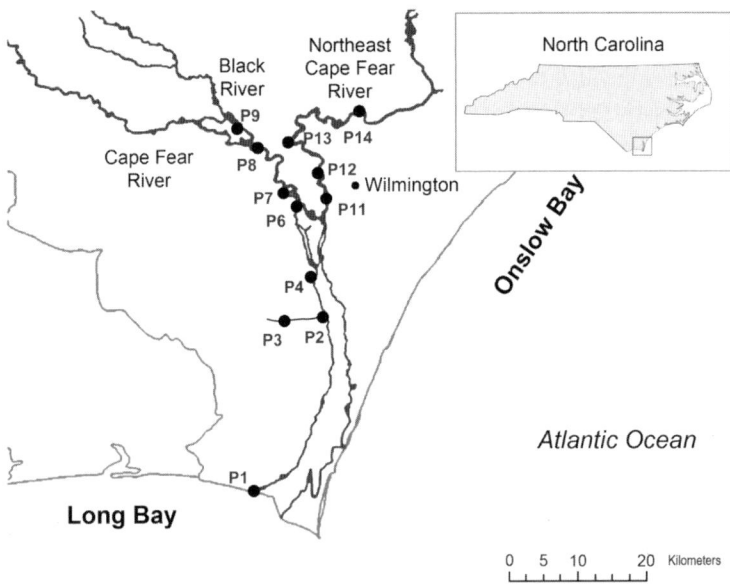

Fig. 8.2. Location of monitoring stations within the tributaries of the Cape Fear River. Stations P1, P2, and P4 do not have associated belt transects and only water level and salinity in the river are collected. Stations P3, P6, and P11 are best characterized as tidal marsh, while P7, P8, P9, P13, and P14 were tidal swamp in 2000. From 2000-2006, P12 has been in a state of transition from tidal swamp to tidal marsh.

Table 8.1. Classification of sampling sites. Yearly mean maximum and minimum salinities (ppt) of river water are shown for the period June 2004-May 2005. The mean discharge in the mainstem CFR over this period was comparable to the 30 year mean ($156.6 \text{ m}^{-3}\text{s}^{-1}$).

Station Number	Environmental Setting	Yearly Mean Max Salinity	Yearly Mean Min Salinity
P1	Open water	31.8	6.8
P2	Marsh	16.7	1.3
P3	Marsh with sparse trees	10.4	0.1
P4	River	12.9	0.7
P6	Marsh	10.7	< 0.1
P7	Swamp with marsh fringe	0.3	< 0.1
P8	Swamp	0.1	< 0.1
P9	Swamp	0.1	< 0.1
P11	Marsh with sparse trees	11.8	0.1
P12	Marsh (sites 1-2); swamp (sites 3-6)	10.6	< 0.1
P13	Swamp	3.1	< 0.1
P14	Swamp	0.1	< 0.1

mum values every 3 minutes. Conductivity and temperature are also sampled by a UNIDATA conductivity instrument and recorded by the Starlogger every three minutes. On wetland surfaces, water levels (relative to soil surface and NAVD88 datum) and salinity were measured every 24 minutes for two-week periods each fall and spring beginning in 2000 using RDS-20 or RDS-40 water level monitors (accurate ± 0.3 cm) at each of six subsites at nine stations as noted in Figure 8.2. Swamp/marsh water level was determined at six subsites distributed from river to upland edge within each belt transect continuously for two weeks, twice each year. Swamp/marsh salinity was collected whenever water was present on the swamp/marsh surface via conductivity and temperature. Both river salinity and swamp/marsh salinity was computed from conductivity and temperature using standard algorithms. Soil metabolism was determined in winter and summer within soils at all six subsites within each belt transect twice each year by monitoring methane, chloride and sulfate in porewaters within the surface 30 cm of soil.

Salt-sensitive vascular vegetation was determined by monitoring the aerial extent of pre-selected stands of species known to be salt sensitive. Data for plant species presence and percent cover are gathered from permanent variable-sized plots (sites P3, P9, P12, P13 and P14) and fixed size plots (sites P7 and P8) that are re-surveyed each August when herbaceous vegetation has reached its full seasonal development. Benthic infauna and epifauna were sampled at eight stations across three of the major tributaries to the lower CFR. For the purpose of benthic characterization, stations were classified according to habitat type on the basis of vegetation because infauna may respond strongly to structural aspects of the habitat as determined by vegetation patterns. Stations P3, P8, and P13 were classified as tidal freshwater swamps even though P3 experiences salinities that regularly exceed >1 ppt. The savannah nature of the vegetation at P3, as well as bank characteristics, suggest a closer relation to swamp habitats structurally. Sites P6, P7, P11, and P12 were classified as mixed tidal swamp/marsh due to the combination of herbaceous and woody vegetation where samples were collected. P2 was classified as an oligohaline/upper mesohaline tidal marsh. Infauna were sampled within the vegetated habitat, along the vegetation/non-vegetated interface, and in the intertidal area immediately adjacent to the marsh or swamp. Samples were collected using cores measuring 0.01 m^2 by 10 cm deep taken 1-2 m from the vegetation edge during summer. Epifauna and nekton were sampled using Breder traps and drop traps during March/April and September/October, which are prime recruitment periods for estuarine nekton. Breder traps (Breder 1960) were placed within the vegetated edge and adjacent unvegetated areas of each habitat on the incoming tide and allowed to fish for 2 hr. Drop traps consisted of a 1m x 1m enclosure deployed from a boom mounted on a shallow draft boat. Drop trap sampling was conducted just seaward of the vegetation and targeted both motile and sedentary shallow-water nekton. Abundances of infauna were calculated on an area basis, while abundances for Breder traps and drop traps were calculated as catch per unit effort (CPUE) with the sampling unit (2 hour catch per Breder trap, per drop catch) as effort. Diversity was calculated using the Shannon Weiner Diversity Index based on yearly faunal abundances and species richness represented as average annual number of species per sampling area.

8.2. Hydrology

Eustatic sea-level change and tectonic uplift throughout the late Quaternary have influenced coastal plain streams in southeastern North Carolina. In the CFR estuary, sea-level rise has been augmented by historic dredging operations that have increased the tidal range and converted freshwater swamp forest to estuarine salt marsh (Hackney and Yelverton 1990). In the CFR basin, the influence of tides extends about 50 km upstream of the estuary, to Lock and Dam 1 on the CFR and beyond Burgaw, North Carolina on the Northeast Cape Fear River. The goals of research in the Cape Fear system has been to quantify tidal patterns in the lower CFR and Northeast Cape Fear River, to identify interannual variability, and to relate changes in tidal patterns to natural events and man-made perturbations.

8.2.1. Tidal patterns

Tides in the river and its tributaries are semi-diurnal and exhibit a strong diurnal inequality in the lower estuary. Within the estuary sub-basin, tidal range is fairly constant, with mean tidal ranges between 1.2 and 1.7 m (Table 8.2). Tidal ranges at upstream stations are generally lower than those within the estuary and range from 0.93–1.31 m in the CFR and 0.7–1.3 in the Northeast CFR. In the CFR, tidal range decreases upstream with the lowest mean tidal range, 0.93 m, observed at station P9. The same pattern exists in the Northeast CFR where the lowest tidal range occurs at the most upstream station, P14 (Table 8.2). Water levels in the CFR basin are strongly affected by regional climatology (e.g., drought and flooding) and these events also affect tidal fluctuations in the river. During periods of reduced rainfall, mean tidal range tends to be higher; especially at the

Table 8.2. Summary of the average difference between high and low tide at each station, time to high tide relative to the river mouth (P1), time to low tide relative to the river mouth, flood tide duration, and ebb tide duration.

Station number	Mean tidal range (m)	Mean flood duration (hr)	Mean ebb duration (hr)	Mean high tide lag from P1 (hr)	Mean low tide lag from P1 (hr)
P1	1.28 ± 21.73%	6.30	6.11	N/A	N/A
P2	1.33 ± 16.01%	5.65	6.75	1.33	1.97
P3	0.82 ± 17.35%	6.20	6.18	2.97	2.98
P4	1.36 ± 15.29%	5.71	6.70	1.62	2.21
P6	1.31 ± 15.44%	5.87	6.53	2.12	2.55
P7	1.16 ± 14.33%	5.78	6.58	2.55	3.02
P8	1.01 ± 16.83%	5.82	6.58	3.17	3.63
P9	0.92 ± 21.23%	5.78	6.61	3.33	3.63
P11	1.29 ± 14.63%	5.82	6.57	2.13	2.58
P12	1.14 ± 14.16%	5.87	6.53	2.53	2.91
P13	0.96 ± 13.98%	6.00	6.40	3.10	3.38
P14	0.67 ± 19.18%	5.91	6.48	4.13	4.50

most upstream sites and the inner Town Creek station (P3). During periods of high river discharge, such as during a period of flooding in spring 2004 (Figure 8.3), water level fluctuations are suppressed leading to reduced tidal ranges upstream and also at site P3 in Town Creek (Figure 8.4). Suppression of tide is not usually associated with less flooding of the swamp/marsh surface as tides propagate on higher water levels in the river. While statistical differences in tidal range have been documented between drought years and non-drought years for sites P1 and P2 in the lower estuary, the overall impact of climatologically driven events on water level remains much less than other up river sites.

Tides in the lower estuary are symmetrical as evidenced by the nearly equal flood and ebb durations at P1 (Table 8.2). Stations in the upper CFR and the Northeast Cape Fear River exhibit more pronounced time asymmetries. At these sites, mean flood durations are shorter than ebb durations and the mean high tide lag relative to site P1 is less than the mean low tide lag time (Table 8.2). This pattern is commonly observed in tidally influenced rivers because the speed of tidal propagation up-estuary depends on water depth which generally decreases upstream in coastal plain systems. Over the period of 2000 to 2005, tidal range for each station was similar among years (Figure 8.5).

8.2.2. River salinity

Tide stage, tidal range, and fluctuations in river discharge result in temporal and spatial variations in salinity in the river system. Near the mouth

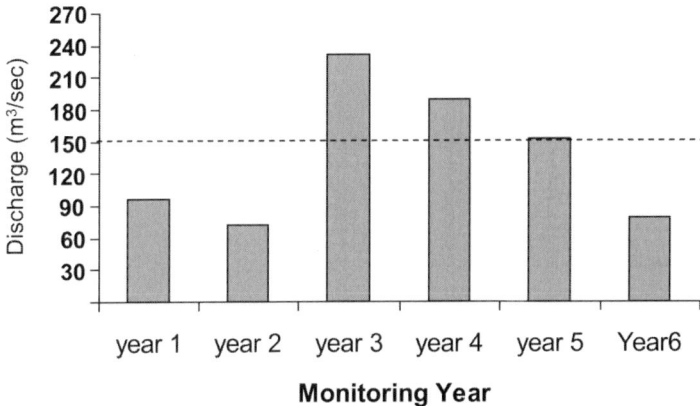

Fig. 8.3. Mean river discharge for each monitoring period. Monitoring year 1 is October 2000 to May 2001; monitoring year 2 is June 2001 to May 2002; monitoring year 3 is June 2002 to May 2003; monitoring year 4 is June 2003 to May 2004; monitoring year 5 is June 2004 to May 2005, and monitoring year 6 is June 2005 to May 2006. The line denotes the long-term mean discharge for the Cape Fear River as measured at Lock 1 by a USGS gauging station.

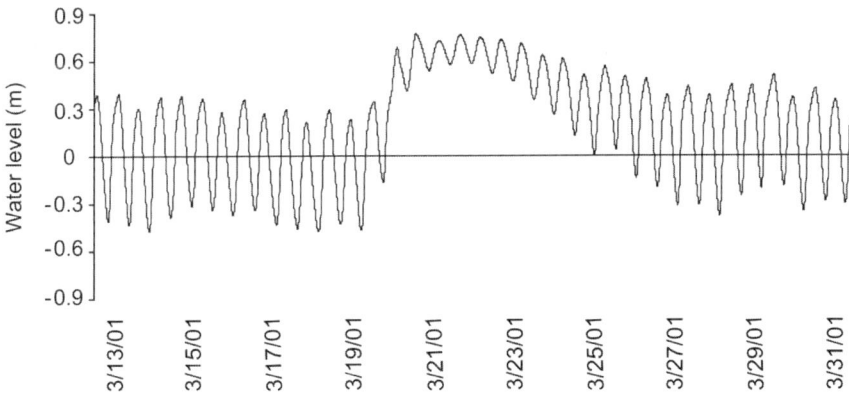

Fig. 8.4. Water level curve for Inner Town Creek (P3). Note the reduction in tidal range and elevation of total water level associated with a period of high rainfall in late March 2001.

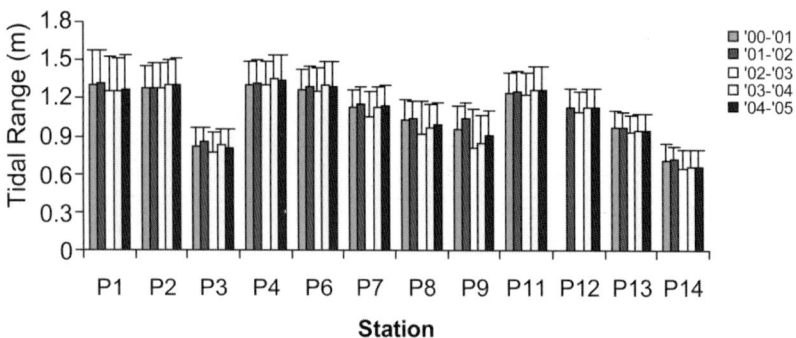

Fig. 8.5. Yearly mean tidal range for each station from 2000 to 2005. All water levels are relative to NAVD88 with the exception of P4, which is relative to MSL. Error bars show one standard deviation.

of the estuary, the near surface salinity ranges from 34 ppt during high water to about 5 ppt during low water. At site P2, also in the lower estuary, salinities are typically lower, ranging from about 25 ppt at high water to less than 1 ppt during low water. This site, however, is located adjacent to the mouth of Town Creek, which discharges low salinity water (<1 ppt) to the estuary on the falling tide and following periods of intense rainfall. In general, salinity decreases upstream as the influence of tides becomes less pronounced and as the influence of upstream discharge increases (Table 8.1). Under normal river discharge conditions, salinity does not exceed 1 ppt at stations upstream of Eagle Island on the CFR because of the continuous release of freshwater upstream. In 2001-2002, a period of intense drought in Southeast North Carolina (Figure 8.3), upstream releases in the CFR had been reduced and salinities as high as 3.5 ppt were measured at P8, while salinities exceeding 14 ppt were measured at Fishing Creek, 12.8 km north of Wilmington, North Carolina in the Northeast Cape Fear River. Drought conditions also existed in summer 2002 and salinities as high as 5.8 ppt and 16.4 ppt, were reported for P8 and P13, respectively. In 2005, a period of more typical flow conditions in the river, maximum salinities for P8 and P13 were 0.2 ppt and 8.6 ppt, respectively. The boundary between saline and non-saline water is critical because brackish marsh replaces tidal freshwater swamp wherever saline waters flood the soil surface on a regular basis.

8.2.3. Hydrology of swamp/marsh surface

Data reported in this section represent only a small subset of the total data collected and are indicative of patterns observed between 2000 and 2006. The general goal of this research was to determine the degree to which tides flooded these wetlands, to determine if flooding patterns coincided with tides, and to measure salinity in floodwaters.

8.2.3.1. Flooding patterns

In general, tidal freshwater swamps and tidal marshes were flooded regularly (Table 8.3). Flooding frequency was related to absolute elevation and distance upstream from the river mouth. As noted in Table 8.1, there is a dampening effect of tidal height with distance upstream resulting in the reduction of tidal range. Clearly, water levels in both swamp and marsh communities are dependent on tides (Figure 8.6) in the river. Note that while water levels within swamp soils may fall below the soil surface when tides do not flood the surface, soils remain saturated because the time between each tidal inundation is generally short.

8.2.3.2. Surface water salinity

In general, the salinity of water flooding the wetland surface decreased with distance upstream. Proximity to the river also influenced the salinity to which each subsite was exposed for any single inundation event for

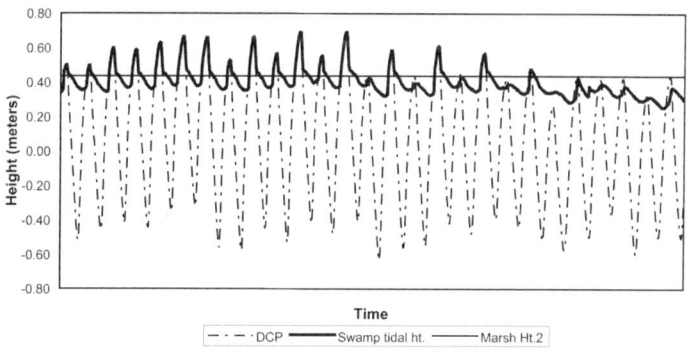

Fig. 8.6. Water level fluctuations relative to the surface of the swamp at station P13, subsite 1 in Spring 2002. Emboldened lines show water levels above and below the marsh surface, while hatched lines show water levels recorded at the adjacent data collection platform (DCP) in the river. The horizontal line indicates the elevation of the marsh surface.

Table 8.3. Flooding frequency, duration of soil surface inundation, and maximum water level above the soil during a typical two week monitoring period.

Station number	Substation number	Season	Start date	End date	# Flood events	Mean flood duration (hr)	Maximum depth (m)	Marsh/ swamp elevation (m)	Actual water level (m)
P3	1	Spr 03	4/24/03	5/8/03	27/27	7.1	0.64	0.20	0.43
	2	Spr 03	4/24/03	5/8/03	26/27	6.8	0.61	0.25	0.37
	3	Spr 03	4/24/03	5/8/03	27/27	7.3	0.64	0.16	0.49
	4	Spr 03	4/24/03	5/8/03	27/27	7.2	0.64	0.45	0.18
	5	Spr 03	4/24/03	5/8/03	26/27	6.5	0.70	0.30	0.40
	6	Spr 03	4/24/03	5/8/03	25/27	5.1	1.28	1.10	0.27
P6	1	Spr 03	3/18/03	4/1/03	26/27	6.7	1.13	0.23	0.88
	2	Spr 03	3/18/03	4/1/03	26/27	5.4	1.10	0.48	0.61
	3	Spr 03	3/18/03	4/1/03	26/27	6.6	0.98	0.26	0.73
	4	Spr 03	3/18/03	4/1/03	26/27	5.7	1.10	1.34	0.76
	5	Spr 03	3/18/03	4/1/03	26/27	5.1	1.10	0.59	0.52
	6	Spr 03	3/18/03	4/1/03	26/27	4.7	0.98	0.53	0.46
P7	1	Spr 03	4/3/03	4/17/03	24/27	7.2	1.37	0.54	0.82
	2	Spr 03	4/3/03	4/17/03	20/27	6.4	1.37	0.68	0.70
	3	Spr 03	4/3/03	4/17/03	18/27	6.1	1.31	0.69	0.61
	4	Spr 03	4/3/03	4/17/03	18/27	5.6	1.37	0.74	0.64
	5	Spr 03	4/3/03	4/17/03	18/27	5.8	1.31	0.70	0.61
	6	Spr 03	4/3/03	4/17/03	17/27	6.0	1.31	0.72	0.58
P8	1	Spr 03	5/13/03	5/27/03	23/27	3.7	0.92	0.65	0.27
	2	Spr 03	5/13/03	5/27/03	26/27	5.1	0.92	0.47	0.46
	3	Spr 03	5/13/03	5/27/03	26/27	5.5	0.95	0.45	0.49
	4	Spr 03	5/13/03	5/27/03	22/27	4.4	0.88	0.60	0.27
	5	Spr 03	5/13/03	5/27/03	16/27	3.7	0.88	0.68	0.21
	6	Spr 03	5/13/03	5/27/03	12/27	3.6	0.85	0.73	0.12
P9	1	Spr 03	5/7/03	5/21/03	27/27	6.9	0.76	0.18	0.58
	2	Spr 03	5/7/03	5/21/03	27/27	4.8	0.85	0.67	0.18
	3	Spr 03	5/7/03	5/21/03	19/27	3.6	0.82	0.37	0.46
	4	Spr 03	5/7/03	5/21/03	9/27	4.5	0.85	0.63	0.21
	5	Spr 03	5/7/03	5/21/03	8/27	4.5	0.85	0.67	0.18
	6	Spr 03	5/7/03	5/21/03	9/27	4.9	0.85	0.59	0.27
P11	1	Spr 03	3/13/03	3/27/03	26/27	6.5	1.10	0.44	0.67
	2	Spr 03	3/13/03	3/27/03	24/27	5.4	1.10	0.56	0.55
	3	Spr 03	3/13/03	3/27/03	25/27	5.9	1.10	0.54	0.55
	4	Spr 03	3/13/03	3/27/03	24/24	5.5	1.10	0.56	0.55
	5	Spr 03	3/13/03	3/27/03	24/24	5.7	1.10	0.58	0.52
	6	Spr 03	3/13/03	3/27/03	23/27	4.9	1.10	0.62	0.49
P12	1	Spr 03	3/27/03	4/10/03	27/27	5.8	0.82	0.27	0.55
	2	Spr 03	3/27/03	4/10/03	21/27	6.1	0.82	0.49	0.34
	3	Spr 03	3/27/03	4/10/03	14/27	4.3	0.79	0.61	0.18
	4	Spr 03	3/27/03	4/10/03	11/27	5.7	0.82	0.58	0.24
	5	Spr 03	3/27/03	4/10/03	15/27	5.9	0.79	0.63	0.15
	6	Spr 03	3/27/03	4/10/03	14/27	6.8	0.79	0.74	0.06
P13	1	Spr 03	2/25/03	3/11/03	25/27	5.6	0.73	0.44	0.31
	2	Spr 03	2/25/03	3/11/03	26/27	6.3	0.69	0.33	0.37
	3	Spr 03	2/25/03	3/11/03	27/27	6.6	0.76	0.23	0.55
	4	Spr 03	2/25/03	3/11/03	27/27	7.0	0.95	0.31	0.64
	5	Spr 03	2/25/03	3/11/03	27/27	5.6	0.73	0.37	0.37
	6	Spr 03	2/25/03	3/11/03	13/27	3.7	0.69	0.50	0.21
P14	1	Spr 03	3/4/03	3/18/03	26/27	8.7	0.70	0.21	0.49
	2	Spr 03	3/4/03	3/18/03	26/27	8.0	0.66	0.27	0.40
	3	Spr 03	3/4/03	3/18/03	26/27	7.1	0.66	0.33	0.34
	4	Spr 03	3/4/03	3/18/03	26/27	6.6	0.66	0.37	0.31
	5	Spr 03	3/4/03	3/18/03	25/27	6.3	0.66	0.39	0.27
	6	Spr 03	3/4/03	3/18/03	23/27	5.5	0.66	0.45	0.21

those sites below the most upstream intrusion of saline water. For those sites where saline water has been detected (>1 ppt), subsites closest to the river experience a greater number of inundations than subsites positioned closer to the upland edge (Table 8.4). The distribution of surface water salinity between sites and across specific sites is mediated by the effects of localized rainfall and upland runoff, river discharge and regional climatology (i.e., drought or flood).

The influence of saline water on wetland community structure was evident at some of the stations examined. Even barely detectable levels of saline water, less than 10% seawater, led to major community change in tidal freshwater swamps if such conditions were chronic. Differences in community type, swamp versus marsh, were clearly related to the penetration of saline water into swamps. Individual or sporadic intrusions of saline water did not immediately lead to a change in vegetation (i.e., tidal swamp to tidal marsh). Chronic exposure to saline water resulted in a shift in vegetation from woody tree species to herbaceous species. In the study area, those stations dominated by herbaceous vegetation (i.e., marsh sub-sites at P3, and P6 and P11) were flooded by saline water >1 ppt almost every tide (Table 8.4). Those stations dominated by large trees, P7, P8, P9, and P14, were either never flooded by saline water or exposed to saline water only during periods of drought. On the infrequent occasions when these swamps were flooded by saline water, the concentration of saline water was generally low (Table 8.4).

Station P12, and to a lesser degree stations P7 and P13, appear to be in various stages of community change from woody swamp to herbaceous marsh (e.g., Figure 8.7). During the study period, there were repeated exposures to saline flood water (Table 8.4), which eventually were linked to increased sulfate reduction in soils (Figure 8.8). Evidence of community change was most profound at station P12 in the Northeast Cape Fear River, where sulfate reduction in soils was chronic (see numerous reports at http://www.saw.usace.army.mil/wilmington-harbor/main.htm) and the vegetation is in the process of transitioning from swamp species to species more characteristic of brackish tidal marsh in the southeastern United States (Figure 8.7). The concentration of saline water necessary to begin the process of community change seems to be approximately 2 ppt based on the long-term data set illustrated by Table 8.4. However, as noted in the next section on biogeochemistry, the causative factor is not likely salinity alone, but microbially-mediated byproducts of remineralization processes associated with the dissolved constituents of seawater.

Table 8.4. Summary of salinity data collected from substations at nine stations along the Cape Fear River and its tributaries in Fall 2002 (ND = no data).

Station	Station name	Substation number	Fall 2002 salinity range (ppt)	Proportion of flood events containing > 1 ppt salinity
P3	Town Creek	1	<1-21	23/27
		2	<1-19	20/27
		3	ND	ND
		4	<1-6	25/27
		5	<1-12	25/27
		6	<1-6	27/27
P6	Eagle Island	1	<1-9	27/27
		2	<1-13	27/27
		3	<1-12	27/27
		4	<1	27/27
		5	<1-6	23/27
		6	<1-7	27/27
P7	Indian Creek	1	<1-2	21/27
		2	<1-1	0/27
		3	<1-1	0/27
		4	1-2	17/27
		5	<1	0/27
		6	<1	0/27
P8	Dollison's Island	1	<1	0/27
		2	<1	0/27
		3	<1	0/27
		4	<1	0/27
		5	<1	0/27
		6	<1	0/27
P9	Black River	1	<1	0/27
		2	<1	0/27
		3	<1	0/27
		4	<1	0/27
		5	<1	0/27
		6	<1	0/27
P11	Smith Creek	1	<1-18	26/27
		2	<1-11	26/27
		3	6-9	27/27
		4	3-13	27/27
		5	<1-18	25/27
		6	<1-18	22/27
P12	Rat Island	1	<1-12	26/27
		2	<1-12	22/27
		3	<1-11	18/27
		4	<1-11	17/27
		5	<1-11	14/27
		6	<1-7	9/27
P13	Fishing Creek	1	<1-2	8/27
		2	<1-2	12/27
		3	<1-2	25/27
		4	<1-2	25/27
		5	<1-2	7/27
		6	<1-1	1/27
P14	Prince George	1	<1-1	1/27
		2	<1-1	0/27
		3	<1-1	3/27
		4	<1-1	10/27
		5	<1	0/27
		6	<1	0/16*

*Data recorded from 10/28/2002 at 11:12 to 11/5/2003 at 13:36.

Fig. 8.7. Station P12 which is in the process of conversion to tidal marsh from tidal swamp. Herbaceous vegetation is present among stands of remnant swamp vegetation at this location (photo by Mark Gay).

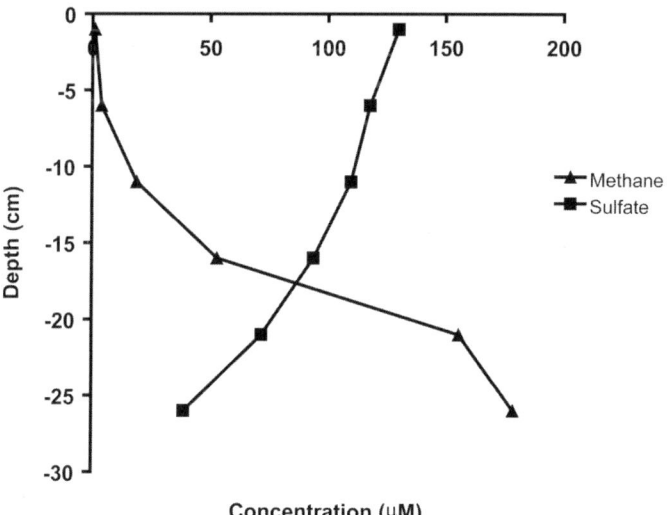

Fig. 8.8. Depth profiles of sulfate and methane porewater concentrations for soils in swamp forest at station P13. Patterns illustrated by this figure are typical across stations and years.

8.2.4. Hydrology summary

Oceanic tides dominate the surface hydrology of the swamps and marshes in the CFR Basin far above the influence of saline water. The boundary between tidal marsh and tidal swamp is not a product of flooding duration or height, but the occurrence of saline water on and in the swamp soil. Once salinity of floodwater exceeds 2 ppt on a regular basis, the tidal swamp begins to convert to oligohaline or brackish tidal marsh.

8.3. Biogeochemistry

8.3.1. Remineralization processes

Biogeochemical conditions in organic rich waterlogged soils are typically governed by microbial remineralization processes (breaking down of organic carbon to CO_2 and methane) and their byproducts. Due to the relatively low solubility of oxygen in water and the high amount of organic matter present in these systems, oxic respiration is often limited to the top few millimeters of soils resulting in anoxic condition just below the surface. In freshwater wetland soils, organic matter is usually remineralized via anaerobic methanogenesis (Wolfe and Higgins 1979). In wetlands exposed to seawater, sulfate present in the sea salts results in remineralization via microbial sulfate reduction. Sulfate reducers out-compete methanogens, so when sulfate is present methanogenesis is shut down (Martens and Berner 1974). During sulfate reduction, hydrogen sulfide and CO_2 are produced by organic matter remineralization instead of methane and CO_2. This has important implications to the community since hydrogen sulfide is toxic and limits both plants and animal species that do not have a behavioral or physiological mechanism to tolerate this bacterial metabolite.

8.3.2. Spatial variations in remineralization processes, depth variations

Byproducts of remineralization processes vary with depth in soils due to competition between microorganisms employing different microbial processes. This biogeochemical zonation results in the most competitive processes at the surface, where oxidants such as sulfate are resupplied by floodwaters, followed by less competitive processes at depth (Froelich et al. 1979). Therefore, when swamp forest sediments contain sufficient sulfate concentrations, sulfate reduction dominates remineralization at the surface, while methanogenesis occurs at depth. Methanogens take over remineralization only after sulfate concentrations reach levels that are too low for sulfate reduction to be energetically favorable. The concentration of sulfate needed to support sulfate reduction (sulfate reduction threshold) has a value of approximately 100mM in wetlands along the CFR (Sexton 2002; Avery and Hackney, unpublished data).

Biogeochemical zonation is common where seawater reaches tidal freshwater swamps, such as the swamp forest at site P13 along the CFR (Figure 8.8). Concentrations of sulfate, the oxidant and reactant for sulfate reduction, decrease with soil depth due to consumption and lack of resupply from the surface. The concentration of methane, the product of methanogenesis, increases steadily with depth after the sulfate concentration falls below the sulfate reducing threshold.

8.3.3. Creek bank to upland variations

Proximity to the river, the source of salinity during high tide for those stations below the most landward intrusion of salt water, impacts the biogeochemistry within adjacent swamp forest wetlands. Our data indicate soils closest to the river have higher salt concentrations while soils closer to uplands often have lower concentrations (Figure 8.9) due to freshwater runoff and subsurface flow of freshwater. The resulting variations in sulfate content of the floodwater impacts the biogeochemical processes and their byproducts in sediments and ultimately fauna and flora inhabiting these swamps. Sulfate reduction can dominate remineralization at substations close to the river when tidewaters contain sufficient sulfate, while methanogenesis can be relatively more important far from the river and adjacent to uplands. For example, the tidal swamp associated with P14 is generally above the influence of saline water. During a historic drought in summer 2001, some saline water penetrated far upriver and salinities at P14 were near 1 ppt, roughly twice the values typical of this site. In the summer of 2002, P12 and P13, sites with salinities typically in the 0.2-0.5 ppt range, also experienced an order of magnitude increase in salinity. Consequently, surface sediment pore water concentrations in substation soils at P14, generally showed a decrease in sulfate concentrations and an increase in methane from creek bank to upland locations (Figure 8.10).

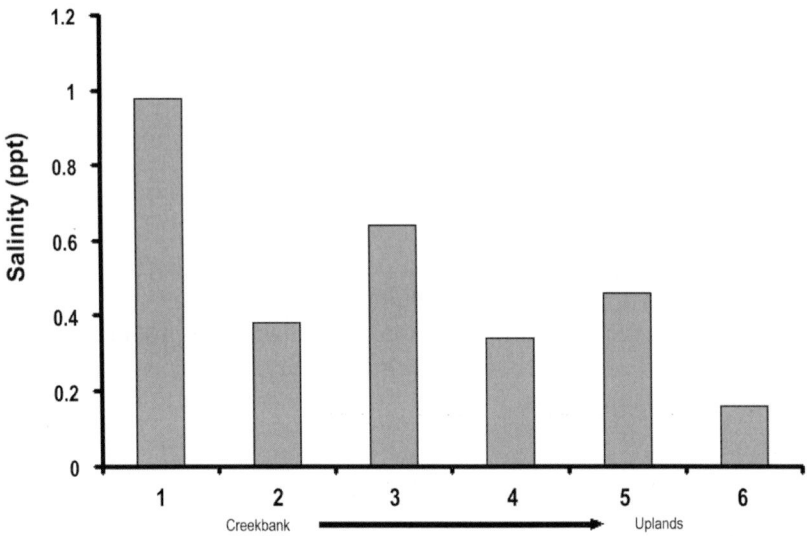

Fig. 8.9. Salinity of surface sediment porewaters for subsites at station P12 during the winter of 2004.

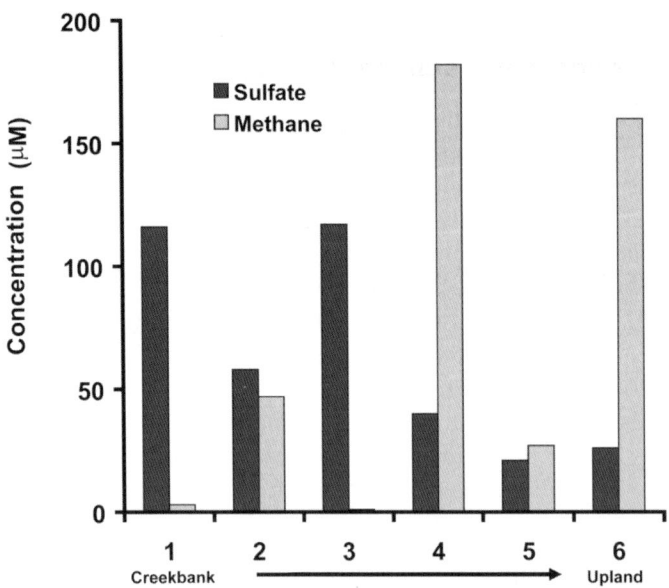

Fig. 8.10. Sulfate and methane concentrations in surface porewaters within swamp forest at P14. This is a pattern typical for tidal freshwater swamps at the upper boundary of saltwater penetration.

8.3.4. Temporal variations

Typically the wetlands along the lower CFR are exposed to highest salinities during the summer when river flows are low. In addition, tidal swamp soils are evacuated by evapotranspiration in the summer, which allows the more rapid penetration of tidal waters into swamp soil pore space. During the winter, river flows are usually higher and evapotranspiration negligible, resulting in lower salinities and penetration of sulfate into soils. It should be noted, however, that this is a general pattern and actual variations can be impacted by events such as drought or hurricanes (Avery et al. 2004). Figure 8.11 shows winter versus summer variations in sulfate and methane concentrations for a typical year in the swamp forest at P13. During the summer, sulfate concentrations were higher indicating that remineralization was likely occurring via sulfate reduction. The importance of sulfate reduction during this time of year was supported by lower methane concentrations indicating inhibition of methanogenesis. Winter conditions were opposite with lower sulfate concentrations and higher methane concentrations since methanogenesis was relatively more important as a remineralization process during fresher times of year. The temporal variations in remineralization processes may have important implications for plants and animals with limited abilities to withstand sulfide exposure.

8.3.5. Biogeochemistry summary

Patterns illustrated in this section are typical of tidal swamp stations in the lower CFR. These tidal freshwater swamps usually experience both spatial and temporal variations in salinity, resulting in shifts between sulfate reduction and methanogenesis as the dominant mode of microbial remineralization. Further, the biogeochemistry in these systems is strongly impacted by climatological events such as drought. As a result, plants and animals in these communities experience varying exposures to sulfide, a toxic substance to many plants and animals.

8.4. Plant community

There are currently no quantitative studies of vegetation in tidal freshwater swamps of North Carolina. The assumption that tidal swamp vegetation is the same as nearby bottomland hardwood communities is not warranted because soils in the tidal systems rarely are aerobic. A list of herbaceous plants and woody shrubs and trees (Table 8.5) was developed

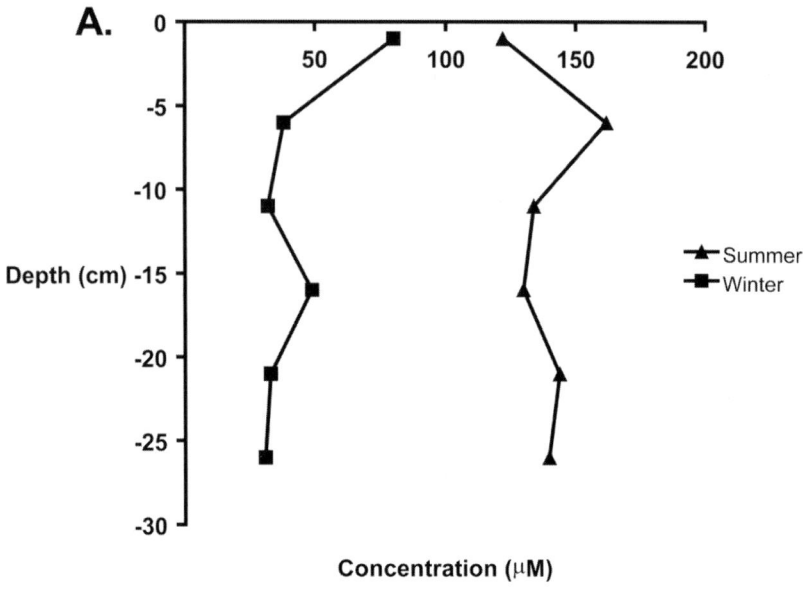

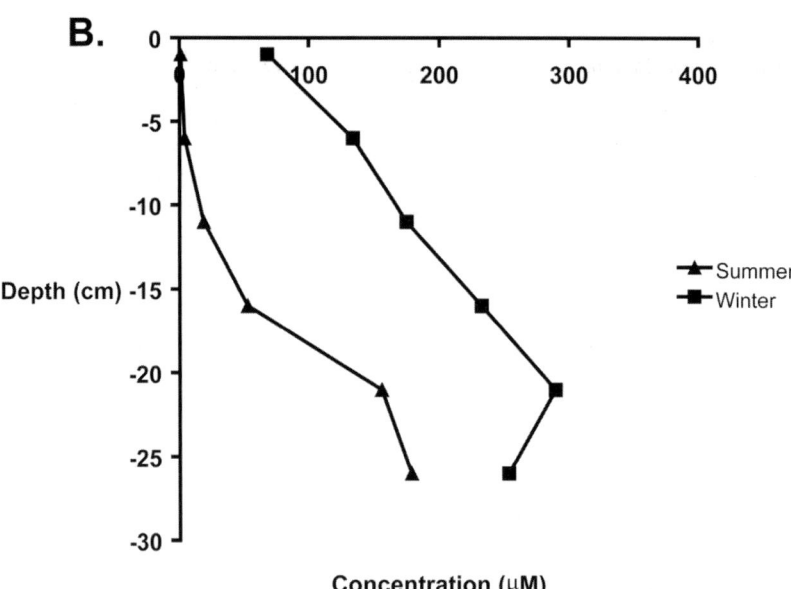

Fig. 8.11. Typical summer and winter depth profiles of sulfate (a) and methane (b) concentrations for swamp forest at station P13.

Table 8.5. A list of vascular plant species from tidal freshwater swamps and marshes in the upper estuary-lower Cape Fear River basin, North Carolina. Species considered sensitive herbaceous species are marked with an asterisk (*).

Scientific Name	Common Name
Acer rubrum L.	red maple
Alternanthera philoxeroides (Mart.) Griseb.*	alligatorweed
Amaranthus cannabinus (L.) Sauer	tidalmarsh amaranth
Apios americana Medik.	groundnut
Aster spp. (probably *Symphyotrichum* spp.*)	
Bidens laevis (L.) B.S.P.	smooth beggarticks
Boehmeria cylindrica (L.) Sw.*	small-spike false nettle
Boltonia asteroides (L.) L'Hér.*	white doll's-daisy
Carex L.*	sedge
Carex crinita Lam.*	fringed sedge
Carex crinita var. *brevicrinis* Fern.*	fringed sedge
Carex hyalinolepis Steud.*	shoreline sedge
Carex lupulina Muhl. Ex Willd.*	hop sedge
Chasmanthium latifolium (Michx.) Yates	broadleaf uniola
Cicuta maculata L.*	common water hemlock
Cinna arundinacea L.*	stout woodreed
Commelina virginica L.*	Virginia dayflower
Decodon verticillatus (L.) Ell.*	swamp loosestrife
Dulichium arundinaceum (L.) Britt.*	threeway sedge
Eryngium aquaticum L.*	rattlesnakemaster
Galium L.*	bedstraw
Hymenocallis floridana (Raf.) Morton*	Florida spiderlily
Impatiens capensis Meerb.	jewelweed
Lilaeopsis chinensis (L.) Kuntze	eastern grasswort
Ludwigia uruguayensis (Camb.) Hara	water primrose
Ludwigia palustris (L.) Ell.*	marsh primrose-willow
Lycopus virginicus L.*	Virginia water horehound
Mikania scandens (L.) Willd.	climbing hempvine
Nyssa aquatica L.	water tupelo
Orontium aquaticum L.*	goldenclub
Osmunda regalis L.	royal fern
Peltandra virginica (L.) Schott*	green arrow arum
Phanopyrum gymnocarpon (Ell.) Nash*	Savannah-panicgrass
Pluchea odorata (L.) Cass.	sweetscent
Polygonum arifolium L.*	halberdleaf tearthumb
Polygonum hydropiper L.*	mild water-pepper
Polygonum punctatum Ell.*	dotted smartweed
Polygonum virginianum L.	jumpseed
Pontederia cordata L.*	pickerelweed
Rhynchospora corniculata (Lam.) Gray*	shortbristle horned beaksedge
Rhynchospora inundata (Oakes) Fern.*	narrowfruit horned beaksedge
Rosa palustris Marsh.	swamp rose
Rumex verticillatus L.*	swamp dock
Sagittaria lancifolia L.*	bulltongue arrowhead
Saururus cernuus L.*	lizard's tail

Table 8.5. Continued

Scientific Name	Common Name
Scutellaria lateriflora L.*	mad dog skullcap
Schoenoplectus americanus (Pers.) Volk. Ex Schinz & R. Keller*	American bullrush
Sium suave Walt.*	common waterparsnip
Spartina cynosuroides (L.) Roth	big cordgrass
Symphyotrichum elliottii (Torr. & Gray) Nesom*	Elliott's aster
Symphyotrichum subulatum (Michx.) Nesom	eastern annual saltmarsh aster
Taxodium ascendens Brongn.	pondcypress
Taxodium distichum (L.) L.C. Rich.	baldcypress
Triadenum walteri (J.G. Gmel.) Gleason*	greater marsh St. Johnswort
Typha latifolia L.*	broadleaf cattail
Zizania aquatica L.*	annual wildrice
Zizaniopsis miliacea (Michx.) Doell & Aschers.*	giant cutgrass

early in our study and includes only plants growing in wet soils. There were many other species growing on hummocks, logs, etc. which elevate roots and stems above the soil surface. The presence of species such as sweetgum (*Liquidambar styraciflua* L.) reflects an abundance of these elevated areas within a tidal swamp. In the study area, tidal swamp forests that have not been previously cleared frequently contain an abundance of hummocks formed on old stumps. Tidal freshwater swamps that had been previously converted into rice fields, however, remain relatively flat and lack hummocks. In general, dominants, both woody and herbaceous plants, within tidal wetlands reflect tolerance to salinity levels of flooding water. We report here the types of plant communities observed in the wetlands of the lower CFR watershed and discuss the impact of salinity variability on these communities.

8.4.1. Sensitive vegetation

Woody vegetation is able to persist for many years after saline water and increased inundation levels have altered more sensitive, shallow-rooted herbaceous plant species. In the upper CFR estuary, baldcypress has been shown to survive, albeit with lower growth (based on tree rings), even after soil salinity had increased above 5 ppt (Harmon 1993; Yanosky et al. 1995). Thus, changes in woody species are not useful in determining the effects of increased salt water exposure or inundation frequency and/or duration.

Herbaceous plant species, especially those with shallow roots, were deemed to be better indicators of short-term changes in hydrology and water chemistry for the purpose of our study. Herbaceous plants that were not tolerant of saline water (i.e., sensitive species) were identified at seven of the 12 stations at the initiation of the study in 1999. No sensitive plant species were found at stations P6 or P11; sites with chronic exposure to saline floodwater (Table 8.4). At remaining stations, the most abundant sensitive, vascular plant species was selected and the extent of one stand of that species was delineated at the initiation of the study. Changes in the aerial extent of coverage by that species were measured every August from 1999 to 2005. A GPS fix (±1 m) was established along the periphery of each resultant hexagon and overlaid on a detailed aerial photograph each year. Inter-annual changes in aerial extent of the polygon and percent cover of species within the polygon were evaluated.

Table 8.6 illustrates the type of data and variations observed for a single defined area of sensitive vegetation. Pickerelweed (*Pontederia cordata* L.) was the sensitive species identified at P12, a site where the plant community is clearly converting from swamp to marsh. However, other species also occurred at this station and their relative contribution to cover noted. Some of the species in these polygons are less sensitive to saline water than others and the observed inter-annual variation in species cover may be a consequence of these differences in salt tolerance (Table 8.6). Most of these species, like pickerelweed, are also perennial and eventually regenerate from rhizomes that survive brief incursions of saline water. For example, after the increased salinities associated with the 2000-2001 drought, a decline in cover of pickerelweed coincided with an increase in cover of the annual, annual wildrice (*Zizania aquatica* L.), another species commonly observed in tidal wetlands in the southeastern United States. With the subsequent return to more typical salinities, however, pickerelweed has rebounded aggressively (Table 8.6).

The aerial extent of sensitive vegetation changes from year to year. These changes, however, are not necessarily related to increased salinity (Table 8.7). In late spring 2004, for example, the upstream stations in the CFR and northeast CFR (P9 and P13) were affected by extremely high water levels resulting from increased river discharge and flooding. Following these flood conditions, the areal extent of salt-sensitive species present at P9 decreased by almost one-half (Table 8.7). This reduction in extent was not associated with an appreciable change in salinity at the sites, but instead appears to reflect a reduction in growth because the vegetation was covered by deep water at a critical time during the growing season. Nonetheless, there are examples of saline water causing changes in sensitive vegetation. In 2000, salinity at P13 (Fishing Creek) was always 0.5 ppt or

Table 8.6. Comparisons of percent cover contributions by sensitive herbaceous species in the polygon for years 2000, 2001, 2002, 2003, and 2004 at the Rat Island (P12) station.

			Year		
Herbaceous species	2000	2001	2002	2003	2004
pickerelweed	20	40	5	30	30
Elliot's aster	<1	--	--	--	--
dotted smartweed	2	1	--	<1	10
common waterparsnip	<1	2	5	1	1
halberdleaf tearthumb	1	3	--	10	15
giant cutgrass	2	<1	<1	5	5
lizard's tail	2	2	--	1	5
common water hemlock	<1	2	--	--	1
bulltongue arrowhead	2	20	5	20	5
goldenclub	<1	--	--	--	--
green arrow arum	<1	1	5	30	12
shortbristle horned beaksedge	<1	<1	--	--	<1
sedge sp.	<1	--	--	--	--
alligatorweed	--	5	<1	<1	--
annual wildrice	--	2	<1	50	<1
white doll's-daisy	--	1	--	--	<1
swamp dock	--	<1	2	1	--
stout woodreed	--	<1	--	<1	<1
rattlesnakemaster	--	<1	5	2	2
American bulrush	--	--	<1	--	--
shoreline sedge	--	--	--	1	--
groundnut	--	--	--	<1 --	<1
Florida spiderlily	--	--	--	2	--
marsh primrose-willow	--	--	--	<1	<1
tidalmarsh amaranth	--	--	--	--	<1
dwarf St. Johnswort	--	--	--	--	<1
small-spike false nettle	--	--	--	--	<1

less. The drought of 2001-2002, however, resulted in exposure to higher salinities during inundation and the extent of salt sensitive, herbaceous, vascular plants declined dramatically. These changes were not permanent, however, and recovery ensued once the drought ended. By 2004, sensitive herbaceous vegetation increased dramatically, adding more than 100 m^2 over the original area in 2000 and quadrupling the previous year's cover (Table 8.7).

Table 8.7. Comparisons of areas (m^2) of sensitive herbaceous vegetation polygons for years 2000, 2001, 2002, 2003, and 2004 at sensitive herbaceous vegetation monitoring stations. The polygon at P3 has split and was considered as three separate units in 2004.

Station name	2000	2001	Year 2002	2003	2004
Inner Town Creek A (P3)	65.90	164.84	121.92	123.32	128.23
Inner Town Creek B Outlier (P3)	--	--	--	--	8.50
Inner Town Creek C Outlier (P3)	--	--	--	--	4.45
Indian Creek (P7)	12.06	12.06	26.21[a]	26.21	26.22
Dollisons Landing (P8)	37.62	37.62	26.61[a]	26.60	26.60
Black River (P9)	40.08	104.16	84.91	52.80	6.45
Rat Island (P12)	49.56	49.56	49.56	50.12	49.60
Fishing Creek (P13)	141.56	153.08	90.39	63.43	243.06
Prince George Creek (P14)	365.58	341.24	482.68	589.65	486.11

[a] Changes in area were caused by an artifact of shift to winter GPS data collection because of excessive interference of GPS signal by the tree canopy in summer.

8.4.2. Bryophytes

Identification of bryophyte species (Tables 8.8 and 8.9) along the Northeast CFR (P11-P14) was undertaken only for those species associated with soils or that occurred near the swamp surface (e.g., on stumps, trees, etc.) (Carroll 2003). A total of 44 genera consisting of 38 moss species (Table 8.8) and 21 liverwort species (Table 8.9) were identified. There was one liverwort recorded, *Cololejeunea setiloba* Evans, that was not previously known from North Carolina. Carroll (2003) examined bryophyte densities and species richness and compared them to water level relative to the marsh surface, salinity, and elevation of the marsh surface at three stations. Each station contained six substations that spanned this distance from river to upland edge. In general, both bryophyte density and species richness increased with distance from the river as water level and salinity decreased.

A Principal Component Analysis using 15 environmental variables, including transect distance upriver, substation distance from river's edge to base of upland, hydrology, elevation, duration of flooding, and salinity, identified a significant correlation between certain species and their tolerance for specific stress-related environmental variables. Bryophyte species clustered and represent groups of species with similar degrees of tolerance or lack of tolerance to both flooding and saline water.

Table 8.8. List of moss species collected on the Northeast Cape Fear River and adjacent swamps and marshes (Carroll 2003).

Scientific name	Common name
Amblystegium Schimp. in B.S.G.	amblystegium moss
Amblystegium serpens (Hedw.) Schimp. in B.S.P.	amblystegium moss
Amblystegium varium (Hedw.) Lindb.	amblystegium moss
Anomodon attenuatus (Hedw.) Hüb.	anomodon moss
Atrichum crispum (James) Sull.	atrichum moss
Brachythecium acuminatum (Hedw.) Aust.	acuminate brachythecium moss
Bryoandersonia illecebra (Hedw.) Robins.	bryoandersonia moss
Bryohaplocladium microphyllum (Hedw.) Wat. & Iwats.	bryohaplocladium moss
Bryum Hedw.	bryum moss
Clasmatodon parvulus (Hampe) Hook. & Wils. ex Sull. in Gray	clasmatodon moss
Climacium americanum Brid.	American climacium moss
Climacium kindbergii (Ren.&Card.) Grout	Kindberg's climacium moss
Entodon seductrix (Hedw.) C. Müll.	seductive entodon moss
Eurhynchium hians (Hedw.) Sande Lac.	eurhynchium moss
Eurhynchium pulchellum (Hedw.) Jenn.	eurhynchium moss
Fissidens fontanus (B.-Pyl.) Steud.	fissidens moss
Fontinalis sullivantii Lindb.	Sullivant's fontinalis moss
Hypnum lindbergii Mitt.	Lindberg's hypnum moss
Isopterygium tenerum (Sw.) Mitt.	isopterygium moss
Leucobyrum albidum (Brid.) Lindb.	
Leucodon bracypus Brid.	leucodon moss
Leucodon julaceus (Hedw.) Sull.	leucodon moss
Leptodictyum humile (P. Beauv.) Ochyra.	eptodictyum moss
Leptodictyum riparium (Hedw.) Warnst.	streamside leptodictyum moss
Lindbergia brachyptera (Mitt.) Kindb.	lindbergia moss
Plagiomnium cuspidatum (Hedw.) T. Kop.	toothed plagiomnium moss
Racomitrium Brid.	racomitrium moss
Schlotheimia rugifolia (Hook.) Schwaegr.	ruggedleaf schlotheimia moss
Sematophyllum adnatum (Michx.) Britt.	sematophyllum moss
Sphagnum affine Ren. & Card.	sphagnum
Sphagnum palustre L.	prairie sphagnum
Sphagnum recurvum P. Beauv.	recurved sphagnum
Steerecleus serrulatus (Hedw.) Robins.	steerecleus moss
Syrrhopodon texanus Sull.	Texan syrrhopodon moss
Taxiphyllum deplanatum (Bruch & Schimp. ex Sull.) Fleisch.	taxiphyllum moss
Taxiphyllum taxirameum (Mitt.) Fleisch.	taxiphyllum moss
Thuidium allenii Aust.	Allen's thuidium moss
Thuidium delicatulum (Hedw.) Schimp. in B.S.G.	delicate thuidium moss

Table 8.9. List of liverwort species collected on the Northeast Cape Fear River and its adjacent swamps and marshes (Carroll 2003).

Scientific name	Common name
Bazzania trilobata (L.) S. Gray	threelobed bazzania
Calypogeia muelleriana (Schiffn.) K. Muell.	
Cephalozia lunulifolia (Dum.) Dum.	cephalozia
Chiloscyphus minor (Nees) Eng. & Schust.	
Chiloscyphus polyanthos (L.) Corda	
Chiloscyphus profundus (Nees)	
Cololejeunea minutissima (Sm.) Schiffn.	
Cololejeunea setiloba Evans	
Frullania kunzei Lehm. & Lindenb.	
Jamesoniella autumnalis (Decandolle) Steph.	
Lejeunea flava (Sw.) Nees	
Lejeunea laetevirens Nees et Mont.	
Leucolejeunea clypeata (Schwein.) Evans	
Leucolejeunea conchifolia Evans	
Metzgeria Raddi sp.	
Odontoschisma prostratum (Sw.) Trev.	
Pallavicinia lyellii (Hook.) Carruth.	
Porella pinnata L.	
Radula complanata L. Dum.	
Riccardia multifida (L.) Gray	
Telaranea nematodes (Gott. Ex Aust.) M. A. Howe	

8.4.3 Plant community summary

The woody vascular plants observed in the tidal swamps of the lower CFR watershed are a subset of vascular plants found within the same region in bottomland hardwood/softwood swamps. Both vascular plants and non-vascular bryophytes along the salinity gradient are limited by incursions of saline water onto the soil surface. Periodic exposure to saline water during events such as drought can exert a temporary affect on plant community structure, but the plant community usually recovers post-drought. Only when the frequency of saline water inundation becomes chronic do tidal freshwater swamps convert to brackish marsh.

8.5. Infauna and nekton

Salt marshes are known to serve a variety of ecosystem functions and are generally well-studied compared to their low salinity counterparts. Low salinity tidal forested wetlands likely also serve as reservoirs for abundant infaunal populations (Rozas and Hackney 1984) similar to higher salinity wetlands (Posey et al. 1997; Alphin and Posey 2000; Posey et al. 2003). In contrast to the recognized importance of saline tidal marshes as infaunal and nekton habitat (Boesch and Turner 1984; Kneib 1984; Baltz et al. 1993; Whaley and Minello 2002), little is known about the faunal communities inhabiting tidal freshwater forests. Tidal freshwater forested wetlands are often dominated by baldcypress along the edge which differ from herbaceous marshes in their belowground structure and tend to contain unconsolidated organic sediments along with woody roots and stems instead of the consolidated sediment and the tight grass root matrix of salt marshes. Tidal freshwater forested wetlands are also often characterized by steep banks transitioning between the forest and adjacent channels – a feature that may or may not occur in oligohaline marsh areas. The following sections describe the dominant infauna and nekton inhabiting tidal freshwater forested wetlands in the lower CFR basin and general faunal patterns are compared to more saline habitats closer to the CFR estuary.

8.5.1. Infauna

Dominance patterns among the three habitat wetland types indicate that salinity, rather than vegetation type, has the strongest influence on infaunal abundance. Freshwater tidal swamp and mixed marsh/swamp were both dominated primarily by oligochaetes, especially Tubificidae (including *Tubificoides*) and Lumbriculidae (Table 8.10). The primary subdominants at both sites were insects, including *Polypedilium*, *Collembola*, and *Bezzia*/*Polpomyia*. The only differences in common infauna between these habitats were juvenile bivalves and the polychaetes *Boccardiella* and *Marenzellaria*. Juvenile bivalves included several facultative estuarine and freshwater species. Their greater abundance within the swamp habitat may reflect the negative impacts of dense marsh root matrices in the mixed marsh/swamp areas on clam burrowing. Both *Boccardiella* and *Marenzellaria* are estuarine polychaetes that do not occur in purely fresh water areas and their dominance at the mixed sites probably reflects the more frequent intrusion of saline water at these sites (Table 8.10). Tidal swamp and mixed swamp/marsh stations had similar overall diversity and species richness. In contrast, the mesohaline marsh habitat (P2) was dominated by a variety of estuarine polychaetes (8 species) and amphipods (4 species), with much lower abundances of oligochaetes. Insects were periodically common, but these represented short-term occurrences of only a few species that peaked immediately after freshets lowered salinity to near zero

Table 8.10. Mean abundance per core (±1 SE) of numerically dominant infauna (at least 1% of total for at least one habitat). Bold entries indicate taxa > 1% of total for specific habitat. Diversity and species richness are based on annual totals. I = insect; C = crustacean, P = polychaete, O = oligochaete.

Species	Tidal Swamp	Mixed Marsh/ Swamp	Mesohaline Marsh
Bivalve sp.	**1.540(0.356)**	0.168(0.035)	0.321(0.081)
Bezzia-Palpomyia gr. [I]	0.330(0.055)	**0.664(0.137)**	**0.625(0.245)**
Boccardiella sp. [P]	0.014(0.008)	**0.441(0.256)**	**4.480(1.730)**
Collembola sp. [I]	**1.300(0.349)**	0.277(0.056)	0.018(0.018)
Corophium acherasicum (C)	0.092(0.045)	0.031(0.021)	**0.893(0.696)**
Corophium acutum [C]	0.004(0.005)	0(0)	**0.553(0.553)**
Corophium lacustre [C]	0.032(0.209)	0.005(0.005)	**2.530(1.140)**
Corophium sp. [C]	0.005(0.005)	0.023(0.016)	**1.050(0.647)**
Dicrotendipes lobus [I]	0.005(0.005)	0.018(0.018)	**1.570(1.340)**
Dicrotendipes sp. [I]	0.055(0.463)	0.068(0.030)	**0.982(0.522)**
Hobsonia florida [P]	0.216(0.064)	0.245(0.102)	**1.890(0.580)**
Lumbriculidae sp. [O]	**1.260(0.506)**	**0.595(0.217)**	0(0)
Marenzellaria viridis [P]	0.018(0.011)	**2.210(0.574)**	0.214(0.179)
Mediomastus ambiseta [P]	0.005(0.005)	0(0)	**0.482(0.334)**
Mediomastus sp. [P]	0.005(0.005)	0.005(0.005)	**0.750(0.345)**
Other Oligochaetes [O]	**22.400(2.960)**	**10.600(1.650)**	**5.410(1.940)**
Parandalia sp.A [P]	0(0)	0.023(0.010)	**1.440(0.339)**
Polydora cornuta [P]	0.009(0.006)	0.009(0.006)	**1.780(1.190)**
Polydora socialis [P]	0.009(0.009)	0.068(0.059)	**0.839(0.516)**
Polypedilium sp. [I]	**0.679(0.115)**	**0.741(0.119)**	**1.160(0.859)**
Streblospio benedicti [P]	0.155(0.055)	0.091(0.040)	**8.000(2.740)**
Hargeria rapax [C]	0.005(0.005)	0(0)	**1.250(0.828)**
Tubificoides heterochaetus [O]	0.087(0.032)	**1.340(0.527)**	0.393(0.166)
Fabriciola spp. [P]	0(0)	0(0)	**2.640(2.640)**
Tubificidae spp. [O]	**12.500(2.290)**	**8.080(1.330)**	**4.140(2.500)**
Balanus improvius [C]	0(0)	0(0)	**0.500(0.431)**
Major Taxonomic Groups:			
Insect	4.870(0.526)	3.850(0.369)	6.020(2.050)
Oligochaete	36.200(3.140)	20.900(2.020)	9.950(3.010)
Polychaete	1.200(0.288)	4.150(0.701)	23.800(4.550)
Amphipod	1.240(0.212)	0.527(0.132)	5.620(1.880)
Decapod	0.160(0.030)	0.145(0.029)	0.178(0.081)
Isopod	0.573(0.098)	0.686(0.179)	0.661(0.398)
Bivalve	0.537(0.356)	0.168(0.035)	0.357(0.082)
Gastropod	0.096(0.029)	0.068(0.019)	0.018(0.018)
Diversity	0.391(0.015)	0.394(0.016)	0.498(0.031)
Species Richness	5.080(0.190)	4.800(0.200)	6.050(0.500)

(e.g., after large amounts of rainfall associated with hurricane events). The presence of a variety of taxa with overall similar abundance, rather than dominance by a few taxa, at the mesohaline marsh station (P2) contributed to a higher diversity. Higher species richness reflects the diversity of polychaetes along with the occasional occurrence of insects and oligochaetes. Overall, oligochaete abundance decreased across habitats with increasing salinity (swamp>mixed>marsh), while polychaetes showed the reverse (Table 8.10).

8.5.2. Epifauna

Species composition for spring epifauna and nekton sampling indicated a general pattern similar to the infauna: the relative salinity to which the site was exposed seemed more important than vegetation types in determining nekton similarity among sites. Spring Breder trap sampling indicated tidal swamp and mixed swamp/marsh stations shared 10 dominant taxa and differed among only 4 taxa, 3 of which were *Uca* species (fiddler crabs) that are not efficiently sampled by Breder traps (Table 8.11). During spring drop trap sampling, the swamp and mixed sites shared 69% of common taxa (Table 8.12). However, the mesohaline marsh site (P2) shared only 36% of common taxa in Breder trap sampling and 42% of common taxa in drop trap sampling with the other two habitat types. Much of the observed difference between the mesohaline site and the other sites involved greater abundance of true estuarine species at the mesohaline marsh site (e.g., blue crabs [*Callinectes*, mostly juveniles in these traps], spot [*Leiostomus*], and grass shrimp [*Palaemonetes*]), and the lack of certain freshwater taxa (e.g., *Gambusia*). Diversity was highest in the tidal swamp habitat, but did not differ appreciably among the other 2 habitat types for spring sampling (Tables 8.11 and 8.12).

In contrast, Fall Breder trap sampling produced a somewhat different pattern, with a gradation occurring among the swamp, mixed, and mesohaline marsh habitat areas. In Fall Breder trap samples, swamp and mixed habitats shared 40% of common taxa; mixed and mesohaline marsh shared 50% of common taxa, while the tidal swamp and mesohaline marsh only shared 23% of common taxa (Table 8.11). A similar pattern occurred for drop traps, with swamp and mixed sites sharing 23% of common taxa (the low number reflecting the extreme abundances of fish larvae at the swamp site, causing few other taxa to approach 1% of total fauna collected), mixed and mesohaline marsh sites sharing 41% of taxa, and swamp and mesohaline marsh sharing only 8% of common taxa (Table 8.12). Diversity also showed a gradient in Breder trap samples, with highest levels in

Table 8.11. Mean annual abundance per trap (±1 SE) of common nekton (>1% total catch) collected in Breder traps. Diversity and species richness are from mean annual catches. Bold indicates taxa > 1% of total for specific habitat.

	Habitat					
	Tidal Swamp		Mixed Marsh/Swamp		Mesohaline Marsh	
Species	Fall	Spring	Fall	Spring	Fall	Spring
Callinectes sapidus	0.004 (0.002)	0 (0)	**0.014** (**0.004**)	**0.029** (**0.006**)	**0.044** (**0.017**)	**0.094** (**0.027**)
Ctenogobius shufeldti	**0.090** (**0.159**)	**0.096** (**0.016**)	**0.157** (**0.240**)	**0.239** (**0.053**)	**0.033** (**0.013**)	**0.117** (**0.032**)
Fundulus heteroclitus	**0.037** (**0.009**)	**0.183** (**0.030**)	0.013 (0.007)	**0.107** (**0.022**)	0.006 (0.006)	**0.128** (**0.031**)
Gambusia affinis	0.003 (0.003)	**0.025** (**0.007**)	0.001 (0.001)	0 (0)	0.006 (0.006)	0 (0)
Gambusia holbrooki	**0.138** (**0.025**)	**0.254** (**0.044**)	0.001 (0.001)	**0.060** (**0.033**)	0 (0)	0 (0)
Lagodon rhomboides	0 (0)	**0.119** (**0.032**)	0 (0)	**0.226** (**0.044**)	0 (0)	0.028 (0.012)
Leiostomus xanthurus	0.001 (0.001)	**0.233** (**0.058**)	0.001 (0.001)	**3.54** (**0.420**)	0 (0)	**2.05** (**0.301**)
Lepomis macrochirus	**0.031** (**0.011**)	0.003 (0.002)	0.007 (0.003)	0.003 (0.002)	0 (0)	0 (0)
Micropogonias undulatus	0.001 (0.001)	0 (0)	0.001 (0.001)	**0.078** (**0.023**)	0.006 (0.006)	0.006 (0.006)
Palaemonetes pugio	**0.197** (**0.029**)	**0.074** (**0.016**)	**0.266** (**0.065**)	**0.157** (**0.048**)	**0.244** (**0.046**)	**1.490** (**0.144**)
Paralichthys dentatus	0.001 (0.001)	**0.032** (**0.013**)	0.001 (0.001)	**0.132** (**0.263**)	0 (0)	0.011 (0.008)
Paralichthys sp.	0 (0)	**0.042** (**0.013**)	0 (0)	**0.639** (**0.013**)	0 (0)	0.022 (0.011)
Penaeus aztecus	**0.047** (**0.013**)	0.001 (0.001)	**0.116** (**0.017**)	0 (0)	**0.789** (**0.160**)	0.006 (0.006)
Symphurus plagiusa	0 (0)	0 (0)	0.003 (0.002)	0 (0)	**0.022** (**0.011**)	0 (0)
unidentified larval fish	0 (0)	0 (0)	0.010 (0.006)	0 (0)	0 (0)	**0.067** (**0.036**)
Uca minax	**0.083** (**0.021**)	**0.100** (**0.018**)	0.011 (0.005)	0.026 (0.015)	0 (0)	0.006 (0.006)
U. pugilator	0 (0)	**0.081** (**0.019**)	0 (0)	0.001 (0.001)	**0.083** (**0.068**)	0 (0)
U. pugnax	**0.058** (**0.011**)	**0.214** (**0.030**)	**0.164** (**0.067**)	0.015 (0.006)	0.006 (0.006)	0 (0)
Uca sp.	0.001 (0.001)	0 (0)	**2.820** (**1.980**)	0.035 (0.012)	**0.050** (**0.045**)	0 (0)
Fundulus diaphanus	**0.059** (**0.022**)	0 (0)	0 (0)	0 (0)	0 (0)	0 (0)
Litopenaeus setiferus	0 (0)	0 (0)	0.010 (0.004)	0 (0)	**0.028** (**0.012**)	0 (0)
Fundulus sp.	0 (0)	**0.020** (**0.008**)	0 (0)	**0.071** (**0.023**)	0 (0)	0.039 (0.016)
Diversity	0.631 (0.036)	0.618 (0.076)	0.439 (1.000)	0.520 (0.082)	0.362 (0.104)	0.450 (0.050)
Species Richness	8.17 (1.17)	8.00 (1.00)	8.83 (0.48)	10.20 (0.98)	4.17 (1.01)	6.50 (0.67)

Table 8.12. Mean annual abundance per drop (±1 SE) of common nekton collected in drop traps. Bold indicates taxa >1% of total for specific habitat.

	Habitat					
	Tidal Swamp		Mixed Marsh/Swamp		Mesohaline Marsh	
Species	Fall	Spring	Fall	Spring	Fall	Spring
Anchoa mitchilli	**0.921(0.626)**	0.003(0.003)	**0.532(0.273)**	0.086(0.047)	**1.250(0.880)**	0.349(0.321)
Brevoortia tyranus	0(0)	**0.210(0.137)**	0(0)	0.002(0.002)	0.028(0.028)	**3.578(3.442)**
Callinectes sapidus	0.059(0.015)	0.140(0.025)	**0.143(0.023)**	**0.234(0.035)**	**0.528(0.110)**	0.349(0.081)
Clupeidae	0(0)	**4.503(2.491)**	0.007(0.007)	**1.837(0.711)**	0(0)	0.037(0.037)
Ctenogobius shufeldti	0.586(0.085)	**0.620(0.124)**	**0.333(0.046)**	**0.352(0.051)**	**0.148(0.043)**	0.064(0.023)
Gambusia holbrooki	0.528(0.133)	**0.228(0.102)**	0.012(0.008)	0(0)	0.009(0.009)	0(0)
Lagodon rhomboides	0(0)	**0.541(0.345)**	0.002(0.002)	**0.090(0.646)**	0.009(0.009)	**0.110(0.046)**
Leiostomus xanthurus	0.035(0.016)	**5.248(1.178)**	0.025(0.012)	**9.469(1.516)**	0(0)	**13.633(6.904)**
Menidia beryllina	0.220(0.130)	**0.725(0.371)**	**0.199(0.047)**	**0.956(0.455)**	**0.102(0.056)**	**1.367(0.592)**
Menidia menidia	0.229(0.139)	0(0)	**0.069(0.060)**	0.014(0.012)	**0.065(0.042)**	**0.752(0.724)**
Minidia sp.	0.003(0.003)	**0.386(0.182)**	0(0)	0.130(0.053)	0(0)	0.147(0.064)
Micropogonias undulatus	0.053(0.021)	**2.655(1.256)**	0(0)	**0.581(0.231)**	**0.083(0.047)**	**4.192(0.955)**
Palaemonetes pugio	0.237(0.077)	**1.140(0.394)**	**1.120(0.419)**	**1.199(0.217)**	**1.657(0.337)**	**13.642(3.492)**
Panopeus herbstii	0.017(0.009)	0.006(0.004)	0.004(0.003)	0(0)	0.092(0.053)	0.009(0.009)
Paralichthys dentatus	0.009(0.005)	**0.301(0.050)**	0.009(0.005)	**1.440(0.273)**	**0.018(0.013)**	0.055(0.025)
Paralichthys sp.	0.009(0.005)	**0.275(0.051)**	0.012(0.008)	**2.324(0.350)**	0(0)	0.293(0.108)
Penaeus aztecus	0.331(0.122)	0.073(0.062)	**0.530(0.195)**	0.011(0.008)	**0.879(0.265)**	**0.706(0.415)**
Penaeus setiferus	0(0)	0(0)	**0.079(0.039)**	0(0)	0.009(0.009)	0(0)
Rhithropanopeus harrisii	0.064(0.029)	0.015(0.008)	0.023(0.010)	0.025(0.009)	**0.065(0.065)**	0.046(0.024)
Sesarma cinereum	0.021(0.008)	0(0)	**0.039(0.017)**	0(0)	0(0)	0(0)
Symphurus plagiusa	0(0)	0.003(0.003)	0.012(0.006)	0(0)	**0.185(0.082)**	0(0)
Trinectes maculatus	0.487(0.122)	0.082(0.022)	**0.178(0.036)**	0.037(0.011)	0.009(0.009)	0(0)
Unidentified larval fish	**61.421 (61.340)**	0.006(0.004)	**0.134(0.075)**	0.004(0.003)	0.009(0.009)	0(0)
Uca pugnax	0.387(0.093)	0.023(0.018)	**0.056(0.025)**	0.002(0.002)	0.018(0.013)	0(0)
Diversity	0.639(0.133)	0.708(0.055)	0.742(0.046)	0.537(0.031)	0.691(0.038)	0.485(0.045)
Species Richness	15.167(1.869)	13.667(1.430)	13.833(0.601)	13.333(1.874)	11.167(0.910)	9.167(0.872)

the tidal swamp, intermediate in the mixed habitat, and lowest in the mesohaline marsh (Table 8.11). However, diversity patterns were more similar among sites for fall drop trap samples (Table 8.12).

8.5.3. Summary of infaunal and epifaunal data

The tidal swamp sites and the mixed swamp/marsh areas sampled in this study shared most common species when averaged over several years of sampling and had similar patterns of diversity and species richness. The few differences occurring among sites could be explained by small differences in salinity (except possibly for juvenile bivalves and certain fish in fall sampling periods). This suggests a broad similarity in vegetation use by both infauna and epifauna despite dramatic differences in vegetation structure and may reflect the opportunistic nature of most species dominating upper estuarine regions. The similarity in these habitats contrasted strongly with the mesohaline marsh sampled, which was dominated by different infauna and epifauna.

8.6. Chapter synthesis

There is a paucity of literature describing the physical, chemical and biological attributes of the tidal freshwater swamps in southeastern North Carolina, and in particular, those systems located in the lower Cape Fear River watershed. Tidal freshwater swamps along the lower CFR clearly differ from more saline habitats within the estuary and from the upstream swamps that are subject to seasonal inundation patterns as opposed to tidal flooding. While the tidal swamps described here flood at about the same frequency as the marshes in the lower estuary, the tidal range is much less and the range is highly suppressed during periods of high river discharge. Because discharge in this river system is highly variable, the frequency and duration of wetland inundation is also highly variable. Nonetheless, the soils of these swamps are almost constantly saturated because of the influence of tides.

While plant species inhabiting tidal freshwater swamps in southeastern North Carolina are poorly documented, the plant communities do appear to be regulated by flooding depth and salinity. The tidal freshwater swamps described here are limited in their downstream extent to near the estuary head due to chronic incursions of saline water. However, these systems appear to be able to resist conversion to tidal marsh if flooding by saline water is infrequent. Periodic pulses of floodwater containing low levels of salt can, however, dramatically affect soil biogeochemistry. Tidal swamp soils typically generate methane, but produce hydrogen sulfide as the dominant microbially-mediated byproduct when saline water is introduced into soil pore space. This phenomenon was observed during periods of regional drought within the watershed. It is likely that both saline water and hydrogen sulfide in the soils following prolonged or chronic exposure to saline water will result in a change of vegetation within tidal freshwater swamps. The data reported here do not conclusively document such change, however, shifts in wetland vegetation from woody to herbaceous have been reported for wetlands where salinity has increased due to anthropogenic modifications to the river. Sea-level rise and anthropogenic activities (e.g., dredging or freshwater removal upstream), will likely lead to the conversion of tidal freshwater swamps to tidal marsh. The data do indicate, however, that even short term pulses of salinity (such as those induced during a drought) can result in temporary plant community changes – such as the shift from perennial to annual species. These changes, however, are short-lived and are typically followed by rapid post-drought recovery.

Little is known with respect to the abundance and diversity of infauna and epifauna in the tidal swamps of the CFR. While structural differences related to woody versus herbaceous vegetation might be expected to have a significant impact in infauna and epifauna, sites dominated by woody vegetation did not differ strongly from those with mixed vegetation of primarily herbaceous plants along the channel edge. Salinity variations are suspected as having a stronger direct impact on these fauna, with possible indirect relations occurring with vegetation type. The data also indicate that tidal swamps provide similar numbers of infauna to epifauna that move onto the swamp surface with each high tide to feed, suggesting that they serve as a trophic link between primary production in swamps and the adjacent aquatic community.

References

Alphin TD, Posey MH (2000) Long-term trends in vegetation dominance and infaunal community composition in created marshes. Wetl Ecol Manage 8:317-325

Avery GB Jr, Kieber RJ, Willey JD, Shank GC, Whitehead RF (2004) The impact of hurricanes on the flux of rainwater and Cape Fear River water DOC to Long Bay, southeastern United States. Global Biogeochem Cycles 18(3): doi: 10.1029/2004GB002229

Baltz DM, Rakocinski C, Fleeger JW (1993) Microhabitat use by marsh-edge fishes in a Louisiana estuary. Environ Biol Fishes 36:109-126

Boesch DF, Turner RE (1984) Dependence of fishery species on salt marshes: the role of food and refuge. Estuaries 7:460-468

Breder CM (1960) Design for a fry trap. Zoologia 45:155-160

Butler RJ, Secor DT Jr (1991) The central Piedmont. In: Horton WJ, Zullo VA (eds) Geology of the Carolinas. Carolina Geological Society Fifteenth Anniversary Volume, University of Tennessee Press, Knoxville, pp 59-78

Carroll DM (2003) Bryophytes as indicators of water level and salinity change along the Northeast Cape Fear River. M.S. thesis, University of North Carolina-Wilmington, Wilmington

Clifton JM (1973) Golden grains of white: rice planting on the lower Cape Fear River. NC Historical Rev 50:356-393

CZR, Inc (1998) A monitoring plan to determine potential effects of increased tidal range on the Cape Fear River ecosystem due to deepening Wilmington harbor, North Carolina. Unpublished Report prepared for the U.S. Army Corps of Engineers, Wilmington District, Wilmington

Froelich PN, Klinkhammer GP, Bender ML, Luedtke NA, Heath GR, Cullen D, Dauphin P, Hammond D, Hartman B, Maynard V (1979) Early oxidation of organic matter in pelagic sediments of the eastern equatorial Atlantic: suboxic diagenesis. Geochim Cosmochim Acta 43:1075-1090

Hackney CT, Yelverton GF (1990) Effects of human activities and sea level rise on wetland ecosystems in the Cape Fear River Estuary, North Carolina, USA. In: Whigham DF, Good RF, Kvet Y (eds) Wetland ecology and management. Kluwer Academic Publishers, Dordrecht, The Netherlands, pp 55-61

Harmon RG (1993) Sodium concentrations in *Taxodium distichum* along an estuarine salinity gradient. M.S. thesis, University of North Carolina-Wilmington

Hubbard RK, Sheridan JM, Marti LR (1990) Dissolved and suspended solids transport from Coastal Plain watersheds. J Environ Qual 19:413-420

Hupp CR (2000) Hydrology, geomorphology and vegetation of coastal plain rivers in southeastern USA. Hydrol Proc 14:2991-3010

Isphording WC, Fitzpatrick JF Jr (1992) Geologic and evolutionary history of drainage systems of the southeastern United States. In: Hackney CT, Adams SM, Martin WH (eds) Biodiversity of the southeastern United States: Aquatic communities. John Wiley and Sons, New York, pp 19-56

Kneib RT (1984) Patterns of utilization of intertidal salt marsh by larvae and juveniles of *Fundulus heteroclitus* (Linaeus) and *Fundulus luciae* (Baird). J Experimental Mar Biol Ecol 83:41-51

Martens CS, Berner RA (1974) Methane production in the interstitial waters of sulfate depleted marine sediments. Science 185:1167-1169

NCDENR (2000) Cape Fear River basinwide water quality plan. North Carolina Department of Environment and Natural Resources, Division of Water Quality, Water Quality Section, Raleigh

Olsen PE, Froelich AJ, Daniels DL, Smoot JP, Gore PJW (1991) Rift basins of early Mesozoic age. In: Horton WJ, Zullo VA (eds) Geology of the Carolinas. Carolina Geological Society Fifteenth Anniversary Volume. University of Tennessee Press, Knoxville, pp 142-170

Phillips JD (1997) Human agency, Holocene sea level, and floodplain accretion in coastal plain rivers. J Coast Res 13:854-866

Posey MH, Alphin TD, Powell CM (1997) Plant and infaunal communities associated with a created marsh. Estuaries 20:42-47

Posey MH, Alphin TD, Meyer DL, Johnson JM (2003) Benthic communities of common reed *Phragmites australis* and marsh cordgrass *Spartina alterniflora* marshes in Chesapeake Bay. Mar Ecol Prog Ser 261:51-61

Renfro AA (2004) Sediment deposition and accumulation in tidal riparian wetlands. M.S. thesis, University of North Carolina-Wilmington

Rozas LP, Hackney CT (1984) Use of oligohaline marshes by fishes and macrofaunal crustaceans in North Carolina. Estuaries 7:213-224

Sexton SG (2002) Rates of carbon remineralization in coastal wetland sediments under sulfate reducing and methanogenic conditions: implications for sea-level rise. M.S. thesis, University of North Carolina-Wilmington

Simmons CE (1993) Sediment characteristics of North Carolina streams, 1970-79. Water-Supply Paper 2364. U.S. Geological Survey, Washington

Soller DR, Mills HH (1991) Surficial geology and geomorphology. In: Horton WJ, Zullo VA (eds) Geology of the Carolinas. Carolina Geological Society Fifteenth Anniversary Volume. University of Tennessee Press, Knoxville, pp 290-308

Steila D (2000) Soils. In: Orr DM, Stuart AW (eds) The North Carolina Atlas. University of North Carolina Press, Chapel Hill, pp 39-46

Stoddard EF, Farrar SS, Horton Jr JW, Butler JR, Druhan RM (1991) The eastern Piedmont in North Carolina. In: Horton WJ, Zullo VA (eds)

Geology of the Carolinas. Carolina Geological Society Fifteenth Anniversary Volume. University of Tennessee Press, Knoxville, pp 79-92

Street MW, Deaton AS, Chappel WS, Mooreside PD (2005) North Carolina coastal habitat protection plan. North Carolina Department of Environment and Natural Resources, Division Marine Fisheries, Morehead City

USACE (1996) Final feasibility report and environmental impact statement, improvement of navigation, Cape Fear-Northeast Cape Fear Rivers comprehensive study. Prepared by US Army Corps of Engineers, Wilmington District

Whaley SD, Minello TJ (2002) The distribution of benthic infauna of a Texas salt marsh in relation to the marsh edge. Wetlands 22:753-766

Wicks EC (2002) Organic content and water holding capacity of wetland soils of the Northeast Cape Fear River and Cape Fear River. Honors Thesis, University of North Carolina-Wilmington

Wolfe RS, Higgins IJ (1979) Microbial biochemistry of methane-a study in contrast. Int Rev Biochem 21:267-353

Yanosky TM, Hupp CR, Hackney CT (1995) Chloride concentrations in growth rings of *Taxodium distichum* in a saltwater intruded estuary. Ecol Appl 5:785-792

Chapter 9 - Ecology of Tidal Freshwater Forests in Coastal Deltaic Louisiana and Northeastern South Carolina*

William H. Conner[1], Ken W. Krauss[2], and Thomas W. Doyle[2]

[1]*Baruch Institute of Coastal Ecology and Forest Science, Clemson University, Box 596, Georgetown, SC 29442*
[2]*U.S. Geological Survey, National Wetlands Research Center, 700 Cajundome Blvd., Lafayette, LA 70506*

9.1. Introduction

Tidal freshwater swamps in the southeastern United States are subjected to tidal hydroperiods ranging in amplitude from microtidal (<0.1 m) to mesotidal (2-4 m), both having different susceptibilities to anthropogenic change. Small alterations in flood patterns, for example, can switch historically microtidal swamps to permanently flooded forests, scrub-shrub stands, marsh, or open water but are less likely to convert mesotidal swamps. Changes to hydrological patterns tend to be more noticeable in Louisiana than do those in South Carolina.

The majority of Louisiana's coastal wetland forests are found in the Mississippi River deltaic plain region. Coastal wetland forests in the deltaic plain have been shaped by the sediments, water, and energy of the Mississippi River and its major distributaries. Baldcypress (*Taxodium distichum* [L.] L.C. Rich.) and water tupelo (*Nyssa aquatica* L.) are the primary tree species in the coastal swamp forests of Louisiana. Sites where these species grow usually hold water for most of the year; however, some of the more seaward sites were historically microtidal, especially where baldcypress currently dominates. In many other locations, baldcypress and water tupelo typically grow in more or less pure stands or as mixtures of

---

* The U.S. Government's right to retain a non-exclusive, royalty-free licence in and to any copyright is acknowledged.

the two with common associates such as black willow (*Salix nigra* Marsh.), red maple (*Acer rubrum* L.), water locust (*Gleditsia aquatica* Marsh.), overcup oak (*Quercus lyrata* Walt.), water hickory (*Carya aquatica* [Michx. f.] Nutt.), green ash (*Fraxinus pennsylvanica* Marsh.), pumpkin ash (*F. profunda* Bush.), and redbay (*Persea borbonia* [L.] Sprengel) (Brown and Montz 1986).

The South Carolina coastal plain occupies about two-thirds of the state and rises gently to 150 m from the Atlantic Ocean up to the Piedmont plateau. Many rivers can be found in the Coastal Plain with swamps near the coast that extend inland along the rivers. Strongly tidal freshwater forests occur along the lower reaches of redwater rivers (Santee, Great Pee Dee, and Savannah) that arise in the mountains and along the numerous blackwater rivers (Ashepoo, Combahee, Cooper, and Waccamaw) that arise in the coastal regions. Most of the tidal freshwater forests were converted to tidal rice fields in the 1700s (Porcher 1995). Canopy members of the present day forests include baldcypress, water tupelo, swamp tupelo (*N. biflora* Walt.), red maple, and Carolina ash (*Fraxinus caroliniana* Miller). Subcanopy and shrub species include Virginia sweetspire (*Itea virginica* L.), dwarf palmetto (*Sabal minor* (Jacquin) Pers.), coastal plain willow (*Salix caroliniana* Michx.), redbay, and water-elm (*Planera aquatica* Gmel.).

9.2. Historical land use

A succession of Native American cultures occupied Louisiana beginning as long as 12,000 years ago (Neuman and Hawkins 1993). Many were local societies sustained by hunting and gathering or subsistence agriculture. During the Mississippian Cultural Period (1200 to 500 years before present), maize (corn) was widely introduced, and Native Americans developed a dependence upon agriculture for their basic food supply (King et al. 2005). Agriculture also gave a greater importance to floodplain areas of moderate and large streams because regular flooding brought a renewal of soil productivity (Kniffen 1968). If a floodplain site became less productive, the site was abandoned for newer, more productive areas that were cleared by girdling the trees and the use of fire (Morse and Morse 1983).

The Louisiana coastal region was possibly visited by Cabeza de Vaca and his fellow survivors of a shipwreck while conducting a Spanish expedition in 1528, and it was certainly seen by some of De Soto's men (1541–42) (Kniffen 1968). In 1682, La Salle reached the mouth of the Mississippi and claimed for France all of the land drained by the river and its tributaries, naming it Louisiana after Louis XIV. Europeans did not permanently settle in the area until around 1700 (Kniffen 1968).

Native Americans were also the first occupants of the coastal region of South Carolina, but war, disease, and slavery took their tolls and many of the tribes had completely disappeared by 1755 (Rogers 1970; Edgar 1998). In 1521, Spanish explorer Lucas Vásquez Ayllón explored the South Carolina coastal area, and returned in 1526 with the intent of establishing a permanent Spanish settlement. San Miguel de Gualdape was established near present day Georgetown and became the first European colony in the United States, although no definitive evidence for this colony has been found to prove this (Rogers 1970). The colony failed due to its swampy location, hostile Native Americans, lack of food, and mutiny. Of the original 600 settlers, only 150 made it safely back to Spain (Edgar 1998).

The first English settlers arrived nearly a century later, but they were successful in establishing a permanent colony at Charleston in 1670 (Rogers 1970), which probably benefited from the great number of Native Americans killed by the deadly diseases introduced by the Spanish explorers (Edgar 1998). Since the rivers provided the main means of transportation, it was important that each plantation or farm fronted on the river. The rivers and bays gave the coastal region a unity making possible the development of a distinct society, which until 1910, had the rice plantation as its center. It was not until the twentieth century that the waterways lost their importance as means of travel to paved roads and bridges (Rogers 1970).

9.2.1. Geologic development

The Mississippi River delta is composed of sediments derived from a river system that drains all or parts of 31 states and 6,086 km^2 before it finally reaches the Gulf of Mexico (Environmental Protection Agency 2006). Since sea level reached its current maximum approximately 7,000 years ago, six natural deltas have been formed and abandoned (Coleman et al. 1998). During the building phase of the delta cycle, the system is dominated by freshwater riverine inputs with the formation of corresponding freshwater marshes and swamps, which then deteriorate during the marine-dominated phase (Roberts 1997). These channel switching events have had a greater impact on sediment dispersal than have waves or currents (National Research Council 2006). The largest areas of Louisiana's coastal wetland forests are swamps in the deteriorating phase of the deltaic plain (Chambers et al. 2005). Currently, much of the sediment of the Mississippi River is diverted by a vast levee system to the Gulf of Mexico.

South Carolina's coastal region developed from continental submergence and emergence with both erosion and deposition of soils (Colquhoun 1974). Soils originated either from the Appalachian Mountains or from coastal processes (McKnight et al. 1981) and are composed of waterborne deposits of sands, silt, or clay and calcareous sediments (Dahl 1999). Unlike the coastal wetland forests of Louisiana, South Carolina wetlands experience strong bidirectional water flows and sediment fluxes with both flowing and ebb tides.

9.2.2. Land use changes

Significant anthropogenic changes to the Mississippi River delta began with European settlement. The Native Americans living in the swamps of Louisiana showed the first European settlers where to settle and how to exploit the natural resources of the wetland forests beginning in the 1700s (Abbey 1979). These early settlers hunted, fished, trapped, and cut baldcypress timber to make a living. Until the 1790s, baldcypress boards and timbers represented the main cash crop of the colonists in the state and remained a stable commodity of the lumber industry into the 1800s because of the wood's durability and workability (Mattoon 1915). Baldcypress lumbering thrived in Louisiana, with the period of highest production occurring between 1890–1925 (Figure 9.1). Baldcypress timber production peaked in 1913 with over 700 million board feet being processed in 94 mills (Mattoon 1915). Depletion of the vast virgin stands of baldcypress timber and the Great Depression caused most of the baldcypress mills to close (Burns 1980), but logging continued in the swamplands of Louisiana to some extent, and the last baldcypress logging operation closed in 1956 (Mancil 1972).

Many logging companies in Louisiana maintained their own dredges to prevent delays in digging access canals (Davis 1975). Canals from 3-12 m wide and 2-3 m deep resulted in partial drainage of many swamps (Mancil 1969, 1980). With the use of pullboat barges, trees could be pulled in from as far as 1,000 m from the canal through runs spaced about 45 m apart in a fan-shaped pattern. Runs were cleared of all trees and stumps, and logs were pulled to the canal. This skidding of timber across the swamp floor damaged and destroyed much young growth, and continual use of a run resulted in a mud-and-water-filled ditch 2.5 m deep for the length of the run (Mancil 1980). This operation left a distinctive wagon wheel-shaped pattern in the swamp forest that can still be seen today in Louisiana swamps. In both states, thousands of km of railway lines were constructed to gain access into swamp areas (Mancil 1969; Fetters 1990).

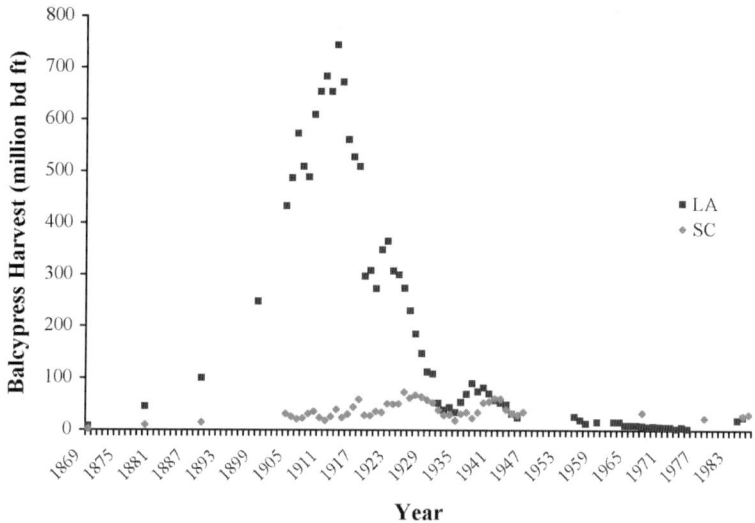

Fig. 9.1. Baldcypress harvest in Louisiana and South Carolina.

During the 19th and 20th centuries, major changes in hydrological patterns occurred as artificial levees were constructed along the Mississippi River. The Mississippi River's natural floodplain has been reduced about 90% in area by levee construction. About 3,620 km of levees exist along both sides of the river within Louisiana (US Army Corps of Engineers 2004), and these levees have reduced the area of seasonally flooded wetlands along the river. Dikes and revetments used to entrain the channel prevent the river from creating new habitats. Another major alteration to Louisiana's coastal wetlands has been the construction of canals and pipelines and their associated spoil banks. Ten major navigation canals and 14,973 km of pipelines exist in the coastal zone (US Army Corps of Engineers 2004). These channels allow saltwater to intrude into freshwater wetland areas (Day et al. 2000) or to permanently impound the swamps (Conner et al. 1981).

Conner and Toliver (1990) showed that baldcypress-dominated ecosystems of coastal Louisiana have experienced widespread hydrological, biogeochemical, and biological changes over the past century from the above-described factors, and declines in forested wetland areas have been apparent. Little is known about the present state of coastal baldcypress ecosystems, and this paucity of information is mainly the result of physical inaccessibility and lack of active forest management following the period of intense logging in the early 20th century. Several factors have raised recent concern regarding the long-term viability of Louisiana's coastal wetland forests, including the increasing number of scientific reports of baldcypress-water tupelo forest death and decline (DeLaune et al. 1987; Conner and Day 1988; Conner and Brody 1989; Pezeshki et al. 1990; Allen 1992; Conner 1993; Krauss et al. 2000) and the recent interest in the harvest of baldcypress in Louisiana for garden mulch (Chambers et al. 2006).

Baldcypress trees in South Carolina were also an important source of timber (Dahl 1999), and swamp ownership considered akin to a gold mine as a source of considerable wealth (Edgar 1998). However, timber production in South Carolina never reached the scale it did in Louisiana (Figure 9.1). By the 1850s there were 50 sawmills operating near the headwaters of the Savannah and Edisto Rivers alone and the largest mill east of the Mississippi River was located in Georgetown, South Carolina. The deep swamps of the Pee Dee and Santee Rivers contained some of the finest timber in the country (Fetters 1990) but were considered inaccessible for timber harvesting until after 1900. Expansion of the railway system into the swamps and the development of steam engines to pull logs from the swamp greatly expanded the loggers' reach. By the 1950s, the majority of the baldcypress stands in the state had been greatly reduced (Dahl 1999).

The greatest impact to coastal tidal forests in South Carolina was the introduction of rice culture. Rice was grown in South Carolina as early as the 1670s (Doar 1936), but the period of greatest expansion occurred in the 1730s. Rice culture first developed in the upper reaches of the rivers. Upper sections of the swamps were diked to form reserve ponds for water storage. The remainder of the swamps was cleared, leveled, and planted. Fields were flooded by releasing water from rain-fed reserve ponds and drained by releasing water downstream of the fields (Gresham and Hook 1982). It was not until the end of the eighteenth century that the broad flat tidal plains were cleared of forests and converted to rice fields (Rogers 1970). Since these rice fields were flooded and drained by controlling the tide, rice planting was confined to stretches of river that were still fresh (Gresham and Hook 1982) which in some cases could extend 48–56 km upstream (Edgar 1998). Rice cultivation grew until 1850 when production peaked at 72,608 metric tons. During the period 1850-1860 there were 227 known plantations in South Carolina with 29,168 ha under cultivation (Doar 1936). Interestingly, a survey of former rice fields conducted in 1974 identified 30,016 ha of former rice fields just within the Ashley-Cooper-Wando, North and South Santee, and Waccamaw-Pee Dee-Black river basins (Gresham and Hook 1982). Of that acreage, 32% was still being managed by landowners (for wildlife usually), 42% was covered with marsh grasses (varying from fresh to saline marsh depending on proximity to the ocean), and only 26% had reverted back to baldcypress-water tupelo forest.

9.2.3. Major sources of water

The Mississippi River is the longest and largest river in North America, flowing 3,705 km from its source in Minnesota, through the midcontinental United States, the Gulf of Mexico coastal plain, and its subtropical Louisiana delta. Louisiana's deltaic plain is dominated by a complex network of distributary channels and natural levees that radiate outward from the Mississippi River main-stem near Baton Rouge and extend southward into the Gulf of Mexico (Frazier 1967; Penland and Boyd 1985).

The coastal plain in South Carolina is nearly flat, and river swamps are a main landscape feature. Winding rivers loop across this flat region of sand, clay, and soft limestone. The northeastern part of South Carolina contains three major water basins. The Waccamaw River/Atlantic Intracoastal Waterway (AIWW) Basin encompasses 11 watersheds and 253,560 ha (20.1% is forested wetland) within the Lower Coastal Plain and Coastal Zone regions (SCDHEC 2000). There are approximately 1,261 stream km in this basin. The Waccamaw River flows across the South Carolina state line from North Carolina and accepts drainage from Kingston Lake and the AIWW before joining the Sampit and Pee Dee Rivers to form Winyah Bay, which drains into the Atlantic Ocean.

The Pee Dee River basin (also referred to as the Great Pee Dee River) encompasses 27 watersheds and 887,071 ha (16.5% is forested wetland) within South Carolina. The Pee Dee River flows across the Sandhills region to the upper and lower coastal plain regions and into the coastal zone region. In the Pee Dee River basin, there are approximately 5,586 stream km. The Pee Dee River receives drainage from Thompson Creek, Crooked Creek, Cedar Creek, Three Creeks, Black Creek, Jeffries Creek, Catfish Creek, Lynches River Basin, Little Pee Dee River, and Black River basin before draining into Winyah Bay (SCDHEC 2000).

The Black River basin encompasses 18 watersheds and 526,500 ha (12.9% is forested wetland) extending from the sandhill region to the upper and lower coastal plains and into the coastal zone. There are approximately 3,532 stream km. The Black River receives drainage from Rocky Bluff Swamp, Pocotaligo River, Pudding Swamp, Kingstree Swamp Canal, and Black Mingo Creek before merging with the Pee Dee River (SCDHEC 2000).

9.3. Climatic drivers and hydrological characterization

9.3.1. Climate

Annual precipitation in coastal Louisiana averages 150 cm, with October tending to be the driest month and July the wettest. Torrential rains are common at any time of the year. High pressure systems from the north and west bring cool, dry air and are easily recognized as cold fronts in the winter although they occur throughout the year (Wax et al. 1978). These fronts are typically followed by atmospheric conditions that bring warm air in from the coast, usually with heavy clouds and rain. Two-thirds of the coastal rainfall is associated with this type of frontal activity, while 10-15% of the annual rainfall is from infrequent tropical storms and hurricanes (Gosselink et al. 1998). Mean monthly temperatures range from a December-January low of about 14°C to a midsummer high of about 30°C. Frost is infrequent with about 300 frost-free days along the coast.

The coastal region of South Carolina is classified as having a humid, subtropical climate (Edgar 1998). The average annual precipitation along the coast is between 122 and 127 cm. Annual rainfall is variable, ranging from a low of 86.6 cm in 1954 to a high of 155.8 cm in 1959. Normally, precipitation is less than potential evapotranspiration from April to September resulting in a slight deficit or dry period during the growing season. The average January temperature is 10°C near the coast, and the average July temperature is 27°C. There are about 290 growing days near the coast. During the winter months, South Carolina is under a continental air mass that is generally cold and dry. Summers are hot and humid, with the moisture for the humidity coming from a Bermuda high-pressure cell in the Atlantic. The clockwise circulation around the Bermuda High causes a prevailing northerly flow of warm, humid air from the Gulf of Mexico.

Hurricanes are common occurrences on the coasts of the southeastern United States, although they are even more common on the Gulf of Mexico Coast (Conner et al. 1989). Over the past 100 years, Louisiana's coastline around Grand Isle has had the highest incidence of major hurricane hits (category 3-5 hurricanes) of any part of the Gulf of Mexico Coast excluding South Florida (Doyle and Girod 1997). The most recent hurricanes to affect the study region were Hurricanes Katrina and Rita in 2005. South Carolina experiences approximately one hurricane every seven years (Dukes 1984), with the last major hurricane affecting the study area in 1989 (Hurricane Hugo).

9.3.2. Hydrology

Tidal amplitude along the south Atlantic coast is more than double that of the northern Gulf of Mexico coast, where South Carolina tides range from 1-2 m, and Louisiana tides never exceed 0.5 m. Except for swamp forest locations alongside larger estuarine lakes, many Louisiana study sites are currently nontidal because of various water control alterations. Most are still subject to periodic salinity pulses from tropical storms. Hence, while South Carolina sites remain strongly tidal, those under comparable salinity regimes in Louisiana are semi-impounded (Figure 9.2).

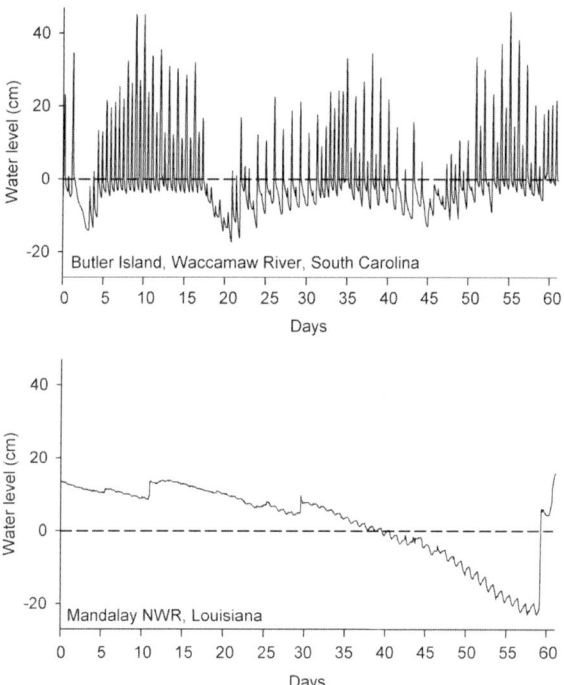

Fig. 9.2. Water level records for coastal wetland forest sites on Butler Island, Waccamaw River, South Carolina (upper) and in Mandalay National Wildlife Refuge, Louisiana (lower) from April 1 to May 31, 2005. Both sites have similar salinity regimes (0.4-0.9 ppt) but represent tidal and nontidal hydroperiods at similar landscape positions. Pulses of water level rise in Mandalay represent rainfall additions. Even though the Mandalay site is semi-impounded, major storm surges and rainfall events can serve as important vectors of hydrologic change. Ground level is represented by the dashed line.

The cumulative effects of eustatic (actual) sea-level rise, crustal sinking, tectonic activity, and sediment consolidation result in high rates of subsidence that dominate the surface elevation and geomorphology of the deltaic plain of Louisiana (Saucier 1994; DeLaune et al. 2004). Subsidence rates for large areas of the deltaic plain range from 30 to 100 cm per century. Relative (eustatic plus subsidence) sea-level rise in the deltaic plain is predicted to range from 50 to 100 cm over the next 100 years (Twilley et al. 2001). South Carolina's coastal region is a relatively stable area built on a 100,000-year-old barrier sand formation. Sea-level rise in this area is much lower than that in the Louisiana deltaic plain region, ranging from 1-2 mm yr^{-1} for South Carolina (Stevenson et al. 1986; Daniels 1992).

9.4. Geospatial description

Study sites in Louisiana are located along two landscape transects in either the Terrebonne/Atchafalaya or Barataria hydrologic basins (Figure 9.3). Along the Terrebonne/Atchafalaya transect, three sites that range from permanently flooded with fresh water to periodically exposed with low salinity (0.9 ± 0.2 ppt) are under study. Sites along the Barataria transect are either permanently flooded with fresh water or microtidal with low (1.3 ± 0.2 ppt) to moderately (2.1 ± 0.1 ppt) saline water. All sites are in the historic Mississippi River floodplain but have been subjected to extensive and permanent alterations to the magnitude and delivery of hydrologic flows over the past century.

Study sites in northeastern South Carolina are organized along two principal landscape transects associated either directly with the Waccamaw River or with a tidal inlet of Winyah Bay at the mouth of this same river at Hobcaw Barony (Figure 9.4). The Waccamaw transect has three swamp sites ranging from tidal, fresh water to tidal with moderately saline water (3.1 ppt). Four sites are located along the Hobcaw transect and range from a permanently flooded depressional swamp to a weakly tidal swamp harboring low salinity levels (0.9 ppt).

9.5. Ecological characterization

9.5.1. Seedling level response

The integrating element of all research on coastal wetland forests has been the potential impacts of sea-level rise. The most direct effects of sea-

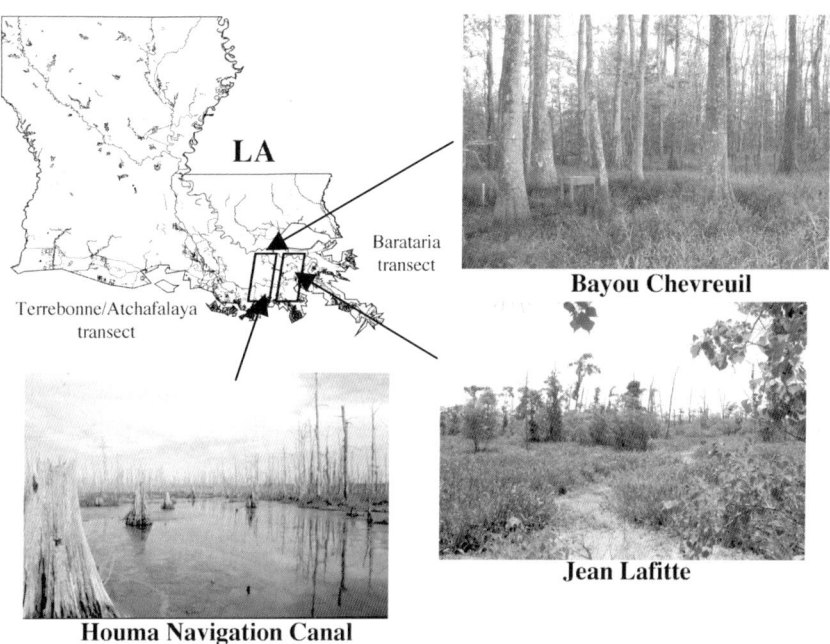

Fig. 9.3. Louisiana map showing location of transects and typical scenes of healthy, stressed, and dead trees (Bayou Chevreuil and Jean Lafittte photos by W. Conner and Houma Navigation Canal photo by J. Allen).

level rise are believed to be increases in flooding of coastal wetlands and increases in salinity levels of soil and surface water. Most of the research has been aimed at elucidating the effects of these stressors on some of the major tree species found in baldcypress-water tupelo swamps. Another factor believed to be important is increased intensity and/or frequency of hurricanes and other large storms (Emmanuel 1987); thus, the potential impacts of storm surges (i.e., large pulses of salt water) have also been investigated. Seedling responses have been studied in species ncluding baldcypress, water tupelo, swamp tupelo, green ash, Chinese tallow tree (*Triadica sebifera* [L.] Small), buttonbush (*Cephalanthus occidentalis* L.), nuttall oak (*Quercus texana* Buckl.), swamp chestnut oak (*Q. michauxii*), overcup oak, water oak (*Q. nigra* L.), loblolly pine (*Pinus taeda* L.), red maple, and redbay.

Experiments with one-year-old and younger seedlings in Louisiana and South Carolina have clearly demonstrated that substantial differences in tolerance to flooding (water 5 cm above surface) and salinity exist among

Fig. 9.4. Location of Waccamaw transect in South Carolina and typical views of healthy, stressed, and dying baldcypress (photos by W. Conner).

species. Oak species, for example, were found to be highly sensitive to flooding with even low level (2 ppt) salinity (Conner et al. 1998). Swamp chestnut oak and water oak showed signs of stress (e.g., leaf necrosis or loss, tip dieback) when flooded with 2 ppt water for only one week. After 11 weeks, both species had 100% mortality. Nuttall and overcup oaks did not show signs of stress until weeks 7 and 9, respectively, but all of these seedlings died by the end of the 5-month experiment. Seedlings of all the oak species died within 7 weeks of flooding with 6 ppt water (Conner et al. 1998). Pezeshki et al. (1990) also found overcup oak to be less flood tolerant than baldcypress. In contrast, baldcypress, water tupelo, and Chinese tallow tree had high survival rates and good growth when flooded with low salinity water (Figure 9.5). When flooded with 2 ppt water, all three species had 100% survival, height growth not significantly different than that for seedlings flooded with fresh water, and only slightly reduced diameter growth (Conner 1994, 1995; McLeod et al. 1996; Conner et al. 1997). Allen et al. (1994) reported small but insignificant reductions in total biomass of baldcypress at 2 ppt.

Measurement of physiological activity of these species demonstrated several response patterns based on their initial and long-term reactions to freshwater flooding in laboratory experiments. Baldcypress and button-

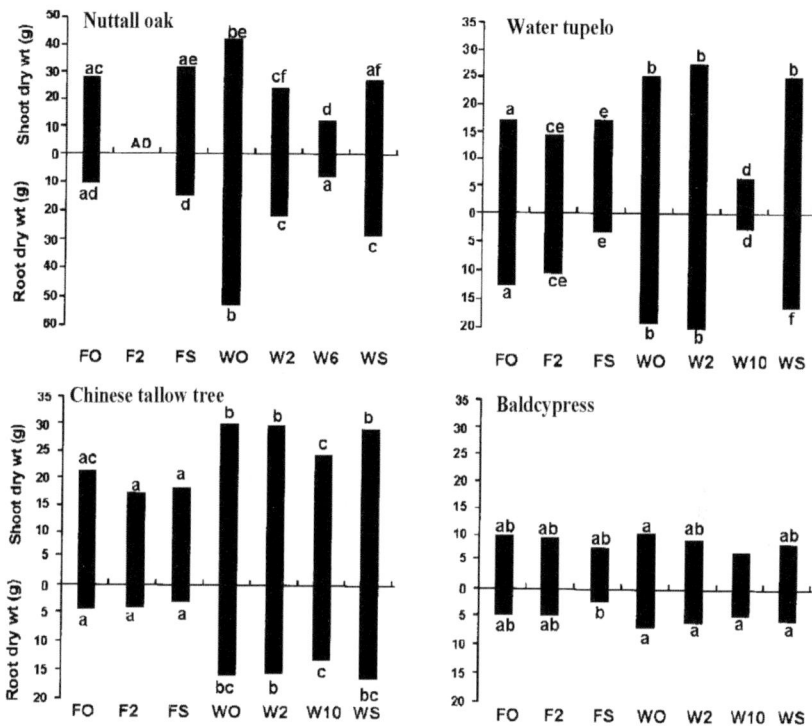

Fig. 9.5. Effects of permanent flooding and salinity on shoot and root biomass of Nuttall oak, baldcypress, water tupelo, and Chinese tallow tree. F0 = flooded with 0 ppt water, F2 = flooded with 2 ppt water, FS = flooded with 0 ppt and surged with 32 ppt water, W0 = watered with 0 ppt water, W2 = watered with 2 ppt water, W10 = watered with 10 ppt water, WS = watered with 0 ppt water and surged with 32 ppt water, AD = all dead, no measurements taken. Unlike letters indicate statistical differences at 5% level for treatments by seedlings (after Allen et al. 1998).

bush showed neither an initial reduction in photosynthesis nor any long-term impact (McLeod et al. 1996; McCarron et al. 1998). Water tupelo and green ash showed a significant initial reduction in photosynthesis but no cumulative effect over time (McLeod et al. 1996). Both of these groups would survive long-term exposure to freshwater flooding. Photosynthesis of overcup oak, nuttall oak, water oak, and Chinese tallow tree was not affected during the first two weeks of flooding, but there was an increasing impact with time beyond their initial exposure (McLeod et al. 1999). Swamp tupelo and swamp chestnut oak photosynthesis was reduced with flooding, and the impact increased with time (McCarron et al. 1998;

McLeod et al. 1999). Since photosynthesis of the latter two groups decreased over time, these species are not adapted to flooding and would eventually perish if flooding continued. This is in contrast to other studieswhere photosynthesis, stomatal conductance, and survival of swamp tupelo was unaffected by flooding and waterlogging (Hook et al. 1970, 1971).

As salinity was increased to 2 ppt as might occur when sea level rises and floods coastal areas, the relative tolerances among the species did not differ greatly, but photosynthesis was reduced in all species. Baldcypress showed no significant reduction in photosynthesis initially, but after four weeks there was a significant reduction (McLeod et al. 1996). Photosynthesis of buttonbush was similar to that of water tupelo and green ash with significant initial reduction and no cumulative effect over time (McLeod et al. 1996; McCarron et al. 1998). The response of the four oak species, swamp tupelo, and Chinese tallow tree to brackish flooding indicated a greater reduction in photosynthesis and ultimate mortality with continued exposure (McLeod et al. 1996, 1999; McCarron et al. 1998).

Exposure to high salinity water (32 ppt) for only two days as is typical of hurricane storm surges resulted in severe reductions in photosynthesis for all species. Buttonbush was the least impacted species but was only slightly more resistant than swamp tupelo, baldcypress, water tupelo, Chinese tallow tree, or green ash, all of which survived for several weeks after the surge (McLeod et al. 1996; McCarron et al. 1998). Photosynthesis of all of the oak species tested was reduced significantly (McLeod et al. 1999), and all died before the beginning of the next growing season (Conner et al. 1998).

In general, these experiments suggest that species already well adapted to long flood durations handle the additional stress of low-level salinity better than do species somewhat less tolerant to flooding. Species such as buttonbush, baldcypress, and water tupelo were among those least affected by flooding with 2 ppt water (Conner 1994; McCarron et al. 1998). This pattern was less clear when trees were flooded with higher salinity water or when the extremely rapid increase in salinity associated with some hurricanes was simulated. Chinese tallow tree, for example, was found to be more tolerant of these treatments than were baldcypress, water tupelo, red maple, and redbay (Conner and Askew 1993; Conner et al. 1997). Based on these findings, a general classification of tolerance to salinity was developed (Table 9.1). These data provide quantitative evidence to support existing rankings that are often based on limited or anecdotal evidence (Allen et al. 1998).

Table 9.1. Flood and salinity tolerance rankings for tree species (after Allen et al. 1998).

Species	Flood tolerance[1]	Salinity Tolerance	
		Low-level+flooding[2]	Storm surge[3]
Baldcypress	Most tolerant	Tolerant	Moderately tolerant
Water tupelo	Most tolerant	Weakly tolerant	Moderately tolerant
Buttonbush	Most tolerant	Weakly tolerant	Moderately tolerant
Swamp tupelo	Most tolerant	Intolerant	Moderately tolerant
Chinese tallow tree	Tolerant	Intolerant	Moderately tolerant
Red maple	Tolerant	Intolerant	Intolerant
Redbay	Not determined	Intolerant	Intolerant
Overcup oak	Tolerant	Intolerant	Intolerant
Loblolly pine	Moderately tolerant	Not determined	Intolerant
Green ash	Moderately tolerant	Weakly tolerant	Moderately tolerant
Nuttall oak	Moderately tolerant	Intolerant	Intolerant
Water oak	Weakly-mod. tolerant	Intolerant	Intolerant
Swamp chestnut oak	Weakly tolerant	Intolerant	Intolerant

[1] Based on rankings by McKnight et al. (1981) and Hook (1984), except for Chinese tallow tree which is from Allen et al. (1998).
[2] Based on responses of seedlings to flooding with 2 ppt water, such as may occur during early stages of saltwater intrusion.
[3] Based on responses of seedlings to simulated storm surge treatments during flooding, such as may occur as a result of hurricanes.

These experiments substantially improve our ability to rank species' tolerances and better predict the impacts of saltwater intrusion. In addition, they provide new insight into how responses to events such as hurricanes may vary depending on site conditions just prior to an event. For example, the effects of a storm surge may be less severe in cases where the soil is already saturated or it may be more severe when the soil is initially well drained. In experiments involving the responses of baldcypress seedlings of differing ages to simulated storm surges, trees showed stress much more quickly and showed less indication of recovery when the soil was well drained prior to surging (Conner, unpublished data). Salinity levels in saturated soils do not become as elevated initially, and they drop to tolerable levels faster following a surge than they do in well-drained soils (Figure 9.6). In areas where storm surge waters become trapped, as in a small baldcypress-swamp tupelo watershed along the Hobcaw transect, high concentrations of salinity can persist for long periods. Williams (1993) found high salinity levels 30 months after the Hurricane Hugo storm surge.

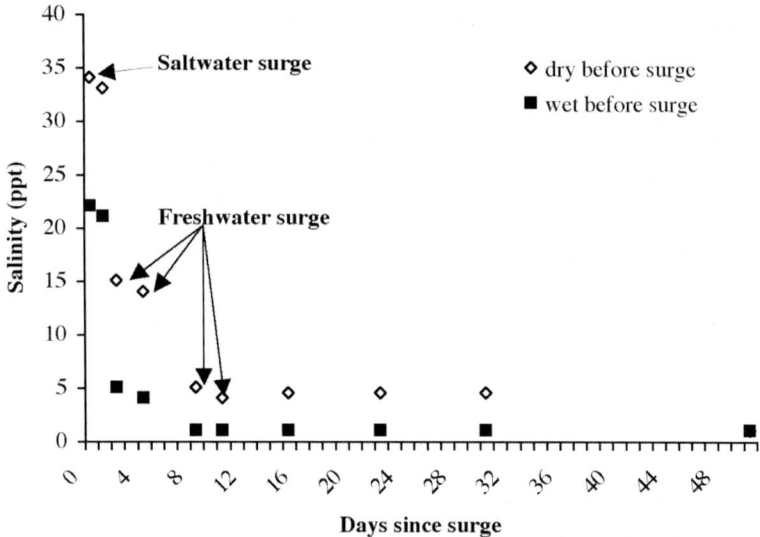

Fig. 9.6. Salinity levels in surge tanks after flooding with saltwater and subsequent freshwater additions (storm surge simulation).

9.5.2. Stand level response

Coastal ecosystems along the Gulf of Mexico and south Atlantic are currently undergoing forest dieback and decline from increasing tidal inundation, saltwater intrusion, and altered freshwater flow attributed to global climate change, variability, and anthropogenic activities (DeLaune et al. 1987; Pezeshki et al. 1990; Allen 1992; Krauss et al. 2000; Chambers et al. 2005). Changes to hydrological patterns in the swamps of Louisiana and South Carolina during the last two centuries can affect tree health and regeneration in these forests. As water levels continue to rise, coastal forests will be subjected to more prolonged and deeper flood events. Even though many of the forest species growing in these areas are adapted to prolonged inundation (Kozlowski 1984), extended flooding during the growing season can cause mortality of these tree species (Hall et al. 1946). Already many of the trees in these areas are showing evidence of severe stress (Conner et al. 1981; Conner et al. 1986; Conner and Day 1987; Shaffer et al. 2003). Even baldcypress and water tupelo, two of the dominant species in Louisiana's coastal forests (Conner and Sasser 1985), slowly die when exposed to prolonged, deep flooding (Penfound 1949; Eggler and Moore 1961; Harms et al. 1980; Brown 1981; Shaffer et al. 2003).

Another important factor to be considered in these coastal forests is the recruitment of new individuals into the forest. Buttonbush, black willow, and cottonwood (*Populus deltoides* Bartr. ex Marsh.) can germinate in standing water, while baldcypress and water tupelo must have dry periods for the seed to germinate and establish (DeBell and Naylor 1972; Hook 1984; Kozlowski 1997). In many cases, this is not happening (Conner et al. 1986; Keim et al. 2006), and if water levels continue to rise, coastal forested areas will eventually be replaced by scrub-shrub stands, marsh, or open water.

Permanent transects and forest research plots in Louisiana and South Carolina were established to assess wetland health and condition in tidal and nontidal areas in order to predict global change impacts on tidal freshwater swamps subject to different hydrogeomorphic settings, tides, droughts, hurricane frequencies, subsidence rates, and varying degrees of riverine influence. From previous studies, we know that saltwater flooding caused by storm surges significantly alters forest communities (Conner 1993; Gresham 1993; Williams 1993; Conner and Inabinette 2003), and it may take years before the forest recovers (Conner 1995). In addition, channel dredging can shorten the distance between the ocean and previously freshwater parts of the estuary (Yanosky et al. 1995).

Specific differences in tree composition between Louisiana and South Carolina coastal swamps include a greater number of species and a greater density of trees in South Carolina, even in more saline areas (Table 9.2). Flooded sites in both states were dominated by baldcypress and water tupelo. Baldcypress was dominant in all South Carolina plots, whereas water tupelo was dominant in 40% of Louisiana plots. South Carolina had higher percentages of swamp tupelo (occurred in only one Louisiana study site, but in 75% of South Carolina sites). Chinese tallow tree, a nonnative invasive in coastal regions of both states, occurred only in one Louisiana site.

Chronic exposure to flooding and saltwater intrusion as experienced in coastal areas because of rising sea level associated with global climate change results in a gradual change in species composition and basal area (Figure 9.7a), as few tree species are capable of surviving increased salinity levels. Furthermore, and as suggested by seedling studies, the number and type of overstory trees decrease as salinity increases from fresh to 1.2 ppt salinity. At that concentration, species such as water tupelo, green ash, and sweetgum (*Liquidambar styraciflua* L.) begin to be excluded, while swamp tupelo and red maple can survive at salinities as high as 2.5 ppt. Baldcypress, however, can remain as a dominant canopy tree and in rela-

Table 9.2. Structural composition of tidal freshwater forests in Louisiana and South Carolina (data from ongoing study; Ozalp 2003; Ratard 2004).

State	Condition	# Species	Density (trees/ha)	Basal Area (m²/ha)	Mean ht (m)	Mean dbh (cm)
LA	no salinity/no tide	3	875	63	24.2	28.1
	no salinity/flooded	3	600	53	17.0	30.9
	low salinity/low tide	2	220	27	19.1	34.5
	high salinity/tidal stress	3	520	36	18.9	27.9
SC	no salinity/no tide	5	825	69	28.5	34.1
	no salinity/low tide	7	1223	61	25.5	23.3
	low salinity/low tide	3	793	52	22.3	26.0
	low salinity/tidal stress	4	465	42	20.5	32.4
	high salinity/tidal stress	2	655	24	14.7	20.5

tively good condition at salinities above 1.8 ppt in natural swamp forests (Figure 9.7a). Stem density of overstory trees shows no apparent relationship with salinity (Figure 9.7b); as larger trees die, smaller trees of more salt tolerant species replace canopy dominants.

In some instances, salt tolerant species have time to disperse into new habitats upslope as environmental conditions change; however, continued development of coastal areas by humans limits the available area for plants to colonize. The current distribution of stem diameters of canopy trees on coastal swamp study sites in Louisiana and South Carolina, for instance, is similar even though South Carolina tended to have a few larger trees (Figure 9.8). Significant impacts from salinity intrusion and invasive species, however, are likely to change these distributions in the future. Overstory species composition is often more similar to understory species composition in freshwater sites than it is in moderately saline swamp sites of Louisiana (Figure 9.9). There is some concern for the invasion of Chinese tallow tree and expansion of wax myrtle (*Morella cerifera* L.) into these saline forests, but the long-term impact of these understory species shifts is uncertain.

The recent occurrence of hurricanes along southeastern and gulf coastal areas has reinforced the importance of understanding the role of acute catastrophic events in shaping the distribution of species and the composition of plant communities. Winds can cause considerable damage to trees although baldcypress and water tupelo are not as severely impacted as other species (Doyle et al. 1995; Conner and Inabinette 2003). More importantly, storm surges can bring large volumes of saline water inland (Figure 9.10). This saline water is deposited on the landscape where it infiltrates into the soil and can result in immediate death of trees or lead to

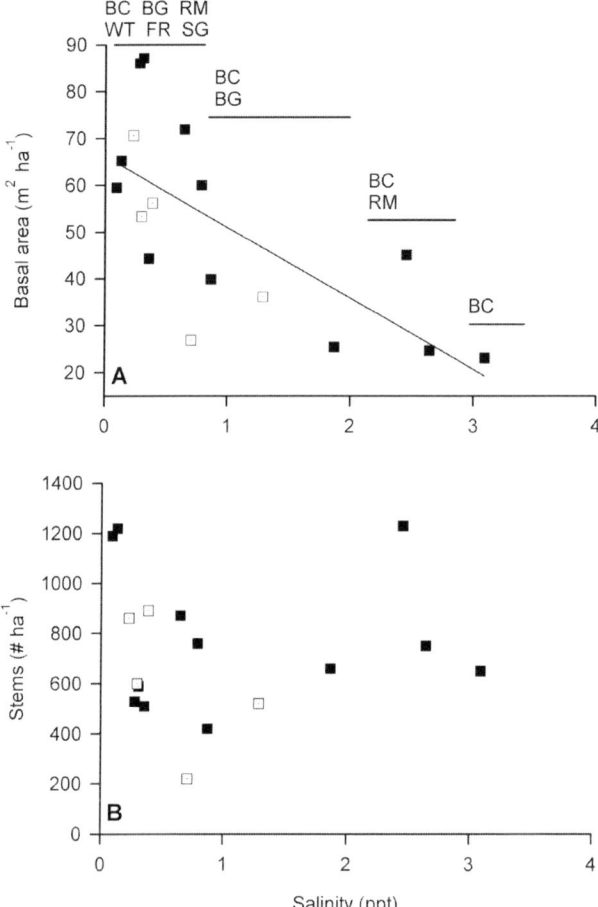

Fig. 9.7. Changes in basal area distribution and species distribution (A) and stem density (B) at different mean salinities for sites along study transects in Louisiana (open squares) and South Carolina (filled squares). Plot basal area decreases significantly with increases in salinity ($r^2 = 0.493$; $P = 0.002$). BC = baldcypress; WT = water tupelo; BG = blackgum; FR = ash sp.; RM = red maple; SG = sweetgum.

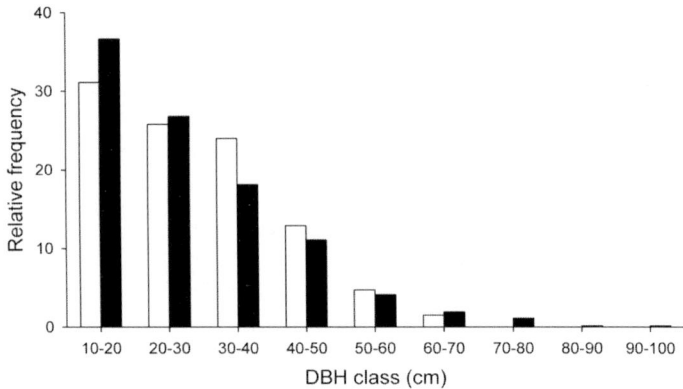

Fig. 9.8. Mean diameter distribution of dominant vegetation (≥ 10 cm DBH) for Louisiana (open bars) and South Carolina (filled bars) study sites.

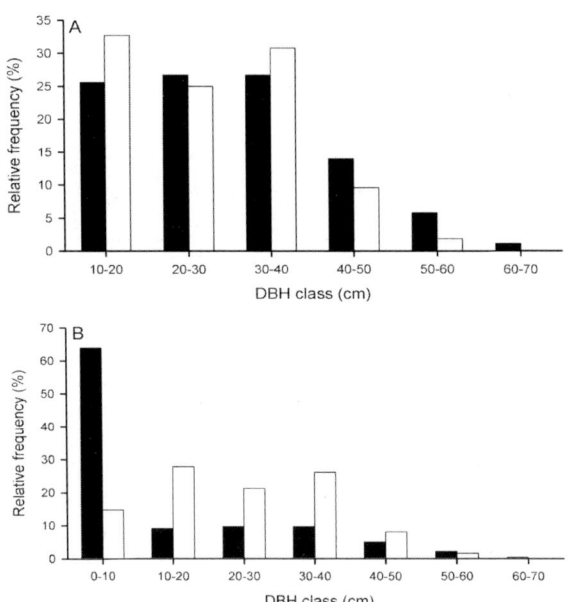

Fig. 9.9. Mean diameter distribution (A) of dominant vegetation (≥ 10 cm DBH) on fresh water (open bars) and moderately saline sites (3.1 ppt: filled bars) in Louisiana and the corresponding mean diameter distribution shift when understory woody vegetation (< 10 cm DBH) is included (B).

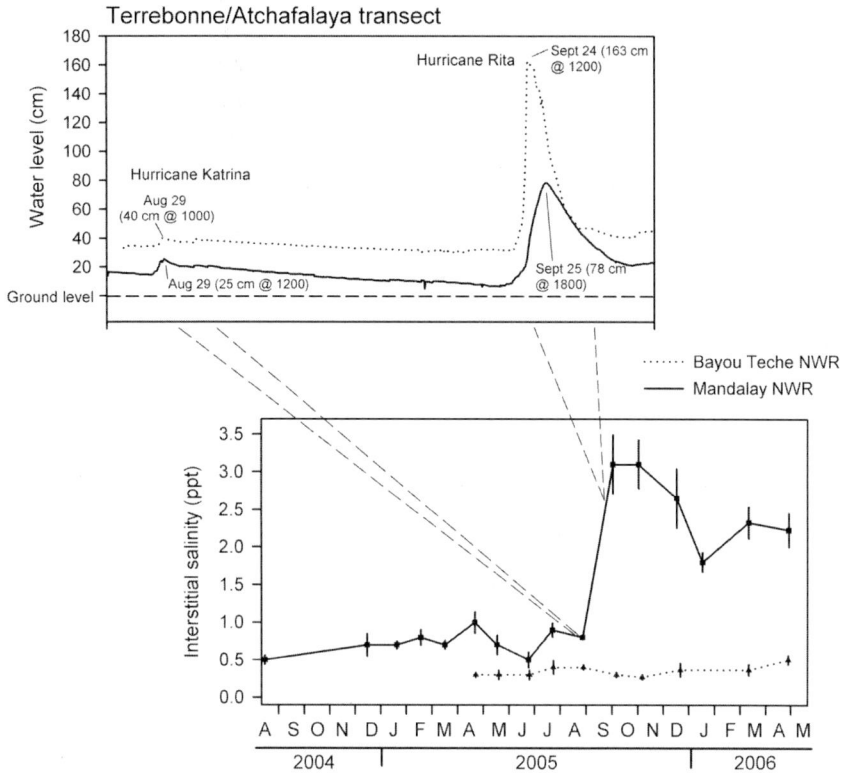

Fig. 9.10. Hydrographs of Louisiana swamp forest sites showing the impacts of two hurricanes (hurricanes Katrina and Rita) on water levels and interstitial salinity concentrations.

continued mortality for many months following the hurricane (Conner 1993). If the saline water can readily drain off, resulting in temporary flooding, trees may be affected only during the surge and immediately following, thus allowing them to survive. This occurred on Louisiana swamp forest sites when the storm surge caused by Hurricane Rita (2005) elevated water levels in Bayou Teche National Wildlife Refuge higher than in Mandalay National Wildlife Refuge. However, salinity entered the porewaters at Mandalay and persisted for at least 9 months (Figure 9.10).

9.6. Research needs

Ongoing, long-term forest inventory and productivity studies will help relate changes in forest structure, growth, and carbon budget to changing environmental conditions, past and present. Hydrologic monitoring of surface water, shallow groundwater (to 1 m), and interstitial porewater in select watersheds is being conducted to determine the degree of coupling with the regional hydrology. Soil analyses and carbon flux dynamics will be used to determine effects of belowground root growth and decomposition on carbon storage and burial. Growth functions and carbon relations on a species and site basis will be established to upgrade existing forest simulation models to predict present and future habitat quality and distribution as related to changing climate, drought and hurricane effects, and freshwater flow alternatives.

Even so, a better understanding of the tolerance of species growing in coastal wetland forests to include additional seedling/sapling studies and mature trees, and their response to increased flooding and salinity is still needed. More attention is necessary concerning the possible consequences of sea-level rise in forested areas so that techniques can be developed for effective management of this resource for the future. Natural regeneration in these areas has been affected by rising water levels (Conner et al. 1986), and planting is difficult because of herbivores (Conner and Toliver 1987). Thus, more work needs to be done on how to ensure adequate survival of planted seedlings. Wetland forest regeneration and sustainability may not be receiving adequate consideration in coastal Louisiana or South Carolina. The loss and rapidly deteriorating condition of coastal forests in Louisiana, interest in managing and restoring this natural resource, and the paucity of information available to accomplish these goals all point to a need to place increased emphasis on the conservation, protection, and study of coastal forests.

Direct manipulation of water levels is an option that should be considered, especially in relation to sediment delivery to subsiding swamp forests. In Louisiana, the numerous canals and spoil banks in the coastal area could be used to pump out areas for one to two years to let natural regeneration occur. This is not a long-term solution, however, because of the trend of continued water level rise and possibility of saltwater intrusion. The introduction of sediments from the Atchafalaya and Mississippi Rivers should also be considered to save these forests. Since the lack of sediment is what is causing the problem, it would be beneficial to these areas if sediment-laden waters were once again directed through these forests. There have been several suggestions to divert fresh water and sediments into wetlands (Gosselink and Gosselink 1985; Templet and Meyer-Ardendt 1988), and two Mississippi River diversions (Caernarvon and Davis Pond) are in place below New Orleans. Templet and Meyer-Ardendt (1988) reported that diverting 11% of the flow of the lower Mississippi

River during high discharge would offset apparent water level rise and subsidence in coastal swamp forests.

9.7. Conclusions

While tidal freshwater forests represent a unique and important ecosystem in the southeastern United States, these forests have experienced widespread hydrological, biogeochemical, and biological changes over the past century. These wetlands are of tremendous economic, ecological, cultural, and recreational value to humans and wildlife. Large-scale and localized alterations of processes affecting these wetlands have caused the complete loss of some coastal wetland forests and reduced the productivity and vigor of remaining areas. This loss and degradation threatens ecosystem functions and the services these systems provide. Unfortunately, there are few data on the value of the specific ecosystem services provided by these forests.

Hydrological patterns in mesotidal landscapes of South Carolina are still fairly intact in comparison to the microtidal landscapes of Louisiana. Although several of the rivers in South Carolina have dams farther inland that can alter flooding patterns, Louisiana wetlands have also been impacted by highways, railroads, channelization, navigation canals, oil and gas exploration canals, flood control structures, levee systems, conversion of forests to urban and agricultural land, and nonsustainable forest practices. In addition, the cumulative effects of sea-level rise, crustal sinking, tectonic activity, and sediment consolidation result in water levels increasing 100 cm/century in Louisiana while South Carolina's more stable coastal region is experiencing a water level increase of only 20 cm/century.

The recent increase in hurricanes striking the southeastern and gulf coastal areas has reinforced the importance of severe weather events in shaping the distribution of species and the composition of plant communities. Winds can cause considerable damage to trees, although baldcypress and water tupelo are not as severely impacted as other species, and baldcypress trees do blow over, although such occurrences are rare. More importantly, storm surges can bring large volumes of saline water into freshwater areas which can infiltrate into the soil, resulting in immediate death of trees or lead to continued mortality for many months following the hurricane. If the saline water temporary floods the landscape, trees may be affected only during the surge and immediately following, thus allowing them to survive.

Alterations in hydrological patterns and increased salinization in coastal areas are of major concern to long-term sustainability of coastal forests. In Louisiana especially, direct manipulation of water levels and the introduction of sediments from the Atchafalaya and Mississippi Rivers might be necessary to save existing forests and restore degraded ones.

9.8. Acknowledgments

This material is based upon work supported by the USGS Global Climate Change Program and by the CSREES/USDA, under project number SC-1700271. Technical Contribution No. 5267 of the Clemson University Experiment Station. The authors thank Dr. Becky Sharitz and Dr. James Luken for their helpful comments on this manuscript.

References

Abbey DG (1979) Life in the Atchafalaya Swamp. The Lafayette Natural History Museum, Lafayette

Allen JA (1992) Cypress-tupelo swamp restoration in southern Louisiana. Restoration Manage Notes 10:188-189

Allen JA, Chambers JL, McKinney D (1994) Intra-specific variation in the response of *Taxodium distichum* seedlings to salinity. For Ecol Manage 70:203-214

Allen JA, Conner WH, Goyer RA, Chambers JL, Krauss KW (1998) Freshwater forested wetlands and global climate change. In: Guntenspergen GR, Vairin BA (eds) Vulnerability of coastal wetlands in the southeastern United States: Climate change research results, 1992-97. Biological Science Report USGS/BRD/BSR-1998-0002. U.S. Geological Survey, Biological Resources Division, Lafayette, pp 33-44

Brown CA, Montz GN (1986) Baldcypress: the tree unique, the wood eternal. Claitor's Publishing Division, Baton Rouge

Brown S (1981) A comparison of the structure, primary productivity, and transpiration of cypress ecosystems in Florida. Ecol Monogr 51: 403-427

Burns AC (1980) Frank B. Williams: cypress lumber king. J For Hist 24:127-134

Chambers JL, Conner WH, Day JW, Faulkner SP, Gardiner ES, Hughes MS, Keim RF, King SL, McLeod KW, Miller CA, Nyman JA, Shaffer GP (2005) Conservation, protection and utilization of Louisiana's coastal wetland forests. Final Report to the Governor of Louisiana from the Coastal Wetland Forest Conservation and Use Science Working Group. Louisiana Governor's Office of Coastal Activities, Baton Rouge

Chambers JL, Conner WH, Keim RF, Faulkner SP, Day JW, Gardiner ES, Hughes MS, King SL, McLeod KW, Miller CA, Nyman JA, Shaffer GP (2006) Towards sustainable management of Louisiana's coastal wetland forest: problems, constraints, and a new beginning. In: Proceedings of international conference on hydrology and management of forested wetlands. American Society of Agricultural and Biological Engineers, St. Joseph, pp 150-157

Coleman JM, Roberts HH, Stone GW (1998) Mississippi River delta: an overview. J Coast Restor 14:698-717

Colquhoun DJ (1974) Cyclic surficial stratigraphic units of the middle and lower coastal plains, central South Carolina. In: Oaks RQ, DuBar JR (eds) Postmiocene startigraphy central and southern Atlantic coastal plain. Utah State University Press, Logan, pp 179-190

Conner WH (1993) Artificial regeneration of baldcypress in three South Carolina forested wetland areas after Hurricane Hugo. In: Brissette JC (ed) Proceedings of the seventh biennial southern silvicultural research conference. General Technical Report SO-93. U.S. Department of Agriculture, Forest Service, Southern Forest Experiment Station, New Orleans, pp 185-188

Conner WH (1994) The effect of salinity and waterlogging on growth and survival of baldcypress and Chinese tallow seedlings. J Coast Res 10:1045-1049

Conner WH (1995) Woody plant regeneration in three South Carolina *Taxodium/Nyssa* stands following Hurricane Hugo. Ecol Engineering 4: 277-287

Conner WH, Askew GR (1993) Impact of saltwater flooding on red maple, redbay, and Chinese tallow seedlings. Castanea 58:214-219

Conner WH, Brody M (1989) Rising water levels and the future of southeastern Louisiana swamp forests. Estuaries 12(4):318-323

Conner WH, Day JW Jr (1987) The ecology of Barataria Basin, Louisiana: An estuarine profile. Biological Report 85(7.13). U.S. Fish and Wildlife Service, Washington

Conner WH, Day JW Jr (1988) Rising water levels in coastal Louisiana: implications for two forested wetland areas in Louisiana. J Coast Res 4:589-596

Conner WH, Inabinette LW (2003) Tree growth in three South Carolina (USA) swamps after Hurricane Hugo: 1991-2001. For Ecol Manage 182:371-380

Conner WH, Sasser CE (1985) Vegetational composition of the upper Barataria basin swamp, Louisiana. Northeast Gulf Sci 7:163-166

Conner WH, Toliver JR (1987) Use of "Vexar" seedling protectors not effective in reducing nutria damage to planted baldcypress seedlings. Tree Planters' Notes 38:26-29

Conner WH, Toliver JR (1990) Long-term trends in the baldcypress (*Taxodium distichum* (L.) Rich.) resource in Louisiana. For Ecol Manage 33/34:543-557

Conner WH, Gosselink JG, Parrondo RT (1981) Comparison of the vegetation of three Louisiana swamp sites with different flooding regimes. Amer J Bot 68:320-331

Conner WH, Toliver JR, Sklar FH (1986) Natural regeneration of baldcypress (*Taxodium distichum* (L.) Rich.) in a Louisiana swamp. For Ecol Manage 14:305-317

Conner WH, Day JW Jr, Baumann RH, Randall J (1989) Influence of hurricanes on coastal ecosystems along the northern Gulf of Mexico. Wetl Ecol Manage 1:45-56

Conner WH, McLeod KW, McCarron JK (1997) Flooding and salinity effects on growth and survival of four common forested wetland species. Wetl Ecol Manage 5:99-109

Conner WH, McLeod KW, McCarron JK (1998) Survival and growth of four bottomland oak species in response to increased flooding and salinity. For Sci 44:618-624

Dahl TE (1999) South Carolina's wetlands – status and trends 1982-1989. U.S. Department of the Interior, U.S. Fish and Wildlife Serice, Washington

Daniels RC (1992) Sea-level rise on the South Carolina coast: two case studies for 2100. J Coast Res 8:56-70

Davis DW (1975) Logging canals: a distinctive pattern of the swamp landscape in south Louisiana. For People 25:14-17, 33-35

Day JW, Shaffer GP, Britsch LD, Reed DJ, Hawes SR, Cahoon D (2000) Pattern and process of land loss in the Mississippi Delta: a spatial and temporal analysis of wetland habitat change. Estuaries 23:425-438

DeBell DS, Naylor AW (1972) Some factors affecting germination of swamp tupelo seeds. Ecology 53:504-506

DeLaune RD, Pezeshki SR, Patrick Jr WH (1987) Response of coastal plants to increase in submergence and salinity. J Coast Res 3(4):535-546

DeLaune RD, Callaway JC, Patrick WH Jr, Nyman JA (2004) An analysis of marsh accretionary processes in Louisiana coastal wetlands. In: Davis DW, Richardson M (eds) The coastal zone: Papers in honor of H. Jesse Walker. Geoscience Publications. Department of Geography and Anthropology, Louisiana State University, Baton Rouge, pp 113-130

Doar D (1936) Rice and rice planting in the South Carolina Lowcountry. Charleston Museum Historic Contribution VIII. The Charleston Museum, Charleston

Doyle TW, Girod G (1997) The frequency and intensity of Atlantic hurricanes and their influence on the structure of south Florida mangrove communities. In: Diaz H., Pulwarty R (eds) Hurricanes, climatic change and socioeconomic impacts: A current perspective, Westview Press, New York, pp 111-128

Doyle TW, Keeland BD, Gorham LE, Johnson DJ (1995) Structural impact of Hurricane Andrew on forested wetlands of the Atchafalaya Basin in coastal Louisiana. J Coast Res 18:354-364

Dukes EK (1984) The Savannah River Plant environment. E.I. du Pont de Nemours & Co., DP-1642. Savannah River Ecology Laboratory, Aiken

Edgar W (1998) South Carolina: A history. University of South Carolina Press, Columbia

Eggler WA, Moore WG (1961) The vegetation of Lake Chicot, Louisiana, after eighteen years of impoundment. Southwestern Naturalist 6: 175-183

Emmanuel KA (1987) The dependence of hurricane intensity on climate. Nature 326:484-485

Environmental Protection Agency (2006) Mississippi River Basin and Gulf of Mexico hypoxia: culture and history. Available online at http://www.epa.gov/msbasin/culture.htm (2 May 2006)

Fetters T (1990) Logging railroads of South Carolina. Heimburger House Publishing Company, Forest Park

Frazier DE (1967) Recent deposits of the Mississippi River, their development and chronology. Trans Gulf Coast Assoc Geol Soc 17:287-311

Gosselink JG, Gosselink L (1985) The Mississippi River delta: a natural wastewater treatment system. In: Godfrey PJ, Kaynor ER, Pelczarski P (eds) Ecological considerations in wetlands treatment of municipal wastewaters. Van Nostrand Reinhold Company, New York, pp 327-337

Gosselink JG, Coleman JM, Stewart RE Jr (1998) Coastal Louisiana. In: Mac MJ, Opler PA, Puckett Haecker CE, Doran PD (eds) Status and trends of the nation's biological resources. U.S. Department of the Interior, U.S. Geological Survey, Reston, pp 385-436

Gresham CA (1993) Changes in baldcypress-swamp tupelo wetland soil chemistry caused by Hurricane Hugo induced saltwater inundation. In: Brissette JC (ed) Proceedings of the seventh biennial southern silvicultural research conference. General Technical Report SO-93. U.S. Department of Agriculture, Forest Service, Southern Research Station, New Orleans, pp 171-175

Gresham CA, Hook DD (1982) Rice fields of South Carolina: a resource inventory and management policy evaluation. Coast Zone Manage J 9(2):183-203

Hall TF, Penfound WT, Hess AD (1946) Water level relationships of plants in the Tennessee Valley with particular reference to malaria control. Rept Reelfoot Lake Biol Sta 10:18-59

Harms WR, Schreuder HT, Hook DD, Brown CL (1980) The effects of flooding on the swamp forest in Lake Ocklawah, Florida. Ecology 61:1412-1421

Hook DD (1984) Waterlogging tolerance of lowland tree species of the South. South J Appl For 8:136-149

Hook DD, Brown CL, Kormanik PP (1971) Lenticel and water root development of swamp tupelo under various flooding conditions. Bot Gaz 131:217-224

Hook DD, Langdon OG, Stubbs J, Brown CL (1970) Effects of water regimes on the survival, growth, and morphology of tupelo seedlings. For Sci 16:304-311

Keim RF, Chambers JL, Hughes MS, Dimov LD, Conner WH, Shaffer GP, Gardiner ES, Day JW Jr (2006) Long-term success of stump sprouts in high-graded baldcypress-water tupelo swamps in the Mississippi Delta. For Ecol Manage 234:24-33

King SL, Shepard JP, Ouchley K, Neal JA, Ouchley K (2005) Bottomland hardwood forests: past, present, and future. In: Fredrickson LH, King SL, Kaminski RM (eds) Ecology and management of bottomland hardwood systems: The state of our understanding. Gaylord Memorial Laboratory Special Publication 10. University of Missouri-Columbia, Puxico, pp 1-17

Kniffen FB (1968) Louisiana: its land and people. Louisiana State University Press, Baton Rouge

Kozlowski TT (1984) Responses of woody plants to flooding. In: Kozlowski TT (ed) Flooding and plant growth. Academic Press, New York, pp 129-163

Kozlowski TT (1997) Responses of woody plants to flooding and salinity. Tree Physiol Monogr No. 1:29 pp

Krauss KW, Chambers JL, Allen JA, Soileau DM Jr, DeBosier AS (2000) Growth and nutrition of baldcypress families planted under varying salinity regimes in Louisiana, USA. J Coast Res 16:153-163

Mancil E (1969) Some historical and geographical notes on the cypress lumbering industry in Louisiana. La Studies 8:14-25

Mancil E (1972) A historical geography of industrial cypress lumbering in Louisiana. Ph.D. thesis, Louisiana State University

Mancil E (1980) Pullboat logging. J For Hist 24:135-141

Mattoon WR (1915) The southern cypress. Bulletin 272. U.S. Department of Agriculture, Washington

McCarron JK, McLeod KW, Conner WH (1998) Flood and salinity stress of wetland woody species, buttonbush (*Cephalanthus occidentalis*) and swamp tupelo (*Nyssa sylvatica* var. *biflora*). Wetlands 18:165-175

McKnight JS, Hook DD, Langdon OG, Johnson RL (1981) Flood tolerance and related characteristics of trees of the bottomland hardwood forests. In: Clark JR, Benforado J (eds) Wetlands of bottomland hardwood forests. Elsevier Scientific Publishing Company, Amsterdam, pp 29-69

McLeod KW, McCarron JK, Conner WH (1996) Effects of inundation and salinity on photosynthesis and water relations of four southeastern coastal plain forest species. Wetl Ecol Manage 4:31-42

McLeod KW, McCarron JK, Conner WH (1999) Photosynthesis and water relations of four oak species: Impact of flooding and salinity. Trees 13:178-187

Morse DF, Morse PA (1983) Archaeology of the central Mississippi Valley. Academic Press, Inc, New York

National Research Council (2006) The historic and existing Louisiana coastal systems. In: Drawing Louisiana's new map – addressing land loss in coastal Louisiana. The National Academies Press, Washington, DC, pp 29-42

Neuman RW, Hawkins NW (1993) Louisiana prehistory, 2nd edn. Anthropological Study Series No. 6. Department of Culture, Recreation and Tourism, Louisiana Archaeological Survey and Antiquities Commission, Baton Rouge

Ozalp M (2003) Water quality, aboveground productivity, and nutrient dynamics during low flow periods in tidal floodplain forests of South Carolina. Ph.D. thesis, Clemson University

Penfound WT (1949) Vegetation of Lake Chicot, Louisiana in relation to wildlife resources. Proc La Acad Sci 12:47-56

Penland S, Boyd R (1985) Mississippi delta shoreline development. In: Penland S, Boyd R (eds) Transgressive depositional environments of the Mississippi River delta plain: A guide to barrier islands, beaches and shoals in Louisiana. Louisiana Geological Survey, Baton Rouge, pp 53–122

Pezeshki SR, DeLaune RD, Patrick WH Jr (1990) Flooding and saltwater intrusion: potential effects on survival and productivity of wetland forests along the U.S. Gulf Coast. For Ecol Manage 33/34:287-301

Porcher RD (1995) Wildflowers of the Carolina Lowcountry and lower Pee Dee. University of South Carolina Press, Columbia

Ratard MA (2004) Factors affecting growth and regeneration of baldcypress in a South Carolina tidal freshwater swamp. Ph.D. thesis, Clemson University

Roberts HH (1997) Dynamic changes of the Holocene Mississippi River delta plain: the delta cycle. J Coast Res 13:605-627

Rogers GC Jr (1970) The history of Georgetown County, South Carolina. The University of South Carolina Press, Columbia (reprinted 2002 by The Reprint Company, Publishers, Spartanburg)

Saucier RT (1994) Geomorphology and quaternary geologic history of the Lower Mississippi Valley, Volume 1. U.S. Army Corps of Engineers, Waterways Experiment Station, Vicksburg

SCDHEC (2000) Watershed water quality management strategy: Pee Dee Basin. Technical Report 015-00. South Carolina Department of Health and Environmental Control, Bureau of Water, Columbia

Shaffer GP, Perkins TE, Hoeppner SS, Howell S, Benard TH, Parsons AC (2003) Ecosystem health of the Maurepas swamp: Feasibility and projected benefits of a freshwater diversion. Environmental Protection Agency, Region 6, Dallas

Stevenson JC, Ward LG, Kearney MS (1986) Vertical accretion in marshes with varying rates of sea level rise. In: Wolfe DA (ed) Estuarine variability. Academic Press, Inc., Orlando, pp 241-259

Templet P, Meyer-Ardendt K (1988) Louisiana wetland loss: A regional water management approach to the problem. Environ Manage 12: 181-192

Twilley RR, Barron EJ, Gholz HL, Harwell MA, Miller RL, Reed DJ, Rose JB, Siemann EH, Wetzel RG, Zimmerman RJ (2001) Confronting climate change in the Gulf Coast region: Prospects for sustaining our ecological heritage. Union of Concerned Scientists, Cambridge, MA and Ecological Society of America, Washington

US Army Corps of Engineers (2004) Louisiana coastal area (LCA), Louisiana – ecosystem restoration study. New Orleans District

Wax CL, Borgengasser MJ, Muller RA (1978) Barataria Basin: Synoptic weather types and environmental responses. Sea Grant Publication LSU-T-78-001. Center for Wetland Resources, Louisiana State University, Baton Rouge

Williams TM (1993) Salt water movement within the water table aquifer following Hurricane Hugo. In: Brissette JC (ed) Proceedings of the seventh biennial southern silvicultural research conference. General Technical Report SO-93. U.S. Department of Agriculture Forest Service, Southern Forest Experiment Station, New Orleans, pp 177-183

Yanosky TM, Hupp CR, Hackney CT (1995) Chloride concentrations in growth rings of *Taxodium distichum* in a saltwater-intruded estuary. Ecol Appl 5:785-792

Chapter 10 - Ecology of the Coastal Edge of Hydric Hammocks on the Gulf Coast of Florida

Kimberlyn Williams[1], Michelina MacDonald[2], Kelly McPherson[3], and Thomas H. Mirti[4]

[1]*Department of Biology, California State University, San Bernardino, 5500 University Parkway, San Bernardino, CA 92407*
[2]*P. K. Yonge, Developmental Research School, 1080 SW 11th Street, Gainesville, FL 32601*
[3]*Alachua County Department of Environmental Protection, 201 SE 2nd Avenue, Suite 201, Gainesville, FL 32601*
[4]*St. Johns River Water Management District, 4049 Reid Street, Palatka, FL 32177*

10.1. Introduction

Hydric hammocks are forested wetlands that can be distinguished from other forested wetlands by their hydrology, soils, and species composition. The distribution and ecology of hydric hammocks have been reviewed by Vince et al. (1989) and Simons et al. (1989). They are patchily distributed across northern and central Florida, entering the tidal reach only at their coastal margins. Hydric hammocks are considered wetland forests because freshwater flooding saturates their soils for substantial periods, excluding species that cannot tolerate saturated soils. However, they flood less frequently and less deeply than baldcypress (*Taxodium distichum* [L.] L.C. Rich.) swamps, and they are not marked by the constant seepage and deep peat soils of bayheads or bay swamps (Vince et al. 1989). Hydric hammocks may not flood at all during some years, but may flood for 1-2 months during other years (Simons 1990; Light et al. 2002). In contrast to bottomland hardwood forests, hydric hammocks do not occur on alluvial soils, but typically occur on sand, loam, or occasionally muck soils where either limestone is close to the surface or springs deliver a high concentration of calcium and other limestone-derived minerals to the forest stand

(Vince et al. 1989; Simons et al. 1989). Ewel (1990) suggested that hydric hammocks might be distinguished from other types of forested wetlands in Florida by the combination of their relatively short hydroperiod, low fire frequency, low organic matter accumulation, and high contribution of deep groundwater to their flooding.

Dominant tree species in hydric hammocks include cabbage palm (*Sabal palmetto* [Walt.] Lodd. ex J.A. & J.H. Schultes), southern redcedar (*Juniperus virginiana* var. *silicicola* [Small] J. Silba), a mixture of hardwood species, and frequently, loblolly pine (*Pinus taeda* L.; Table 10.1). Hydric hammocks often grade into other forest types. They may be found at the drier edges of other swamp types and at the wetter edges of mesic hammocks and other upland forest types (Vince et al. 1989; FNAI 1990). There is substantial overlap in species composition between hydric hammocks and other types of swamps (Ewel 1990). However, hydric hammocks are characterized by the dominance of cabbage palm in combination with a mixture of hardwood species and/or southern redcedar. Baldcypress, which is common in many other swamp forests, is rare in hydric hammocks (Vince et al. 1989; Ewel 1990).

Table 10.1. Common dominants of hydric hammock.

Scientific name	Common name	Abundance[1]	FNAI[2]	Type[3]
Sabal palmetto (Walt.) Lodd. ex J.A. & J.H. Schultes	cabbage palm	A*	X	CLS
Juniperus virginiana var. *silicicola* (Small) J. Silba	southern redcedar	A*	X	C
Quercus virginiana P. Mill.	live oak	A*		CIL
Quercus nigra L.	water oak	A	X	IL
Quercus laurifolia Michx.	laurel oak	A*	X	IS
Liquidambar styraciflua L.	sweetgum	A*		IL
Pinus taeda L.	loblolly pine	A*		L
Acer rubrum L.	red maple	C*	X	S
Magnolia virginiana L.	sweetbay	C*	X	S
Morella cerifera (L.) Small[4]	wax myrtle	A	X	
Persea palustris (Raf.) Sarg.	swamp bay	C		
Carpinus caroliniana Walt.	American hornbeam	A*		
Ulmus americana L.	American elm	C*		

[1]Species listed as "abundant" (A) and a subset of those listed as "common" (C) in hydric hammocks by Vince et al. (1989). Those marked with an asterisk constituted > 10 percent of the relative importance value in at least one of the 12 stands studied.
[2]A subset of the species listed as typically occurring in hydric hammocks by the Florida Natural Areas Inventory (FNAI 1990).
[3]Dominance in hydric hammock subtypes recognized by Simons et al. (1989): coastal (C), inland (I), loblolly pine (L), and seepage (S).
[4]A shrub: common, but contributing little to the basal area in a stand.

The most extensive stands of hydric hammock occur along the west coast of Florida, from the Hernando/Pasco County line in the south to the St. Marks River in the Florida panhandle (Vince et al. 1989). Along this coast, they extend to the edge of the tidal marshes on the Gulf of Mexico. A large fraction of the hammocks in this region are subject to periodic storm surges (Figure 10.1). Variation in salt exposure influences forest composition, with live oak (*Quercus virginiana* P. Mill.), cabbage palm and southern redcedar becoming more dominant toward the edge of the salt marsh (Vince et al. 1989; Williams et al. 1999a). These hammocks exist in a dynamic environment, with composition changing as sea level rises and exposure to tidal water increases. Both storms and reductions in freshwater supply can modify species composition and accelerate canopy loss as these forest stands are converted to coastal marsh by rising seas.

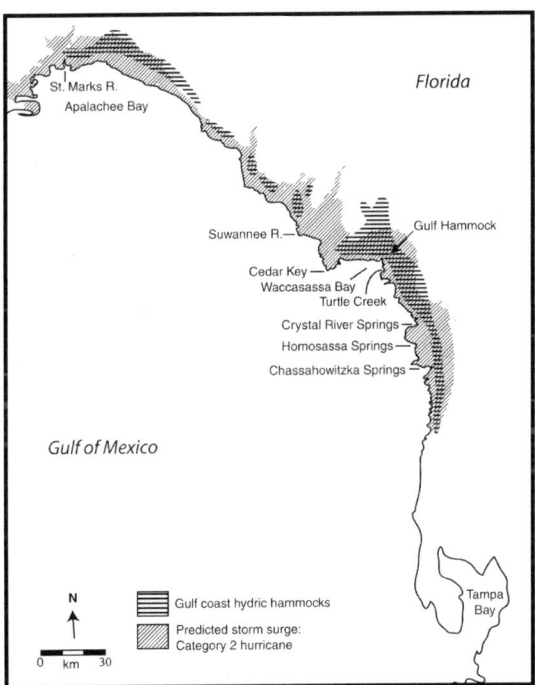

Fig. 10.1. General locations of the most extensive stands of Gulf Coast hydric hammock as mapped by Vince et al. (1989). Hydric hammocks located in other parts of Florida are not shown. Conversion to pine plantation and other uses has reduced the extent of coastal hydric hammocks from that shown to a more narrow coastal band in areas such as Gulf Hammock (see Sect. 10.2.2). The predicted extent of a storm surge generated by a Category 2 hurricane along the coastline occupied by hydric hammock is indicated (Florida Division of Emergency Management 2005).

10.2. Historical land use and geologic history

Humans have occupied the Gulf Coast of Florida since the late Pleistocene (Clausen et al. 1979; Milanich 1994; Faught and Carter 1998). The low elevational gradient of this coast has influenced human occupation, with sea-level rise having drowned prehistoric settlements (as evidenced by underwater archeological sites off this coast; Dunbar et al. 1988, 1992), and hurricane flooding having caused abandonment of coastal towns, such as Atsena Otie, after European occupation. Conversely, effects of humans on topography have affected the rate at which coastal forest succumbs to sea-level rise. Ancient shell middens contribute to the patchwork of topographic highs that sustain forest stands in areas where the surrounding forest has succumbed to rising seas. Today, effects of humans on both topography and hydrology may hasten or delay the rate at which sea-level rise eliminates coastal forest.

10.2.1. Geology

The hydric hammocks along the west coast of Florida grow on a platform of limestone and dolomite that formed in shallow seas during the Tertiary (Vernon and Puri 1964; Brown et al. 1990). The topography and porosity of the limestone influence both the distribution and hydrology of the Gulf Coast hydric hammocks. The region of coastline occupied by these hydric hammocks (the Big Bend region of Florida) coincides with a region where the extensive upper Floridan aquifer comes to the surface and is unconfined (Bush and Johnson 1988). Discharge from that aquifer contributes to forest flooding and influences the distribution of hydric hammocks.

This limestone platform also helps support extensive salt marshes adjacent to the forest. The Big Bend region of Florida is a low-energy, microtidal coast with a tidal range of less than 1 m and mixed semi-diurnal tides (Stumpf and Haines 1998). It has such low wave energy that it has been described as a "zero-energy" coastline (Montague and Wiegert 1990). The most extensive coastal salt marshes in Florida occur in this region, despite its low sediment supply (Montague and Wiegert 1990). These extensive marshes may buffer tidal action, reducing tidal flooding frequency at the coastal edge of hydric hammocks (Williams et al. 1999a).

The very shallow slope of the limestone platform renders coastal forest susceptible to effects of sea-level rise despite the tectonic stability of the region. The rate of relative sea level rise at Cedar Key, near the center of distribution of coastal hydric hammocks, is comparable to estimates of global eustatic sea-level rise. Best estimates of global sea-level rise fall between 1.0 and 2.0 mm/yr (IPCC 2001), whereas relative sea-level rise at Cedar Key has risen an average of 1.6 mm/yr since 1938 (Figure 10.2), and 1.87 mm/yr if one includes an earlier period of record from 1914-1925 (Zervas 2001). Despite the low rate of relative sea-level rise in this region, the flatness of the platform on which these hammocks exist causes minor increases in sea level to affect large expanses of land. Minor topographic highs in the limestone and mounds formed from ancient shell middens near the coast enhance tree regeneration, affect community composition and delay conversion of forest to coastal marsh in the face of rising seas (Cooke 1939; Kurz and Wagner 1957; Williams et al. 1999a). The low topography exposes many of the Gulf hydric hammocks to occasional storm surges: approximately half of the area potentially supporting hydric hammock on the Gulf Coast is currently within the surge zone of tropical storms, and most is within reach of tidal surges predicted to accompany Category 2 hurricanes (Figure 10.1).

10.2.2. Land use changes

The Big Bend coastline of western Florida is relatively undeveloped, but the region has a long history of commercial forestry and use for animal

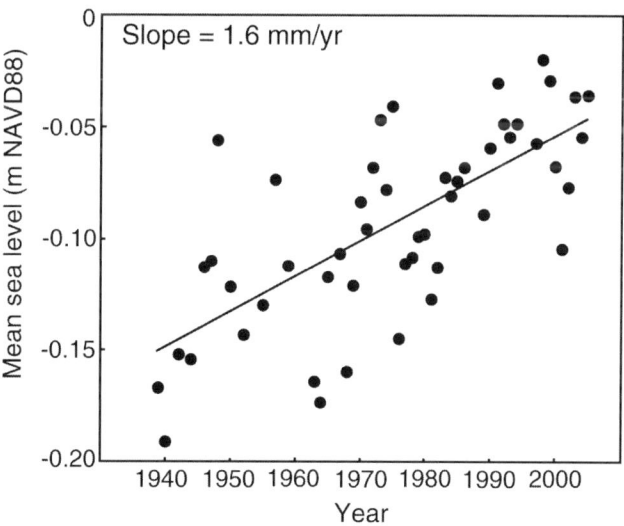

Fig. 10.2. Mean sea level record from Cedar Key, Florida. Annual means are shown, and years with incomplete records have been omitted.

production. Most of the dominant tree species of hydric hammock have been exploited for commercial use at some point during the past two centuries. Selective harvest of live oak, southern redcedar, cabbage palm, loblolly pine, and sweetgum (*Liquidambar styraciflua* L.) has occurred at different times since the early 1800s (Simons et al. 1989).

Intensive harvesting of southern redcedar occurred during the latter half of the 19th century, peaking in the 1880s. Slat mills on the west coast of Florida supplied wood for the pencil factories of New York and Europe, shipping (by one account) 28,000 cubic meters of trimmed cedar wood in one year (Monk 1987; Simons et al. 1989). The largest cedar slat mills were at Cedar Key, and Gulf Hammock, the largest expanse of hydric hammock in Florida, supplied much of the southern redcedar for these mills. By the mid-1890s, southern redcedar of useable size was growing scarce, and the mills were in decline. A strong hurricane in 1896 destroyed the mills, marking the end of large-scale harvesting for pencil wood in these hammocks (Monk 1987).

Harvesting of cabbage palm on a commercial scale followed the peak years of southern redcedar harvest, providing fiber for broom and brush manufacture. A fiber factory in Cedar Key operated from 1910 to 1950 and processed palms harvested from coastal regions between the Suwannee and Chassahowitzka Rivers. Bud cutters, seeking the terminal "bud" of the tree for fiber (the terminal ca. 1.5 m), targeted primarily shorter trees (B. Hudson, Inglis, FL and R.K. Runnels, Crystal River, FL, personal communication). A good bud-cutter could cut 100 buds per day (R.K. Runnels, Crystal River, FL, personal communication), and the fiber factory required a minimum of 600 buds per day (Simons et al. 1989). Fiber production peaked in the 1940s, and the fiber factory closed following Hurricane Easy in 1950. Today, palms are harvested primarily for horticultural purposes, extracting whole trees, typically larger than those previously preferred by the bud-cutters.

Despite the exploitative nature of these early harvests, hydric hammock appears to have regenerated well where fire, herbicide, and deliberate conversion to agriculture, plantation forestry, and urban development did not occur (Simons et al. 1989). Palms are still common where subsequent tree removals did not occur, and many forest stands with high densities of old southern redcedar stumps still have high densities of southern redcedar. These early harvests probably had their most permanent effects in stands at the coastal margin where sea-level rise had already eliminated tree regeneration (see Sect. 10.4.3). These nonregenerating stands, which were already succumbing to sea-level rise, could not have recovered.

Today, most of the coastal fringe of hydric hammock on the Gulf Coast of Florida is in conservation land. A number of federal wildlife refuges combine with several coastal state preserves to cover over 85 percent of the coastline where hydric hammock meets coastal marsh (Florida DEP 2006). Many of these conservation areas are relatively narrow, however, typically extending less than 5 km inland from the coast. This narrow swath places almost all of the hydric hammocks south of Crystal River in protected lands, but leaves many of those farther north in private ownership. Hydric hammocks in private ownership have been subjected to extensive modification. Most of Gulf Hammock, the largest historical expanse of hydric hammock on the Gulf Coast, has been converted to loblolly pine plantations (Simons 1990). Similarly, hydric hammock is the forest type that has experienced the greatest conversion to agriculture and urban development in the lower reaches of the Suwannee River watershed (Light et al. 2002). Because private land and geomorphological features such as thick sand deposits prevent hydric hammock from migrating inland along most of the coastline, many authors consider sea-level rise to constitute the greatest threat to coastal hydric hammock (e.g, Simons et al. 1989; Williams et al. 1999a; Doyle et al. 2003; Raabe et al. 2004).

10.2.3. Major sources of water

Hydric hammocks typically receive much of their fresh water from the Floridan aquifer. Along the region of the Gulf Coast that supports hydric hammocks, this aquifer comes to the surface, with water flowing out through the karstic limestone bedrock, supplying forest stands and rivers. In some regions, large volumes of water come to the surface as major springs. The Springs Coast, near the southern limit of coastal hydric hammock, contains numerous high-volume springs (eg., Crystal River, Weeki Wachi Springs, and Homosassa Springs). In regions near such high-volume rivers and springs, hydric hammocks may experience tidal waters that are less saline than those in other areas and border marshes dominated by jamaica sawgrass (*Cladium mariscus* ssp. *jamaicense* [Crantz] Kükenth.). In contrast, where freshwater discharge is lower and tidal waters more saline, coastal hydric hammocks tend to abut a marsh dominated by needlegrass rush (*Juncus roemerianus* Scheele.), a more salt-tolerant marsh species. Discharge from the Floridan aquifer, both into rivers that supply freshwater to the coast and into the forest stands themselves, interacts with ocean water and local rainfall to influence both the flooding regime and salinity of soils in coastal hydric hammocks.

10.3. Climate drivers and hydrological characterization

10.3.1. Climate

The stretch of Florida coastline occupied by the Gulf Coast hydric hammocks has a climate that is considered subtropical. Mean maximum temperatures are slightly over 32°C in July, dropping to 17-22°C in January (Tanner 1992). Average annual precipitation ranges from approximately 125 cm to 150 cm per year (Jordan 1984). Rainfall in this region is seasonal, with over half of the precipitation falling between June and September, and a second minor peak in precipitation occurring in March with the passage of frontal storms (Figure 10.3; Jordan 1984).

Hurricanes strike the Big Bend coast of Florida less frequently than they hit the south tip of Florida or the Apalachicola delta and points west (Jordan 1984). However, storm winds and storm surges still occur frequently enough to affect coastal hydric hammock stands many times during the life of a tree. HURASIM, a model that estimates storm surges based on hurricane tracks and wind speeds (Doyle 1994), calculates that hurricanes have produced storm surges that struck Cedar Key at least 26 times between 1852 and 2004, an average return interval of less than six years. Based on storm surge records between 1939 and 1975, Ho and Tracey (1975) estimated that Cedar Key experienced storm tides reaching 1.7 m above sea level an average of once every ten years, and storm tides reaching 4.0 m above sea level an average of once every 50 years. The coastal edge of Gulf Hammock, along the embayment south of Cedar Key, may experience storm tides of these magnitudes more frequently (Hine et al. 1987), and coastal hydric hammocks at the southern end of their distribution may experience them less frequently (Ho and Tracey 1975).

10.3.2. Hydrology

Although soil inundation may not occur in hydric hammocks every year, such flooding is more likely to occur during some seasons than others. Seasonal patterns of rainfall, groundwater level, river discharge, tide heights, and storm incidence all interact to affect patterns of flooding in hydric hammocks. Near the coast, these interactions also affect soil and groundwater salinity. On a longer time-scale, multi-year droughts and groundwater withdrawals reduce freshwater supply to the coast, affecting both flooding and salinity in coastal hydric hammocks.

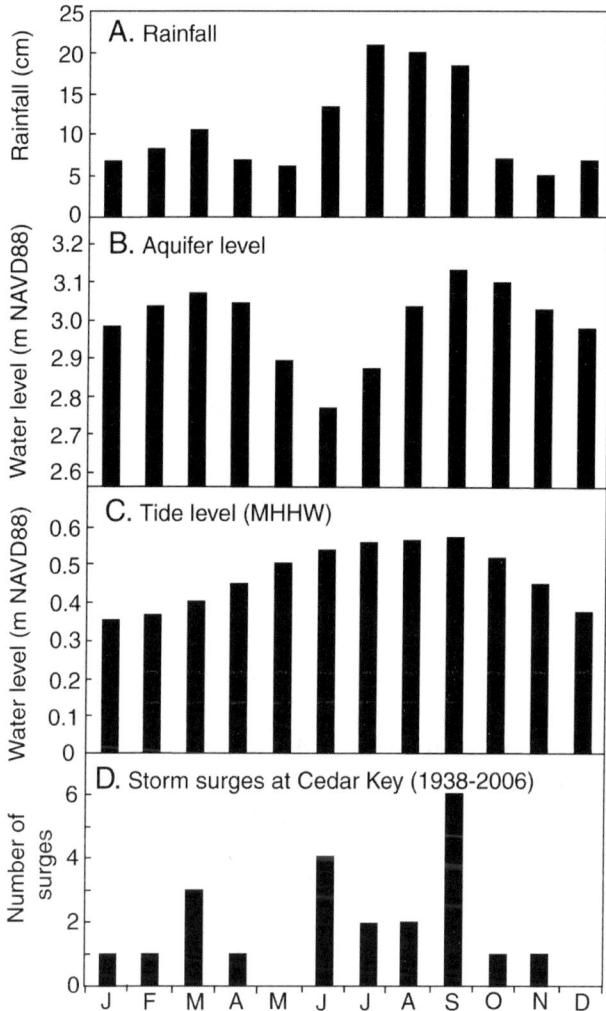

Fig. 10.3. Seasonal variation in hydrological inputs to coastal hydric hammock. (a) Mean monthly precipitation (1941-1970) along the Big Bend coastline of Florida (after Jordan 1984). (b) Mean monthly elevation of the top of the Floridan aquifer near Waccasassa Bay, Florida (Data for October 1982–September 2005, from well S141429001 operated by the Suwannee River Water Management District, located slightly above and within 1-2 km of the edge of the hydric hammocks that ring Waccasassa Bay. Data delivered from SRWMD, March 7, 2006). (c) Mean high tide level (MHHW, mean higher high water) at Cedar Key, Florida (1982-2005, COOPS 2006). (d) Number of tides estimated to have flooded all hydric hammock plots studied by Williams et al. 1999a between October 1938 and July 2006. (Estimated as tides exceeding 1.15 m NAVD at Cedar Key, Florida; COOPS 2006 and R.P. Stumpf, unpublished data).

10.3.2.1. Seasonality of hydrologic inputs and flooding

The Floridan aquifer, which supplies water to most if not all hydric hammocks in this region, may exhibit two seasonal peaks in water level: one following the convective storms of summer and one around March (Figure 10.3). Potential evapotranspiration is relatively low from October through March and high from May through August (Wolfe 1990). This helps depress the level of the aquifer during the summer and elevate it during late winter/early spring. Most springs and rivers along the Big Bend of Florida are linked to the Floridan aquifer, and their flows also tend to peak in late winter/early spring (February and March) or in late summer and fall (August through October, e.g., Yobbi and Knochenmus 1989; Yobbi 1992; Light et al. 1993, 2002). In general, rivers in the northern range of coastal hydric hammocks experience peak flows in late winter, those toward the southern end of the range experience peak flows in late summer, and those in the middle of the range (e.g., the Waccasassa River and some reaches of the Suwannee River) experience bimodal flow patterns (Kelly 2004). High water levels in the Floridan aquifer and in the connected springs and rivers may simply saturate soils of hydric hammocks for extended periods or flood them more deeply (Vince et al. 1989).

Although few studies have been conducted on the flooding regime in hydric hammocks, those that exist generally indicate seasonal patterns that are consistent with seasonal fluctuations in the Floridan aquifer. Water levels measured in an inland hydric hammock in 1986 (Vince et al. 1989) were highest in late winter and spring, low in June, and high again in the fall, similar to seasonal patterns generally found for ground water in central Florida. Light et al. (2002), estimating flooding in a hydric hammock from river levels in the lower reaches of the Suwannee River, showed a similar pattern: with the exception of hurricane flooding, water levels that would have flooded this hammock for substantial periods occurred primarily during February and March with some incidences of flooding occurring in late summer and fall. Where water levels have been measured in hydric hammocks, authors have noted that rises in water level often exceed that which can be explained based on local rainfall (Vince et al. 1989; Williams et al. 1999a). Seepage from surrounding uplands and from the underlying aquifers following regional rainfall events have been hypothesized to account for such a pattern.

Near the coast, tides affect the flooding of hydric hammock stands. Water levels in wells along the coast often display a tidal signal and lack the summer depression that typifies water levels in wells farther from the coast. Instead, groundwater levels near the coast (including that in forest stands near coastal marsh) are often lowest during the winter months

(Trommer 1993; Williams et al. 1999a), reflecting seasonal changes in tide levels in the Gulf of Mexico (Figures 10.3 and 10.4). Tides that flood the coastal fringe of hydric hammock are most likely to occur during the summer and fall, when tides are relatively high. Similarly, hurricane surges that flood great expanses of coastal hydric hammock occur between June and November (hurricane season). Tide gauge data from Cedar Key, however, indicate that storm tides and wind tides can flood coastal hydric hammock during almost any month (Figure 10.3).

10.3.2.2. Salinity

As in all coastal systems, soil salinity in coastal hydric hammock results from interactions between ocean water and freshwater sources. These interactions affect the salinity of tidal waters, the salinity of groundwater, and the dynamics of soil salinity in coastal hydric hammock. In general, tidal waters near large springs and spring-fed rivers are less saline than those further away (e.g., Perry and Williams 1996). However, local interactions between bedrock and water flows may modify soil salinity in tidally flooded areas and, therefore, vegetation patterns. Some springs emit salt water along this coast (E. Raabe, personal communication), undoubtedly affecting vegetation. Conversely, local freshwater seeps have been hypothesized to account for both the existence of small sawgrass patches in areas dominated by the more salt-tolerant black needlerush (Williams et al. 1999a; T. Doyle, personal communication) and for enhanced tree regeneration in stands flooded frequently by tidal water (Williams et al. 1999a).

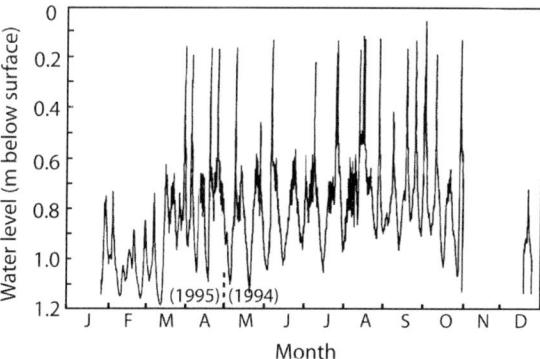

Fig. 10.4. Groundwater level in a coastal remnant of hydric hammock at Waccasassa Bay, Florida (after Williams et al. 1999a). Data from 1994 and 1995 are combined to show patterns in water level over a calendar year. Missing data during November, December, and January represent periods when groundwater levels dropped below the level of the sensors in the shallow monitoring well.

Groundwater salinity near the coast often displays rapid and extreme fluctuations. Rainfall, tide level, and changes in the potentiometric surface of the upper Floridan aquifer all affect groundwater salinity (Trommer 1993; Williams et al. 1999a). High tides can depress the instantaneous outflows of coastal springs, changing water quality rapidly (Hine et al. 1987; Yobbi 1992; Yobbi and Knochenmus 1989). In remnants of coastal hydric hammock, however, some stands appear to experience greater fluctuations in groundwater salinity than others, perhaps reflecting differences in the structure of karstic channels in the underlying limestone (Williams et al. 1999a).

Longer-term changes in tidal flooding and freshwater supply can change the soil salinity in a forest stand, causing major changes in species composition. Increases in tidal flooding frequency, linked to sea-level rise, correlate with increased soil salinity and forest stand decline in some areas (Williams et al. 1999a; see Sect. 10.4). Drought may involve both a reduction in local rainfall and a reduction in freshwater supply through the aquifer (see Sect. 10.4), again increasing soil salinity. Other factors that affect freshwater supply to coastal hydric hammocks (e.g., groundwater withdrawal, climate change, freshwater diversion, etc.) may all affect salinity in coastal stands, with ecological consequences for the stand.

10.4. Forest change at the coastal margin

10.4.1. Historical rates of forest conversion to salt marsh

Forest conversion to salt marsh in the Big Bend region of Florida has been extensive. Tree stumps, tree roots, and forest peats in and under salt marsh in this region have been interpreted as evidence for the drowning of coastal forest by rising seas (e.g., Cooke 1939; Vernon 1951; Kurz and Wagner 1957). Additionally, records of aerial photography have captured the shrinking of forested islands in salt marsh and the retreat of the coastal edge of forest during the 20th century (e.g., Figure 10.5).

Rates of coastal forest retreat have not been constant. Goodbred et al. (1998), studying sediment cores from Waccasassa Bay, described a swamp forest that developed at the mouth of the Waccasassa River between ~4,400 yrs B.P. and 1,800 yrs B.P., then retreated rapidly (10 to < 20 m/yr) around 1,800 yrs B.P. They linked development of the forest peats there to a period of relatively slow sea-level rise and linked the subsequent rapid retreat of the forest to a period of rapidly rising sea level. Raabe et al. (2004), studying forest retreat since the 1800s, compared coastal surveys made between 1852 and 1886 to satellite imagery from 1995. These comparisons indicated that the forest/marsh boundary moved up to 0.5 km inland along most of the Big Bend coast during that period, with less change in marsh/ocean boundary. At Waccasassa Bay, however, the forest margin (which bordered the largest historical expanse of hydric hammock) retreated 1 km or more in most areas, an average retreat of 7 m/yr or more. This does not imply that the migration rate of the forest/marsh boundary has been constant since the mid-1800s. Studies of recent change using satellite imagery indicate that conversion of forest to salt marsh continues (Castaneda and Putz 2007) and that the disappearance of forest canopy may occur in pulses (see Sect. 10.4.4).

10.4.2. Forest zonation and changes in composition near the coastal margin

Most of what is known about forest zonation and dynamics in coastal hydric hammock comes from studies conducted in Gulf Hammock, which borders Waccasassa Bay (e.g., Vince et al. 1989; Simons et al. 1989; Perry

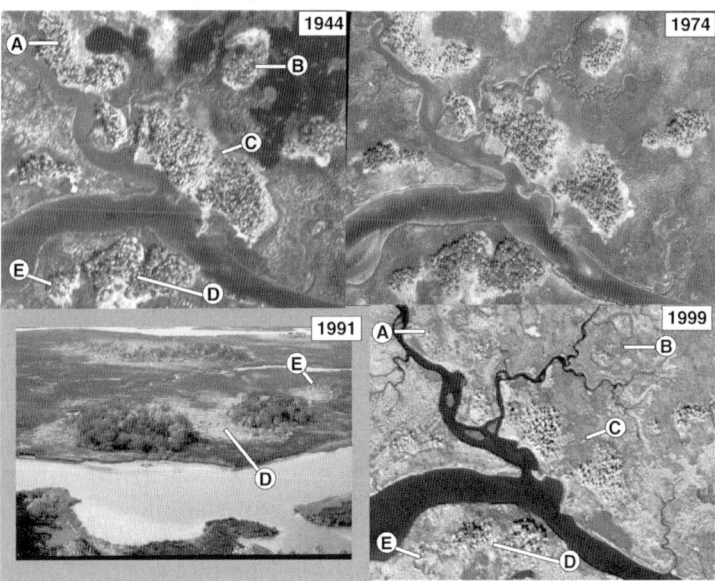

Fig. 10.5. Progressive conversion of coastal forest to salt marsh at Waccasassa Bay. Forested islands A, B, and E almost disappeared between 1944 and 1999. A band of marsh between two forested islands (C) expanded as marsh replaced forest, and a remnant patch of forest that existed as a single island in 1944 became two separate islands with the death of trees in a trough that ran through the center of the island (D).

and Williams 1996; Williams et al. 1998; Williams et al. 1999a; Williams et al. 2003). The coastal edge of this forest lies within the Waccasassa Bay State Preserve, where remnant forest patches persist on topographic highs in the limestone bedrock, surrounded by marsh. Forest within the Preserve contains only small areas of basin swamp and mesic to scrubby flatwoods; most is hydric hammock (Abbott and Judd 2000). This area has a modest freshwater supply compared to some of the other coastal areas that support hydric hammock (e.g., the southern Springs Coast and the mouth of the Suwannee River), and potential differences between the forest zonation found at Waccasassa Bay and that found in areas with higher freshwater discharge should be recognized. Freshwater outflow to Waccasassa Bay is diffuse, with numerous small local streams, springs, and the Waccasassa River emptying into the bay. Freshwater supply to the bay from the Waccasassa River is modest: at least five waterways that empty into the Gulf of Mexico between the St. Marks River and the Hernando/Pasco county line discharge more water than the Waccasassa River (Fernald and Patton 1984, Nordlie 1990, USGS 2005). As a result, tidal waters that influence hydric hammock are fairly salty, though extremely variable (3-30 ppt in Turtle Creek, Figures 10.1 and 10.6; Perry and Williams 1996; Williams et al. 1999a).

At Waccasassa Bay, there is a distinct zonation of coastal hydric hammock with distance from the adjoining salt marsh. At elevations above approximately 1.0 m NAVD or at distances a few hundred meters from the marsh edge, hydric hammock contains a wide variety of tree species (Table 10.1). Closer to the coast, however, the forest is dominated by three species: live oak, southern redcedar, and cabbage palm. At the saltmarsh edge, only two species (cabbage palm and southern redcedar) persist (Vince et al. 1989; Williams et al. 1999a). Carcasses of cabbage palm and southern redcedar, standing in salt marsh, mark former stands of coastal forest (Figure 10.6). In many (but not all) areas, cabbage palm is the last surviving tree species as hydric hammock is converted to marsh.

Forest zonation appears to result from attrition, with certain tree species being eliminated as one approaches the saltmarsh edge. Although cabbage palm and southern redcedar often increase in density toward the edge of the salt marsh, they are also present in inland stands of hydric hammock. Change in forest composition over time may be inferred from patterns of tree regeneration. Studies of stand structure and tree demography in a series of forest plots across a tidal-flooding gradient near Turtle Creek (Figure 10.6) revealed that each of the tree species that existed across a substantial portion of this gradient existed as relict, non-regenerating stands at the coastal margin of its distribution (Figure 10.7; Williams et al. 1999a).

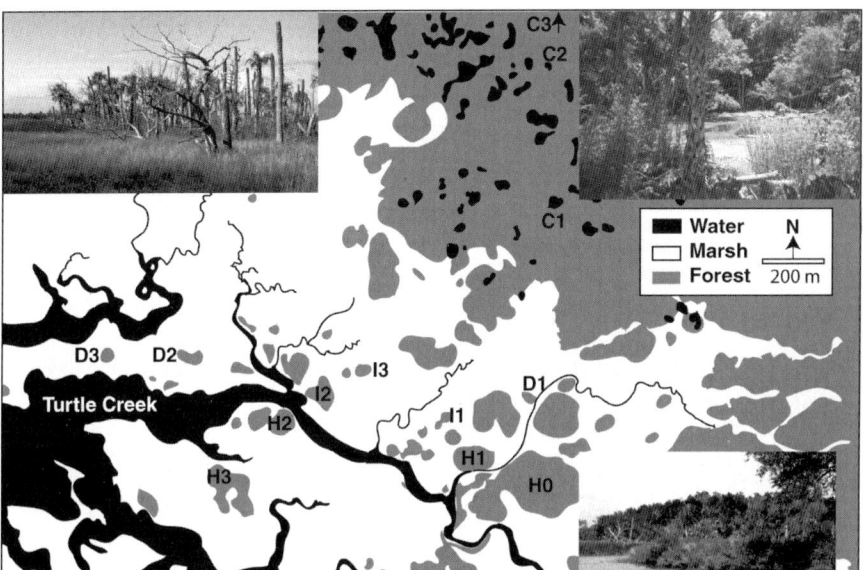

Fig. 10.6. Map of Turtle Creek estuary, near the southern end of Waccasassa Bay. The forest plots (each 400 m^2) studied by Williams et al. 1999a are indicated. Plots C1-3 were in unbroken coastal hydric hammock, plots H0-3 were "healthy" stands on forested islands in salt marsh that supported small seedlings of cabbage palm. Plots I1-3 were intermediate stands with healthy canopy palms but low densities of trunkless palms, and plots D1-3 were stands in advanced stages of decline, with high densities of standing dead palms. Photos: K. Williams and L. Perry.

Such a pattern suggests that changes in forest composition over time in this system are caused largely by failure of regeneration of some species as tidal flooding increases, followed by attrition of their canopy trees. Estimated age structures of relict stands of cabbage palm suggest that canopy trees may persist many decades after regeneration ceases (Williams et al. 1999a).

10.4.3. Patterns of tree regeneration

10.4.3.1. Effects of tidal flooding and salinity on tree regeneration

The lack of tree regeneration in relict stands appears to result from the inability of seedlings to survive environmental conditions in those stands, rather than any other limitation, such as seed availability. Seeding and transplanting seedlings into relict stands have failed to enhance regeneration (Perry and Williams 1996). Studies following planted seeds of cabbage palm and transplanted seedlings of both cabbage palm and southern redcedar across a tidal flooding gradient have confirmed that seedling emergence fails and seedlings cannot survive in stands where no natural seedlings are found (Table 10.2). Patterns of emergence and survival of

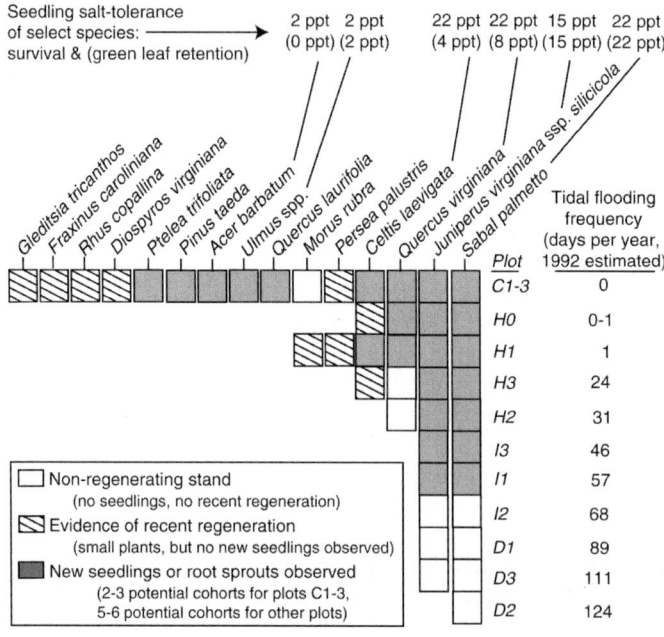

Fig. 10.7. Stand structures of tree species found across a gradient of tidal flooding frequency at Waccasassa Bay (Williams et al. 1999a). For each species a stand is shown as a relict stand, with canopy trees but no recent regeneration (open bars), a regenerating stand, with seedlings and/or vegetative reproduction documented during the study (shaded bars), or a stand on the brink of regeneration failure, with small plants present in the stand, but no seedling regeneration observed during the study (hatched bars). Estimates of tidal flooding frequency for each plot in 1992 were calculated from the tide gauge record at Cedar Key and the relationship between that record and documented weekly high-water levels in each plot (Williams et al. 1999a). (The accuracy of the estimate was better for plots with low flooding frequency than for plots with high flooding frequency, where the relationship between tide height at Cedar Key and plot flooding was slightly poorer.) Two measures of salt tolerance of seedlings of select species are indicated. The first number is the highest salinity that seedlings survived in a greenhouse experiment that subjected seedlings to 0, 2, 4, 8, 15, and 22 ppt sea salt (Williams et al. 1998). The number in parentheses is the highest salinity under which seedlings maintained green leaf tissue. Discrepancies between the two values indicate that plants appearing dead resprouted following freshwater flushing of the soil.

Table 10.2. Tree seedling emergence and survival across a gradient of tidal flooding frequency.

Plot[1]	Elevation[2]	Tidal flooding[3] (days/yr)	Volunteer seedlings of cabbage palm				Planted seeds and seedlings of cabbage palm		Southern red cedar transplants[7]
			Number emerged per plot per year[4]	First-year survival	Second-year survival[5]	Cumulative 2-year survival	Percent emergence from seed[6]	One-year survival of transplants[6]	Cumulative 2-year survival
C1	1.10	0.6	11 (n=3)	86%	83%	71%	-	-	-
C2	1.09	0.6	312 (n=2)	67%	65%	44%	-	-	-
C3	0.69	?[8]	24 (n=2)	53%	37%	20%	-	-	-
H0	0.96	0.8	125 (n=5)	84%	74%	62%	66%	70%	-
H1	0.93	1.5	46 (n=6)	81%	76%	62%	46%	80%	70%
H2&H3[9]	0.79	10 & 17	9 (n=6)	16%	31%	5%	12%	65%	5%
I1,I3	0.62	28 & 58	1 (n=6)	29%	25%	7%	50%	15%	5%
I2	0.67	68	0 (n=6)	-	-	-	0	0%	0%
D1	0.52	79	0 (n=6)	-	-	-	-	0%	0%
D2&D3	0.61	119 & 139	0 (n=6)	-	-	-	0	0%	0%

[1] Plots located on Figure 10.6.
[2] Median elevation for plot(s) in m NAVD (Williams et al. 1999a).
[3] Number of days each plot flooded over a 5-year period (November 1991-October 1996), estimated from tide gauge records at Cedar Key and a temporary tide gauge near the study site. Comparison of the two records suggest that these values, estimated primarily from the Cedar Key tide gauge, may have missed a number of wind-driven tides and could be up to 30% higher than shown.
[4] Mean number of new seedlings per year. Plots were censused for different lengths of time. "n" gives the number of cohorts included in the average.
[5] Percent of first-year survivors that survived the second year
[6] Source: Perry and Williams (1996). No seeds were planted near plots C1, C2, C3, or D1. Fifty seeds were planted near each of the others in the autumn of 1991. Ten small seedlings were planted near each of the ten lower plots in July 1993.
[7] Ten seedlings (<10 cm tall) were planted near each of the lowest nine plots in June 1992. Much of the mortality in plots H2 and H3 was associated with storm damage that occurred in March 1993.
[8] Plot C3 may have been more frequently flooded by tidal water than plots C1 and C2. It was the lowest and wettest of the three plots in continuous forest, with seedlings surviving primarily in small elevated microsites.
[9] Plots H2 and H3 sustained heavy storm damage in March of 1992. Consequently, part of the low emergence and survival of volunteer palm seedlings and part of the low survival of southern red cedar transplants in these plots may be attributable to direct and indirect effects of storm damage.

volunteer palm seedlings parallel those found in experimental plantings. A monitoring study that followed several cohorts of volunteer palm seedlings over several years at Waccasassa Bay showed that emergence and survival declined drastically as tidal flooding frequency exceeded 10-20 days per year (Table 10.2).

The flooding frequency at which regeneration fails differs for different tree species, and the order in which tree species are eliminated from a stand appears to be related to the relative salt-tolerances of their seedlings (Figure 10.7; Williams et al. 1998). Greenhouse experiments have shown that seedlings of tree species confined to areas that experience only the occasional storm surge die at salinities exceeding 2 ppt. This does not imply that these species can regenerate under salinities produced by storm surges, only that periods with low salinities between surges are sufficiently long to allow occasional regeneration of these species. Seedlings of species that persist into the most frequently flooded areas (i.e., cabbage palm and southern redcedar) maintain green leaf tissue at relatively high salinities. Seedlings of species that are excluded from the most frequently flooded plots, but persist in areas that experience the occasional high tide (i.e., live oak and sugarberry [*Celtis laevigata* Willd.]) have seedlings that experience shoot death at high salinity, but resprout after salt is flushed from the soil. While this characteristic would allow seedlings to establish in zones where tides are infrequent, freshwater flushing of soils is adequate, and periods between high tides are sufficiently long to allow seedling recovery, it would not help a seedling survive in zones with persistently high salinity.

Factors other than tidal flooding (e.g., drought and local variation in freshwater outflow) may influence the salinity of tidal water and soils, thereby affecting the position in the landscape where seedling regeneration of various tree species ceases. Comparative transplant studies at Waccasassa Bay and at Chassahowitzka Springs (Figure 10.1), a site with fresher tidal waters, have shown that cabbage palm seedlings tolerate tidal flooding better where the tidal waters are less saline (2-5 ppt in contrast to 3-23 ppt; Perry and Williams 1996). Much more work is required to determine the relative effects of flooding and salinity on patterns of tree regeneration across the entire coast occupied by hydric hammock. So far, however, studies of species zonation, seedling salt tolerance, and comparison of sites with tidal waters of different salinities, indicate that increasing exposure to salt is a major (if not the major) factor that eliminates tree regeneration and creates relict stands in hydric hammock as sea level rises (Perry and Williams 1996; Williams et al. 1998; Williams et al. 1999a).

10.4.3.2. Effects of storms on tree regeneration

Storms can directly damage seedlings with incursions of salt water, disturbance of soil, and deposition of wrack. They may indirectly affect seedlings by modifying their competitive environment. Undoubtedly, storm surges periodically introduce salt water into most of the coastal hydric hammocks on the Gulf Coast of Florida (Figures 10.1 and 10.3). The fact that hydric hammock trees continue to reproduce (except near the coastal fringe) suggests that flushing of salt from soil in this system has historically been sufficient to allow periodic tree recruitment over most of this area. Wrack deposition and erosion may be expected to occur primarily at the coastal edge of hydric hammock where the more salt-tolerant species exist. Data from a storm that struck the west coast of Florida during an ongoing study (the March 1993 "Super Storm", "No-Name Storm". or "Storm of the Century") suggested that seedlings of the most salt tolerant species, cabbage palm, are also very resistant to storm damage. Although small seedlings with simple, strap-shaped leaves suffered some mortality that could be attributed to wrack deposition, survival was still relatively high (30% survival two years after the storm for seedlings buried by > 1 cm of wrack, compared to 68% for seedlings in the same plots that were surrounded by a shallower layer of wrack; $X^2=10.1$, $p<0.01$, K. Williams and M. MacDonald, unpublished data). Larger palm regeneration (trunkless plants with bifid leaves and more numerous pinnae) suffered little mortality at all: of 109 such plants recorded in three "healthy" island plots prior to the storm (H1, H2, and H3; Figure 10.6), 108 were still living two years later. These plots included two plots (H2 & H3) that sustained heavy storm damage, primarily associated with southern redcedar blow-down.

Storm surges can produce long-lasting increases in soil salinity that alter communities and suppress tree regeneration for years (e.g., Conner and Inabinette 2003), or they may produce only transient increases in salinity. The degree and duration of surge-induced increases in salinity often depends on the hydrology and geomorphology of the site and on the water content of the soil prior to the surge (e.g., Gardner et al. 1992). The porous nature of the limestones that support coastal hydric hammock, and the general outwelling of water from the Floridan aquifer in the region, suggest that flushing of salts from soil after a storm surge may be more effective in these coastal hammocks than in other systems. Rapid changes in groundwater salinity in forest stands near this coast attest to the ability of salt and fresh water to move rapidly in and out of this system. Such conclusions are speculative, however, and little formal information exists on the potential duration of storm surge effects on regeneration in coastal hydric hammocks.

Storms can also affect tree regeneration by altering ground level, and changing the susceptibility of the site to sea-level rise. Erosive effects of storms in forests along this coast are not well studied. The March 1993 storm caused little erosion of the salt marsh and deposited sediment in the marsh (Goodbred and Hine 1995). The same storm deposited a thin layer of sediment in forest plots near the marsh. This deposition in forest stands appeared minor, however, and unlikely to agrade soils to elevations more conducive to tree regeneration. Erosion generally cannot change elevation much in these forest stands, because of the shallowness of the soil and the proximity of the limestone bedrock. However, a loss of even 10 cm can change the frequency of tidal flooding in low-lying sites. The March 1993 storm did not result in obvious soil loss from forest plots at Waccasassa Bay, but large fans of soil were ripped from the bedrock and left dangling in the air, as southern redcedars tipped over, carrying soil aloft on its roots. The long-term net effect of this disturbance is unclear. Although tree regeneration is likely to be low in areas with no soil, mounds of soil created by the tip-ups may provide raised microsites that provide refugia for tree seedlings and prolong tree regeneration in a stand.

The most permanent effects of storms on tree seedling recruitment may result from changes in the community composition of a stand and consequent changes in the seedlings' competitive environment. Plant community changes following storms, particularly those involving increases in exotic species, have received much attention (e.g., Smith et al. 1997; Kwit et al. 2000; Horvitz and Koop 2001). Toward the southern end of the range of coastal hydric hammock, Brazilian peppertree (*Schinus terebinthifolius* Raddi) is displacing the native vegetation, and efforts are underway to eradicate it. It quickly colonizes disturbed areas (Abrahamson and Hartnett 1990), and storm damage may favor it. Further north, other species that are not common components of hydric hammocks have been observed to invade stands following storm damage. At Waccasassa Bay, big sandbur (*Cenchrus myosuroides* Kunth) rapidly colonized openings in two previously healthy forest stands (H2 & H3) that suffered major canopy damage during the March 1993 storm. This grass formed dense thickets where an open forest understory once existed (Figure 10.8). Although big sandbur is native to the southern United States, it is an opportunistic species that is not commonly found in hydric hammock. These tall, dense stands of sandbur persist today, twelve years after the storm, and undoubtedly suppress the regeneration of hydric hammock species.

Fig. 10.8. Colonization of a storm-damaged stand by big sandbur. The right foreground shows a section of the stand where wrack was deposited, but the canopy was left intact. The left side shows a thicket of big sandbur that colonized areas where storm damage had removed the tree canopy.

10.4.3.3. The role of plant competition in the failure of tree regeneration

In the absence of other disturbances, competition seems to play little role in the failure of tree regeneration and in the conversion of forest to salt marsh in the tidal reaches of coastal hydric hammock. Analysis of plant community changes across a gradient of tidal flooding frequency at Turtle Creek (Williams et al. 1999a) indicated that tree regeneration of all tree species, including the most salt-tolerant trees, failed before typical salt marsh species became common in the understory. As noted above, however, the introduction of exotic plant species and/or disturbances that damage or remove the tree canopy may allow colonization of forest stands by species that do exert strong competitive pressure on tree seedlings. Exotic species such as cogon grass (*Imperata cylindrica* [L.] Beauv.) and Brazilian peppertree are expanding into new areas and have been observed on forested islands in the southern range of coastal hydric hammock. Competition from such exotics may play an increasing role in eliminating tree regeneration and hastening the elimination of coastal hydric hammock.

10.4.4. Patterns and dynamics of canopy tree loss

Canopy trees may be lost from a stand through gradual attrition following regeneration failure, or they may die in pulses. Variation among species in longevity or in susceptibility to stressors may cause differences in how quickly individual species are lost from stands following regeneration failure. Such variation contributes to changes in species composition of the forest. The rapidity with which the last surviving tree species dies dictates how long forest structure remains on a site after increased tidal flooding has eliminated all tree regeneration. Temporal and spatial variation in environmental conditions, such as freshwater supply and salinity, influences the health of coastal forest trees and may affect the length of time that relict, non-regenerating stands persist in the tidal reaches of hydric hammock.

10.4.4.1. Effects of storms on canopy composition and loss

Trees growing at the coastal margins of forests, where periodic hurricanes and storm surges occur, are often both salt-tolerant and wind-tolerant (Williams et al. 1999b and references therein). Variation in susceptibility to wind damage may contribute to tree species zonation near the coast, with storms removing canopy trees of more susceptible species near the forest/marsh boundary. Some studies suggest that two of the tree species found near the coastal margin of hydric hammock (live oak and cabbage palm) are more resistant to storm damage than the other tree species (Touliatos and Roth 1971; Williams et al. 1999b). Although southern redcedar may persist near the saltmarsh boundary longer than many other tree species because of its high salt tolerance (Touliatos and Roth 1971), storms can remove canopy trees, leaving the overstory dominated by more storm-resistant species such as cabbage palm. The storm that struck the west coast of Florida in March 2003 selectively removed southern redcedar trees from two of 13 study plots at Turtle Creek (plots H2 & H3, Figure 10.6). Mortality among southern redcedars was tightly linked to visible storm damage, and mortality continued for two years after the storm. Palms, in contrast were unaffected (Williams et al. 2003). This difference in susceptibility to storms contributes to cabbage-palm dominance in coastal forest stands.

10.4.4.2. Effects of freshwater supply on canopy composition and loss

Reductions in freshwater supply to the coast appear to hasten the elimination of canopy trees from the relict, non-regenerating stands at forest/marsh boundary. At Waccasassa Bay, increased mortality among southern redcedar trees and cabbage palms coincided with a historic multi-year drought that began in 1998 and lasted through 2002 (Figure 10.9). This drought, which affected the entire southeastern United States, was the most severe drought to affect the Big Bend region of Florida since the mid-1950s (Verdi et al. 2006). Low flows from rivers and low groundwater levels resulted both from low rainfall during this period and from increased groundwater withdrawals that were linked to dry conditions and, consequently, higher demand (Verdi et al. 2006). Increased mortality was detected first among southern redcedar trees in the most fre-

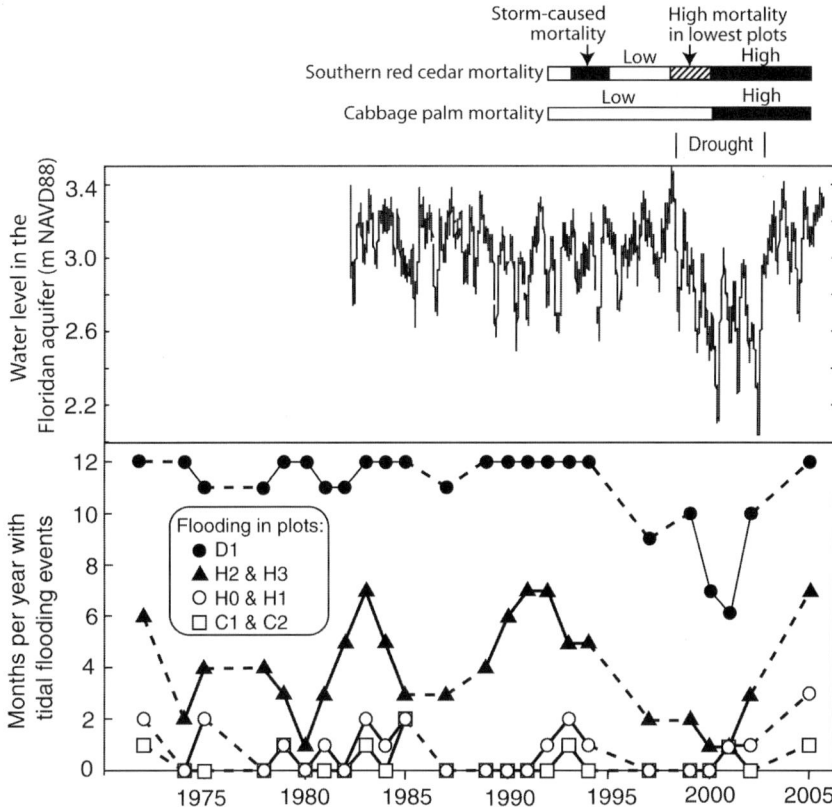

Fig. 10.9. Changes in the level of the Floridan aquifer, tidal flooding frequency, and tree mortality in the most frequently flooded plots at Waccasassa Bay. Record includes the drought of 1998-2002. The aquifer level was measured at Rosewood, 5-6 km north of Waccasassa Bay (Suwannee River Water Management District, well S141429001; data provided in March 2006). Months with tidal flooding were estimated from tide gauge records at Cedar Key Florida, and the relationship between those records and tidal flooding measured in plots in 1992 (Williams et al. 1999a). Years with fewer than 11 months of good data were omitted.

quently flooded stands they occupied (Williams et al. 2003). Cabbage palms did not suffer increased mortality until slightly later in the drought (sometime between censuses conducted in 2000 and in 2005). For both species, increased mortality was restricted to the stands most frequently flooded by tidal water, where regeneration had already ceased or fallen to very low levels (S. Bhotika, L. R. Grawe DeSantis, and F. E. Putz, University of Florida, unpublished data).

Although tree mortality is expected to increase with increased tidal flooding (given the earlier regeneration failure and consequent older average tree age in those stands), mortality patterns observed during the drought suggest that these relict stands are also more vulnerable to reductions in freshwater supply. Although the drought of 1998-2002 reduced rainfall across the region, increased canopy tree mortality was only detected in plots flooded frequently by tidal water (S. Bhotika, L. R. Grawe DeSantis, and F. E. Putz, University of Florida, unpublished data). Increased tidal flooding itself, however, could not account for this observed pulse of mortality: tide gauge records indicated that tidal flooding frequency during the drought was lower than normal, rather than higher (Figure 10.9). In the relict stands at the coastal fringe of hydric hammock, therefore, reduced tidal flushing may have combined with reduced freshwater outflow to increase soil salinity, a factor that has previously been associated with tree mortality in this system (Williams et al. 1999a, 2003).

Reductions in freshwater supply to the coast can also change the species composition of coastal forest stands. Because southern redcedars began to die earlier in the drought than cabbage palms (Figure 10.9), a shorter or less severe drought may have selectively removed southern redcedar, leaving cabbage palm more dominant near the saltmarsh edge. Cabbage palms are often considered to be more salt-tolerant than southern redcedars (e.g., Touliatos and Roth 1971), and the large quantities of water stored in their stems may allow them to survive long periods of low water availability (Holbrook and Sinclair 1992). Thus, dominance of the coastal forest fringe by cabbage palms may result partly from their greater resistance to storm damage (above) and partly from their greater tolerance of salinity and water stress.

High freshwater discharge at the coast may slow the rate at which canopy trees are lost in the tidal reaches. In coastal areas where freshwater discharge is relatively high and tidal waters are low in salinity, canopy trees may remain healthy much longer than those in other areas. Palms in intertidal zones of areas with low freshwater outflows often appear more stunted than those in areas with high rates of freshwater discharge, reflecting the negative impact of salinity on canopy tree health (Figure 10.10). Moreover, non-regenerating stands of palms in less saline areas appear to maintain healthy, closed canopies for some time (Figure 10.11). The possibility that adequate freshwater discharge from the coast may prolong forest health and delay forest retreat in the face of rising seas finds some support in the mapping studies conducted by Raabe et al. (2004). Their assessment of forest retreat along the Big Bend coastline since the mid-1800s showed little movement of the forest/marsh boundary in areas with high freshwater discharge (e.g., near the mouths of the Suwannee River and the Weeki Wachi River). Good quantification of the effect of fresh-

Fig. 10.10. Palms in coastal marsh (A) near Turtle Creek, on Waccasassa Bay, and (B) near Chassahowitzka River. Trees in marsh at Turtle Creek tend to have smaller canopies and appear more stunted than those near Chassahowitzka River where tidal waters are less saline. Photos: K. Williams and L. Perry.

Fig. 10.11. A tidally flooded stand of cabbage palms with a healthy closed canopy near the Chassahowitzka River. Some small palms are visible on raised microsites in the background, but most of the stand is not regenerating.

water discharge on rates of forest canopy loss and coastal forest retreat requires further study.

10.4.5. Future changes in Gulf coast hydric hammocks

Many of the environmental factors that shape coastal hydric hammock and cause its conversion to marsh are changing or are predicted to change. Sea level continues to rise and may accelerate (IPCC 2001). Most of the hydric hammocks that are currently protected are in a narrow band of conservation land near the coast and are, therefore, subject to effects of sea-level rise. Digital elevation models predict that Waccasassa Bay State Preserve, which supports the remnants of the once extensive Gulf Hammock, will lose 20 percent of its forest cover over the next 50 years to sea-level rise (Castaneda and Putz 2007), and that St. Marks National Wildlife Refuge will lose thousands of hectares of coastal forest, including expanses of hydric hammock, to sea-level rise by the year 2100 (Doyle et al. 2003). Intense hurricanes, which occasionally flood even the inland reaches of the Gulf Coast hydric hammocks, may become more common as sea-surface temperatures rise (e.g., Webster et al. 2005; Hoyos et al. 2006). Damage caused by high hurricane winds should reinforce palm dominance of the coastal margin and may open greater expanses of hydric hammock to colonization by invasive, opportunistic species. High storm tides, produced by the combination of more intense hurricanes and higher sea level, should flood hydric hammocks more frequently. It is difficult to predict, however, the extent to which these hurricanes will increase the frequency of tidal flooding toward the inland edge of coastal hydric hammocks, eliminate the more salt-sensitive species, and shift zones of forest composition inland.

Rainfall, which serves to raise the level of the aquifer, flush salts from the soil, and lower tidal salinity, is difficult to forecast. Global Change Models, while generally predicting increased average precipitation across the earth, are currently poor at predicting precipitation change at the regional scale (Gleick 2000). These models and the models that use them variously predict increasing or decreasing rainfall and runoff for the southeastern United States (Gleick 2000). Increases in rainfall may help counteract any increases in groundwater consumption that occur as the region's human population grows, whereas reductions in rainfall will exacerbate groundwater depletion. Changes in precipitation, while currently difficult to predict, have important consequences for the fate of coastal hydric hammocks. Our best information to date, while limited, suggests that drought hastens loss of coastal forest in tidal areas, and that high freshwater

ter outflow at the coast may prolong both seedling regeneration and canopy tree health as sea level rises.

10.5. Research needs

Quantification of the relative roles that groundwater seepage, local rainfall, and overland flow play in delivering fresh water to coastal hydric hammock stands, flushing salt from soil, and repelling saltwater intrusion is needed for informed management of these stands. A reduction in all freshwater sources, observed during the drought of 1998-2002, was associated with a pulse of tree death at the margin of coastal hydric hammock. However, water consumption by a growing human population may draw down the aquifer independently of any changes in rainfall. Determining the relative roles that local rainfall and groundwater seepage play in reducing salinity at the coastal margin and maintaining forest health is needed to inform water management policies about ecological impacts of groundwater depletion. Other human activities, such as ditching and mounding for drainage and access roads, may alter hydrology, but the extent to which such modifications change flooding in a hydric hammock is not clear. The impacts may vary with distance from seeps and springs. Beyond knowing that hydric hammocks derive much of their water from the Floridan aquifer, that they flood occasionally, that increasing salinity causes compositional change and forest loss at the coastal margin, and that increasing salinity is partially caused by rising sea level, too little is known about the controls on water and salt dynamics in this system for useful management application.

Better quantification is needed of how freshwater outflow (1) affects the elevation at which tree regeneration can occur in tidally flooded areas, and (2) affects how long mature trees persist and remain healthy in non-regenerating stands. Digital elevation models that are used to predict the rate and pattern of forest retreat as sea level rises are excellent tools for planning and management. However, if freshwater outflow changes the elevation at which tree regeneration occurs and prolongs the existence of tree canopy cover, parameters in such models may need to be adjusted to account for variation in freshwater supply at the coast. Furthermore, if high supplies of freshwater maintain healthy forest canopies long after regeneration ceases, aerial surveys of coastal forest may not detect the extent of relict, non-regenerating stands. In areas with only modest freshwater outflows, the tree canopy may thin soon after regeneration failure of the last remaining tree species. These relict stands, appearing as a "transitional scrub" of halophytic shrubs with thin tree cover, can be detected through remote sensing (Raabe et al. 2004). In areas with higher freshwater outflows, relict stands that maintain a healthy canopy may be more difficult to detect. Since tree regeneration is important in the long-term fate of the forest, knowing the extent of relict stands has important implications for management and planning in coastal reserves.

Dendrochronology may prove a useful tool in reconstructing stand histories, defining environmental tolerances, and approaching questions of how long mature trees of various species persist in hydric hammock after regeneration failure. Tree rings can yield histories of both salt-exposure (Yanosky et al. 1995) and tree growth, potentially shedding light on the relative salt-tolerances of different tree species and yielding data on site-to-site differences in the history of salt exposure of forest stands. Unfortunately, cabbage palm, the most persistent tree species in hydric hammock, does not form growth rings and is not amenable to such studies. However, as sea level rises, most hydric hammock species are lost from coastal forest long before cabbage palm and the last vestiges of forest structure disappear. These tree species include some considered rare in Florida (Abbott and Judd 2000), and tree ring studies may elucidate the dynamics and timing of this earlier phase of hydric hammock simplification and loss.

Finally, better understanding of how different forest types in the Big Bend region of Florida react to rising sea level is needed. Although this chapter describes what we know about processes driving the conversion of hydric hammock to salt marsh, and other swamp forests are treated elsewhere in this book, additional forest types exist along Florida's Big Bend coastline. These contribute to the pattern of coastal forest retreat detected in aerial surveys and remote sensing studies. Some of these forests may have quite different hydrological relationships to the Floridan aquifer. Pine flatwoods, for example, with their characteristic hard pans that limit water exchange between vegetation and underlying ground water, abut salt marsh in some areas, and these hardpans extend into the marsh (Kurz and Wagner 1957). The interplay between salt water and fresh water in these forests may differ somewhat from that in hydric hammocks, potentially affecting rates and patterns of forest conversion to salt marsh.

10.6. Conclusions

Both natural and anthropogenic factors shape the structure, dynamics, and extent of coastal hydric hammock. While these forests have been largely converted to other uses on private lands, intact stands persist in a narrow band of conservation lands near the coast. The predominant forces influencing these remaining stands appear to be those affecting the interplay between fresh water and salt water, punctuated by periodic wind damage from storms. Factors affecting salinity in these stands include frequency of tidal flooding, storm surges, droughts, proximity to high-volume springs and rivers, and, potentially, groundwater withdrawal that reduces freshwater efflux from the Floridan aquifer. Because of the low topographic relief of the regions supporting coastal hydric hammock and because of barriers to forest migration inland, sea-level rise is generally considered to be the greatest threat to these forests. Freshwater supplies may ameliorate impacts of tidal flooding, however, and prolong tree regeneration and tree health as sea level rises.

In general, rising sea level increases the frequency of tidal flooding in coastal hydric hammock stands, sequentially eliminating the regeneration of different tree species. Mature trees may persist long after regeneration ceases, leaving relict stands at the coastward margin of their distribution. Because seedlings of different tree species have different tolerances of saltwater flooding, this leaves relict stands of more susceptible species within regenerating stands of more tolerant species.

Attrition of canopy trees is expected to follow regeneration failure, changing the species composition of the forest as sea level rises and trees in relict stands reach the end of their lifespan. However, variation in the tolerance of mature trees to coastal stressors may selectively remove certain species from relict stands, contributing to forest zonation. While less is generally known about tolerance limits of mature trees than about tolerance limits of seedlings, the relatively rapid and drastic response of southern redcedar to drought and storms suggests that drought and storms may play roles in eliminating southern redcedar from relict coastal stands, leaving cabbage palm the sole survivor.

Progressive conversion of coastal forest to marsh has been documented along the coastline occupied by hydric hammock, moving the forest/marsh boundary more than 1 km inland in places since the mid-1800s. It is likely that the rate of this boundary movement has not been constant. Both droughts and storms produce episodes of rapid canopy loss in this intertidal zone. The movement of this boundary, however, represents primarily death of canopy trees in stands that have long since ceased to regenerate. It also represents only the final loss of a greatly simplified forest community. By the time forest structure is finally lost and converted to coastal marsh, most hydric hammock tree species have long since succumbed to encroaching coastal influences and been eliminated from the stand.

10.7. Acknowledgments

We would like to thank Norm Mangrum of the National Wetlands Inventory for providing information on coastal wetland types along the Springs Coast, Rick Stumpf of NOAA for providing detailed tide gauge records for Cedar Key, Ellen Raabe of the U.S. Geological Survey for providing information on ongoing studies and useful comments on an earlier draft of this manuscript, staff at the Water Resources Institute at California State University San Bernardino for providing reports and resources, Linda Perry for providing photographs, Tom Doyle for providing storm surge estimates and helpful comments, and Jack Putz, Smriti Bhotika, Larisa Grawe DeSantis, and Hector Castaneda for providing unpublished data of recent studies at Waccasassa Bay.

References

Abbott JR, Judd WS (2000) Floristic inventory of the Waccasassa Bay State Preserve, Levy County, Florida. Rhodora 102:439-513

Abrahamson WG, Hartnett DC (1990) Pine flatwoods and dry prairies. In: Myers RL, Ewel J (eds) Ecosystems of Florida. University of Central Florida Press, Orlando, pp 103-149

Brown RB, Stone EL, Carlisle VW (1990) Soils. In: Myers RL, Ewel J (eds) Ecosystems of Florida. University of Central Florida Press, Orlando, pp 35-69

Bush PW, Johnson RH (1988) Ground-water hydraulics, regional flow, and ground-water development of the Floridan aquifer system in Florida and in parts of Georgia, South Carolina, and Alabama. Professional Paper 1403-C. U.S. Geological Survey, Washington

Castaneda H, Putz FE (2007) Predicting sea-level rise effects on a nature preserve on the Gulf Coast of Florida: a landscape perspective. Florida Scientist 70:166-175

Clausen CJ, Cohen AD, Emiliani C, Holman JA, Stipp JJ (1979) Little Salt Spring, Florida: a unique underwater site. Science 203:609-614

Conner WH, Inabinette LW (2003) Tree growth in three South Carolina (USA) swamps after Hurricane Hugo: 1991-2001. For Ecol Manage 182:371-380

Cooke CW (1939) Scenery of Florida interpreted by a geologist. State Geological Survey, Tallahassee

COOPS (2006) Observed water levels. Center for Operational Oceanographic Products and Services, National Ocean Service, National

Oceanographic and Atmospheric Administration, Silver Springs, http://tidesandcurrents.noaa.gov/

Doyle TW (1994) Field and modeling studies of hurricane disturbance on Gulf coastal forest ecosystems. In: Global strategies for environmental issues: NAEP 19th annual conference proceedings. NAEP Publications, Washington, pp 167-173

Doyle TW, Day RH, Biagas JM (2003) Predicting coastal retreat in the Florida Big Bend region of the Gulf Coast under climate change induced sea-level rise. In: Ning ZH, Turner RE, Doyle T, Abdollahi KK (eds) Integrated assessment of climate change impacts on the Gulf Coast region. GCRCC, Baton Rouge, pp 201-209

Dunbar JS, Webb SD, Faught MK (1988) Page/Ladson (8 Je 591): an underwater Paleo-Indian site in northwestern Florida. Florida Anthropologist 41:442-452

Dunbar JS, Webb SD, Faught MK (1992) Archaeological sites in the drowned Tertiary Karst region of the eastern Gulf of Mexico. In: Johnson L, Stright M (eds) Paleo-shorelines and prehistory: An investigation in method. CRC Press, Boca Raton, pp 117-146

Ewel KC (1990) Swamps. In: Myers RL, Ewel J (eds) Ecosystems of Florida. University of Central Florida Press, Orlando, pp 281-323

Faught M, Carter B (1998) Early human occupation and environmental change in northwestern Florida. Quaternary International 49/50:167-176

Fernald EA, Patton DJ (1984) Water resources atlas of Florida. Florida State University, Tallahassee

Florida DEP (2006) Land boundary information system. Florida Department of Environmental Protection, Tallahassee (http://data.labins.org, accessed April 12, 2006)

Florida Division of Emergency Management (2005) Surge zones. Florida Division of Emergency Management, GIS Section, Tallahassee, http://floridadisaster.org/PublicMapping/

FNAI (1990) Guide to the natural communities of Florida. Florida Natural Areas Inventory and Florida Department of Natural Resources, Tallahassee

Gardner LR, Michener WK, Williams TM, Blood ER, Kjerve B, Smock LA, Lipscomb DJ, Gresham C (1992) Disturbance effects of Hurricane Hugo on pristine coastal landscape: North Inlet, South Carolina, USA. Netherlands J Sea Res 30:249-263

Gleick PH (lead author) (2000) Water: The potential consequences of climate variability and change for the water resources of the United States. A report of the National Water Assessment Group for the U.S.

Global Change Research Program. Pacific Institute for Studies in Development, Environment and Security, Oakland

Goodbred SL, Hine AC (1995) Coastal storm deposition: salt-marsh response to a severe extratropical storm, March 1993, west-central Florida. Geology 23:679-682

Goodbred SL, Wright EE, Hine AC (1998) Sea-level change and storm-surge deposition in a Late Holocene Florida salt marsh. J Sed Res 68:240-252

Hine AC, Evans MW, Mearns DL, Belknap DF (1987) Effect of Hurricane Elena on Florida's marsh-dominated coast: Pasco, Hernando, and Citrus Counties. Florida Sea Grant College Technical Paper 49. University of South Florida, St. Petersburg

Ho FP, Tracey RJ (1975) Storm tide frequency analysis for the Gulf Coast from Cape San Blas to St. Peterburg Beach. NOAA Technical Memorandum, National Weather Service, HYDRO-20

Holbrook NM, Sinclair TR (1992) Water balance in the arborescent palm, *Sabal palmetto*. II. Transpiration and stem water storage. Plant Cell Environ 15:401-409

Horvitz CC, Koop A (2001) Removal of nonnative vines and post-hurricane recruitment in tropical hardwood forests of Florida. Biotropica 33:268-281

Hoyos CD, Agudelo PA, Webster PJ, Curry JA (2006) Deconvolution of the factors contributing to the increase in global hurricane intensity. Science 312:94-97

IPCC (2001) Climate change 2001: The scientific basis. Contribution of Working Group I to the third assessment report of the Intergovernmental Panel on Climate Change. Houghton JT, Ding Y, Griggs DJ, Noguer M, van der Linden PJ, Xiaosu D (eds), Cambridge University Press, UK

Jordan CL (1984) Florida's weather and climate: implications for water. In: Fernald EA, Patton DJ (eds) Water resources atlas of Florida. Florida State University, Tallahassee, pp 18-35

Kelly M (2004) Florida river flow patterns and the Atlantic Multidecadal Oscillation. Draft report. Southwest Florida Water Management District, Brooksville

Kurz H, Wagner K (1957) Tidal marshes of the Gulf and Atlantic coasts of northern Florida and Charleston, South Carolina. Florida State University Studies Number 24. Florida State University, Tallahassee

Kwit C, Platt WJ, Slater HH (2000) Post-hurricane regeneration of pioneer plant species in south Florida subtropical hardwood hammocks. Biotropica 32:244-251

Light HM, Darst MR, Lewis LF (2002) Hydrology, vegetation, and soils of riverine and tidal floodplain forests of the lower Suwannee River, Florida, and potential impacts of flow reductions. Professional Paper 1656A. U.S. Geological Survey, Tallahassee

Light HM, Darst MR, MacLaughlin MT, Sprecher SW (1993) Hydrology, vegetation, and soils of four North Florida river flood plains with an evaluation of state and federal wetland determinations. Water-Resources Investigation Report 93-4033. U.S. Geological Survey, Tallahassee

Milanich JT (1994) The archaeology of Precolumbian Florida. University Press of Florida, Gainesville

Monk JF (1987) The cedar keys and the pencil industry. Unpublished manuscript, Cedar Key Historical Society, Cedar Key

Montague CL, Wiegert RG (1990) Salt marshes. In: Myers RL, Ewel J (eds) Ecosystems of Florida. University of Central Florida Press, Orlando, pp 481-516

Nordlie FG (1990) Rivers and springs. In: Myers RL, Ewel J (eds) Ecosystems of Florida. University of Central Florida Press, Orlando, pp 392-425

Perry L, Williams K (1996) Effects of salinity and flooding on seedlings of cabbage palm (*Sabal palmetto*). Oecologia 105:428-434

Raabe EA, Streck AE, Stumpf RP (2004) Historic topographic sheets to satellite imagery: A methodology for evaluating coastal change in Floridal's Big Bend tidal marsh. U.S. Geological Survey Open-File Report 02-211. U.S. Geological Survey, Center for Coastal and Regional Marine Studies, St. Petersburg

Simons RW (1990) Chapter 5. Terrestrial and freshwater habitats. In: Wolfe SH (ed) An ecological characterization of the Florida Springs Coast: Pithlachascotee to Waccasassa Rivers. Biological Report 90(21). U.S. Fish and Wildlife Service, Washington

Simons RW, Vince SW, Humphrey SR (1989) Hydric hammocks: A guide to management. Biological Report 85(7.26 Supplement). U.S. Fish and Wildlife Service, Washington

Smith GF, Nicholas NS, Zedaker SM (1997) Succession dynamics in a maritime forest following Hurricane Hugo and fuel reduction burns. For Ecol Manage 95:275-283

Stumpf RP, Haines JW (1998) Variations in tidal level in the Gulf of Mexico and implications for tidal wetlands. Estuar Coastal Shelf Sci 46:165-173

Tanner WF (1992) Natural environment. In: Fernald EA, Purdum ED (eds) Atlas of Florida. University Press of Florida, Tallahassee, pp 12-77

Touliatos P, Roth E (1971) Hurricanes and trees – ten lessons from Camille. J For 69:285-289

Trommer JT (1993) Description and monitoring of the saltwater-freshwater transition zone in aquifers along the wet-central coast of Florida. Water-Resources Investigations Report 93-4120. U.S. Geological Survey, Tallahassee

USGS (2005) Water resources data for Florida – Water Year 2005. U.S. Geological Survey, Tallahassee

Verdi RJ, Tomlinson SA, Marella RL (2006) The drought of 1998-2002: Impacts on Florida's hydrology and landscape. Circular 1295. U.S. Geological Survey, Tallahassee

Vernon RO (1951) Geology of Citrus and Levy Counties, Florida. Geological Bulletin No. 33. Florida Geological Survey, Tallahassee

Vernon RO, Puri HS (1964) Geologic map of Florida. Division of Geologic Map Services No 18. U.S. Geological Survey in cooperation with Florida Board of Conservation, Tallahassee

Vince SW, Humphrey SR, Simons RW (1989) The ecology of hydric hammocks: A community profile. Biological Report 85(7.26). U.S. Fish and Wildlife Service, Washington

Webster PJ, Holland GJ, Curry JA, Chang H-R (2005) Changes in tropical cyclone number, duration, and intensity in a warming environment. Science 305:1844-1846

Williams K, Meads MV, Sauerbrey DA (1998) The roles of seedling salt tolerance and resprouting in forest zonation on the west coast of Florida, USA. Am J Bot 85(12):1745-1752

Williams K, Ewel KC, Stumpf RP, Putz FE, Workman TW (1999a) Sea-level rise and coastal forest retreat on the west coast of Florida, USA. Ecology 80:2045-2063

Williams K, Pinzon ZS, Stumpf RP, Raabe EA (1999b) Sea-level rise and coastal forests on the Gulf of Mexico. Open-File Report 99-441. U.S. Geological Survey, Center for Coastal Geology, St. Petersburg

Williams K, MacDonald M, Sternberg LDSL (2003) Interactions of storm, drought, and sea-level rise on coastal forest: a case study. J Coast Res 19(4):1116-1121

Wolfe SH (1990) Chapter 4. Hydrology and water quality. In: Wolfe SH (ed) An ecological characterization of the Florida Springs coast: Pithlachascotee to Waccasassa Rivers. Biological Report 90(21). U.S. Fish and Wildlife Service, Washington

Yanosky TM, Hupp CR, Hackney CT (1995) Chloride concentrations in growth rings of *Taxodium distichum* in a saltwater-intruded estuary. Ecol Appl 5:785-792

Yobbi DK (1992) Effects of tidal stage and ground-water levels on the discharge and water quality of springs in coastal Citrus and Hernando Counties, Florida. Water-Resources Investigations Report 92-4069. U.S. Geological Survey, Tallahassee

Yobbi DK, Knochenmus LA (1989) Effects of river discharge and high-tide stage on salinity intrusion in the Weeki Wachee, Crystal, and Withlacoochee River estuaries, Southwest Florida. Water-Resources Investigations Report 88-4116. U.S. Geological Survey, Tallahassee

Zervas C (2001) Sea level variations of the United States 1854-1999. NOAA Technical Report NOS CO-OPS 36. National Oceanic and Atmospheric Administration, Silver Spring

Chapter 11 - Ecological Characteristics of Tidal Freshwater Forests Along the Lower Suwannee River, Florida*

Helen M. Light[1], Melanie R. Darst[1], and Robert A. Mattson[2]

[1] *U.S. Geological Survey, Florida Integrated Science Center, 2010 Levy Avenue, Tallahassee, FL 32310*
[2] *St. Johns River Water Management District, 4049 Reid Street, Palatka, FL 32177*

11.1. Introduction

An integrated understanding of hydrology, topography, and soils can help explain the composition and distribution of tree species in wetland forests along coastal plain rivers. Hydrologic factors, which are typically the key forcing functions in wetlands, can be especially complex in tidally-influenced forests of river floodplains. River flooding from upstream as well as coastal influences including storm surges, tides, and salinity play important roles in determining the structure and composition of floodplain forests near the mouth of a river. This chapter uses an interdisciplinary approach to describe ecological characteristics of tidal freshwater swamps of the lower Suwannee River.

The Suwannee River flows from its headwaters in the Okefenokee Swamp in southeastern Georgia to the Gulf of Mexico in the Big Bend region of north-central Florida (Figure 11.1). The river drainage divides the southern temperate panhandle region of Florida from the subtropical peninsular region of the State. Biotic richness in the Suwannee River Basin is high, because many northern and some subtropical taxa of flora and fauna reach the geographical limits of their distribution in this area (Clewell 1985; Carr 1994). In the lower Suwannee River from the confluence of the

---

* The U.S. Government's right to retain a non-exclusive, royalty-free licence in and to any copyright is acknowledged.

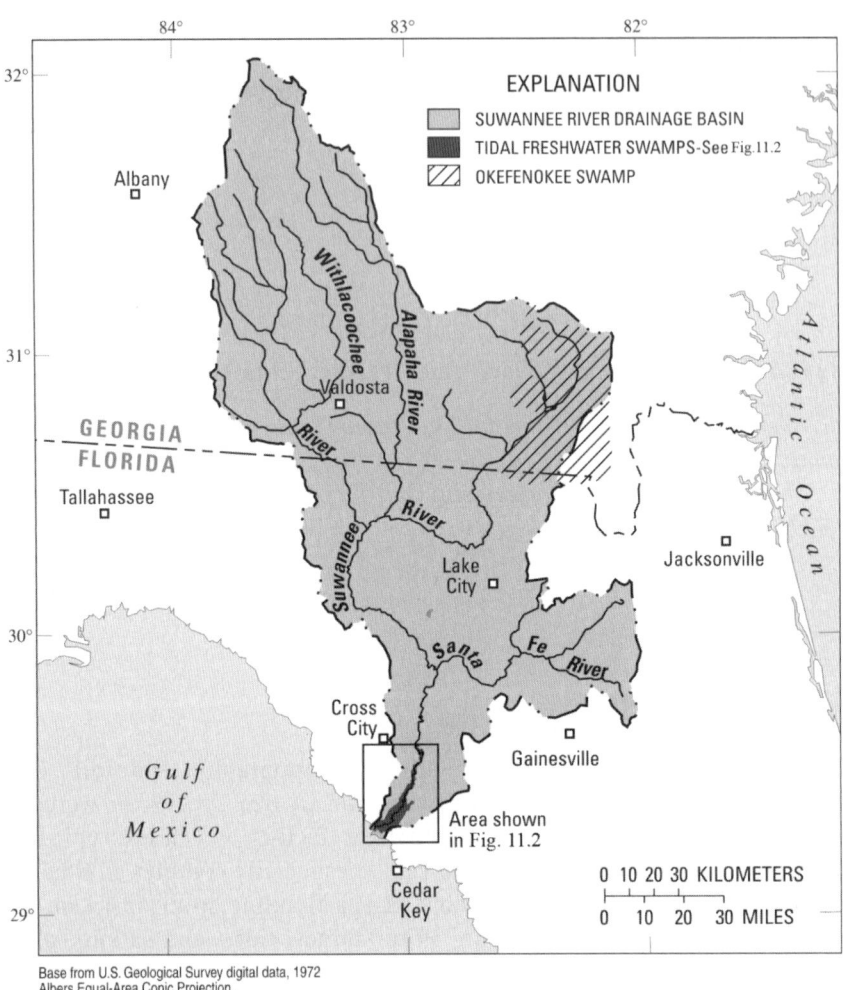

Fig. 11.1. Drainage basin of the Suwannee River in Florida and Georgia.

Santa Fe River to the Gulf of Mexico, species richness of canopy and subcanopy trees and woody vines in the river floodplain (77 species) is among the highest of river floodplains in North America (Brinson 1990; Light et al. 2002).

Tidal freshwater forests, covering 8,160 ha in the lower Suwannee River floodplain, change in composition along a gradient from river kilometer (rkm) 45 downstream to rkm 2, where forests give way to coastal marshes at the "treeline" (Figures 11.2-11.5). Although this transition is gradual, variations in hydrology, vegetation, and soils generally support a division of the lower Suwannee River floodplain into three reaches: nontidal, up-

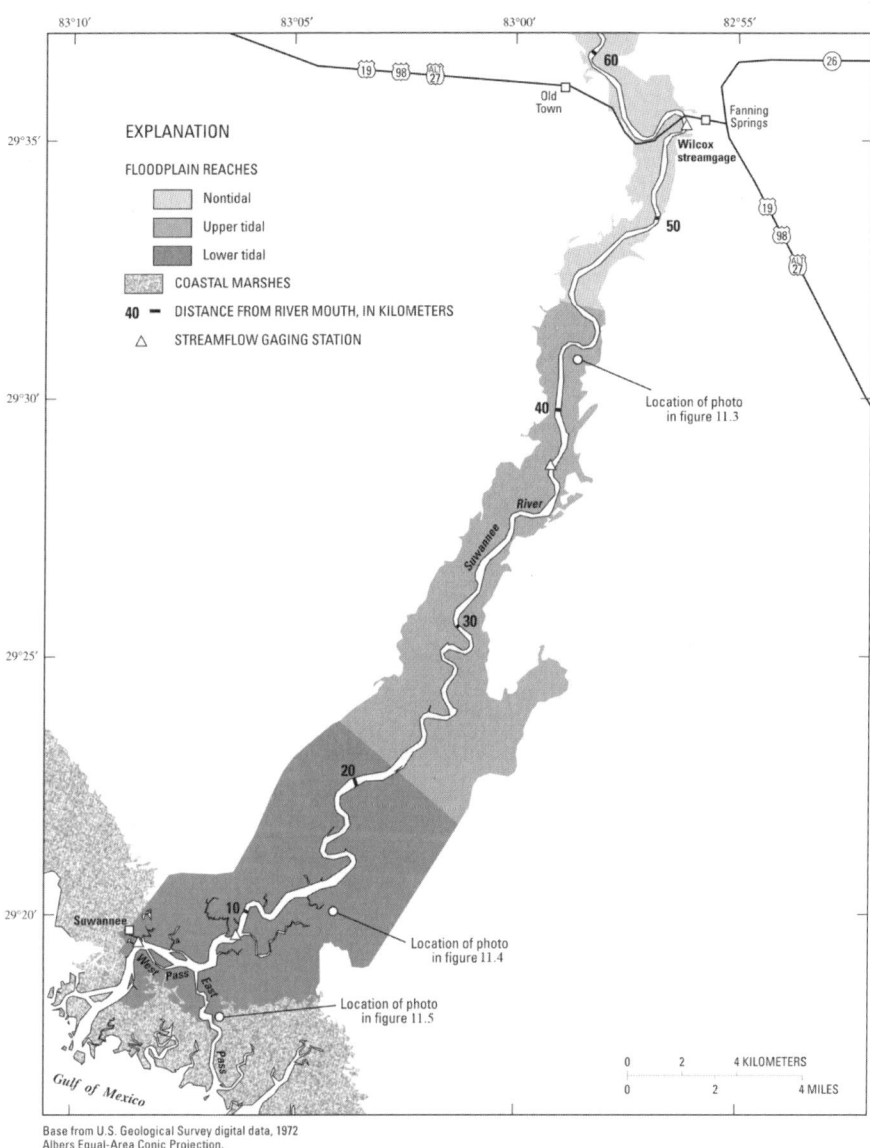

Fig. 11.2. Study area showing major reaches of the lower Suwannee River, Florida.

Fig. 11.3. Swamp in the upper tidal reach of the lower Suwannee River floodplain, Florida. Swollen bases of water tupelo and baldcypress have formed in response to deep flooding at this UTsw1 forest.

Fig. 11.4. Mixed forest in the lower tidal reach of the Suwannee River floodplain, Florida. Both swamp and hammock species are found in mixed forests. This site is located near a tidal creek and is inundated almost daily by high tides.

Fig. 11.5. Stunted stand of pumpkin ash trees in the lower tidal reach near the tree line of the lower Suwannee River floodplain, Florida. Trees in this stand are less than 6 m tall.

per tidal (UT), and lower tidal (LT). Seven tidal forest types described in this chapter are named using both the reach names (UT and LT) and the general forest type: swamp (sw), mixed forest (mix), or hammock (ham) (Table 11.1). Swamps (UTsw1, UTsw2, LTsw1, and LTsw2) are found in the lowest elevations of the floodplain. Bottomland hardwood or hammock species are more important in the canopy of sw2 forests than in sw1 types. Mixed forests (UTmix and LTmix) are dominated by a mixture of swamp and bottomland hardwood or hammock species. Hydric hammocks (LTham), a unique wetland forest type that is rare outside Florida, support a mixture of evergreen and semi-evergreen trees (Vince et al.1989). Forest types in the upper tidal reach that are dominated by bottomland hardwoods are not addressed in this chapter because they are not tidally influenced. Much of the information in this chapter was collected and analyzed as part of a comprehensive study of Suwannee River floodplain by the U.S. Geological Survey (USGS) in cooperation with the Suwannee River Water Management District (SRWMD) in Florida from 1996 to 2000 (Darst et al. 2002; Lewis et al. 2002; Light et al. 2002; Darst et al. 2003).

11.2. Geology, physiography, and historical land use

The Suwannee River and its three major tributaries, the Alapaha, Withlacoochee, and Santa Fe Rivers, lie entirely within the southeastern Coastal Plain (Berndt et al. 1996). The river is 394 km long and drains an area of about 25,770 km^2. An important geologic feature known as the Cody Scarp crosses the Suwannee River Basin from northwest to southeast about 80 km inland from the coast (Figure 11.6) (Ceryak et al. 1983). This karst escarpment has been highly modified by marine processes (Randazzo and Jones 1997). Active sinkhole formation, "disappearing" streams (streams captured by a sink feature), and "poljes" (karstic lakes) are present in the vicinity of the scarp. The upper Suwannee River above the Cody Scarp lies in the Okeefenokee Basin physiographic region, and most of the drainage basins of the Alapaha and Withlacoochee Rivers lie in the Tifton Uplands physiographic region (Figure 11.6). In both the Tifton Uplands and Okefenokee Basin, sandy soil over subsurface clay layers create conditions favorable to surficial drainage, resulting in numerous creeks and tributary streams that drain the upper part of the Suwannee River Basin.

Downgradient from the Cody Scarp, in the Ocala Uplift physiographic region, the landscape is highly karstic, influenced by the dissolution of limestones that lie at or near land surface (Crane 1986). Dozens of springs

Table 11.1. Tidal forest types of the lower Suwannee River floodplain, Florida.

Reach	Forest type name and abbreviation	Area, in ha	Dominant canopy species
Upper tidal	Swamps type 1 (UTsw1)	950	Water tupelo
	Swamps type 2 (UTsw2)	1,150	Baldcypress Pumpkin ash
	Mixed forests (UTmix)	400	Baldcypress Pumpkin ash Laurel oak
Lower tidal	Swamps type 1 (LTsw1)	1,400	Swamp tupelo
	Swamps type 2 (LTsw2)	1,390	Pumpkin ash Baldcypress
	Mixed forests (LTmix)	1,080	Pumpkin ash Swamp tupelo Sweetbay
	Hydric hammocks (LTham)	1,790	Cabbage palm Loblolly pine
Total		8,160	

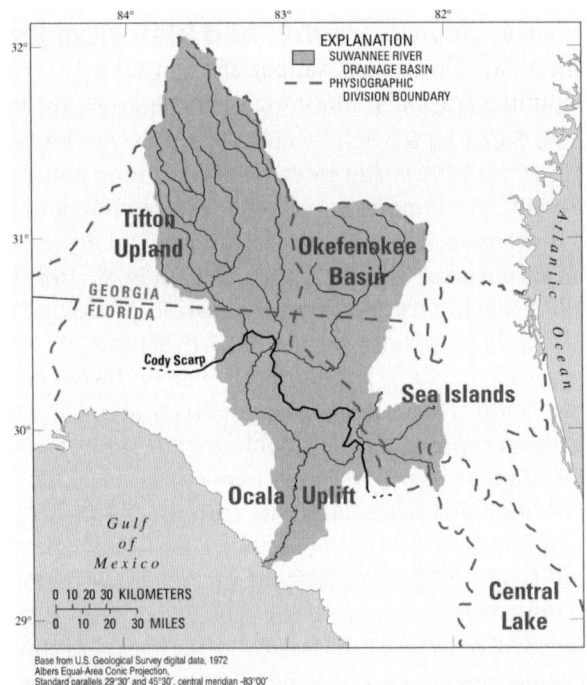

Fig. 11.6. Physiographic regions of the Suwannee River Basin in Florida and Georgia (from Clark and Zisa 1976; Brooks 1981).

along the riverbank and in the adjacent floodplain discharge large amounts of groundwater to the Suwannee. Numerous relict marine terraces in the Ocala Uplift region were established by different sea-level stands during the Pleistocene (and possibly Pliocene) Epoch. The transition between different forest types in the tidal floodplain of the lower river may be influenced in part by the presence of these terrace features.

In the lower Suwannee River and adjacent tidal freshwater swamps, a m layer of sands 15-30 thick overlies limestones. These limestones form the "container" for the water in the Floridan aquifer system; a major regional aquifer underlying Florida, Georgia, and parts of South Carolina and Alabama (Crane 1986; Berndt et al. 1996). The many springs along the lower river are fed by water from the Floridan aquifer system (Crane 1986).

The limestones of the Floridan aquifer system are less compact (less dense and more conductive of water) in the Suwannee River estuary than they are in areas both north and south of the estuary (Rupert and Arthur 1990). Thus, a combination of freshwater inflow from the river and diffuse groundwater inflow modifies the plant communities of the region with mesohaline and oligohaline brackish marshes grading into tidal freshwater swamp and hydric hammock communities. North and south of the Suwannee estuary, where limestones are denser, plant communities are more saline, with intertidal saltmarshes, saltbarrens and salinas, and coastal upland forest communities (usually maritime hammock or scrubby flatwoods). The existence of an unusually freshwater-influenced flora in the Suwannee estuary was recognized nearly a century ago (Harper 1910).

The river estuary is a deltaic type with a coastline regarded by marine geologists as "sediment starved" due to the low sediment load of the Suwannee River (Crane 1986; Wright 1995). During the Eocene Epoch, limestones that underlie the region were formed when most of the region was under a shallow sea. During the Oligocene Epoch, the area was infilled by sands derived from the Appalachian region. These sands were subsequently reworked by numerous transgressions of sea level during glacial and interglacial periods of the Pleistocene Epoch. Currently, the estuarine delta and the upstream tidal wetland forests are dominated by mucks and sandy soils (Clewell et al. 1999; Light et al. 2002). Available evidence indicates that vertical accretion in the marshes is keeping pace with sea-level rise (Wright 1995).

Native people inhabited the Suwannee River Basin for about 12,000 years, but apparently did not rely on domesticated crops, such as corn and beans, until about 800 years ago (Milanich 1994). The impact of native inhabitants on the basin was slight, and ceased altogether in the mid-1800s when they were relocated from the basin to reservations in the West

(Wright 1986). Settlers began clearing land for agriculture and timber in the 1800s, initiating a process of large-scale deforestation of the basin that continued into the 1900s with the addition of phosphate mining in the upper basin (Tebeau 1971; Ewel JJ 1990). Virtually every swamp in Florida was logged between the late 1800s and 1950, including cypress stands on the Suwannee River floodplain (Ewel KC 1990). Once-thriving lumber towns declined in populations because the harvest of virgin stands of timber was unsustainable (Anderson et al. 1998). After the 1950s, modern silviculture practices were introduced.

Agriculture and commercial forestry are presently the dominant land uses throughout the drainage basin, with urbanized areas covering a minor part of the land area (Berndt et al. 1996). Modern land uses appear to be affecting water quality in the Suwannee River, with a substantial increase in nitrate concentrations and loads in the past 40 years (Katz et al. 1999). Concerns have been raised about the effects of elevated nitrate concentrations on aquatic communities in the river and estuary, but effects on the tidal freshwater forests may not be significant.

Recent population increases throughout the basin have been substantial, with an estimated 41 percent increase from 1980-2005 (Bureau of Economic and Business Research 2006; Carl Vinson Institute of Government 2006; U.S. Census Bureau 2006). The Georgia part of the basin had slightly more than half of the total basin population in 2005. The rate of population increase from 1980-2005 in the Florida part of the basin was more than twice that in the Georgia part.

Tidal freshwater forests of the lower Suwannee River floodplain are largely intact, with much of the land presently under public ownership. Because these forests are dependent on adequate freshwater inflow, however, increasing demands for water throughout the drainage basin could potentially affect floodplain habitats in the future.

11.3. Hydrology and climate

The Suwannee River is the second largest river in Florida with an average discharge of 285 m^3 s^{-1} (1931-2004 average at Wilcox streamgage, Figure 11.2). River discharge is usually highest in March-April and lowest in November-December. The median tidal range at the mouth of the Suwannee River is about 1 m (Tillis 2000). Tides are mixed semidiurnal, typically with two unequal high tides and two unequal low tides each day. Tides affect river stages at low and medium flows in the upper tidal reach, and at all flows in the lower tidal reach.

Hydrographs of daily high stages from 1985-99 are shown in Figure 11.7 for the upstream end (A) and downstream end (B) of the tidal reach, along with river stage values noted on the right side of each graph for median monthly high (MMH), median daily high (MDH), and median daily low (MDL) for this period. Traditional tidal datums could not be calculated for locations throughout the tidal reach because of the mixed influence of riverflow and tide.

From the upstream end of the tidal reach downstream to the Gulf, the influence of river flooding on river stage gradually decreases, and the influence of tides and storm surges on river stage gradually increases (Figure 11.7). At rkm 43, river flooding dominated the hydrograph about one-third of the time from 1985-99, and storm surge levels were minimal (less than the 2-year flood peak). At rkm 5, major river flood levels were not much higher than the MMH, with the highest storm surge (2.2 m) exceeding the peak of a 25-year flood event by 1 m. Coastal areas with relatively shallow offshore areas, such as the Big Bend and the south Florida Gulf Coast, will generally experience a higher storm surge than areas with deep offshore waters (Ho and Tracey 1975; Chen and Gerber 1990; Federal Highway Administration 2006). The height and inland extent of storm surge depends not only on the maximum sustained wind speed of the storm, but also the direction of storm movement, the speed at which the storm approaches the coast, the location of storm landfall, and the tidal cycle and riverflow at the time of landfall. Two unnamed tropical storms, for example, produced storm surge elevations nearly as high as hurricanes during the 1985-99 period (Figure 11.7B).

Although hurricanes occur less frequently in the Big Bend than in other regions along the Gulf Coast to the west (Henry 1998), storm surges in the Suwannee River region occur frequently enough (two to three times per decade) to have important effects on tidal freshwater swamps in the lower tidal reach. Between 1871-1995, an estimated 14 unnamed tropical storms and 16 hurricanes affected the Suwannee River region (Williams and Duedall 1997). Particularly strong hurricanes appear to have occurred in 1842 and 1882, with storm surges of 5.5 and 4.9 m, respectively, at Cedar Key (Ho and Tracey 1975). A storm surge of 3.1 m at Cedar Key occurred during Hurricane Alma in June 1966, and again during Hurricane Agnes in June 1972. Most storm surges occur during the hurricane season from June through November. A rare, nontropical winter storm on March 12-14, 1993, caused a storm surge of 2.9 m at Cedar Key (National Oceanic and Atmospheric Administration 1994).

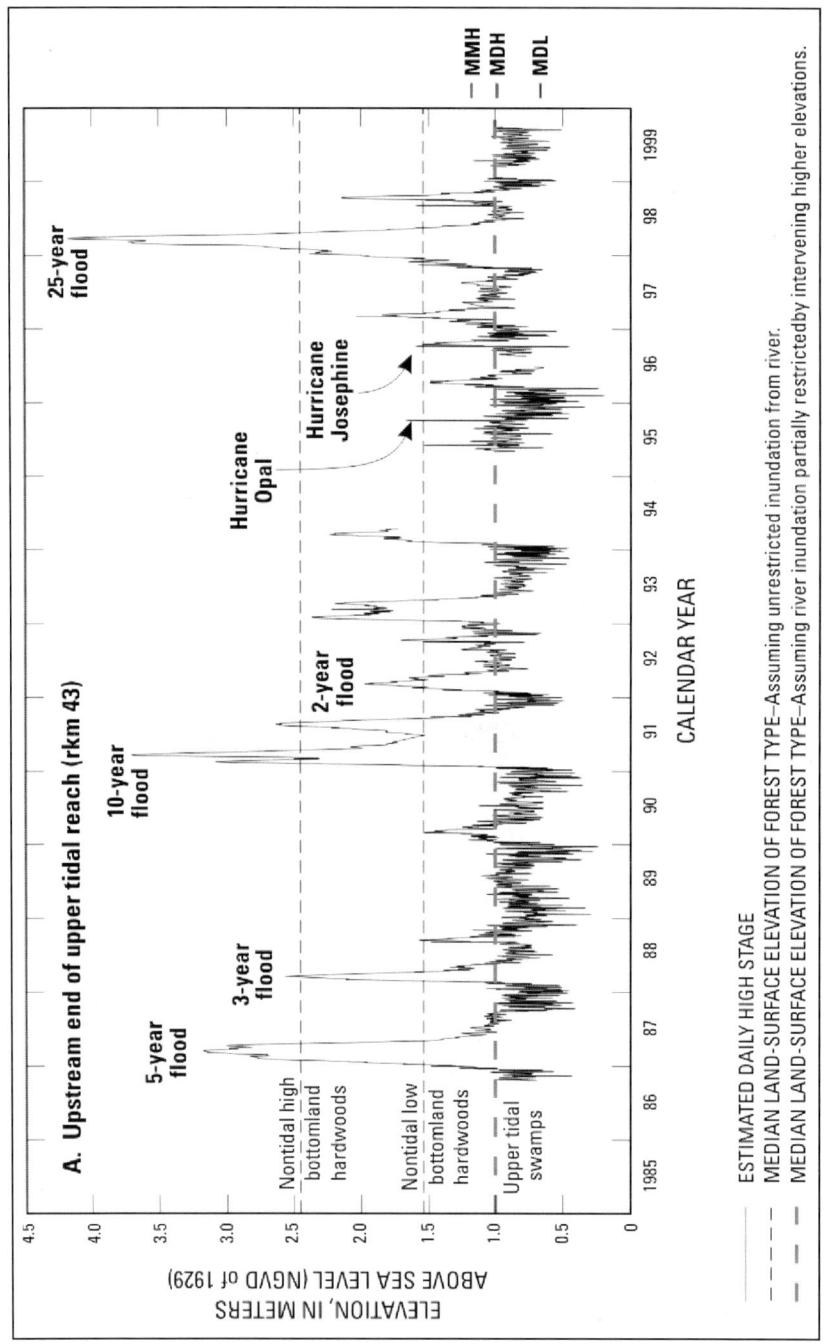

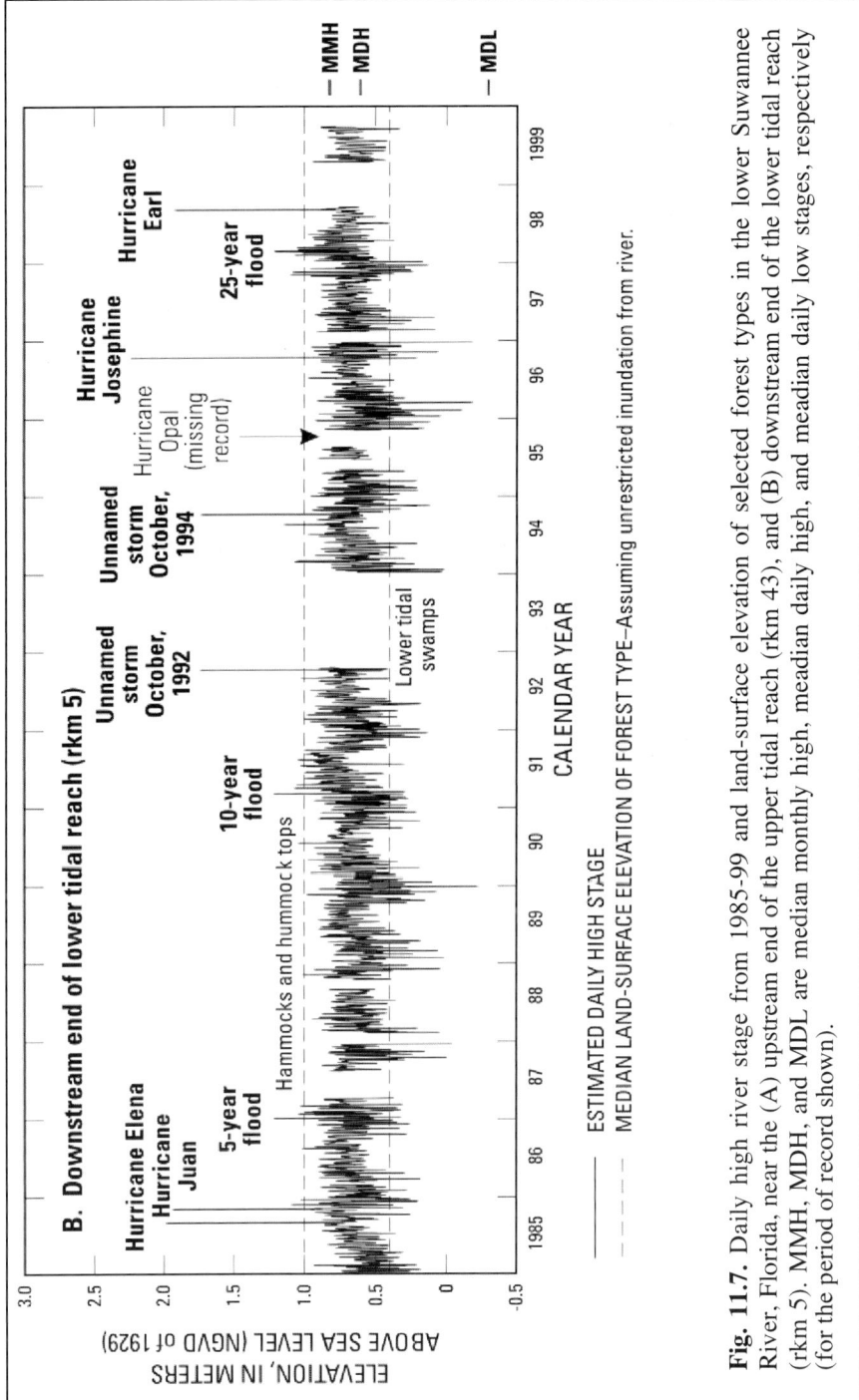

Fig. 11.7. Daily high river stage from 1985-99 and land-surface elevation of selected forest types in the lower Suwannee River, Florida, near the (A) upstream end of the upper tidal reach (rkm 43), and (B) downstream end of the lower tidal reach (rkm 5). MMH, MDH, and MDL are median monthly high, meadian daily high, and meadian daily low stages, respectively (for the period of record shown).

The warm, temperate climate in the lower Suwannee River region is characterized by long, humid summers. The growing season (50 percent probability freeze-free period) varies from 259 days at the confluence of the Suwannee and Santa Fe Rivers to 283 days near the river's mouth. Average summer air temperature (June, July, and August) is 26.4 °C and average winter air temperature (December, January, and February) is 12.1 °C. Average annual precipitation (1961-90) at Cross City is 146 cm (Owenby and Ezell 1992).

Sea-level rise of 30.5 cm since 1852 was estimated for the Gulf Coast in the vicinity of the mouth of the Suwannee River (Raabe 2000). Evidence that the Suwannee River estuary and nearby areas are experiencing sea-level rise or coastal forest retreat has been presented by a number of investigators (Kurz and Wagner 1957; Carlton 1977; Clewell et al. 1999; Williams et al. 1999).

11.4. Ecological characterization

The gradual transition in ecological characteristics of floodplain forests from the upstream to the downstream ends of the tidal reach is primarily a response to the changing flood levels, storm surge heights, tidal fluctuations, and salinity in the river. All of these hydrologic factors interact in complex ways with landscape features, topographic relief, and soil characteristics to determine the distribution and composition of biological communities in tidal freshwater swamps.

11.4.1. Study design

Land-surface elevations, hydrologic conditions, soil characteristics, and vegetation (canopy, subcanopy, and ground cover) of floodplain forests were measured at nine transects throughout the tidal reach (Light et al. 2002). Hydrologic observations in tidal forests were made during a variety of flow conditions during the 3-year data collection period (1996-99), which included a normal year, a wet year, and a dry year. Long-term hydrologic conditions in floodplain forests were estimated using riverflow and stage data at a series of continuous recording streamflow gaging stations throughout the lower Suwannee River. A total of 46 soil borings were made in upper and lower tidal forests with soil profiles described to a depth of 1.5 to 2 m at most locations, and 32 soil samples collected from 21 locations in the lower tidal reach were analyzed for electrical conductivity. A total of 4,517 canopy and subcanopy plants in the tidal reach were identified to species. Algal and faunal species were observed or sampled in separate studies.

Visual signatures of forest types within reaches of the lower Suwannee River floodplain were identified on infrared aerial photographs of the floodplain (Darst et al. 2003). Vegetative sampling on the ground was used to define species composition, which was based on relative basal area of trees with diameter at breast height (dbh) of 10 cm or more. Forest types were defined (Table 11.1), mapped, and the mapping was verified by vegetation surveys at many additional plots. The area and distribution of each forest type was determined from the forest map. The boundary between reaches was not a visible feature on aerial photographs, although signatures changed gradually from upstream to downstream. The boundary between the nontidal, upper tidal, and lower tidal reaches was based on vegetative data, using characteristic tree species of each reach as indicators.

11.4.2. Floodplain topography and hydrology

Typical land-surface elevations of tidal swamps and mixed forests generally decrease from about 1 m above sea level (NGVD of 1929) in the upper tidal reach to about 0.5 m near the treeline. Hydric hammocks in the lower tidal reach are slightly higher, with land-surface elevations ranging from 1.0-1.2 m above sea level even as far downstream as the treeline. Much of the variation in land-surface elevations in swamps and mixed forests is due to the presence of microtopographic features called hummocks, which are elevated mounds of land around the base of a tree or a clump of trees (Figure 11.8). The trees that grow on these hummock tops are the same species found in hydric hammocks. Hummocks are present throughout the tidal reach, but are more common in the lower tidal reach where they are generally larger and more well defined. Lower tidal mixed forests, in particular, have a distinct hummock/mud-floor microtopography that supports hammock species on the hummocks and swamp species on the mud floor.

The depth of river flooding gradually decreases from the upper end of the tidal reach downstream to the treeline. Figure 11.9 shows depth of continuous flooding for 14 days in tidal and nontidal swamps in relation to distance from river mouth, for 2-year, 5-year, and 25-year recurrence interval floods. Fourteen days of continuous flooding was chosen for this analysis because total submergence of leaves and stems that continues for 10-20 days is fatal to seedlings of many swamp and bottomland hardwood species (Demaree 1932; Hosner 1960). For the 5-year recurrence interval

Fig. 11.8. Hummock in the lower tidal reach of the Suwannee River floodplain, Florida. Hammock tree species can grow in swamps on the elevated soil of hummocks, but cannot survive on the surrounding mud floor.

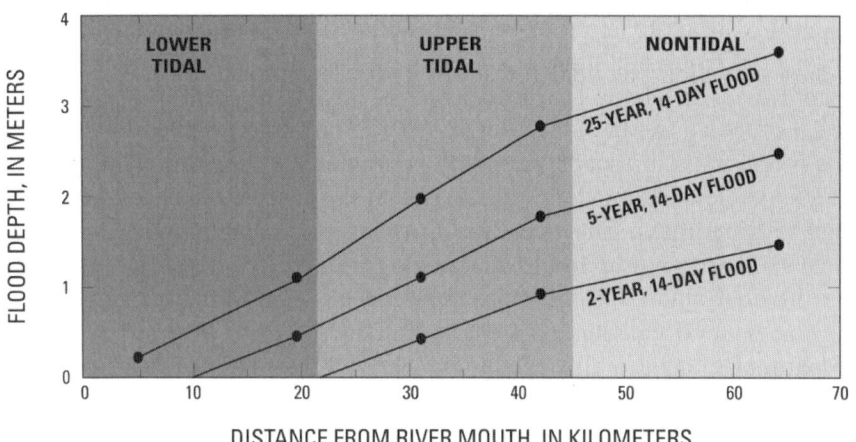

Fig. 11.9. Depth of continuous flooding for 14 days in swamps of the lower Suwannee River, Florida, in relation to distance from river mouth. Depths during 2-year, 5-year, and 25-year recurrence interval floods were based on 67 years of streamflow records (1933-99) at the head of the study area, along with stage-discharge ratings and median land-surface elevation of swamps at each tidal and nontidal sampling location.

flood, for example, depth of continuous flooding for 14 days ranges from 1.8 m at the upper end of the tidal reach to zero at the lower end. Most lower tidal swamps are flooded nearly every day at high tide, but flooding is not continuous for 14 days because the ground is usually exposed every day at low tide. At the downstream end of the tidal reach, for example, mean lower low tide, which is similar to the elevation of the MDL shown on Figure 11.7B, is 0.7 m below the median land-surface elevation of the lower tidal swamps.

Near the upstream end of the upper tidal reach, land-surface elevations of tidal swamps are close to the elevation of the MDH (Figure 11.7), but standing water in these forests (Figure 11.3) does not fluctuate with the tidal cycle because these swamps are isolated from regular tidal inundation by prominent natural riverbank levees. Downstream from that location, however, riverbank levees are very low or nonexistent and tidal creeks are common, especially in the lower tidal reach, allowing daily high tides to move freely into adjacent swamps. Because high tides last only an hour or two, daily tidal inundation may not reach the more isolated swamps that are relatively far from the main channel and tidal creeks. During the study period, water levels measured in shallow pools of isolated swamps near the outer edges of the floodplain and distant from any tidal creeks did not fluctuate with the tides except when river levels were above the MMH. The highest elevations of tidal forests in hydric hammocks or on the tops of hummocks are above the MMH but are submerged by large floods and storm surges (Figure 11.7).

11.4.3. Salinity of tidal creeks and isolated ponds

Under normal low-flow conditions (not during storm surge), tidal creeks and the main river channel are fresh (<0.5 ppt salinity) throughout the tidal reach upstream from rkm 9. Downstream from that location during normal low-flow conditions, vertically averaged salinity is about 5 ppt at rkm 5, and both surface and vertically averaged salinity are typically greater than 10 ppt at the treeline near rkm 2 (Tillis 2000; Bales et al. 2006). Salinities are much higher during most storm surges. At rkm 9, where the river is usually fresh throughout the water column, a bottom salinity of 11 ppt was measured during a minor storm surge (Jerry Krummrich, Florida Fish and Wildlife Conservation Commission, written communication, 2000).

Major hurricanes with large storm surges probably deposit saline water over floodplain forests of the entire lower tidal reach and the downstream end of the upper tidal reach. Storms abate quickly (in less than 24 hours), but saline water is retained in isolated ponds in the interior of the floodplain. In the low rainfall months that commonly follow the storm season, water trapped in isolated ponds increases in salinity as ponds shrink from evaporation. This probably occurred following Hurricane Josephine in two ponds near the downstream end of the lower tidal reach for which salinity measurements were made intermittently from 1997-99 (Figure 11.10). Salts accumulate in the soils, so that even after ponds are flushed with freshwater during major river flooding, such as during the 25-year flood event in March 1998, pond salinity gradually increases after floods recede.

11.4.4. Soil textures and conductivity

Soils in upper tidal and lower tidal forests are predominantly organic on the surface, with organic or mineral subsurface textures. With the exception of high spots in the landscape, such as narrow riverbank levees and hummock tops, upper tidal swamps and lower tidal swamps and mixed forests have muck surface soils that are continuously saturated. Saturated conditions persist in these areas despite seasonal and annual fluctuations in riverflow because land-surface elevations are close to the elevation of the MDH (Figure 11.7). Slightly better drainage provided by sandy subsoils

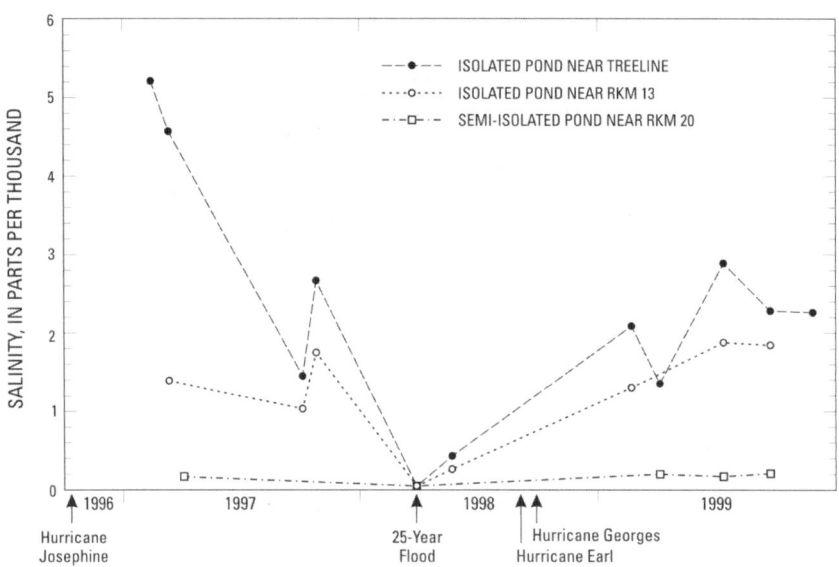

Fig. 11.10. Salinity of isolated ponds in the lower tidal reach of the lower Suwannee River floodplain, Florida. Salinity near zero during the 25-year flood event (March 25, 1998) occurred when the entire floodplain was covered with fresh river water moving downstream through the forest as sheetflow.

below surface muck in lower tidal mixed forests, such as in the forest shown in Figure 11.4, may partially explain why this forest type supports a drier mix of species than lower tidal swamp forests. Soil profiles of upper tidal mixed forests and lower tidal hammocks consist of a thin layer (<10 cm) of muck over mineral subsoils (usually sand). Although most of these slightly higher areas were not continuously saturated, a high water table probably exists for long enough periods to maintain thin surface mucks. Limerock was encountered at a depth of 80 cm at one location, which was not unexpected considering that the entire lower Suwannee River Basin is underlain by limestone, most of which is at or near land surface (Crane 1986).

Soils with an electrical conductivity greater than 4 mmhos/cm are generally considered to be saline; however, reductions in the yield of some crop species can occur in soils with conductivities of 2 mmhos/cm (Richards 1954; Hanlon et al. 1993; U.S. Department of Agriculture 1996). Average surface soil conductivities in the upstream part of the lower tidal reach (rkm 12-20) were less than 2.0 mmhos/cm, but maximum conductivities indicate that sensitive tree species could be affected at some sites (Table 11.2). At a downstream site near the treeline where saline water was retained in isolated ponds after storm surges, conductivities of surface soils were high enough to exclude many nontolerant tree species.

The maximum conductivity of 5.4 mmhos/cm in subsurface soils at an upstream site was high considering the location of that particular measurement (rkm 20), and indicates that exceptionally large storm surges probably affected the floodplain this far upstream at some time in the past. In a similar study, Coultas (1984) reported that conductivity of subsurface soils from forested wetlands in the lower Apalachicola River (6-22 km upstream from the mouth) were higher than that of surface soils. The increase in conductivity at greater depth was expected, because the surface tends to be flushed frequently with rain and river water resulting in a dilution of salts (Coultas 1984).

11.4.5. Important tree species and distribution in tidal reaches

Baldcypress (*Taxodium distichum* [L.] L.C. Rich.), pumpkin ash (*Fraxinus profunda* [Bush] Bush), swamp tupelo (*Nyssa biflora* Walt.), and water tupelo (*Nyssa aquatica* L.) are the dominant canopy species by basal area in tidal forests of the lower Suwannee River (Table 11.3). Pumpkin ash, Carolina ash (*Fraxinus caroliniana* P. Mill.), American hornbeam (*Carpinus caroliniana* Walt.), and waxmyrtle (*Morella cerifera* L.) are the

Table 11.2. Soil conductivity in lower tidal forests of the Suwannee River floodplain, Florida. Soil samples were collected January-March 1999 during normal tidal conditions. No river flooding or storm surges occurred during that period or in the two months prior to sampling.

Depth of soil sample	Site type	Electrical conductivity of soil, in mmhos/cm					
		Upstream sites (rkm 12 to 20)			Downstream sites (<1.5 km from tree line)		
		n	Avg.	Range	n	Avg.	Range
Surface (0-30 cm)	Sites inundated by daily or monthly tides	6	1.0	0-2.0	4	3.2	1.5-4.6
	Isolated or semi-isolated sites (tidal inundation rare)	8	1.5	0-4.5	3	8.1	5.2-13.7
Subsurface (variable, 50-200 cm)	All site types	6	2.6	0.6-5.4	5	6.1	4.7-7.9

most common subcanopy tree species in tidal forests. Many tidal canopy species with high relative basal areas also have high relative densities. Baldcypress was heavily logged in the 1900s, but is recovering dominance, especially on publicly owned lands in the floodplain.

The distribution of tree species along a downstream gradient in the lower Suwannee River floodplain (Figure 11.11) is determined primarily by the depth and duration of river floods (Figure 11.9) and the salinity of soils (Table 11.2). The seedlings of some slow growing species, such as sweetbay (*Magnolia virginiana* L.), southern redcedar (*Juniperus virginiana* var. *silicicola* [Small] J. Silba), and pumpkin ash, cannot survive in floodplain locations where inundation is too deep and too long in duration. The distribution of cabbage palm (*Sabal palmetto* [Walt.] Lodd. ex J.A. & J.H. Schultes) is limited by deep flooding as well, but is found on some levees in the nontidal reach, where floods are shallower and recede more quickly than in the lower-elevation forests behind the levee. In contrast, water tupelo can withstand deep flooding but is known to be sensitive to salinity (Penfound and Hathaway 1938), and this sensitivity probably limits its occurrence in the lower tidal reach. The downstream limit of distribution of species such as water locust (*Gleditsia aquatica* Marsh.) and river birch (*Betula nigra* L.) is probably due to variations in salinity tolerance. Pumpkin ash and cabbage palm appear to be the most salt-tolerant species in the lower Suwannee River floodplain, and stands of stunted trees of these species are found along the treeline.

Table 11.3. Important canopy and subcanopy species in tidal forests of the lower Suwannee River floodplain, Florida. Tidal wetland forests include all species in upper and lower tidal reaches. All canopy species with relative basal area of 2 percent or more and all subcanopy species with relative density of 2 percent or more are listed. Canopy tree species have diameter at breast height (dbh) of 10 cm or greater; subcanopy species have dbh of 2 to 9.9 cm.

Canopy species	Relative basal area, in percent	Relative density, in percent	Subcanopy species	Relative density, in percent
Baldcypress	19.4	16.6	Pumpkin ash	22.2
Pumpkin ash	16.1	18.6	Carolina ash	16.1
Swamp tupelo	14.3	13.8	American hornbeam	11.0
Water tupelo	10.6	6.1	Waxmyrtle	8.9
Cabbage palmetto	6.7	6.3	Buttonbush	6.4
Laurel oak	5.6	3.9	Sweetbay	4.5
Sweetgum	4.7	5.4	Baldcypress	4.1
Red maple	4.7	4.0	Stiff dogwood	3.7
Sweetbay	3.6	3.2	Swamp bay	2.8
Loblolly pine	2.0	0.9	American elm	2.2
32 other species	12.3	21.0	Sweetgum	2.0
			28 other species	16.1
Total	100.0	98.8		100.0

11.4.6. Characteristics of forest types

Water tupelo is the most important tree in upper tidal swamps, and comprised over 75 percent of total basal area at some sites. Upper tidal swamps have greater basal area than upper tidal mixed forests (Table 11.4). Pumpkin ash is more important in upper tidal mixed forests than in upper tidal swamps. Areas of slightly higher elevations in upper tidal mixed forests have lower depths of inundation, enabling seedlings of pumpkin ash to survive floods. The depths and duration of river flooding, along with the size of canopy trees and subsequent shading of forest floors, are probably responsible for the lower density of ground cover in upper tidal swamps than in lower tidal swamps. Variable panicgrass (*Dichanthelium commutatum* [J.A. Schultes] Gould) and seven sisters (*Crinum americanum* L.) are commonly dominant ground-cover species in upper tidal forests.

The density of lower tidal swamps is high, with an average of 1,198 trees/ha for both types (LTsw1 and LTsw2). Lower tidal swamps have the highest basal area of all lower tidal forests, but low tree species richness. Swamp tupelo is the most important canopy species in LTsw2 forests, and pumpkin ash is the most important canopy species in LTsw1 and LTmix

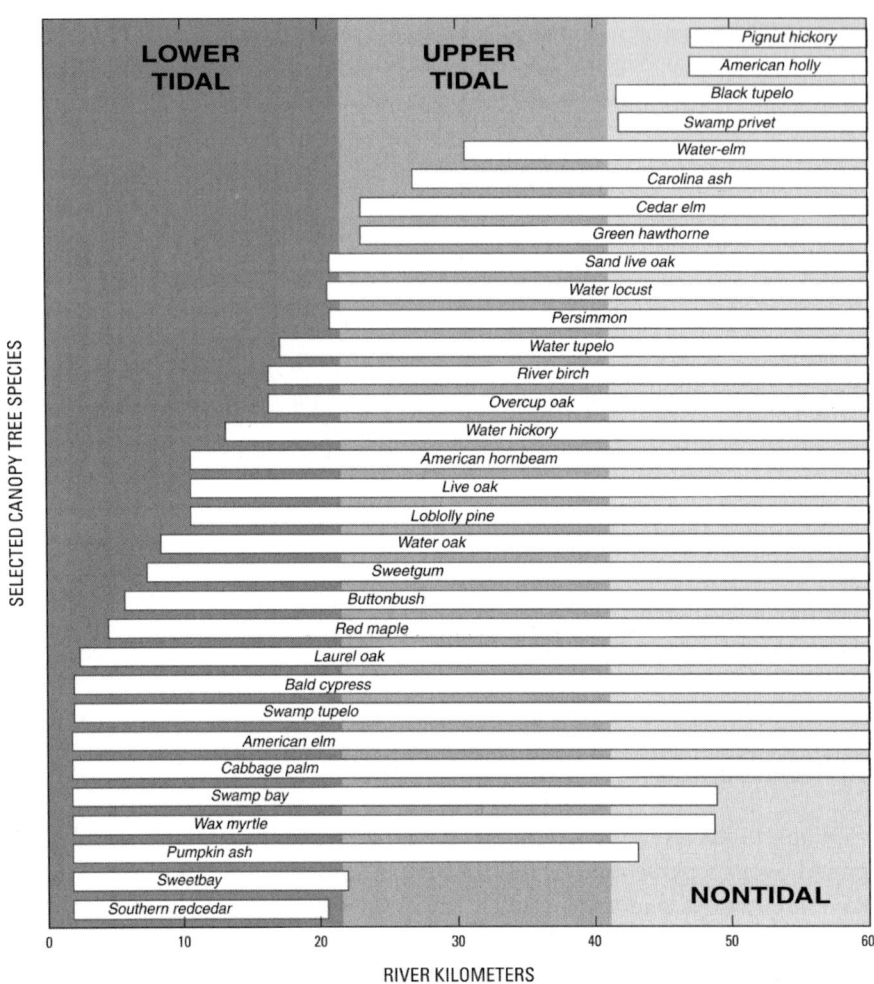

Fig. 11.11. Distribution of selected canopy tree species in relation to distance from the mouth of the Suwannee River, Florida.

forests. About 27% of LTsw1 are stunted stands of pumpkin ash (Figure 11.5) that have a distinctive signature on aerial photographs. Trees within stunted stands appear to have uniform heights and may be even-aged, growing where previous forests have been destroyed by a storm surge. Annual rings counted on a stunted tree of 11 cm dbh at rkm 4 indicated that the tree was about 122 years old (Clewell et al. 1999). Tree stunting is barely perceptible at rkm 18, gradually increases downstream, and is severe in stands at the treeline where canopy heights average 5-6 m. Ground-cover density is higher in lower tidal swamps than in upper tidal swamps, probably because of greater light reaching the swamp floor, shallower

Table 11.4. Vegetative characteristics of tidal forest types of the lower Suwannee River floodplain, Florida. Includes all trees with diameter at breast height of 10 cm or greater. See Table 11.1 for description of the forest types.

Forest type	Canopy			Ground cover	
	Number of tree species	Basal area, in m^2/ha	Density, in trees/ha	Number of species	Density, in percent cover
UTsw1	16	70.0	962	64	20
UTsw2	23	57.6	958	81	21
UTmix	19	51.3	668	64	34
LTsw1	17	59.0	1,238	101	41
LTsw2	17	57.6	1,157	106	49
LTmix	18	52.1	1,086	100	28
LTham	23	26.4	575	101	48

flood depths, and shorter durations of inundation. Lizard's tail (*Saururus cernuus* L.) and species of sedge (*Carex* spp.) are commonly dominant ground-cover species in lower tidal swamps.

Lower tidal mixed forests include swamps with numerous hummocks (Figure 11.8). Sweetbay is a characteristic species of lower tidal mixed forests and is usually present on hummocks and streambanks. Sweetbay trees have a unique signature that distinguishes lower tidal mixed forests on aerial photographs. Lower tidal hammocks are dominated by cabbage palm and loblolly pine (*Pinus taeda* L.). The prominence of loblolly pine in hammocks may be due to attempts to grow this species commercially on many hammock areas of the lower tidal reach. The ground cover of hammocks is dominated by cinnamon fern (*Osmunda cinnamomea* L.) and slender woodoats (*Chasmanthium laxum* [L.] Yates). Ground-cover species richness is greater in lower tidal than in upper tidal forests because of the inclusion of two groups of species unique to that reach: obligate brackish-marsh or saltmarsh species, such as eastern grasswort (*Lilaeopsis chinensis* [L.] Kuntze) and southern quillwort (*Isoetes flaccida* Shuttlw. ex A. Braun) and species tropical in distribution and sparsely distributed along the northern Gulf Coast of Florida such as bramblefern (*Hypolepis repens* [L.] K. Presl.).

11.4.7. Algal communities

Euryhaline marine red algae (*Bostrychia rivularis* Harvey, *Catenella repens*, and *Caloglossa leprieurii* Montagne) dominated the algal community at three sites in lower tidal forests of the Suwannee River floodplain in a limited qualitative survey by Mattson and Krummrich (1995). Although quantitative studies of algae in these forests have not been conducted, algae are known to be an important component of other tidal ecosystems and may be a sensitive environmental indicator, responding to water-quality changes (e.g., salinity) more rapidly than woody vegetation. Studies in saltmarsh ecosystems have shown that algae can account for a substantial fraction of the primary production and authochthonous food base (Gallagher and Daiber 1974; Sullivan and Moncreiff 1990). Odum et al. (1984) summarized the scant literature on the algae of tidal freshwater marshes and reported that green (Chlorophyta) and blue-green algae (Cyanobacteria) are major groups. Algae are an important component of the primary productivity in mangrove swamps (Dawes 1981), and additional studies of algal communities in tidal freshwater swamps are needed.

11.4.8. Faunal communities

The macroinvertebrate and small vertebrate communities in forests of the Suwannee River floodplain are dominated by crayfish, salamanders, frogs, and freshwater snails in the nontidal and upper tidal reaches, and by fiddler crabs and estuarine snails in the lower tidal reach, according to observations by Wharton et al. (1982). The abundance of redjointed fiddler crabs (*Uca minax*) and squareback marsh crabs (*Amrases cinereum*) in lower tidal swamps and in the downstream end of the upper tidal reach has been confirmed by the authors and others (Golder Associates, Inc. 2000). Redjointed fiddler crabs are characteristically found far up estuaries in tidal freshwater regions, whereas squareback marsh crabs are found in a variety of tidal communities throughout the estuary, including high salinity habitats.

Benthic macroinvertebrate communities in main channel habitats adjacent to the tidal forests are dominated (in terms of species richness and abundance) by various aquatic insect groups, primarily Chironomidae (midges), Ephemeroptera (mayflies) and Trichoptera (caddisflies)

(Mattson 1992). Estuarine taxa include euryhaline species such as olive nerite (*Neritina usnea*), mysids (*Taphromysis bowmani*), and blue crab (*Callinectes sapidus*). Similar invertebrate species may exist in tidal creeks that penetrate into the swamps, but would be influenced by factors in the creeks, such as drying during low tide and reduced dissolved oxygen. Sampling of macroinvertebrates in East Pass indicated that invertebrate communities on wood habitat were primarily dominated by freshwater taxa, whereas the communities of the littoral soft-bottom areas included many estuarine taxa.

Fish communities collected by boat electrofishing in East Pass adjacent to tidal forests were dominated by freshwater taxa, including largemouth bass (*Micropterus salmoides*), sunfish (*Lepomis* spp.), gar, and killifish (*Fundulus* spp.) (Mattson and Krummrich 1995). Estuarine taxa such as striped mullet, gray snapper, spot, and croaker were present seasonally. Similar fish communities have been reported for tidal freshwater marshes (Odum et al. 1984). Some of these are known to utilize the larger tidal creeks that penetrate the tidal forests. Although no comparable work has been done in tidal freshwater forests, the fisheries habitat values of other tidal wetlands have been well established (Day et al. 1989), indicating that tidal forests in the upstream areas of the Suwannee River estuary may have high value as fish nursery habitat.

Few studies of the wildlife habitat value of tidal forests have been made. The abundance of fiddler and shore crabs may make these forests important foraging areas for wading birds such as yellow-crowned night heron and green-backed heron, and other animals such as raccoons. Flocks of wild turkey have been observed foraging in upper tidal forests, and deer and wild hogs are known to feed on acorns and other mast in these swamps. Predators such as bobcat, fox, owls, hawks, and other raptors also forage in these swamps. Sykes et al. (1999) found that swallow-tailed kites use swamp forests in the lower tidal reach as nesting habitat.

11.5. Research needs

Biological communities in tidal forests vary along a downstream gradient that is maintained in part by riverflow; thus, long-term changes in flow will affect forest ecology. Regular monitoring of consumptive water use throughout the drainage basin is important, especially considering the substantial increases in population that have occurred in the basin since 1990. In addition, flow-climate modeling is needed to understand the relative contribution of various natural and anthropogenic factors to changes in flow. Natural temperature swings of the Atlantic multidecadal oscillation have resulted in long-term cycles of higher and lower than normal rainfall that have "alternately disguised and accentuated" anthropogenic effects on streamflow (Enfield et al. 2001).

Based on preliminary continuous discharge measurements made by the USGS over a 37-day period in 2002, the river downstream from the Wilcox stream gage appears to have considerable gains and losses in flow that are not fully explainable by spring inputs or losses to tidal distributaries (Stewart A. Tomlinson, USGS, written communication, 2006). Although developing accurate stage-discharge ratings is difficult in this reach because of wind and tidal effects, a better understanding of flow changes in the tidal reach of the lower Suwannee River is needed to assist in the development of models to support water-management decisions.

Long-term monitoring of forest composition will be needed if increased water consumption in the basin causes river flows to decrease. Potential impacts of flow reductions anticipated by Light et al. (2002) include: tidal forests moving upstream and replacing riverine forests, upper tidal forests becoming more like lower tidal forests, salt-tolerant species increasing and salt-intolerant species decreasing in lower tidal forests, and marshes replacing forests at the treeline. Although some of these changes may occur as a result of sea-level rise, they could be accelerated or intensified by decreased freshwater inflow. Permanent transects established in the floodplain forest (Lewis et al. 2002) can be used for long-term monitoring of ecosystem changes resulting from flow reductions or sea-level rise, although plans for such monitoring have not yet been developed.

The position of the treeline determined from a time series of aerial photographs could be used to estimate the historic rate of movement of the treeline. Studies that clarify how salinity from storm surges builds up in soils in the tidal floodplain could be useful in predicting shifts between tidal forests and tidal marshes. Research is needed to determine if potential flow reductions could decrease soil moisture in tidal forests and increase the amount of salt absorbed from saline water deposited by storm surges.

Increased study and monitoring of algal communities are needed, particularly with regard to their usefulness as ecological indicators of water quality. Additional research of faunal communities (invertebrates, fish, and wildlife) is needed to better understand the ecological roles of tidal forests in this unique transition area between riverine and estuarine ecosystems.

11.6. Summary

From the upstream end of the tidal reach of the lower Suwannee River (rkm 45) downstream to the Gulf of Mexico, the influence of river flooding on river stage gradually decreases, and the influence of tides and storm surges on river stage gradually increases. Although this transition is gradual, variations in hydrology, vegetation, and soils along a downstream gradient generally support a division of the lower Suwannee River floodplain into three reaches: nontidal, upper tidal, and lower tidal. Tidal freshwater forests, which cover 8,160 ha of the river floodplain, include swamps, mixed forests, and hydric hammocks.

Median tidal range at the mouth of the Suwannee River is about 1 m with storm surges 2-3 m above sea level occurring about two to three times a decade. Saline water is limited to the downstream-most 9 km of the river except during storm surges. At the treeline, which is the downstream limit of tidal forest about 2 km from the mouth of the river, salinity typically exceeds 10 ppt at low flow. Major hurricanes with large storm surges probably deposit saline water over floodplain forests of the entire lower tidal reach and the downstream end of the upper tidal reach.

The most important species in swamps and mixed forests are pumpkin ash and baldcypress, with water tupelo, important in the upper tidal reach, replaced by swamp tupelo, a more salt-tolerant species, in the lower tidal reach. Mixed forests and swamps have continuously saturated muck soils; however, mixed forests have numerous hummocks that are slightly elevated above the saturated muck floor. Hummocks provide microhabitats for slow growing species such as sweetbay, which is found only in the lower tidal reach because of its intolerance of deep flooding. Hydric hammocks, dominated by cabbage palm and loblolly pine, are found on higher elevations of the lower tidal reach. Hammocks are affected only by the highest river floods and storm surges, although ground-water levels are high year-round. Pumpkin ash and cabbage palm appear to be the most salt-tolerant species in the lower Suwannee River floodplain, and stands of stunted trees of these species are found along the treeline.

The tidal freshwater forests of the lower Suwannee River floodplain are largely intact, with most of the land under public ownership. Population throughout the drainage basin is increasing, however, and potential flow reductions from increased water withdrawals could change conditions in tidal forests. Both water-use monitoring and flow-climate modeling are needed to understand the relative contribution of natural and anthropogenic factors to changes in flow. Studies designed to better understand the effects of storm surges on the salinity of soils and the effects on forests, particularly at the treeline, may help predict impacts of potential flow reductions. Baseline data on algal and faunal communities, which have received little study, are needed to better understand key biotic components in this transition area between riverine and estuarine ecosystems.

11.7. Acknowledgments

John C. Good, SRWMD, contributed to analysis and field work in the 1996-99 study, and provided review comments for this chapter. All soils data in this chapter were based on profile descriptions provided by David A. Howell, Natural Resources Conservation Service, U.S. Department of Agriculture. Extensive contributions to field work and analysis were made by Lori J. Lewis, USGS. Grayal E. Farr, volunteer, assisted with literature review. Palmer Kinser, St. Johns River Water Management District, provided review comments for this chapter, and assistance in preparation of the manuscript was received from Ronald S. Spencer, illustrator, and Twila D. Wilson, editor, USGS.

References

Anderson S, Austin R, Arcuri A (1998) Suwannee River valley environmental and historical narrative and inventory. Report prepared by Janus Research for Suwannee River Water Management District, Live Oak

Bales JD, Tomlinson SA, Tillis G (2006) Flow and salt transport in the Suwannee River estuary Florida, 1999-2000. Analysis of data and three-dimensional simulations. Professional Paper 1656-B. U.S. Geological Survey, Tallahassee

Berndt MP, Oaksford ET, Darst MR, Marella RL (1996) Environmental setting and factors that affect water quality in the Georgia-Florida Coastal Plain study unit. Water Resources Investigations Report 95-4268. U.S. Geological Survey

Brinson MM (1990) Riverine forests. In: Lugo AE, Brinson M, Brown S (eds) Ecosystems of the world—15: Forested wetlands. Elsevier, pp 87-141

Brooks HK (1981) Guide to the physiographic divisions of Florida. Center for Environmental and Natural Resources Programs, University of Florida, Gainesville

Bureau of Economic and Business Research (2006) Florida estimates of population 2005. URL http://www.bebr.ufl.edu

Carlton JM (1977) A survey of selected coastal vegetation communities of Florida. Florida Marine Research Publication Number 30. Florida Department of Natural Resources, St. Petersburg

Carl Vinson Institute of Government (2006) Georgia counties. URL http://www.cviog.uga.edu/Projects/gainfo/county.htm

Carr A (1994) A naturalist in Florida. Yale University Press, New Haven

Ceryak R, Knapp MS, Burnson T (1983) The geology and water resources of the upper Suwannee River basin. Report of Investigations 87. Florida Bureau of Geology, Tallahassee

Chen E, Gerber JF (1990) Climate. In: Myers RL, Ewel JJ (eds) Ecosystems of Florida. University of Central Florida Press, Orlando, pp 11-34

Clark Jr WZ, Zisa AC (1976) Physiographic map of Georgia. Georgia Department of Natural Resources, Atlanta

Clewell AF (1985) Guide to the vascular plants of the Florida Panhandle. Florida State University Press, Tallahassee

Clewell AF, Beaman RS, Coultas CL, Lasley ME (1999) Suwannee River tidal marsh vegetation and its response to external variables and endogenous community processes. Suwannee River Water Management District, Live Oak

Coultas CL (1984) Soils of swamps in the Apalachicola, Florida estuary. Fla Scientist 47:98-107

Crane JJ (1986) An investigation of the geology, hydrogeology, and hydrochemistry of the lower Suwannee River basin. Report of Investigations 96. Florida Bureau of Geology, Tallahassee

Darst MR, Light HM, Lewis LP (2002) Ground-cover vegetation in wetland forests of the lower Suwannee River floodplain, Florida, and the potential impacts of flow reductions. Water-Resources Investigations Report 02-4027. U.S. Geological Survey, Tallahassee

Darst MR, Light HM, Lewis LJ, Sepulveda AA (2003) Forest types in the lower Suwannee River floodplain, Florida-A report and interactive map. Water-Resources Investigations Report 03-4008. U.S. Geological Survey, Tallahassee

Dawes CJ (1981) Marine botany. John Wiley, New York

Day Jr JW, Hall CAS, Kemp WM, Yanex-Arancibia A (1989) Estuarine ecology. John Wiley, New York

Demaree D (1932) Submerging experiments with *Taxodium*. Ecology 13(3):258-262

Enfield DB, Mestas-Nunez AM, Trimble PJ (2001) The Atlantic multidecadal oscillation and its relation to rainfall and river flows in the continental U.S. Geophysical Research Letters 28(10):2077-2080

Ewel JJ (1990) Introduction. In: Myers RL, Ewel JJ (eds) Ecosystems of Florida. University of Central Florida Press, Orlando, pp 3-10

Ewel KC (1990) Swamps. In: Myers RL, Ewel JJ (eds) Ecosystems of Florida. University of Central Florida Press, Orlando, pp 281-323

Federal Highway Administration (2006) HEC 25 – Tidal hydrology, hydraulics, and scour at bridges, appendix E ADCIRC surge estimates. U.S. Department of Transportation, URL http://www.fhwa.dot.gov/engineering/hydraulics/hydrology/hec25appe.cfm

Gallagher JL, Daiber FC (1974) Primary production of edaphic algal communities in a Delaware salt marsh. Limnol Oceanogr 19(3):390-395

Golder Associates, Inc (2000) Mapping low-salinity submerged aquatic vegetation beds in the lower Suwannee River. Final report submitted to the Suwannee River Water Management District, Live Oak

Hanlon EA, McNeal BL, Kidder G (1993) Soil and Container Media Electrical Conductivity Interpretations. Florida Cooperative Extension Service, Circular 1092. University of Florida, Gainesville

Harper RM (1910) Preliminary report on the peat deposits of Florida. In: Sellards EH (ed) Third annual report, 1909-1910. State Geological Survey, Tallahassee, pp 197-375

Henry JA (1998) Weather and climate. In: Fernald EH, Purdum ED (eds) Water resources atlas of Florida. Florida State University Press, Tallahassee, pp 16-37

Ho FP, Tracey RJ (1975) Storm tide frequency analysis for the Gulf Coast of Florida, from Cape San Blas to St. Petersburg Beach. Technical Memorandum NWS HYDRO-20. National Oceanic and Atmospheric Administration, Office of Hydrology, Silver Spring

Hosner JF (1960) Relative tolerance to complete inundation of fourteen bottomland tree species. For Sci 6(3):246-251

Katz BG, Hornsby HD, Bohlke JF, Mokray MF (1999) Sources and chronology of nitrate contamination in spring waters, Suwannee River Basin, Florida. Water-Resources Investigations Report 99-4252. U.S. Geological Survey, Tallahassee

Kurz H, Wagner K (1957) Tidal marshes of the Gulf and Atlantic Coasts of northern Florida and Charleston, South Carolina. Studies Number 24. Florida State University, Tallahassee

Lewis LJ, Light HM, Darst MR (2002) Location and description of transects for ecological studies in floodplain forests of the lower Suwannee River, Florida. Open-File Report 01-410. U.S. Geological Survey, Reston

Light HM, Darst MR, Lewis LJ (2002) Hydrology, vegetation and soils of riverine and tidal floodplain forests of the lower Suwannee River, and potential impacts of flow reductions. Professional Paper 1656-A. U.S. Geological Survey, Washington

Mattson RA (1992) Characteristics of benthic macroinvertebrate communities of the tidal freshwater region of the Suwannee River. Fla Scientist 55(1):11

Mattson RA, Krummrich J (1995) Determination of salinity distributions in the upper Suwannee River estuary. Cooperative project of the Florida Game and Freshwater Fish Commission and the Suwannee River Water Management District. Unpublished

Milanich JT (1994) The Archaeology of Precolumbian Florida. University Press of Florida, Gainesville

National Oceanic and Atmospheric Administration (1994) Superstorm of March 12-14, 1993. Natural Disaster Survey Report. National Weather Service, Silver Spring

Odum WE, Smith III TJ, Hoover JK, McIvor CC (1984) The ecology of tidal freshwater marshes of the United States east coast: A community profile. FWS/OBS-83/17. U.S. Fish and Wildlife Service, Washington

Owenby JR, Ezell DS (1992) Monthly station normals of temperature, precipitation, and heating and cooling degree days 1961-90. Climatography of the United States No. 81. U.S. Department of Commerce, National Oceanic and Atmospheric Administration, National Climatic Data Center, Asheville

Penfound WT, Hathaway ES (1938) Plant communities in the marshlands of southeastern Louisiana. Ecol Monogr 8(1):3-5

Raabe E (2000) Coastal forest retreat on Florida's Big Bend Gulf Coast. U.S. Geological Survey, Center for Coastal Geology, URL http//coastal.er.usgs.gov/wetlands/forest-retreat/

Randazzo AF, Jones DS (eds) (1997) The geology of Florida. The University Press of Florida, Tallahassee

Richards LA (ed) (1954) Diagnosis and improvement of saline and alkali soils. Handbook 60. U.S. Department of Agriculture, Washington

Rupert FR, Arthur JD (1990) The geology and geomorphology of Florida's coastal marshes. Open-File Report 34. Florida Geological Survey, Tallahassee

Sullivan MJ, Moncreiff CA (1990) Edaphic algae are an important component of salt marsh food-webs. Evidence from multiple stable isotope analysis. Mar Ecol Prog Series 62:149-159

Sykes PW Jr, Kepler CB, Litzenberger KL, Sansing HR, Lewis ETR, Hatfield JS (1999) Density and habitat of breeding swallow-tailed kites in the lower Suwannee ecosystem, Florida. J Field Ornithology 70(3):321-336

Tebeau, CW (1971) A history of Florida. University of Miami Press, Coral Gables

Tillis GM (2000) Flow and salinity characteristics of the upper Suwannee River estuary, Florida. Water-Resources Investigations Report 99-4268. U.S. Geological Survey, Tallahassee

U.S. Census Bureau (2006) Population estimates. URL http.//www.census.gov/popest

U.S. Department of Agriculture (1996) Soil survey laboratory methods manual. Soil Survey Investigations Report 42, version 3.0. Natural Resources Conservation Service, Washington

Vince SW, Humphrey SR, Simons RW (1989) The ecology of hydric hammocks: A community profile. Biological Report 85. U.S. Fish and Wildlife Service, National Wetlands Research Center, Washington

Wharton CH, Kitchens WM, Pendleton EC, Sipe TW (1982) The ecology of bottomland hardwood swamps of the southeast: A community profile. FWS/OBS-81/37. U.S. Fish and Wildlife Service, Biological Services Program, Washington

Williams JM, Duedall IW (1997) Florida hurricanes and tropical storms. University Press of Florida, Gainesville

Williams K, Ewell KC, Stumpf RP, Putz FE, Workman TW (1999) Sea-level rise and coastal forest retreat on the western coast of Florida, USA. Ecology 80:2045-2063

Wright EE (1995) Sedimentation and stratigraphy of the Suwannee River marsh coastline. Ph.D. thesis, University of South Florida

Wright Jr JL (1986) Creeks and Seminoles. University of Nebraska Press, Lincoln

Chapter 12 - Community Composition of Select Areas of Tidal Freshwater Forest Along the Savannah River

Jamie Duberstein[1] and Wiley Kitchens[2]

[1]*Baruch Institute of Coastal Ecology and Forest Science, Clemson University, Department of Forestry and Natural Resources, PO Box 596, Georgetown, SC 29442*
[2]*USGS Florida Cooperative Fish and Wildlife Research Unit, Building 810, University of Florida, Gainesville, FL 32611*

12.1. Introduction

12.1.1. Regional location and extent

The Savannah River serves as the jurisdictional state line between Georgia and South Carolina. In the lower Savannah River basin there are approximately 3900 ha of tidal freshwater forest, with an additional 500 ha of seasonally flooded tidal forest, and 150 ha of temporarily flooded tidal forest (National Wetlands Inventory [USFWS 1993a, 1993b, 1999a, 1999b] geographical information systems coverages). Although freshwater tidal forests occur as several community types along the Savannah River, this chapter will focus on describing 4 communities within two designated forest areas (East and West; Figure 12.1). The first at river kilometer 43.5 of the main branch of the Savannah River, and the second is located at river kilometer 41.8 of the Little Back River, a tributary of the main Savannah River.

12.1.2. Major tree and shrub species

In general, tree species that are tolerant to longer hydroperiods were found. The majority of trees include water tupelo (*Nyssa aquatica* L.),

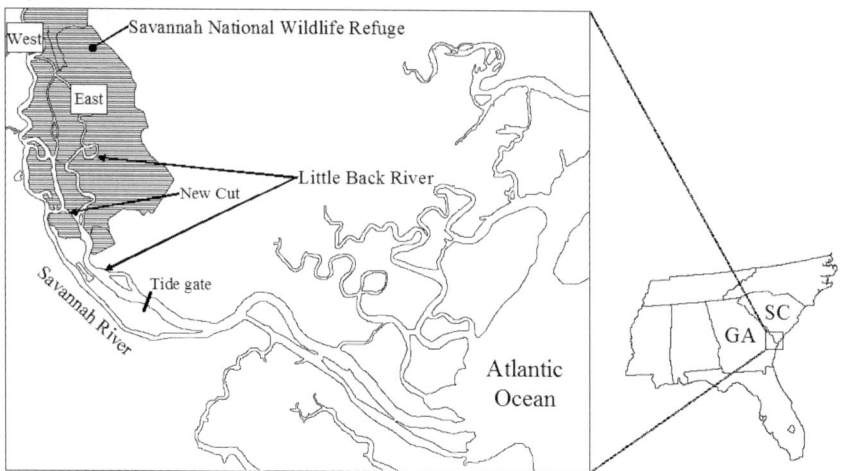

Fig. 12.1. Savannah River, Georgia and South Carolina. The eastern study area is located off of the Little Back River, and the western study area is located off of the main channel of the Savannah River.

swamp tupelo (*Nyssa biflora* Walt.), and water oak (*Quercus nigra* L.). Other abundant trees include ash (*Fraxinus* spp.), sweetgum (*Liquidambar styraciflua* L.), red maple (*Acer rubrum* L.), baldcypress (*Taxodium distichum* [L.] L.C. Rich.), and American hornbeam (*Carpinus caroliniana* Walt.). Shrubs found primarily include alder (*Alnus serrulata* [Ait.] Willd.), dahoon (*Ilex cassine* L.), wax myrtle (*Morella cerifera* L.), swamp doghobble (*Leucothoe racemosa* [L.] Gray), and swamp bay (*Persea palustris* [Raf.] Sarg.).

12.2. Geologic history and historical land use

12.2.1. Soil and underlying bedrock

The underlying bedrock is geologically recent coastal plain sedimentary rocks composed of marsh and lagoon deposits from the Pleistocene and Holocene epochs (Quaternary Period). Technically referred to as the Pamlico Shoreline Complex, the underlying bedrock is composed predominantly of sand and sandy clay with marsh and lagoonal facies which were deposited at former high sea levels (GA DNR 1976, 1977).

General soil descriptions indicate the soils in the eastern area are Levy Soils, which are very poorly drained, nearly level soils on the lower coastal plain. The surface layer is very dark gray silty clay loam 20 cm thick. The underlying material, to a depth of 152 cm, is gray silty clay over silty clay loam (USDA 1980). Descriptions of soils in the western area indicate that they are composed of Angelina and Bibb soils, also frequently flooded and poorly to very poorly drained. These two soil series occur together, in approximately a 4:2:4 ratio of Angelina:Bibb:other ("other" being Chipley, Kershaw, and Ocilla soils). They have been formed in recent deposits of sediments washed from soils on the coastal plain. Surface layers are very dark gray loam about 8 cm thick (Angelina) or light brownish gray loamy sand about 46 cm thick (Bibb). The underlying areas are black to light-gray sand to silty clay loam (Angelina) or mottled light-gray to greenish-gray coarse sand to sandy loam (Bibb). The clay content between depths of 25 and 102 cm within the Bibb series is less than 18 percent (USDA 1974).

12.2.2. Land use changes

The lower Savannah River basin has been severely altered to facilitate a variety of anthropocentric benefits. In the mid 1700's much of the tidal portions of marsh and forest along the Savannah River were converted to rice cultivation. Through this deforestation process, trees were harvested and moved out of the way or burned, and the stumps largely removed (Doar 1936). The presence of some remaining large stumps in Rifle Cut, a man-made tidal creek, suggests that the tidal forest once extended at least 8 km further downstream, relative to their current position (2006). After the Civil War ended in 1865, rice cultivation in the tidal marshlands failed and much of the land was abandoned (McKenzie et al. 1980).

It is likely that the study areas were previously logged. Both tracts of land, now owned by the U.S. Fish and Wildlife Service (USFWS), were owned by logging interests prior to federal acquisition. The Wylly family owned the study area south of Interstate 95 (depicted as West in Figure 12.1), where the southern portion of the western study area is located. Conversations with John Wylly, the previous landowner, indicated that his family had owned the land for the past 100 years, and that the tract was probably cut in the 18th century (William Webb, USFWS, personal communication). The Argent Swamp Tract, where the eastern study area is located, was purchased in 1978 from Union Camp Lumber Company, who had purchased it from Argent Swamp Lumber.

Since both areas were previously owned by logging interests, it is assumed that they were used for lumber production during the boom of the early 20th century. Though there is little direct evidence of past logging operations (only one small patch of three- and four-stemmed water tupelo trees was found in the eastern area), the lack of baldcypress as a codominant in the canopy of the study areas seems to support the notion that baldcypress logging was indeed done prior to USFWS acquisition of the land. The presence of dramatic hummock and hollow topography may even be a relic of previous cypress removal given the spacing and consistency of the hummocks.

12.2.3. Major source of water

The Savannah River is an alluvial river that arises in the southern Appalachian Mountains. Its discharge (Figure 12.2) is fifth largest in the southeastern United States next to the Mississippi, Alabama, Apalachicola, and Altamaha Rivers. The Savannah River basin encompasses approximately 2.7 million ha, originating in the Blue Ridge Mountains near the intersection of the North Carolina, South Carolina, and Georgia borders, coursing through the Piedmont and coastal plain to its mouth at the Atlantic Ocean near Savannah, Georgia. In route it flows southeast through a series of three "peaking" hydropower stations (Figure 12.2) in the Peidmont before eventually reaching the ocean.

The first hydropower project built on the Savannah River was the J. Strom Thurmond Dam and Lake, completed in 1954. The dam is located 35 km above Augusta, Georgia, approximately 385 km above the mouth of the Savannah River. It impounds Thurmond Lake, which at normal pool elevation (101 m, msl) covers 28,773 ha with a shoreline of 1931 km (USACE 2006a).

The second hydropower project built on the Savannah River was Hartwell Dam and Lake, completed and put into operation in 1963. The dam is located 496 km from the mouth of the Savannah River, which is 11 km below the point at which the Tugaloo and Seneca Rivers join to form the Savannah River. The dam impounds Hartwell Lake, which at normal pool elevation (200 m, msl) covers 22,662 ha with 1548 km of shoreline (USACE 2006b).

The third hydropower station built on the Savannah River was Richard B. Russell Dam and Lake, initiated in 1974, completed in 1983, and put into operation generating power in 1985. This dam is located 48 km downstream of the Hartwell Dam and 60 km upstream of the J. Strom Thurmond Dam. The dam impounds Lake Russell, which covers 10,785 ha with 869 km of shoreline at normal pool (145 m, msl) (USACE 2006c).

Calculations published in 1976 state that the J. Strom Thurmond Dam and Hartwell Dam combined are capable of trapping 90% of incoming sediment (Meade 1976), based on reservoir size and calculated from a pre-

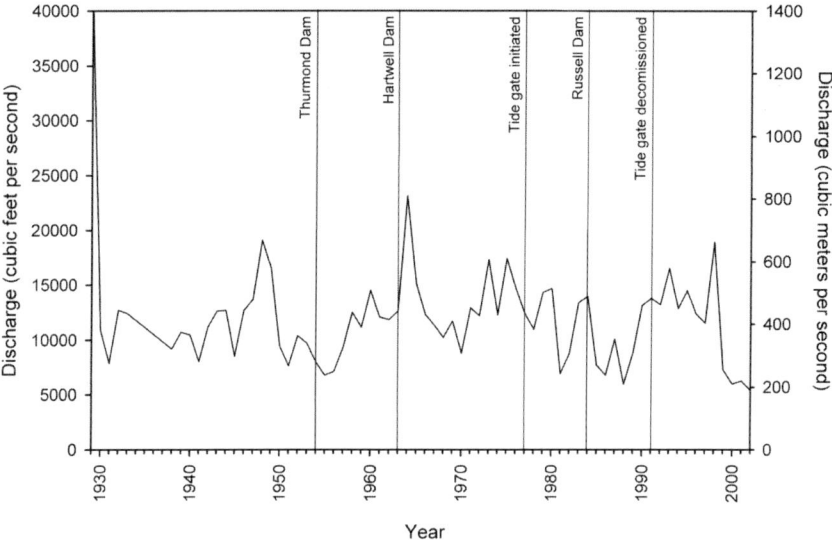

Fig. 12.2. Mean annual discharge of the Savannah River taken at USGS monitoring station #02198500 near Clyo, Georgia. Specific anthropogenic alterations to natural flow are shown.

vious study (Brune 1953). The J. Strom Thurmond Dam and Lake alone is credited with reducing the amount of sediment carried by the river into Savannah Harbor by 22% (USACE 1979 as cited in USACE 2006c).

The navigation channel between the Atlantic Ocean and the Savannah harbor, now extending 34.6 km from the mouth of the river, has been modified many times. In the early years between 1733 and 1850, a minimal amount of dredging was done out of necessity (Granger 1968, as cited in ATM 2001). Between the years of 1874-1890, the navigation channel between the city of Savannah and the Atlantic Ocean was formed to a depth of 4.7 m MLW (ATM 2001). Since that time it has been deepened and widened many times, including installation of several turning basins and curve wideners. The most recent depth increases came in 1992, and is currently being maintained at those levels (Alan Garrett, USACE, personal communication May 15, 2006):

> Depth of the channel from the ocean bar extending to the mouth of the Savannah River is 13.4 m MLW. The inner harbor channel, running from the mouth of the river to Georgia Ports Authority Terminal in Garden City 30 km upstream, is 12.8 m deep; the next 2 km segment extending to the Savannah Sugar Refinery is 11 m deep. The remainder of the channel is 9.1 m deep up to a point 457

m below the Atlantic Coastal Highway Bridge, approximately 34 km from the mouth (USACE, 2006d).

For a detailed list of past dredging activities see Plachy (2002).

In 1977, a one-way tidal flap gate was installed on the Back River, approximately 23 km from the river mouth, as a mechanism for minimizing the amount of maintenance dredging in the shipping channel (Front River) of the Savannah River. In-flowing water was allowed to pass upstream through the gate during flood tides. The one-way flap gate was shut at slack high, and the entire volume of entrained water was forced to flow through New Cut, a diversion channel constructed through Argyle Island (Figure 12.1), and out the main channel during ebb tide, thereby increasing the velocity and scour through the harbor area. However, the blockage caused saltwater intrusion into the Little Back River and Middle River portions of the Savannah River. With each tidal cycle, the salt wedge was pushed further upstream, resulting in a dramatic shift in marsh vegetation from freshwater species to those that are more tolerant of oligohaline conditions (Georgia Ports Authority 1998). Salinity projections by Pearlstine et al. (1993) indicate portions of the tidal forest were impacted from operation of the tide gate, though the projected salinities in the tidal freshwater swamp area were minor (< 1 ppt). In 1991 the tide gate was taken out of operation, with the subsequent closure of New Cut in 1992. The freshwater and oligohaline marsh communities present before the Tide Gate era had significantly recovered by 1995, only to be altered by a deepening in 1996 (Dusek 2003).

Hydropower stations, the tide gate, as well as past and continuing navigation channel dredging have been documented as changing the natural conditions along many points of the Savannah River. Some key effects are decreased sediment deposition at the harbor and measurable interstitial salinity increases (on the order of 2-fold increases) in the freshwater marshes (Latham and Kitchens 1995; Dusek 2003). Although data from the freshwater tidal forests have not been collected and analyzed at this point, it is very likely that the management upstream and downstream is causing change in these areas as well.

12.3. Climatic drivers and hydrological characterization

12.3.1. Climate

Savannah, Georgia has an average annual temperature of 19°C, with the highest monthly average of 27.8°C in July and lowest monthly average of

9.6°C in January (NOAA 2002). The average frost free season is 226 days long (90% confidence), occurring between March 30–October 31(NOAA 1988). Average annual precipitation is 126 cm with an average monthly high of 14 cm in June (NOAA 2002). The mean sea-level trend for Fort Pulaski, Georgia (located at the mouth of the Savannah River) is +3.05 mm/year with a standard error of 0.2 mm/yr based on monthly mean sea-level data from 1935 to 1999 (NOAA 2001).

12.3.2. Tidal hydrology

The lower portion of the Savannah River undergoes a regular, semidiurnal flooding regime and is a salt-wedge type estuary (Hansen and Rattray 1966). Tidal range of the Savannah River is in excess of 3 m with flow reversals 45 km upstream of the river mouth. Tidal ranges at the two areas studied, however, are lesser: 1.5-2 m during non-drought periods (Figure 12.3). Flooding of the surface layer of the soil in the freshwater tidal forests was only in the range of a few centimeters to 0.5 m during 2002. However, hydrologic conditions resulting from the range and consistency of the semidiurnal tides keep soils saturated for the entire year in most areas of the tidal forest, even during the extended drought conditions of 1999-2002.

12.4. Ecological characterization

Historic rice cultivation converted much of the former freshwater tidal forest areas to what is now freshwater marsh. Manipulations of the Savannah River that occurred since the 1860's have certainly resulted in alterations to the tidal wetlands. Specifically, the modifications mentioned previously have changed the freshwater marsh vegetation communities to those of brackish marsh communities (Pearlstine et al. 1993) and caused drastic declines in the number of striped bass eggs found at historic spawning grounds (Van Den Avyle et al. 1990; Van Den Avyle and Maynard 1994). It was these changes, coupled with the rarity of the tidal freshwater marshes in the southeastern United States, that resulted in the removal of the tide gate in 1992. Since this time, the Georgia Ports Authority has strived to not cause decline of the Savannah River freshwater marshes. As part of the effort to monitor these areas, tidal freshwater forest areas were also contracted for monitoring, thus bringing about the need for this study. One objective of the study was to identify the tree communities in two areas projected as having different salinities (<0.1 ppt in the west vs. 1.0 ppt

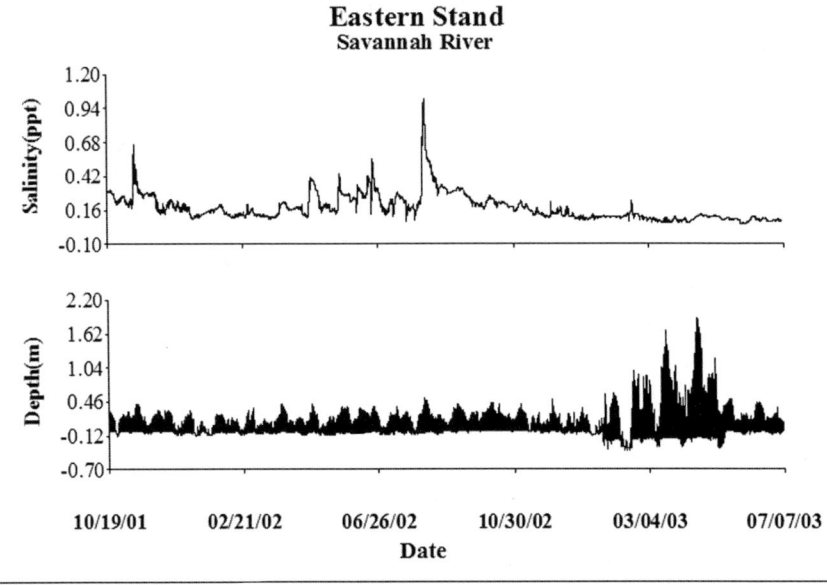

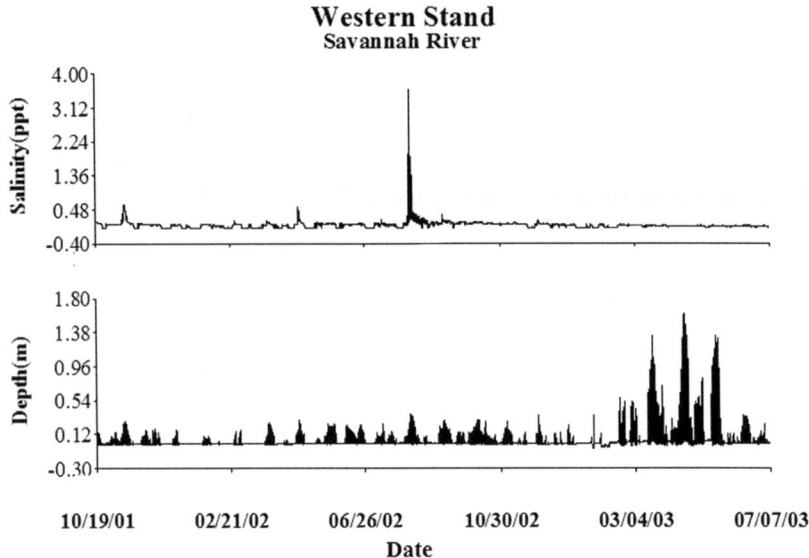

Fig. 12.3. Hydrographs of freshwater tidal forest areas studied in the Savannah NWR. Data were taken from interstitial measurements recorded at 15 minute intervals. Drought conditions persisted from summer 1999–spring 2003.

in the east) during the operation of the tide gate (see Pearlstine et al. 1993). However, as the areas were investigated it became clear that previous salinity effects could not be inferred due to the over-riding differences of soil type and drainage capabilities. A second objective was to ordinate and describe the current tree and shrub communities in the freshwater tidal forest of the lower Savannah River basin, and to identify soil chemical properties that determine plant community differentiation. Finally, an attempt was made to document the present conditions in anticipation of accelerated saltwater intrusion resulting from future shipping channel deepening.

12.4.1. Methods

Two forest areas were chosen, each approximately 140 ha. The eastern area is located off the Little Back River, a tributary of the Savannah River, and the western area is located directly off the main channel of the Savannah River (Figure 12.1). Both areas are roughly 40 km from the Atlantic Ocean. In 2001 a pilot study was done to determine the tree species richness and diversity within the study areas. It was quickly evident that the eastern area has greater diversity, so pilot study efforts were focused within its boundaries. In all, ten nested quadrats were cataloged for information regarding species and diameter at breast height (DBH). The smallest reasonable area to be quantified was assumed to be 25 m^2 (5 m x 5 m). By lengthening each quadrat by 5 m on two ends, each nested quadrat, then, consisted of one of each quadrat size: 25, 100, 225, and 400 m^2. With this information, a species-area curve (Cain 1938; Kent and Coker 1992) was developed to determine the minimum quadrat area that would capture the community structure and composition. Unlike the traditional species-area curves that use a progressive doubling of the quadrat size (Kent and Coker 1992), our species-area curve necessitated an algorithm that could incorporate several samples of the same quadrat size. To accomplish this, the computer program Sigma Plot 8.02 (SPSS Inc. 2001) was used to perform a nonlinear regression, resulting in an optimal quadrat (i.e., plot) size of 100 m^2 (10 m x 10 m), which results in 0.1% of the areas being sampled.

12.4.1.1. Data collected

In each area, 16 plots (10 m x 10 m) were established using a stratified random design, resulting in 32 sample plots total. In 2002 each plot was flagged off, and all tree species at least 1.38 m tall (breast height) were identified and measured for DBH. Additionally, two soil samples were taken at each plot to a depth of 12.6 cm using a 6.9 cm diameter soil corer. Soil samples were placed in a freezer upon return from the field to minimize decomposition. Once thawed, samples were oven dried at 50°C for at least 2 weeks to remove all moisture. From each sample, material (mineral and organic matter) was passed through a 2 mm sieve, homogenized, and sent to the Analytical Research Laboratory at the University of Florida for analysis. Concentrations of phosphorous (P), potassium (K), calcium (Ca), magnesium (Mg), zinc (Zn), manganese (Mn), sodium (Na), and iron (Fe) were determined by Mehlich extraction. Electrical conductivity and chloride ion (Cl$^-$) concentrations were determined using a 2:1 water to soil ratio with 250 cm^3 soil. The second soil sample was homogenized, weighed to determine bulk density, and combusted to determine organic matter content by the loss on ignition method (Klawitter 1962).

12.4.1.2. Data preparation

Due to inherent differences of plot structure, a method of representing the competitive interactions at each plot was needed. For example, a given plot may be comprised of many small, shrubby trees whose collective basal area is small. Conversely, a plot may be made up of relatively few big trees with a large cumulative basal area. The freshwater tidal forests along the Savannah River floodplain have structures described in both scenarios. For this reason, a standardization transformation of the data was performed. Standardizations of this type are widely used in gradient analyses because it increases the strength of the relationship between species dissimilarity and ecological distance for moderate or long gradients (Faith et al. 1987). For this study, importance values were computed in a manner similar that of Curtis and McIntosh (1950, 1951), with the elimination of the relative frequency term. The value is then the average of two components: relative density and relative dominance.

Rare species were removed from the analyses in an effort to tighten patterns and enhance the detection of relationships between community composition and soil variables. Using a method offered by McCune and Grace (2002), those species that were present in fewer than 5% of the plots (i.e., 2 plots or fewer) were removed from the analyses. Though deletion of rare species is considered inappropriate when examining patterns in species diversity (Cao et al. 1998), it is often helpful for multivariate analysis of community structure (McCune and Grace 2002) such as Nonmetric Multidimensional Scaling (NMS) ordination. In total, 8 species were removed: inkberry (*Ilex glabra* [L.] Gray), highbush blueberry (*Vaccinium corymbosum* L.), sweetbay (*Magnolia virginiana* L.), eastern baccharis (*Baccharis halimifolia* L.), common winterberry (*Ilex verticillata* L.), black willow (*Salix nigra* Marsh.), water-elm (*Planera aquatica* J.F. Gmel.), and laurel oak (*Quercus laurifolia* Michx.). In addition to the rare species removal, an outlier analysis (done with PC-ORD, McCune and Mefford 1999) indicated that one plot located in the western area was an outlier, and it was removed from the analyses. This plot was comprised of mostly canopy and subcanopy trees, including (predominantly) swamp tupelo, with some baldcypress and ash of similar canopy position. Relatively few shrubs were cataloged in this plot, likely resulting in the outlying nature. A single water-elm sapling (DBH < 1cm) and a laurel oak sapling (DBH 4.1 cm) were also found in this plot only. Edaphic properties were not dissimilar to other plots.

Soil nutrient values were analyzed both as a concentration, and on a per-volume basis. For each plot, nutrient concentrations were multiplied by bulk density to produce a per-volume value (i.e., nutrients present). All parameters were then scaled to reflect the same scale as the species data. To accomplish this task, soil nutrient values were multiplied or divided by orders of 10. Following the relativization, NMS procedures were used to determine correlations between soil parameters and ordination axes (and, therefore, study plots). To increase parsimony and strengthen relationships between tree and shrub communities and soil properties, soil parameters that had a coefficient of determination (r^2) less than 0.392 to any of the axes for any rotation were removed from the analyses (Table 12.1).

12.4.1.3. A Priori grouping

In addition to the *a priori* grouping of plots into areas (east and west), each plot was placed into one of three categories based on their landscape position and assumed hydrogeomorphologic differences:

1) Plots proximal to either the main channel of the Savannah River or a large distributary. These plots are likely to be of higher elevation and have higher mineral content of soil since they are associated with the natural levee of the river.
2) Plots associated with tidal creeks and drainages. Lower in elevation than the first group, the proximity of these plots to tidal rivulets in the floodplain leads to intermediate drainage conditions and soil mineralization, as compared to the other two groups.
3) Plots relatively far removed from tidal creeks and drainages. These are essentially backswamp plots furthest removed from the main rivers, experiencing increased residence time, and decreased water flux with each tidal cycle. Relative isolation leads to very poor drainage, ponding, and accumulation of organic matter and nutrients.

Table 12.1. Soil parameters collected in each plot. Only variables with a Pearson's correlation (r^2) of at least 0.392 were retained for further analyses.

Variables retained		Variables removed
Organic matter	pH	Zn present
Ca concentration	P concentration	Mn Concentration
Mg concentration	K concentration	Mn present
Electrical conductivity	K present	Cu concentration
Na concentration	Ca present	Fe concentration
Cu present	Mg present	Fe present
Bulk density	Zn concentration	Na present
P present		

12.4.1.4. Statistical analyses

Multivariate approaches were used to determine the community structure of the freshwater tidal forest communities in the two areas, as well as aid in their description. Plots were subject to hierarchical agglomerative cluster analysis, followed by indicator species analyses for various numbers of clusters (i.e., communities). Multi-response permutation procedures (MRPP) used species data to test for differences at the various community levels, landscape level (i.e., a priori group), as well as differences between the two areas. Nonmetric Multidimensional Scaling (NMS) was used to visualize sample plot grouping based upon species data, and examine trends in soil characteristics through biplot overlays. As a final cross-validation step in determining the appropriate number of communities to interpret, classification and regression tree (CART) analysis was used to group communities based solely on soil parameters. Once the appropriate number of communities was determined, indicator species analysis assisted in determining a naming convention.

A hierarchical, polythetic, agglomerative clustering was done on study plots based upon the importance value of each species in each plot. The clustering routine utilizes the Sørensen distance measure in combination with a flexible beta ($\beta = -0.25$) linkage method (McCune and Grace 2002). Group memberships from the cluster analysis were written to a secondary data matrix, and later used as categorical variables in subsequent analyses.

To assist with pruning of the cluster dendrogram, several indicator species analyses were performed based on species composition, and using Dufrene and Legendre's (1997) method. For each iteration (level of grouping), the groups to which each plot belonged were used as categorical variables. A requisite of this analysis is that each group must be comprised of at least two or more plots, therefore the maximum number of groups that could be analyzed with data from this study was eight. Logically, the minimum number of groups was two, since placing all plots in one group leaves nothing to compare and contrast. A total of seven separate indicator species analyses were performed, ranging from 2 to 8 groups. For each iteration, an indicator value is computed for each species in each group ranging from 0 (no indicator) to 100 (perfect indicator) with a perfect indicator being faithful (always present) and exclusive to all plots in that group. The largest indicator value for a given species across all groups was recorded as the indicator value for that species. A Monte Carlo test using 1000 randomized runs tested significance. Each of the seven analyses resulted in different *p*-values for species as indicators for a given grouping level. The *p*-values were then summed across all species for each of the analysis.

MRPP used species data to test for differences between groups at various scales. For this data set, we tested the difference between: (1) forest areas (eastern vs. western), (2) *a priori* landscape grouping, and (3) groups defined by the cluster analysis, of which the seven levels mentioned above were tested.

The software package PC-ORD (McCune and Mefford 1999) used species data to perform indirect gradient analysis using NMS ordinations based on Sørensen distances, also known as the Bray-Curtis coefficient (Faith et al. 1987, as cited in McCune and Grace 2002). The first NMS run utilized PC-ORD's autopilot mode to determine the appropriate number of axes to interpret, and compute correlations with soil variables. A random number was generated for the starting configuration during this particular ordination. While in autopilot mode, PC-ORD recommends dimensionality by comparing stress values among the best solutions for each of the 6 dimensional possibilities it investigates. Once the optimal dimensionality is determined, the autopilot mode does a final run with the appropriate dimensionality. Biplots of soil variables are overlaid onto the ordinations of plots in species space, and correlations of the soil variables to the axes are output. These correlations were used for the removal of insignificant soil variables. Subsequent NMS ordinations used a random starting configuration, were restricted to 3 dimensions, and did 100 runs using real data. A Monte Carlo test of significance used 100 runs of randomized data.

A CART analysis using the soil properties data (without species data) was performed to classify plots into communities (groups) using the statistical program S-Plus 2000 Professional Release 3 (Mathsoft 2000) with the TreesPlus add-in (De'ath 2002). This approach was chosen as the final step in determining how many groups to interpret (as communities) due to its predictive and descriptive ability to model community composition with environmental correlates. It was also chosen for its ability to handle correlations among variables since only the single best predictor is selected at each branch, while different predictors are still free to be selected at other branches of the tree (Urban 2002).

12.4.2. Statistical results

The clustering routine produced a dendrogram (Figure 12.4) with only 3.52 percent chaining using the flexible beta linkage method, which was ultimately pruned at the point where 50 percent of the information was remaining. This pruning is the key step in determining how many freshwater tidal forest communities exist in Savannah NWR. As noted previously, the entire suite of statistical analyses were carried out at several grouping levels to determine where to prune. The starting point for each iteration was determining the group membership based upon this cluster analysis.

For all indicator species analyses, the lowest total p-values were 0.1667 and 0.1657, found in cluster sizes 4 and 2, respectively (Figure 12.5). The number of significant indicator species ($\alpha < 0.05$) for each iteration (grouping level) was also tallied (Figure 12.5) and used as an aid for pruning the dendrogram. With 5 or more groups the average p-value increases sharply, while the number of significant indicators drops, indicating that 5

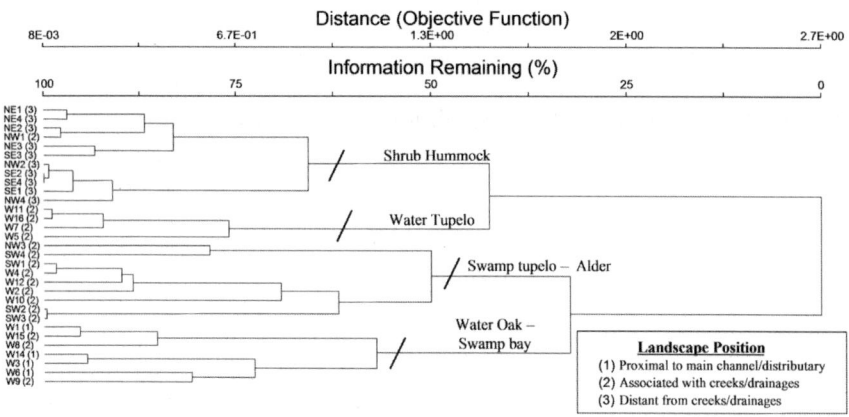

Fig. 12.4. Cluster dendrogram resulting from analysis of 20 species in 31 plots describing community clusters for tidal freshwater forests in Savannah NWR. Plot names are listed on the left. Those beginning with W are from the western area whereas the NE, NW, SE, and SW plots are from the eastern area. A priori grouping of landscape position is indicated by parentheses and defined in the inset legend. Pruning of the dendrogram is indicated by the /, and community names are given for each of the 4 groups.

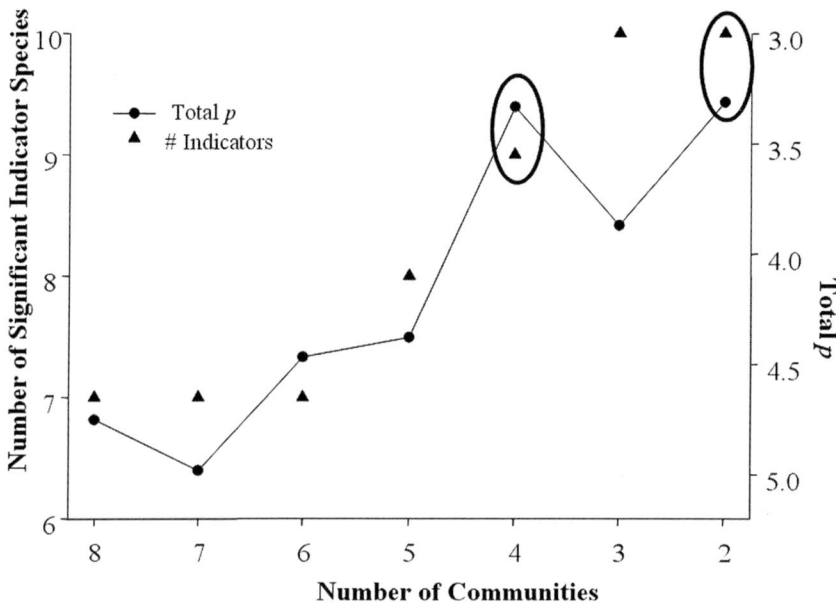

Fig. 12.5. Summary of the 7 indicator species analyses. *P*-values are based on Monte Carlo randomization, then summed over all species for each grouping level (x axis). Circles denote best alternatives: lowest total *p*-values, highest number of indicator species, and logical interpretation.

or more distinct communities probably do not exist in the study areas. Although grouping levels of 3 and 2 resulted in the highest number of significant indicator species and had low total *p*-values, the cluster size of 4-community option was chosen due to the fact that it still has a very low total *p*-value, a high number of significant indicators, and gives optimal interpretation. Later analyses, including MRPP, NMS ordinations, and CART, produced additional support of this community level. After analysis was complete, names were given to communities based upon the significant indicator species in each.

Results of the MRPP tests determined that community composition is significantly different (in general and in pairwise comparisons) between the 2 areas ($p<0.01$), the 3 landscape positions ($p=0.01$), and the 4 communities ($p<0.01$).

The NMS autopilot run indicated that a 3-dimentional solution was optimal, with a stress value of 6.62 ($p=0.02$) for 82 iterations and 50 Monte Carlo runs. This is well within the acceptable range (Kruskal 1964), especially when statistics of this sort are applied to ecological community data

(McCune and Grace 2002). Additional ordinations run to fit a 3-D solution resulted in a final stress of 6.62 (p<0.01) and a final instability of 0.00001 for 93 iterations and 100 Monte Carlo runs. Axes 2 and 3 represent the largest proportion of variance explained by the ordinations (Table 12.2), and the plots naturally separate out when viewing these axes.

Though the soil parameters have no affect on how the dissimilarity matrix is calculated, nor where plots are positioned in the display of the NMS ordination (Figure 12.6), it is helpful to see how they are correlated to ordination axes and, therefore, to plots. It is also helpful to see how they correlate to other soil variables. Organic matter content, electrical conductivity, Ca concentration, Mg concentration, and Na concentration are all closely correlated with axis 2 (Figure 12.6). At the same time, the values for bulk density, K present, and Cu present are also correlated to axis 2, but negatively correlated to the other soil parameters. Note that longer vectors indicate stronger relationships to plots in which they are directed toward. For example, the biplot overlays of organic matter content, electrical conductivity, Ca concentration, Mg concentration, and Na concentration are all associated with high values in shrub hummock.

The NMS ordination gives good visualization of 2- and 4-community grouping levels, as indicated by the larger solid line circles around the plots in Figure 12.6. This helps explain the 2 vs. 4 group results given by the indicator species analyses (Figure 12.5). However, knowledge of the areas proves to be an important interpretive tool, and demonstrates how these exploratory statistical analyses were meant to point the ecologist in the proper direction rather than dictate interpretation. There are essentially 2 broader groups that are comprised of 2 subgroups, all of which are significantly different from each other. The primary differences between the 2 larger groups are the relative abundances of water tupelo vs. swamp tupelo

Table 12.2. Proportion of variance represented by axes based on the r^2 distance in the NMS ordination space and distance in the original space.

	r^2	
Axis	Increment	Cumulative
1	0.025	0.025
2	0.407	0.433
3	0.527	0.960

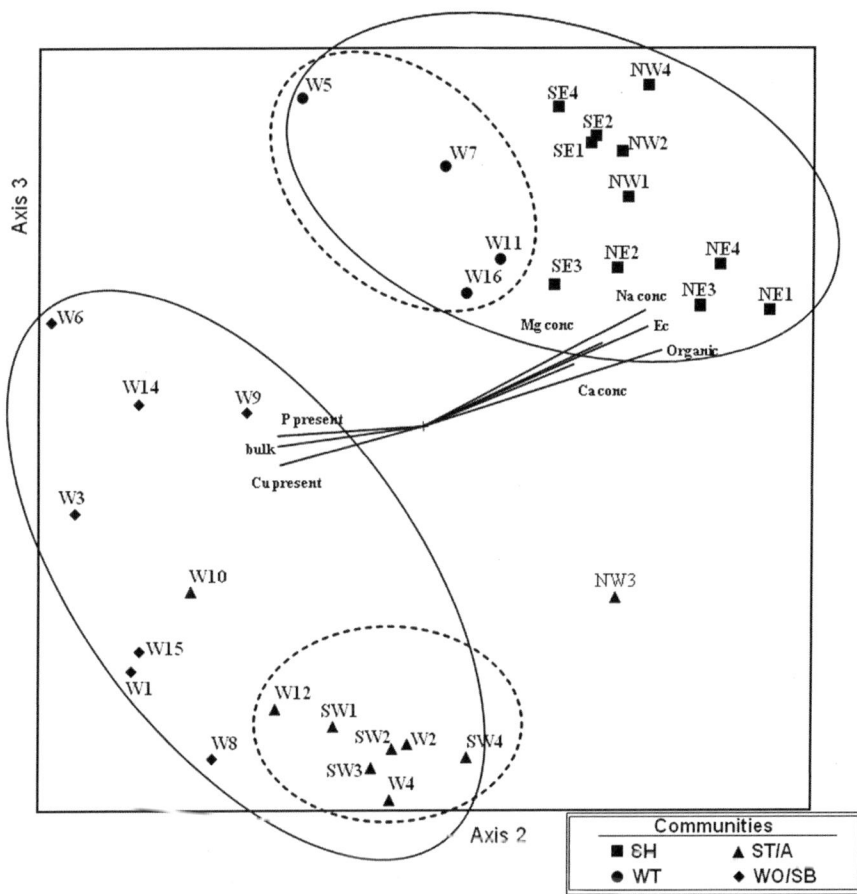

Fig. 12.6. NMS oridination of plots in species space. Communites include: shrub hummock (SH), water tupelo (WT), swamp tupelo – alder (ST/A), water oak – swamp bay (WO/SB). Biplot overlays only include soil parameters with $r^2 > 0.392$ to either axis. Longer lines indicate stronger relationships.

(Table 12.3). The finer differences are pointed out in the community descriptions.

The best fit CART model (Figure 12.7) explained 55% of the variation in the edaphic data, which was accomplished when four communities were classified, reinforcing the earlier analyses that 4 communities exist in the study area. Results indicate that the occurrence of these four communities can be predicted with soil values for organic matter, Na, and electrical conductivity with an error rate of approximately 50%. When soil organic matter is greater than 78%, the CART models show a clear split, indicating the most likely community to be found would be the shrub hummock. This

Table 12.3. Averages species' importance values and soil values within given community types. Values in parentheses indicate how many plots are in each community and, therefore, were used in averaging.

Species	Community type			
	Shrub (11)	Water tupelo (4)	Swamp tupelo/alder (9)	Water oak/ swamp bay (7)
red maple	4.07	1.96	4.76	5.15
alder	12.77	18.59	23.71	1.87
buttonbush	0.08	0.00	0.07	0.00
ash	4.86	11.04	4.32	11.49
dahoon	4.67	0.18	0.44	0.00
Virginia sweetspire	0.43	0.00	0.00	0.00
swamp doghobble	10.22	0.00	1.97	0.26
wax myrtle	12.69	1.08	1.84	0.39
water tupelo	38.54	49.01	1.99	12.79
swamp tupelo	2.77	7.41	32.64	30.87
baldcypress	5.72	3.26	14.12	9.19
possumhaw viburnum	0.13	0.00	0.04	0.00
possumhaw	0.14	0.00	0.22	1.04
fetterbush lyonia	0.26	0.00	0.00	0.00
swamp bay	0.17	0.35	2.79	6.50
arrowwood	0.15	0.00	0.07	0.00
sweetgum	0.99	0.00	0.82	1.69
water oak	0.32	6.55	7.62	16.03
stiff dogwood	0.10	0.20	0.96	0.13
American hornbeam	0.00	0.37	1.35	2.34
Soil parameters				
Organic matter (%)	86.26	49.76	52.98	42.62
Bulk density (g/cm^3)	0.08	0.18	0.20	0.27
Electrical conductivity (dS/m)	7.81	3.54	3.92	2.48
Na concentration (mg/kg)	1033.00	315.30	435.49	261.91
Ca concentration (mg/kg)	7326.00	4454.50	4746.44	3714.57
Mg concentration (mg/kg)	1817.15	539.00	787.96	510.43
P present (μg/cm^3)	5.61	13.76	9.72	13.19
Cu present (μg/cm^3)	0.37	1.15	1.28	1.50
Plot statistics				
Density (stems/ha)	18209	6425	9611	4043
Basal area (m^2/ha)	54	69	56	59

was the strongest and most consistent result, even when 2, 3, and 4 communities were run in the model (not presented graphically). Further splitting (which undoubtedly increased the error rate up to 50%) was done based upon Na concentration, followed by electrical conductivity.

Most plots in the swamp tupelo – alder community have a soil organic matter content less than 78 percent and Na concentration greater than 353

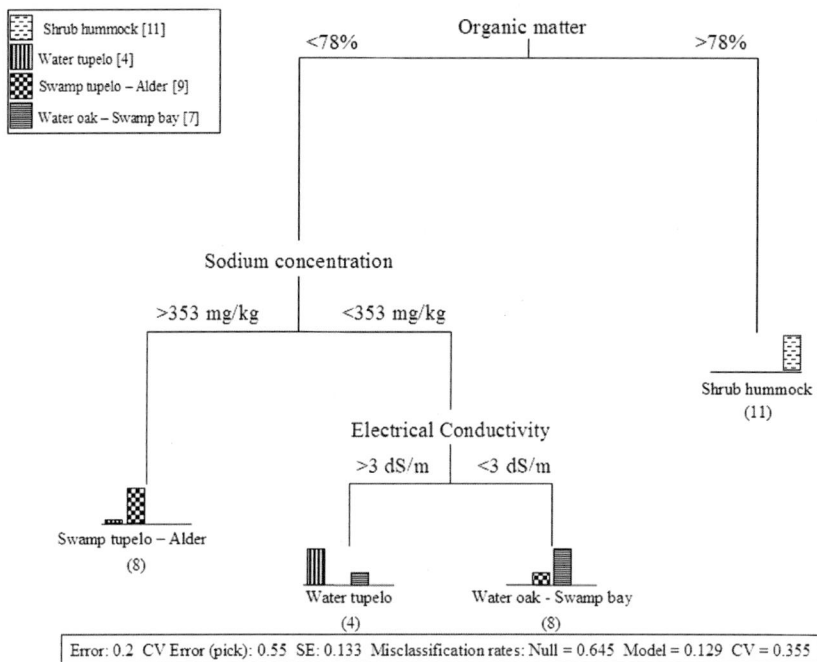

Fig.12.7. Classification and regression tree (CART) depicting amounts of organic matter, Na concentration, and electrical conductivity used to categorize communities. Brackets in legend indicate total number of each community, while parentheses indicate number of plots each (labeled) community contained in CART leaves.

mg/kg, while most plots in the water tupelo community have a soil organic matter content less than 78%, Na concentration less than 353 mg/kg, and electrical conductivity greater than 3 dS/m. Finally, most plots in the water oak – swamp bay community have soil properties similar to those of the water tupelo community, with the exception of electrical conductivity, being less than 3 dS/m.

12.4.3. Community descriptions

12.4.3.1. Shrub hummock

The shrub hummock is the most distinct of all the communities described in this study. It is characteristically lacking many tall trees, and occurred in all backswamp plots (i.e., distant from creeks and drainages) in the eastern area, except for one. Within these areas, the diverse microtopography is a feature that makes them quickly distinguishable. Of all trees inventoried in this community type, 90% of them occurred on the microtopographic highs (hummocks). Of those that did grow in the unconsolidated hollows, 54% of the individuals inventoried were alder. The shade-producing canopy trees are noticeably shorter in this community, as is the overall average height of all trees. Larger trees are found, but are infrequent and, given their height relative to the rest of the canopy, their presence does not inhibit sunlight from penetrating to the forest floor. These supra-canopy individuals are primarily water tupelo and baldcypress (rarely swamp tupelo, sweetgum, and red maple) rooted almost exclusively (95%) on hummocks. Their large basal area relative to the other individuals resulted in large importance values for the statistical analyses, which influenced the results of the cluster analysis by causing plots with this plant community to closely resemble the water tupelo community (Figures 12.4 and 12.6) when, in fact, they are quite different structurally (Figure 12.8).

Shrub hummocks have the highest diversity of all the communities described in this study, as well as the highest stem density of individuals smaller than 5 cm DBH (Figure 12.8). Significant indicator species are da-

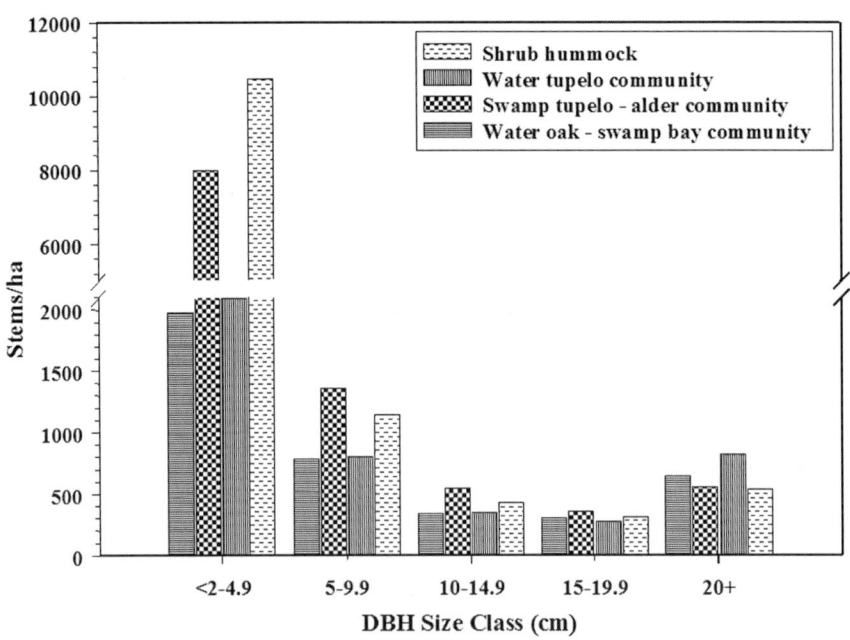

Fig. 12.8. Number of stems per hectare, broken into DBH size classes, for each community type.

hoon, Virginia sweetspire, fetterbush, and wax myrtle. Other common species are alder and regeneration-sized (sapling and subcanopy) water tupelo and red maple. This community also contains the most uncommon of the species analyzed: arrow wood (*Viburnum dentatum* L.), fetterbush lyonia (*Lyonia lucida* [Lam.] K. Koch), and possumhaw viburnum (*Viburnum nudum* L.). Diversity in the shrub hummocks was so high that some plants were removed from the analyses due to their rareness: inkberry, highbush blueberry (*Vaccinium corymbosum* L.), sweetbay (*Magnolia virginiana* L.), eastern baccharis (*Baccharis halimifolia* L.), common winterberry (*Ilex verticillata* L.), and black willow (*Salix nigra* L.).

Soil organic matter content is very high (average = 86.26 percent), as indicated by the long biplot overlay line pointing toward plots with this community (Figure 12.6), the CART model (Figure 12.7), and the means presented in Table 12.3. The inverse relationship between soil weight and organic matter results in soil bulk density values that are the lowest of any community described in this study. Shrub hummock soils also have the lowest amount of P present based on a per-volume basis, while concentrations of Na, Ca, and Mg are the highest. It follows that soil electrical conductivity is also highest.

12.4.3.2. Water tupelo community

This community was found in only 4 plots, all of which were associated with tidal creeks and drainages in the western area (Figure 12.9). The lower densities of trees (Table 12.3) leave a relatively open canopy, allowing sunlight to penetrate to the forest floor. The result is a well developed herbaceous layer, which is typical of the western area. Water tupelo is the defining species of this community, but occurs in neither highest density nor greatest basal area here than in any other community. However, relative to other species in each plot (all 31 plots included, not only those with this community type), water tupelo has the highest importance value in this community type. This is partially due to the fact that shrub layer is poorly developed, evidenced by the fact that swamp doghobble and possumhaw (*Ilex decidua* Walter) were not documented in any of the plots. There were many regeneration-aged stems of water oak, ash, red maple, stiff dogwood (*Cornus foemina* P. Mill.), and baldcypress, species that can eventually make it to the canopy and subcanopy, given significant disturbance. Oddly, sweetgum was not found in this community.

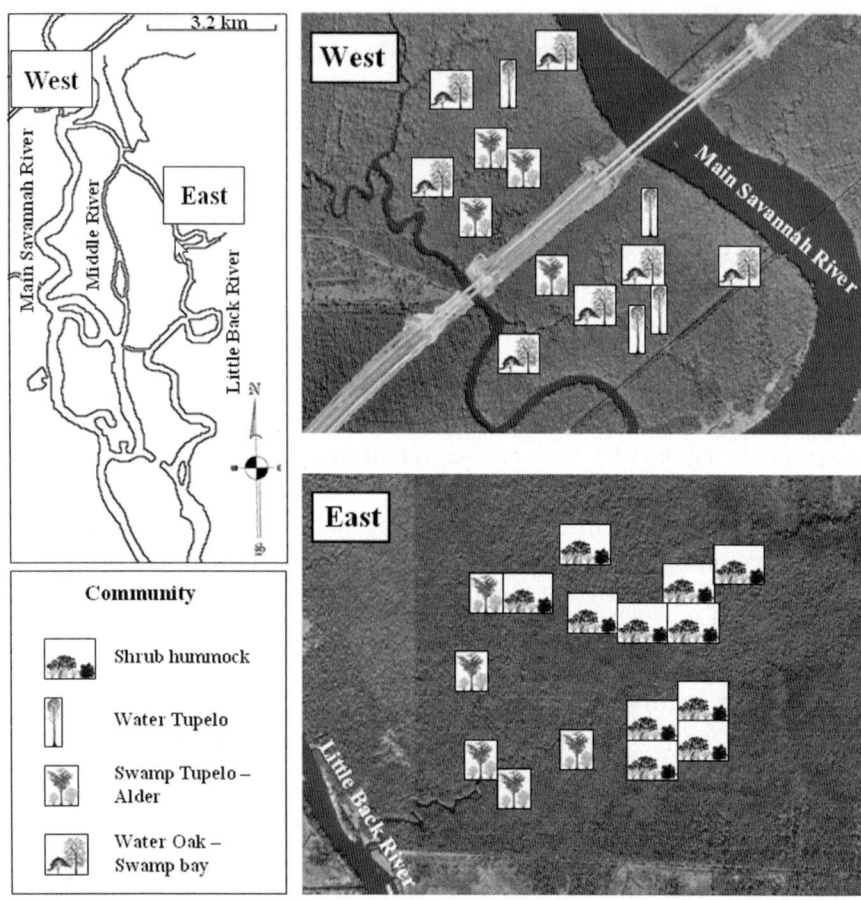

Fig. 12.9. Locations of communities sampled.

12.4.3.3. Swamp tupelo – alder community

This is likely the most abundant community throughout the study areas, and probably best represents the "typical" tidal freshwater forest community in Savannah NWR. It occurs in proximity to tidal rivulets and drainages, and is the only community to be found in both the eastern and western areas. The canopy is tall, well developed, and dominated by swamp tupelo, along with the highest abundance of baldcypress found in the areas examined (Table 12.3).

Floral diversity is high, in part attributed to a well developed shrub layer, as evidenced by the abundance of smaller size-classed trees and shrubs (Figure 12.8). Though the shrub layer is dominated by alder, stiff dogwood and buttonbush (*Cephalanthus occidentalis* L.) are also relatively abundant. The overall canopy position of most species are equally distributed. The following species are found in all layers (canopy, subcanopy, and shrub): red maple, ash, dahoon holly, fetterbush lyonia, wax myrtle, possumhaw viburnum, possumhaw, swamp bay, sweetgum, water oak, and American hornbeam. It should be noted, however, that determination of a species' canopy-class position was relative to other trees in the immediate area, not to heights of trees in other areas visited.

12.4.3.4. Water oak – swamp bay community

This community was found in all plots associated with the main channel of the Savannah River, as well as those associated with the largest tidal creeks and drainages in the western area. The canopy and subcanopy is similar to the swamp tupelo – alder community. It is tall, well developed, and contains many of the same species, which explains why the plots are relatively close in both the cluster analysis (Figure 12.4) and the NMS ordination (Figure 12.6). With the exception of swamp tupelo, there are no strong dominants in this community, in contrast to the water tupelo and swamp tupelo – alder communities. Water oak and swamp bay are significant indicators, but are found in all layers. A key compositional difference comes from the shrub layer. It is not dominated by "shrub" species such as stiff dogwood and wax myrtle, but rather regeneration-sized "tree" species that are also in the canopy and subcanopy. Again, similar to the water tupelo community, this community is unique in its absence of a very common species. Alder, perhaps the most widely distributed of all the species, is completely absent from 4 of the 7 plots exhibiting this community.

Soil organic matter is relatively low in this community (Table 12.3), ranging from 17 percent in very clayey soils near the main channel of the river to 62 percent at a plot next to a tidal stream (Figure 12.9). Soil electrical conductivity, Na concentration, and Ca concentration are also, on average, the lowest found in any of the community types (Table 12.3).

12.5. Discussion

There are key differences between the 2 freshwater tidal forest areas, evidenced by the fact that only 1 of the 4 communities (swamp tupelo – alder) is found in both areas. The most obvious difference is landscape position: the western area is located adjacent to the main channel of the Savannah River, and the eastern area is located adjacent to the Little Back vannah River, and the eastern area is located adjacent to the Little Back River, a distributary of the Savannah River. This results in less topographical gradient, lower water velocity in the river and tidal rivulets and, therefore, less flushing of the soil in tidal forests adjacent to the Little Back River. The dams situated upstream likely compound this situation by minimizing annual freshwater pulses to the river system (see 1929 discharge in Figure 12.2). The decreased flushing helps explain the 2-fold higher values for soil electrical conductivity, Na, Cl, and Ca in the shrub community (Table 12.3). This situation was exacerbated by the operation of the tide gate from 1977-1991, which further decreased flushing in the eastern area (by design), pushed the salt-wedge and its associated ions further upstream, and increased the aforementioned soil nutrient values in the eastern area. River flow surrounding and covering the western area, in contrast, was not altered during the operation of the tide gate, allowing for continuous flushing in that era.

The higher water velocity and proximity to the main channel of the Savannah River, and its sediments, may be a factor in the recovery of the area after cypress logging was done. The western area does not show the dramatic hummock and hollow topography that exists in the eastern area, though it does exist to a lesser extent. Perhaps this sediment load allowed for the hollows to fill in, eventually nearing the tops of the stumps. Since the eastern area is covered with lower velocity water, it is likely that the high organic matter content [of the shrub hummock] in the backswamp areas is a result of reduced sediment input. In fact, that area likely experiences flooding attributed more to tidal forcing of the water table rather than overland flow. The dramatic hummock and hollow topography in the eastern area may be a result of the baldcypress harvest of the 1800s followed by a period of sediment deprivation. The baldcypress stumps likely served as nurse logs for the large water tupelo trees that now contribute to the community composition of the current shrub hummock, and the lack of sediments gave the area dramatic microtopography and highly organic soil.

12.6 Research needs

Many research goals can be obtained by studying sites that are easily accessible by truck, boat, or foot. However, the difficultly accessing interior regions of freshwater tidal floodplain forests such as those presented in this chapter makes them unlikely candidates for studies involving many repeat visits. Remoteness of a study site increases cost dramatically as field crew time is increased to several days (versus several hours for bankside sites), crew lodging becomes necessary for overnight stays out of town, and specialized equipment becomes required for crew safety and longer term datalogging.

The remoteness of interior study sites also provides potentially desirable habitat for many wildlife species that are not commonly found along edges. More than a dozen warbler species, avian families Sylviidae and Parulidae, were visually identified in the shrub community during a single 15 minute event. As wildlife habitat continues to be fragmented by an array of anthropomorphic forces, this forest community may play an important role in providing habitat to some wildlife species that are sensitive to forest edge effects. Further, given the semi-aquatic nature of the ground substrate, it may also be found that less mobile wildlife such as amphibians and reptiles use these areas as sources for metapopulations.

Water flux in and out of the interior areas is likely quite slow. Yet, during the 2001 and 2002 years of data collection when the southeastern United States was undergoing an extended drought, the substrate in the hollows of the interior areas remained completely saturated having the consistency of pea soup to a depth of approximately a meter or more in most areas. Since little overland flow was observed, the question of water source becomes intriguing. It seems likely that there is a tidal forcing of the groundwater table during flood tides, even during times of drought. This seemingly ever-present water source, combined with the periodic drying times required to establish long hydroperiod species such as baldcypress and water tupelo, may explain the restriction of large trees to the hummocks. If there is a tidal forcing of the water table to the interior zones of the floodplain, there may be cause for concern as ocean levels rise due to global climate changes.

12.7. Acknowledgments

This material is based upon work supported by the Florida Cooperative Fishand Wildlife Research Unit, the Georgia Ports Authority, the USGS Global Climate Change Program and by the CSREES/USDA, under project number SC-1700271. Technical Contribution No. 5280 of the Clemson University Experiment Station. Comments by three anonymous reviewers and the book editors were helpful in completing this manuscript.

References

Advanced Technologies and Management Inc (ATM) (2001) Tidal amplitude study. Savannah harbor expansion project. Final report by ATM for Georgia Ports Authority, Charleston

Brune GM (1953) Trap efficiency in reservoirs. Trans Am Geophysical Union 34:407-418

Cain SA (1938) The species-area curve. Am Midl Nat 19:573-581

Cao Y, Willliams DD, Williams NE (1998) How important are rare species in aquatic community ecology and bioassessment? Vegetatio 42:1403-1409

Curtis JT, McIntosh RP (1950) The interrelations of certain analytic and synthetic phytosociological characters. Ecology 31:434-455

Curtis JT McIntosh RP (1951) An upland forest continuum in the prairie-forest border region of Wisconsin. Ecology 32:476-496

De'ath G (2002) Multivariate regression trees: a new technique for modeling species-environment relationships. Ecology 83:1105-1117

Doar D (1936) Rice and rice planting in the South Carolina low country, Second printing May 1970. Contributions from the Charleston Museum VIII. Milby Burton (ed) The Charleston Museum, Charleston

Dufrene M, Legendre P (1997) Species assemblages and indicator species: the need for a flexible asymmetrical approach. Ecol Monogr 67:345-366

Dusek ML (2003) Multiscale spatial and temporal compositional change of the tidal marshes of the lower Savannah River delta. M.S. thesis, University of Florida

Faith DP, Minchin PR, Belbin L (1987) Compositional dissimilarity as a robust measure of ecological distance. Vegetatio 69:57-68

Georgia Department of Natural Resources, Geologic and Water Resources Division (1976) Geologic Map of Georgia. Williams & Heintz Map Corporation, Washington

Georgia Department of Natural Resources, Geologic and Water Resources Division (1977) Geologic Map of Georgia. Williams & Heintz Map Corporation, Washington

Georgia Ports Authority (1998) Environmental impact statement: Savannah harbor expansion feasibility study. http://www.sysconn.com/harbor/

Granger ML (1968). Savannah Harbor: Its origin and development 1733-1890. Annex B. U.S. Army Corps of Engineers, Savannah District

Hansen DV, Rattray Jr M (1966) New dimensions in estuarine classification. Limnol Oceanogr 11:319-326

Kent M, Coker P (1992) Vegetation description and analysis: A practical approach. John Wiley & Sons Ltd, West Sussex, UK

Klawitter RA (1962) Sweetgum, swamp tupelo, and water tupelo sites in a South Carolina bottomland forest. Ph.D. thesis, Duke University

Kruskal JB (1964) Multidimensional scaling by optimizing goodness of fit to a nonmetric hypothesis. Psychometrika 29:1-27

Latham PJ, WM Kitchens (1995) Changes in vegetation and interstitial salinities following tide gate removal on the lower Savannah Rivera: 1986-1994. Final report by Florida Cooperative Fish and Wildlife Research Unit to U.S. Army Corp of Engineers, Savannah District

Mathsoft Inc (2000) S-Plus Professional 2000 Professional. Release 3. Lucent Technologies Inc., Murray Hill

McCune B, Mefford MJ (1999) PC-ORD. Multivariate Analysis of Ecological Data, Version 4.27. MjM Software Design, Gleneden Beach

McCune B, Grace JB (2002) Analysis of Ecological Communities. MjM Software Design, Gleneden Beach

McKenzie MD, Miglarese JV, Anderson BS, Barclay LA (1980) Ecological characterization of the Sea Island coastal region of South Carolina and Georgia. Vol 2: Socioeconomic features of the characterization area. U.S. Fish and Wildlife Service, Washington

Meade RH (1976) Sediment problems in the Savannah River Basin. In: Dillman BL, Stepp JM (eds) The future of the Savannah River. Water Resources Research Institute, Clemson University, pp 105-129

National Oceanic and Atmospheric Administration (NOAA) (1988) Climatology of the US No. 20, suppliment No. 1: Freeze/Frost data

National Oceanic and Atmospheric Administration (NOAA) (2001) Sea level variations of the United States 1854-1999. Technical Report NOS CO-OPS 36

National Oceanic and Atmospheric Administration (NOAA) (2002) Climatological data annual summary for Georgia 106:13

Pearlstine LG, Kitchens WM, Latham PJ, Bartleson RD (1993) Tide gate influences on a tidal marsh. Water Resour Bull 29:1009-1019

Plachy DH (2002) Savannah Harbor Expansion Project Management Plan. (Appendix C: Project Authorizations of the online report by US Army Corps of Engineers. http://www.sas.usace.army.mil/shexpan/ Appendix%20C.pdf. Updated 02/09/2002)

SPSS Inc (2001) SigmaPlot 2002 for Windows, Version 8.02. Systat Software Inc, Richmond

Urban DL (2002) Classification and regression trees. In: McCune B, Grace JB (2002) Analysis of ecological communities. MjM Software Design, Gleneden Beach, pp 222-232

U.S. Army Corps of Engineers (USACE) (1979) Water resources development in Georgia. Savannah

U.S. Army Corps of Engineers (USACE) Savannah District (2006a) J. Strom Thurmond Dam and Lake. (Webpage http://www.sas.usace.army.mil/lakes/thurmond/index.htm. updated 04/25/2006)

U.S. Army Corps of Engineers, (USACE) Savannah District (2006b) Hartwell Dam and Lake. (Webpage http://www.sas.usace.army.mil/lakes/hartwell/index.htm. updated 05/11/2006)

U.S. Army Corps of Engineers (USACE) Savannah District (2006c) Richard B. Russell Dam and Lake. (Webpage http://www.sas.usace.army.mil/lakes/russell/index.htm. updated 04/25/2006)

U.S. Army Corps of Engineers (USACE) Savannah District (2006d) Savannah Harbor Expansion Project Management Plan. (Webpage www.sas.usace.army.mil/shexpan/description.htm. No update reported.)

U.S. Department of Agriculture (USDA) (1974) Soil survey of Bryan and Chatham Counties, Georgia. Soil Conservation Service, Washington

U.S. Department of Agriculture (USDA) (1980) Soil survey of Beaufort and Jasper Counties, South Carolina. Soil Conservation Service, Washington

U. S. Fish and Wildlife Service (USFWS) (1993a) National Wetlands Inventory website: Hardeeville, SC. U.S. Department of the Interior, Fish and Wildlife Service, St. Petersburg. http://www.nwi.fws.gov

U. S. Fish and Wildlife Service (USFWS) (1993b) National Wetlands Inventory website: Limehouse, SC. U.S. Department of the Interior, Fish and Wildlife Service, St. Petersburg. http://www.nwi.fws.gov

U. S. Fish and Wildlife Service (USFWS) (1999a) National Wetlands Inventory website: Port Wentworth, GA. U.S. Department of the Interior, Fish and Wildlife Service, St. Petersburg. http://www.nwi.fws.gov

U. S. Fish and Wildlife Service (USFWS) (1999b) National Wetlands Inventory website: Rincon, GA. U.S. Department of the Interior, Fish and Wildlife Service, St. Petersburg. http://www.nwi.fws.gov

Van Den Avyle MJ, Maynard MA (1994) Effects of saltwater intrusion and flow diversion on reproductive success of striped bass in the Savannah River estuary. Trans Am Fisheries Soc 123:886–903

Van Den Avyle MJ, Maynard MA, Klinger RC, Blazer VS (1990) Effects of Savannah Harbor development on fishery resources associated with the Savannah National Wildlife Refuge. Final Report by Georgia Cooperative Fish and Wildlife Research Unit to U.S. Fish and Wildlife Service, Athens

Chapter 13 - Ecology of the Maurepas Swamp: Effects of Salinity, Nutrients, and Insect Defoliation

Rebecca S. Effler[1], Gary P. Shaffer[2], Susanne S. Hoeppner[3], and Richard A. Goyer[4]

[1] University of Georgia Marine Institute, Sapelo Island, GA 31327
[2] Department of Biological Sciences, Southeastern Louisiana University, Box 10736, Hammond, LA 70402
[3] Louisiana State University, Department of Oceanography and Coastal Sciences, Energy, Coast & Environment Building, Baton Rouge, LA 70803
[4] Louisiana State University, 404 Life Sciences, Department of Entomology, Baton Rouge, LA 70803

13.1. Introduction

Historically, the Maurepas swamp was what most people envisioned when they thought of Louisiana: majestic baldcypress draped with Spanish moss, Cajun locals fishing and hunting, and scenic bayous winding slowly to the Gulf of Mexico. The swamp was, and still is, rich with wildlife, drawing in commercial and recreational fishermen and hunters. Its beauty captivates artists and photographers, boaters, and naturalists.

The Maurepas swamp is located in the Lake Pontchartrain Basin in southeastern Louisiana (Figure 13.1). The lake from which the swamp derives its name is 21 km in diameter and lies between New Orleans and Baton Rouge. The dominant tree species are baldcypress (*Taxodium distichum* [L.] L.C. Rich.) and water tupelo (*Nyssa aquatica* L.) (Table 13.1). The midstory consists mostly of drummond maple (*Acer rubrum* var. *drumundii* [Hook. & Arn. ex Nutt.] Sarg.) and ashes (*Fraxinus* spp.). This wetland system complex is the second largest contiguous coastal forest in Louisiana, and consists of 77,550 ha of swamp and 5,204 ha of fresh/intermediate marsh (Coast 2050 1998), much of which experience both meteorological and wind-driven tides.

Table 13.1. Common plants in the Maurepas swamp.

Scientific Name	Common Name
Trees and Shrubs	
Acer rubrum var. *drummondii* (Hook. & Arn. ex Nutt.) Sarg.	drummond maple
Cephalanthus occidentalis L.	buttonbush
Fraxinus pennsylvanica Marsh.	green ash
Morella cerifera L.	wax myrtle
Nyssa aquatica L.	water tupelo
Nyssa biflora Walt.	swamp tupelo
Quercus laurifolia Michx.	laurel oak
Salix nigra Marsh.	black willow
Triadica sebifera (L.) Small	Chinese tallow tree
Taxodium distichum (L.) L.C. Rich.	baldcypress
Herbaceous	
Alternanthera philoxeroides (Mart.) Griseb.	alligatorweed
Amaranthus australis (Gray) Sauer	southern amaranth
Aster spp.	aster
Baccharis halimifolia L.	eastern baccharis
Cyclospermum leptophyllum (Pers.) Sprague ex Britt. & Wilson	marsh parsley
Echinochloa walteri (Pursh) Heller	coast cockspur
Eleocharis spp.	spikerush
Galium tinctorium L.	stiff marsh bedstraw
Hydrocotyle spp.	hydrocotyle
Iris virginica L.	Virginia iris
Panicum hemitomon J.A. Schultes	maidencane
Peltandra virginica (L.) Schott	green arrow arum
Polygonum punctatum Ell.	dotted smartweed
Pontederia cordata L.	pickerelweed
Sabal minor (Jacq.) Pers.	dwarf palmetto
Sagittaria lancifolia L.	bulltongue arrowhead
Vigna luteola (Jacq.) Benth.	deer pea

Unfortunately, anthropogenic alterations combined with sea-level rise are causing this valuable system to transition into marsh and open water at alarming rates. This rapid loss of swamp habitat has captured the attention of government agencies, conservation groups, scientists, and the public at large. The Maurepas swamp is under consideration for restoration through reintroductions of Mississippi River water. Much of the research discussed in this chapter describes the current conditions of woody and herbaceous vegetation of areas in various stages of decline, and discusses the potential for restoration of these sites.

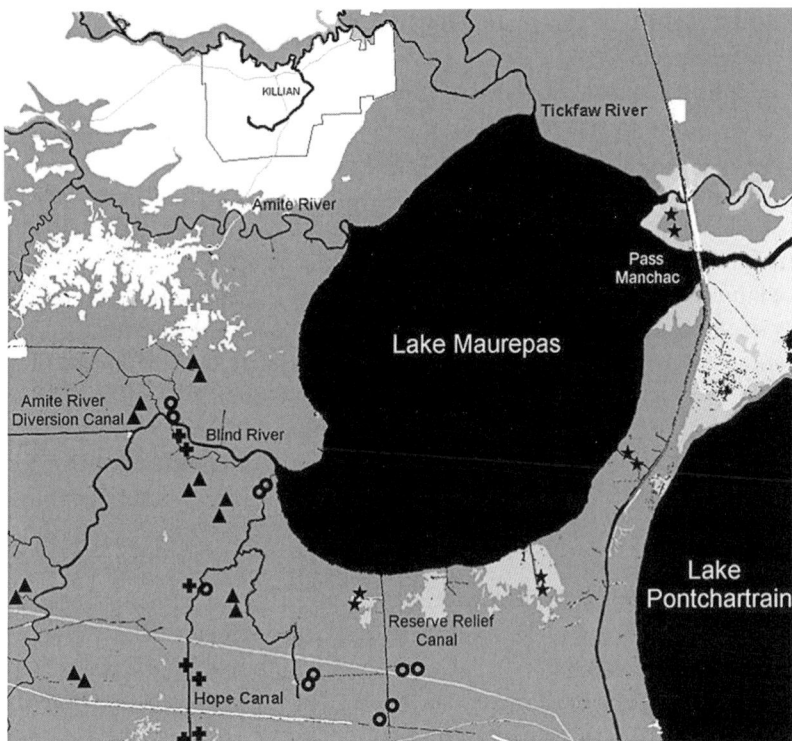

Fig. 13.1. The Maurepas swamps (dark grey) rim the lake margin of Lake Maurepas in southeastern Louisiana, and are bounded by urban development (white) to the North, West, and South. The largest field study in this area was conducted by Shaffer et al. (2003) and included four swamp habitat types: Throughput sites (✚), Interior sites (Δ), Edge sites (O), and Lake sites (★). Existing marshes within the study area are mostly located in the eastern part of the basin and are shown in light grey.

13.2. Historical land use and geologic history

13.2.1. Exploration and geologic development

The geologic development of most of coastal Louisiana occurred through a series of delta lobe shifts of the Mississippi River, which occurred roughly once every 1,000 years (Scruton 1960). Lakes Pontchartrain and Maurepas were formed during the evolution of two separate delta lobes over the last 4,000 years. The Cocodrie lobe expanded over the area where New Orleans presently resides, forming the lakes' southern shore.

The St. Bernard Lobe then completed the lakes eastern shore line (Saucier 1963; Frazier 1967).

The first European explorer to find the Mississippi River was Alfons Alvarz de Pinda in search of a route to the Pacific for Spain in 1519 (Hoover 1975). De Pinda was not able to navigate the coastline because of extensive mudflats at the mouth of the river, and thus was not able to explore the Mississippi River. Over a hundred years later, Robert Cavelier, Sieur de La Salle again located the mouth of the Mississippi River in April 1682, and claimed the Mississippi River Valley for France. Again, few explorations of the surrounding terrain were attempted, and the Maurepas swamps remained largely unexplored by European colonists and explorers. It was not until 1699 when Pierre Le Moyne, Sieur d'Iberville and his brother, Jean-Baptiste Le Moyne, Sieur de Bienville arrived to explore Louisiana and the Mississippi River. Iberville and Bienville travelled the Mississippi River up to the mouth of the Red River, and named lakes Pontchartrain and Maurepas. Relationships were established mainly with the Chitimacha, the oldest Native American tribe of southern Louisiana, which has been traced back as far as 12,000 B.C. Baldcypress was regarded as a sacred tree and the Chitimacha believed god had come down from heaven on a cypress tree and taught them how to build huts and canoes. The Chitimacha defined their land as "areas where cypress trees grew" – a description that captures the essence of the Maurepas swamps to this day.

13.2.2. Land use changes

Following European colonization, the vast virgin stands of baldcypress in the deltaic plain were quickly recognized by plantation owners and other colonists as an invaluable resource (Mattoon 1915). Baldcypress wood became the dominant cash crop in Louisiana during the 18th century. It was not until the 1890s, however, that large-scale clear-cutting of the baldcypress swamps became possible due to the invention of pullboats and skidders, which increased the accessibility of interior swamps (Mancil 1972). The depletion of timber resources and the Great Depression eventually caused most baldcypress mills to close (Burns 1980).

Most areas in the Pontchartrain Basin regenerated as second-growth baldcypress-tupelo swamps. By the 1960s, however, many areas in the southeastern portion of the basin started to convert to marsh and open water (Barras et al. 1994). The dominant land uses today are recreational and commercial fishing, as well as hunting. The most popular gamefish are catfish, largemouth bass, perch, and crappie, although recently redfish and speckled trout have increased in abundance and popularity. The most sought after species for hunters in this area are whitetail deer, squirrels, and bullfrogs, with a lesser emphasis placed on rabbits and raccoons. Contract trapping for alligators is allowed each year, and bounty hunting for nutria is encouraged and financed by the Coastal Wetlands Planning, Protection, and Restoration Act. Bird watching, sightseeing, and boating are also common. Despite the doubtful regeneration capacity of most of the Maurepas swamp, timber harvesting is still at the discretion of landowners. This topic has recently come to public attention due to renewed interest in the timber value represented by the maturing second-growth swamps.

Until recently, land ownership has been predominantly private. However, in the summer of 2001, 25,300 ha of the Maurepas swamp were donated to the Louisiana Department of Wildlife and Fisheries (LADWF) by the Richard King Mellon Foundation. This area is contiguous with the 3,370 ha Manchac Wildlife Management Area (WMA) already maintained by the LADWF. Further negotiations to enlarge the existing WMA are in progress. Plans for the WMA include the construction of a board walk for public access, the deployment of 200 duck-boxes to enhance breeding habitat, and the establishment of a program to control invasive plant species.

13.2.3. Major sources of water

Although Lake Maurepas is 21 km wide, it is no more than 3 m deep. This shallow lake system is fed by three rivers: the Amite, Tickfaw, and Blind (Figure 13.1). The Blind River is a former tributary of the Mississippi River and has a highly modified, inconsequential watershed. By contrast, the watersheds of the Amite and Tickfaw Rivers are largely natural to this day. The Amite River watershed is 4,822 km^2 in size and is 58% forested, whereas the Tickfaw River watershed encompasses 1,896 km^2, of which 66% are forested (Xu and Wu, in press). A further supply of freshwater reaches the lake through three storm-water drainage canals: the Amite River Diversion Canal, Hope Canal, and Reserve Relief Canal. Each of these canals channels storm-water run-off from nearby urban areas surrounding the western half of the Maurepas swamps. Another major hydrological feature of the Maurepas swamps is Pass Manchac, a natural tidal inlet, 10 to 20 m deep, through which saltwater enters the system from the adjoining Lake Pontchartrain during drought periods and meteorological high tides, hurricanes, and tropical storms.

13.2.4. Anthropogenic drivers

Anthropogenic alterations in the Maurepas swamps began in the early 1700s with the initiation of baldcypress logging. However, it was not until the invention of pullboat logging and associated train exploitation at the turn of the 20th century that allowed for mass harvest of baldcypress swamps (Mancil 1980). These changes in logging technology caused considerable damage to the landscape, as the ground-skidded trees ripped wide linear groves in the soil.

After the great flood of 1927, the construction of the Mississippi River levee system began isolating much of coastal Louisiana from the riverine inputs that had heretofore built and sustained it. The separation from the Mississippi River changed the Maurepas swamps from a phase of deltaic maintenance to one of degradation. This process was sped up by the construction of the Mississippi River Gulf Outlet (MRGO) that was completed in 1965. MRGO is a 106 km channel that provides a shorter route between the Gulf of Mexico and the inner harbor of New Orleans. The outlet was intended to facilitate deep-draft shipping to the port of New Orleans. Because of erosion, however, the original 200-m wide canal has broadened to 455 m in width, and now serves as a major conduit for saltwater intrusion directly into Lake Pontchartrain.

Further anthropogenic alterations that impacted ecosystem health in the Lake Pontchartrain Basin include oil and gas exploration, as well as shell dredging. Oil and gas exploration in the basin started as early as 1931. Several canals were dug into the swamps near the lake margins, causing direct wetland loss. More dramatically, however, were the long-term effects of the spoil banks associated with each canal. Fortunately, very little oil and gas exploration was conducted in the Lake Pontchartrain Basin because of a scarcity of oil and gas deposits in the relatively thin sedimentary layers in the area. No new oil and gas drilling leases have been allowed since 1992. Shell dredging of rangia clam shells (*Rangia cuneata*) was conducted in both Lakes Maurepas and Pontchartrain for several decades prior to a moratorium in 1989. The clam shells were primarily sought after as fill materials, and no thought was given to the devastating effect the removal of these clams had on the lake ecosystems. Due to a complete disruption of the lake benthos, the removal of a solid and consolidating lake bottom, as well as the lack of water filtration through the clams, water quality in both lakes dramatically deteriorated. The resulting increase in turbidity caused a complete loss of all submerged aquatic vegetation from the lakes, severely disrupted fisheries production, and completely altered the lake ecosystem structure. Since the moratorium in 1989, rangia clam populations have rebounded in the lakes (Abadie and Poirrier 2000), water quality has improved, and submerged aquatic vegetation is slowly returning.

Aside from abiotic alterations, people have also introduced several invasive species to the Maurepas swamps. First and foremost of troublesome invasive species is the rodent nutria, or coypu (*Myocastor coypus*), which was introduced along the northern Gulf of Mexico around 1937. Introductions reportedly originated in Louisiana through intentional and accidental releases of animals farmed for the fur industry. By the late 1950s, approximately 20 million nutria inhabited coastal Louisiana. Nutria eat and devastate marsh vegetation, kill swamp seedlings, and girdle mature trees (Myers et al. 1995). Other problematic invasive species in coastal Louisiana include water hyacinth (*Eichornia crassipes* (Mart.) Solms) and watermoss (*Salvinia* spp.), both of which degrade open water habitat.

13.3. Climatic drivers and hydrological characterization

13.3.1. Climate

The Gulf Coast enjoys a climate uncharacteristic of its latitude, which typically hosts warm, arid, and semi-arid climates. The Gulf of Mexico, Caribbean Sea, and Atlantic Ocean substantially influence the region's climate and moderate winter temperatures. Occasionally, however, these mild winters are punctuated by cold air masses reaching far south from the northern Pacific or the Arctic, bringing low temperatures and freezing conditions. This situation arises when the midlatitudinal jet stream that governs the tracks of storm systems shifts from a more east-west direction into north-south meanders, allowing cold air and winter storms to penetrate southern regions. Summers along the northern Gulf of Mexico tend to be hot and humid (Twilley et al. 2001).

El Niño/Southern Oscillations (ENSO) also influence climate in this Gulf Coast region. El Niños significantly decrease temperatures in winter (especially in Florida) and spring. During La Niña, fall and winter seasons in the Gulf Coast are warmer, especially in Louisiana, followed by little change in spring, and higher temperatures in summer. El Niño is also linked to substantial increases in winter rainfall in the Gulf Coast region. La Niña events, by contrast, are associated with regional droughts such as occurred in the central and western Gulf Coast from 1998 to 2000 (NOAA 2001). Moreover, these events strongly influence the number of Gulf Coast hurricanes. During La Niña events, the average number of hurricanes making landfall along the northern Gulf of Mexico is typically higher than during El Niño or non-ENSO years (Bove et al. 1998; Twilley et al. 2001).

13.3.2. Hydrology

The three tributaries that feed Lake Maurepas are flashy rivers prone to brief, high-intensity floods that can occur at almost any time of the year. Mean monthly Amite River discharge did not exceed 28 m^3/s during a severe drought in 2000, though it can reach 283 m^3/s in wet years (Day et al. 2004). The long-term annual rainfall average in the Pontchartrain Basin is 1,600 mm/yr, ranging from 1,108 mm/yr to 2,178 mm/yr over the past half century (Xu and Wu, in press). Along with the rivers, the storm-water drainage canals carry only a small fraction of the sediment and nutrients that the Mississippi River historically delivered to the Maurepas swamp as recent as the 1890s (Saucier 1963). Tragically, canal spoil banks ranging in height from 0.5 to 13 meters confine the flow and deliver urgently needed nutrients and sediments directly to the lake, thus bypassing the adjacent swamps. In addition, logging ditches and abandoned railroad tracks from historic logging activities further alter the hydrology of the swamp ecosystem, causing increased flashiness in some areas, while impounding others.

Tidal pulses are introduced to the system through Pass Manchac. Lunar tidal fluxes average about 30 cm, but are generally overwhelmed by meteorological tidal fluxes. During the summer and early fall, storms and prevailing winds from the southeast raise water levels and push Gulf water into the system. During the winter months, continental fronts with prevailing winds from the northeast lower water levels and push water out into the Gulf. Lake Maurepas salinities near Pass Manchac typically range from 0 to 3 ppt, but have been observed to reach 12 ppt near the Pass and 6 ppt on the other side of the lake at the Blind River entrance, respectively, during a drought event in 2000 (Lane et al. 2003). In general, there are decreasing salinity gradients from Pass Manchac to the western Maurepas, and from anywhere on the lake margin to the interior swamps.

13.3.3. Relative sea-level rise

Regional estimates place relative sea-level rise (RSLR) between 3.6 and 4.5 mm yr^{-1} in this basin (Penland and Ramsey 1990). Holocene sediments are roughly 16 meters thick. Flooding duration has been reported to have doubled in the Manchac Wildlife Management Area adjacent to the Maurepas swamp since 1955 due to sea-level rise and subsidence (Thomson et al. 2002). Similar increases in flooding are expected to have taken place in most wetlands in the Pontchartrain Basin.

13.4. Ecological characterization

13.4.1. Swamp structure and productivity

In an effort to characterize the Maurepas swamp, Shaffer et al. (2003) selected twenty field sites with two replicates each throughout a 350 km^2 area in the southern wetlands of Lake Maurepas (Figure 13.1). Field research was conducted from 2000-2004, a time period that included the end of a severe drought event (1998-2000) in the southeastern United States. The study sites were chosen to represent four swamp habitat types with different hydrologic regimes. Lake sites were located close to the margin of Lake Maurepas and were most susceptible to saltwater intrusion events. Edge sites were located along major bayous and canals connected to Lake Maurepas. Deep inside the swamp, interior sites were chosen to represent swamps largely isolated from lake hydrology. Lastly, several throughput sites were chosen to characterize intact swamp that received non-point source runoff from nearby urban areas.

13.4.1.1. Abiotic characterization

The abiotic environment in the Maurepas field study was characterized by measuring soil water salinity, pH, and bulk density, as well as water concentrations of nitrate, ammonia, and phosphorus (Shaffer et al. 2003; Day et al. 2004). Soil salinity levels at the Lake sites were higher than anywhere else in the study area with a mean of 4.15 ± 0.29 ppt (mean ± SE). At all sites, salinity was highest during the drought year (2000) and steadily decreased in subsequent years (Figure 13.2). Soil salinity was also found to decrease with increasing distance from Pass Manchac, as well as with increasing distance from the margin of Lake Maurepas into the interior swamp.

With the exception of the throughput sites, soil strength, as measured by bulk density, was very low throughout the study areas, and in the range of those typically found in fresh and intermediate marshes (e.g., Hatton 1981). The highest bulk densities in the study area were found at the throughput sites (mean = 0.194 ± 0.016 g/cm^3). Interior and edge sites had bulk densities of 0.132 ± 0.011 g/cm^3 and 0.154 ± 0.021 g/cm^3, respectively, and did not differ from lake sites (mean = 0.117 ± 0.013 g/cm^3).

The pH values measured throughout the Maurepas swamp are slightly acidic and thus are also indicative of organic soils with low bulk densities.

Nutrient levels in surface water samples indicated that the system is nutrient poor (Lane et al. 2003). Nitrate levels in the swamp were less than 1% of those found in Mississippi River water. Ammonium levels also were lower than those in the Mississippi River, but not as dramatically. Total nitrogen in the Maurepas swamp was almost comparable to total nitrogen levels in river water, which indicates the presence of high concentrations of organic nitrogen in the Maurepas swamp (Lane et al. 2003). Phosphate and total phosphorous levels in swamp waters were comparable to those found in the river. Surface water ratios of nitrate/ammonium to phosphate were close to 2:1, a further indication that the Maurepas swamp is nitrogen limited.

13.4.1.2. Swamp structure and tree mortality

Overall, the Maurepas swamps are typical of other coastal Louisiana swamps in that their canopies are dominated by either water tupelo, baldcypress or both, while their midstories are largely dominated by numerous smaller maples and ashes, both of which are more shade tolerant than ei-

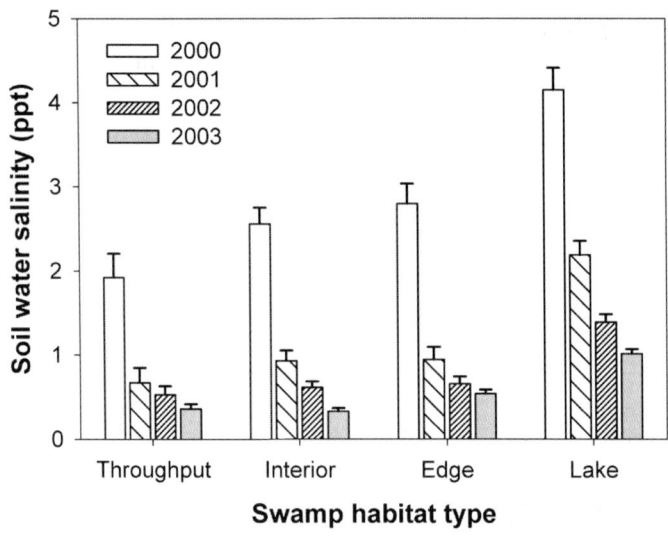

Fig. 13.2. The average observed soil salinity in the Maurepas swamps was greatly elevated during a severe drought in 2000 and declined gradually in subsequent years. This effect was greatest at the Lake sites located near Pass Manchac, which is the main salt water conduit into the system.

ther of the dominants. Except for throughput sites, all of the Maurepas swamp sites have low stem densities and small basal areas that are at best reminiscent of those reported as impounded or continuously flooded swamps (e.g., Conner et al. 1981; Dicke and Toliver 1990; Conner and Day 1992). For example, stem densities and basal areas in 2004 were lowest at the lake sites, averaging only 298 ± 55 stems ha^{-1} and 9.96 ± 3.08 m^2 ha^{-1}, respectively. Interior and edge sites had slightly greater stem densities (mean = 753 ± 54 stems ha^{-1} and mean = 915 ± 90 stems ha^{-1}, respectively) and basal areas (mean = 28.25 ± 1.84 m^2 ha^{-1} and mean = 27.59 ± 3.76 m^2 ha^{-1}, respectively), and did not differ from one another. Throughput sites had the greatest number of trees (mean = 1154 ± 77 stems ha^{-1}) and the highest basal areas (mean = 52.44 ± 6.13 m^2 ha^{-1}).

Both throughput and edge sites were numerically dominated by midstory species, primarily ash and maple, which made up 50-60% of the total number. In both of these habitat types, baldcypress was less abundant than its co-dominant canopy competitor water tupelo. However, despite their high abundance, midstory species contributed only 10-20% of the basal area at the throughput and edge sites. At interior sites, water-tupelo and midstory species were approximately equally abundant, accounting for 80-90% of the trees present. Baldcypress dominated the lake sites, accounting for roughly 60% of the total trees as well as 75% of the basal area. Wax myrtle (*Morella cerifera* L.), Chinese tallow tree (*Triadica sebifera* [L.] Small), and black willow (*Salix nigra* Marsh.) gain in importance at lake sites, most likely because neither maple, water tupelo, nor ash have salt tolerances to 2-5 ppt found at these sites. Shrub-scrub habitats are often observed on the transitional edges between marshes and forested wetlands or uplands (White 1983; Barras et al. 1994). Laurel oak (*Quercus laurifolia* Michx.) and green ash (*Fraxinus pennsylvanica* Marsh.) were found in greater abundance at sites characterized by higher bulk densities, which were indicative of increased throughput and generally less flooding (Shaffer et al. 2003). These observations support similar findings from wetland plant ordinations by White (1983) in the Pearl River, Louisiana, and Rheinhardt et al. (1998) in the forested riverine wetlands of the inner coastal plain of North Carolina. As Chinese tallow has been found to be more shade, flood, and salt tolerant (Jones et al. 1989; Conner and Askew 1993) than native wetland tree species, this invasive tree may become more dominant in the coastal wetlands of the southeastern United States (Conner and Askew 1993).

Despite steady decreases in salinity from 2000-2003, tree mortality remained high in the Maurepas swamps throughout the study period (Figure 13.3). The relatively stress-resistant baldcypress that experienced zero

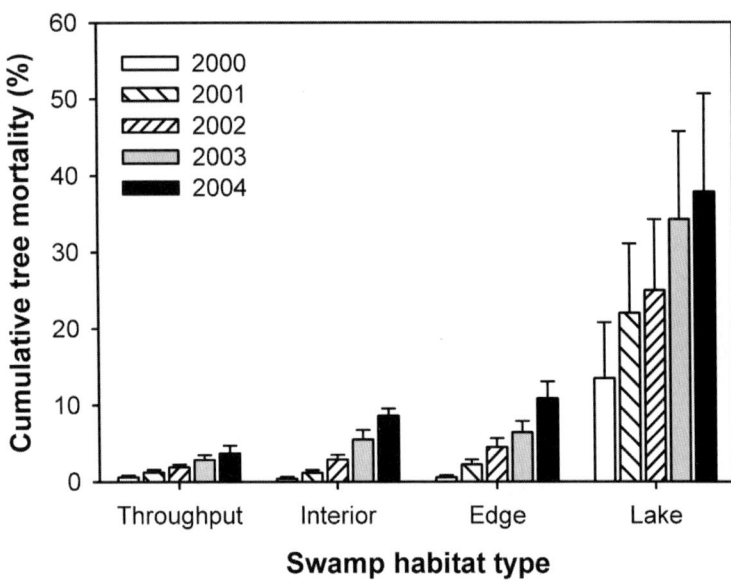

Fig. 13.3. Cumulative tree mortality reached an average of nearly 40% over five years at the Lake sites in the Maurepas swamps. Intact swamp sites (i.e., Throughput sites) had the lowest mortality throughout the study period.

mortality during the drought (2000) started experiencing mortality in subsequent years. Cumulatively, up to 40% of all trees alive at the lake sites in 2000 had died by 2004, compared with roughly 10% at all other sites. Adjusting the overall percent mortality for the time span elapsed, yearly average mortality rates were at or below roughly 2% at throughput, interior, and bayou sites. Lake site trees were dying at a rate of roughly 10% per year, with mortality rates as high as 25% per year near Pass Manchac. The size class information of the dominant tree species at the Maurepas swamp sites (Figure 13.4), further showed a lack of recruitment of canopy-species saplings. This lack of recruitment was particularly pronounced at interior and lake sites. Together, the mortality and size class distribution data indicated that large areas of the Maurepas swamps are relic swamps in steady decline punctuated by extreme climatic conditions such as drought.

13.4.1.3. Primary productivity of trees and herbaceous vegetation

Annual tree biomass production in the Maurepas swamps varied greatly between years and swamp habitat types. In general, tree biomass production was higher at sites with high bulk densities, low salinity, and large

Chapter 13 - Ecology of the Maurepas Swamp

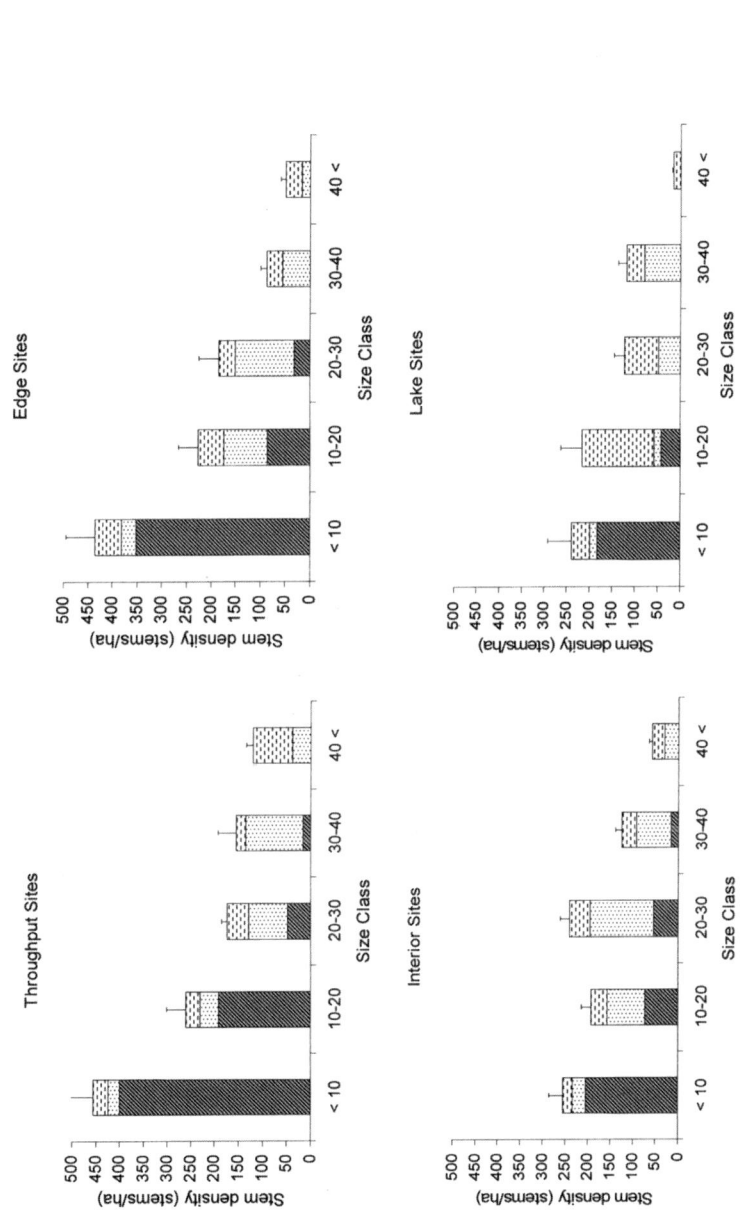

Fig. 13.4. Maurepas swamp habitat type differences in stand structure by size class (mean ± standard error). Horizontal hatching denotes baldcypress, dots mark water tupelo, and diagonal hatching shows other trees. The size class of trees less than 10 cm in diameter is dominated by other trees throughout the study area. Large size classes of 30 cm diameter trees and larger are dominated or exclusively represented by the canopy trees baldcypress and water tupelo.

basal areas. Temporally, tree biomass production was highest during the end of a severe drought (2001) and declined during subsequent wet years with extended periods of hurricane induced deep flooding (Figure 13.5). The drought-related temporal effect suggested that most of the Maurepas swamps are flood-stressed during normal (wet) years. Tree primary production at the Interior sites was not as strongly impacted by the drought as other sites, and remained low throughout the study period. This could be an indication that swamp interior sites were chronically stressed by near-constant, stagnant flooding even during the drought (Shaffer et al. 2003).

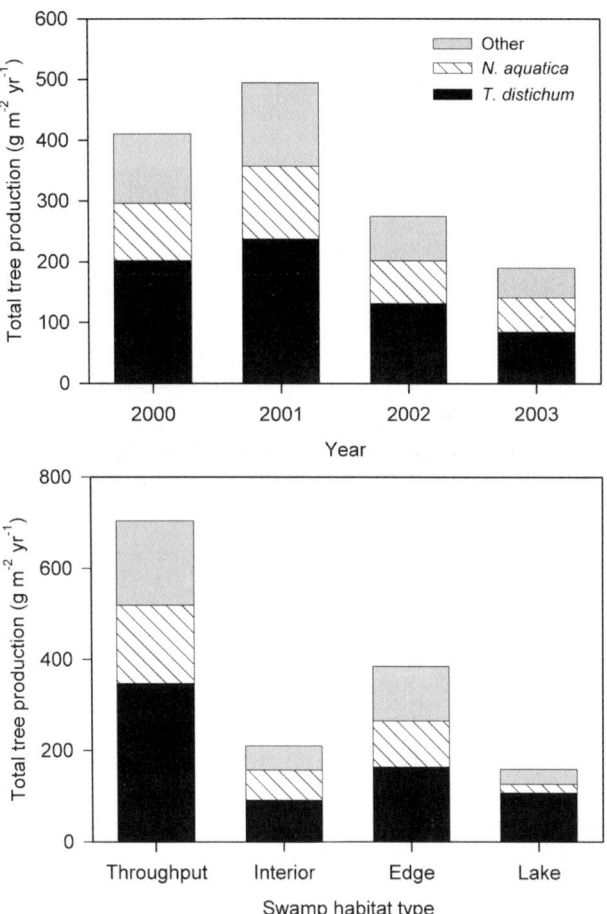

Fig. 13.5. During the 4-year Maurepas swamp field study, the ratios of cypress:tupelo:midstory tree species (other) tree biomass production maintained relatively constant, even though biomass production generally declined over time (a). Water tupelo and midstory tree species produced less biomass annually than bald-cypress at the Lake sites (b).

Overall, the highest rates of annual tree primary production were measured at the throughput sites (approximately 700 g·m^{-2}·yr^{-1}). Edge sites produced more tree biomass (approximately 400 g·m^{-2}·yr^{-1}) than interior sites (about 220 g·m^{-2}·yr^{-1}), and lake sites were the least productive sites (about 170 g·m^{-2}·yr^{-1}). Most of the biomass was produced by baldcypress, which accounted for nearly 50% of all tree primary production throughout the study (Shaffer et al. 2003). Water tupelo, as the only other dominant canopy species, contributed substantial proportions of the total site productivity at most sites, but almost completely dropped out at Lake sites with elevated salinity levels.

Combining total tree biomass production with herbaceous biomass production, Shaffer et al. (2003) found a roughly compensatory increase in annual herbaceous biomass production when tree biomass production was low (Figure 13.6); herbaceous biomass production showed the reverse pattern to tree biomass production. While the compensatory biomass production is not surprising due to light limitation on herbaceous growth in dense swamps, the amount of the compensatory growth was surprising. Most of the herbaceous biomass production could be attributed to 15 dominant ground-cover species, which together represented 97% of the total herbaceous cover throughout the study areas. Alligatorweed (*Alternanthera philoxeroides* [Mart.] Griseb.), dotted smartweed (*Polygonum punctatum* Ell.), and green arrow arum (*Peltandra virginica* [L.] Schott) appeared to be the most ubiquitous species in the swamps of southern Maurepas. Pickerelweed (*Pontederia chordata* L.) decreased in abundance as habitats became saltier and more open, whereas bulltongue arrowhead (*Sagitarria lancifolia* L.) and fall panicgrass (*Panicum dichotomiflorum* Michx.) became more abundant. Maidencane (*Panicum hemitomon* J.A. Schultes) and spikerushes (*Eleocharis* spp.) were generally only present at the interior sites and in ponding areas of lake sites, and might be indicator species of degrading soil strength. Over time, the overall ratio of tree to herbaceous biomass production decreased, reflecting a successional trajectory from forested to marsh sites (Figure 13.6).

In general, for total aboveground net primary production, only the most productive sites of the Maurepas swamp compare well with natural, periodically flooded cypress-tupelo swamps (e.g., Carter et al. 1973; Conner and Day 1976; Conner et al. 1981; Megonigal et al. 1997). The vast majority of the Maurepas swamp, including interior, edge, and lake sites, range in total productivity (including herbaceous productivity) between swamps that have been identified as either nutrient-poor, stagnant, or near-continuously flooded swamps (e.g., Schlesinger 1978; Taylor 1985; Mitsch et al. 1991; Megonigal et al. 1997). Spatial patterns of differences in pro-

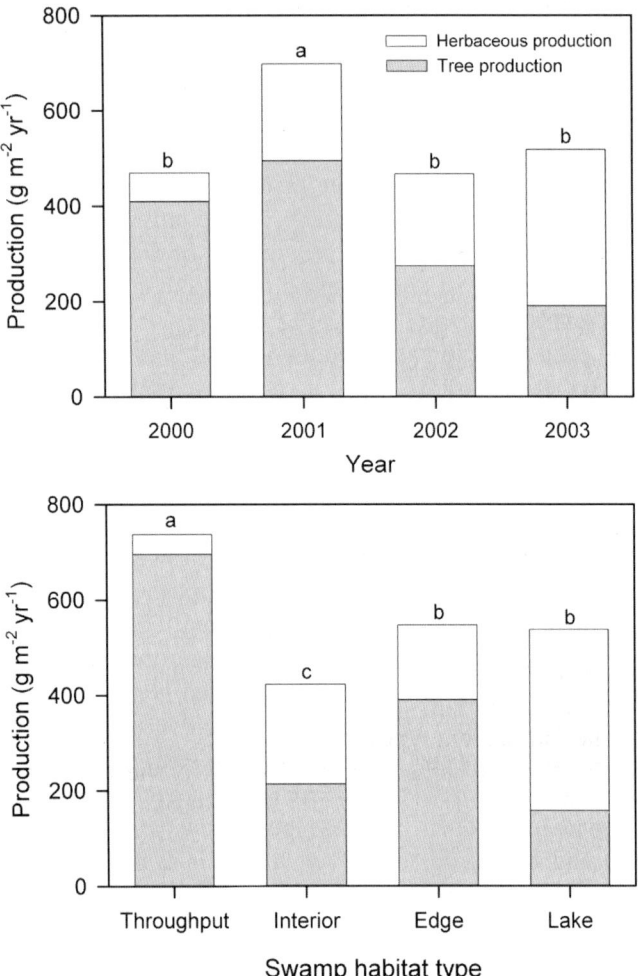

Fig. 13.6. Overall annual primary production in the Maurepas swamps remained relatively steady over the 4-year study period (a), but showed changes in the ratio of herbaceous to tree primary production. In habitat types where trees are producing only little biomass, herbaceous primary production is increased (b).

ductivity among individual sites within the study area are comparable to the spatial patterns of mean yearly salinity, indicating that salt stress is a major factor in influencing tree productivity at all but the swamp Interior sites. The increasing proportions of herbaceous productivity at many sites could be interpreted as indications that these forested wetlands are converting to marshes, primarily as a result of prolonged stagnant flooding at Interior sites and salt water intrusions at the Lake sites.

13.4.2. Experimental manipulations: simulation of river reintroduction via fertilization

Ecological restoration of the Maurepas Basin is partially dependent upon introductions of river water through diversions along the Mississippi River. Seventy 16-m^2 study plots were differentially fertilized to simulate the effects of nutrient enrichment from Mississippi River reintroductions. Dosages of timed-release fertilizer simulated loading rates of 11.25 g N m^2/year and 22.5 g N m^2/year, similar to what a 43.4 m^3/second ('1x') and a 86.4 m^3/second ('2x') river reintroduction would carry to the Maurepas swamp. Plots were fertilized in the spring and again late-summer to simulate river water input throughout the growing season. Cages were installed around half of the plots to exclude nutria and deer. Clip plots were obtained in early summer and again at the end of the growing season to estimate biomass production.

In general, biomass production of the caged and fertilized plots increased with dosage. When plots were not caged, the standing crop remained at low levels regardless of fertilizer dosage, implicating extensive browsing. On average, the standing crop of the caged control plots was 1.53x higher than that of control plots without cages, and yields an estimate of browsing pressure; this ratio was used to adjust annual herbaceous production estimates obtained from biannual clip plots from all 40 sites representing the four habitat types.

It appears that the low herbaceous production for the southern Maurepas swamps as a whole is largely attributable to nutrient limitation. Caged plots augmented with the highest dosage of fertilizer yielded standing crops greater than double that of the caged control plots. As a whole, the nutrient enrichment study indicates that reintroducing water from the Mississippi River into the Maurepas swamp would greatly stimulate primary production. Furthermore, a large reintroduction of as much as 460 m^3/second would be more beneficial than the small reintroduction currently planned (Coast 2050 1998).

13.4.3. Experimental manipulations: baldcypress and water tupelo fertilization and insect defoliation

In addition to the multiple abiotic stresses in the Lake Maurepas swamp, the two dominant tree species, baldcypress and water tupelo, are defoliated periodically by lepidopteran herbivores. In 1983, an outbreak of the baldcypress leafroller (*Archips goyerana* Kruse; BCLR) occurred in and around the Atchafalaya Basin and has since spread to southern Lake

Maurepas and Lake Verret becoming a serious pest on baldcypress in Louisiana (Goyer and Lenhard 1988; Goyer and Chambers 1996). Baldcypress, renowned for its lack of serious insect and disease problems (Brown and Montz 1986), presently experiences annual repeated defoliation, often complete, by the immature stages of this insect resulting in continuous and significant reduction in radial growth and dieback of the crowns (Goyer and Lenhard 1988; Braun et al. 1989).

Water tupelo, the other dominant wetland tree, has been infested by the forest tent caterpillar (*Malacosoma disstria* Hubner; FTC) in Louisiana and Alabama, and seemingly regular outbreaks have been reported since 1948 (Nachod and Kucera 1971). At the turn of the 20th century, "During certain years, the [forest tent] caterpillars were so bad that the railroads would lose traction on the tracks. Sawdust needed to be carried from the mill and placed on the tracks to get adequate traction for the railroad train to pull itself" (Rudy Sparks, Williams Land Inc, Patterson, Louisiana, pers comm.). Although the FTC is a native insect throughout the United States and Canada (Fitzgerald 1995), water tupelo is its preferred host in the south (Smith and Goyer 1986). In Louisiana, widespread, complete defoliation by the FTC has occurred over as much as 250,000 ha during a single season (Nachod 1977). Abrahamson and Harper (1973) found that up to 45% of the annual incremental growth of tupelo was lost due to severe FTC defoliation during a 5-year impact study in Alabama.

Traditional studies often address defoliators as pests, yet recent studies indicate that arthropods affect forest ecosystems in complex ways. Trees are capable of compensatory growth following defoliator outbreaks, often more than replacing growth loss during defoliation (Alfaro and MacDonald 1988; Alfaro and Shepherd 1991; Lovett and Tobiessen 1993; Trumble et al. 1993). Insect herbivory also can cause changes in amount and quality of precipitation reaching the forest enhancing other nutrient cycling processes. Schowalter et al. (1991) found that defoliators removing 20% of the foliage from young Douglas-fir doubled the amount of precipitation reaching the forest floor and increased N, K, and Ca flow to the forest floor by 20-30%. They also reported that three taxa of litter arthropods were more abundant under defoliated trees, suggesting that defoliation could increase species diversity and stimulate litter decomposition. Similarly, Progar et al. (2000), Schowalter (1992), and Schowalter et al. (1992, 1998) demonstrated that saprophytic insects and fungi, including bark beetles and some pathogenic fungi, contribute significantly to nutrient cycling from decomposing wood, and enhance the fertility of underlying soil.

Forest defoliation by insects can also lead to severe disruptions in nutrient cycling by causing premature litterfall (in the form of foliage, insect tissues, and insect excrement) and increasing nutrient content and biomass of litter products reaching the forest floor causing disruption in detrital processes (Klock and Wickman 1978; Schowalter 1981; Swank et al. 1981; Schowalter and Crossley 1983; Seastedt and Crossley 1984; Schowalter et al. 1991). Importantly, nitrate-impoverished streams in the southeastern United States show significant increases in NO_3-N concentrations during periods of defoliation (Swank et al. 1981). This may be particularly important for declining, nutrient-limited Louisiana swamps, as litterfall associated with insect defoliation may be an important nutrient source for primary and secondary productivity in the Spring when biological emergence is at its peak. Baldcypress-tupelo swamps are known to be long-term sinks for both nitrogen and phosphorous via burial of partially decomposed organic and inorganic matter under reduced soil conditions (Kemp et al. 1985).

However, little research has been conducted on nutrient/biological interactions of the trees and defoliators in these swamps, with two exceptions (Meeker 1992; Johnson 2004). During the baldcypress leafroller larval activity, nutrient concentration (N, P, K, and Mg) of baldcypress foliage decreased as feeding progressed; and constituents representing tree defense mechanisms increased (Meeker 1992; Johnson 2004), rapidly creating foliage characteristic unfavorable for the BCLR. Meeker (1992) also found that a flooded swamp may have more suitable foliage for the larvae, as suggested by having higher total nonstructural carbohydrates in the foliage than in an unflooded swamp.

Mature baldcypress and water tupelo trees in three forested stands varying in different degrees of decline were fertilized in one study to mimic nutrient changes that occur during river diversions (Souther-Effler 2004; Effler et al. 2006). Within each of the three sites, both tree species randomly received three nutrient regimes [no nutrients, a 1x dose of 59 g m^{-2} of Osmocote® 18-6-12 (B.W.I. Jackson, Mississippi, USA), and a 2x dose] each Spring from 2001-3. The 1x rate is equivalent to a N loading rate of 43.4 m^3/s springtime effluent of a Mississippi River diversion (Lane et al. 1999). Stem growth and litter productivity for each tree was determined. After refoliation occurred in June, leaves from the crown on the trees were sampled (clipped leaves). Clipped (spring) and fallen leaves (natural abscission) and insect excrement were analyzed for N and P content (Table 13.2).

13.4.3.1. Basal area growth

In the absence of fertilizer, stem growth of each species did not differ across sites (Figure 13.7). At the slight decline site (Throughput Site-Hope

Table 13.2. Overall means (± 1 SE) of nitrogen (%) and phosphorous (%) content of frass and clipped and fallen leaves and frass nitrate content of baldcypress and water tupelo trees using collection periods from the Lake Maurepas basin ranging from 2001-2003.

Variable	Baldcypress	Water tupelo
Nitrogen (%)		
Frass (2001-03)	1.171 ± 0.026	1.757 ± 0.033
Clipped leaves (2002-03)	1.524 ± 0.032	1.632 ± 0.033
Litterfall (2002-03)	1.353 ± 0.042	1.120 ± 0.042
Nitrate (as N in ppm)		
Frass (2002)	137.18 ± 8.11	74.69 ± 13.40
Phosphorous (%)		
Frass (2003)	0.090 ± 0.002	0.186 ± 0.010
Clipped leaves (2002-03)	0.131 ± 0.004	0.108 ± 0.004
Litterfall (2002-03)	0.091 ± 0.003	0.062 ± 0.002

Canal), fertilization did not increase stem growth. However, both species at the moderate decline site (Edge Site, Reserve Relief Canal) yielded a higher growth than the unfertilized trees. At the severely declined site (Edge Site, Blind River), only baldcypress grew more with the fertilizer applications, while water tupelo did not. Basal area growth of baldcypress was not correlated to defoliation.

The increased basal area growth of both tree species with fertilizer application indicates that moderately stressed areas surrounding Lake Maurepas should benefit the most from nutrient enhancement associated with a river diversion. Even in the severely declining swamp, growth of baldcypress trees should increase. However, water tupelo trees, many of which have severely degraded canopies, are not likely to grow in response to nutrient influxes. In the healthiest and densest site, competition with neighboring tree species may limit basal area growth, masking the effects of nutrient augmentation. Increased nutrients may have benefited the trees in other ways that were not detectable by aboveground productivity, for example through increased nutrient storage, photosynthetic rates, belowground biomass, etc. (Webb 1978). Further, this study ascertained that defoliation is more likely to affect basal area growth of water tupelo than baldcypress. However, the effects of long-term defoliation may yield different findings. Also, nutrient augmentation, if high enough, may compensate for defoliation losses to basal area growth for both species.

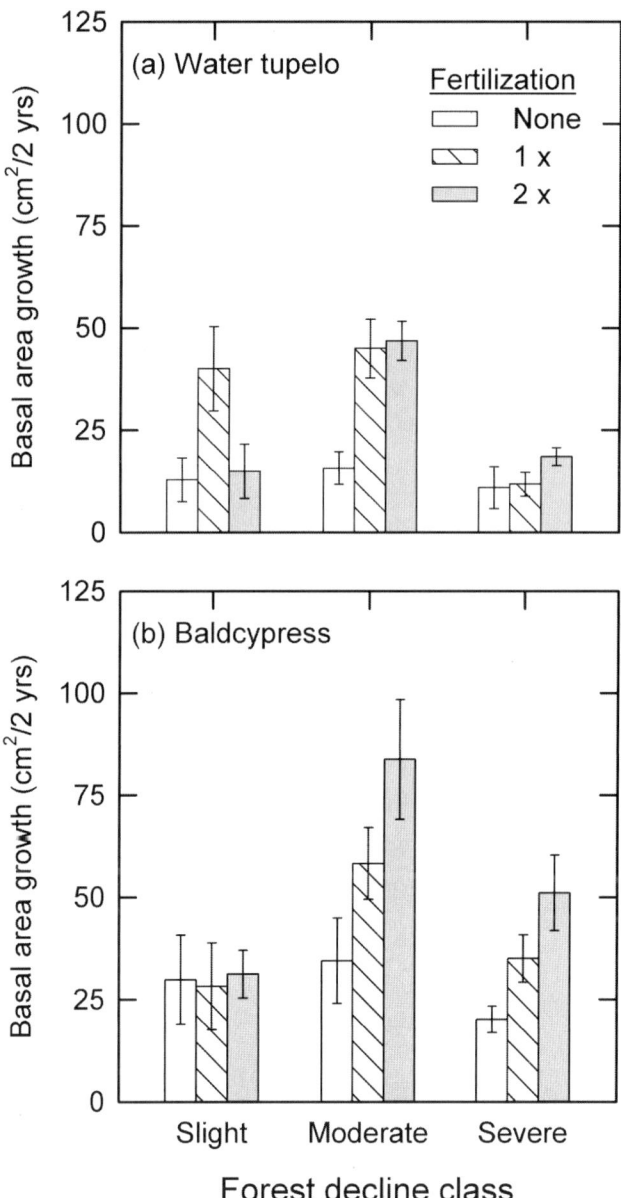

Fig. 13.7. Cumulative basal area growth of (a) water tupelo and (b) baldcypress trees in 2002 and 2003 for three sites in the Lake Maurepas Swamp varying in health after receiving three levels of fertilization (1x=236 g/m^{-2}, 2x=472 g/m^{-2}, n=7 trees for each bar, ±1SE). Reprinted from Forest Ecology and Management, Vol 236, Effler et al. (2006) with permission from Elsevier.

13.4.3.2. Litter productivity

An increase in annual leaf litter productivity was not detected in response to fertilizer, but some interesting correlations were observed with respect to species and defoliation. Individual baldcypress trees produce more litter than tupelo trees (Figure 13.8). Also, as defoliation increased, biomass of leaf fall at the end of the year decreased. However, when frass biomass collected in the spring was added to leaf biomass and the data were re-analyzed, total litter productivity increased with increased defoliation, especially for baldcypress. The increased litter reaching the forest floor as a result of defoliation observed in this study could be important in aquatic food webs particularly in nutrient limited systems. Collectively, this suggests that changes in baldcypress abundance and productivity may be more influential in nutrient cycling processes in this system than other tree species.

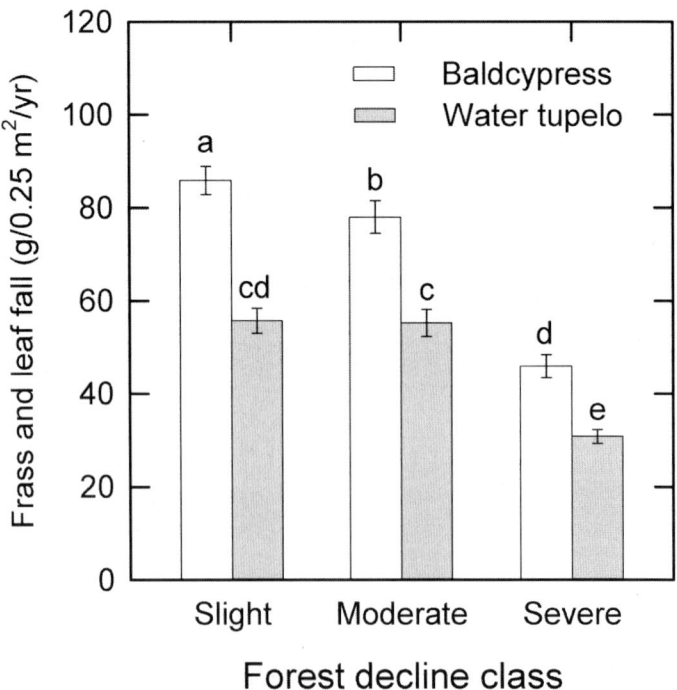

Fig. 13.8. Mean annual frass and leaf fall (g/0.25m²/year) collected in litterfall traps (2002-2003) under (a) water tupelo and (b) baldcypress trees in the Lake Maurepas swamp and correlated to defoliation (%) of each tree (n=126/species, 95% C.I.).

13.4.3.3. Foliar nitrogen content

Nitrogen levels increased as fertilization increased in both clipped leaves and litterfall for both species and in all forest condition classes and fertilization did not interact with any other treatments. As expected, spring clipped leaves had a higher nitrogen content than those in fall leaves. Nitrogen content in clipped leaves was highest at the healthiest, and decreased at both of the more degraded sites, while the nitrogen content of fallen leaves was similar at all sites (Figure 13.9). Clipped water tupelo leaves were higher in nitrogen content than clipped baldcypress leaves on a dry weight basis, but baldcypress had a higher nitrogen content in fallen leaves than water tupelo. For water tupelo, frass (insect feces) nitrogen content was higher than clipped leaves and litter fall. The opposite was true for baldcypress frass, where the nitrogen content was lower in frass than in clipped leaves and fallen leaves. Furthermore, foliar nitrogen content increased in clipped leaves and litterfall for both species as defoliation increased (Figure 13.10).

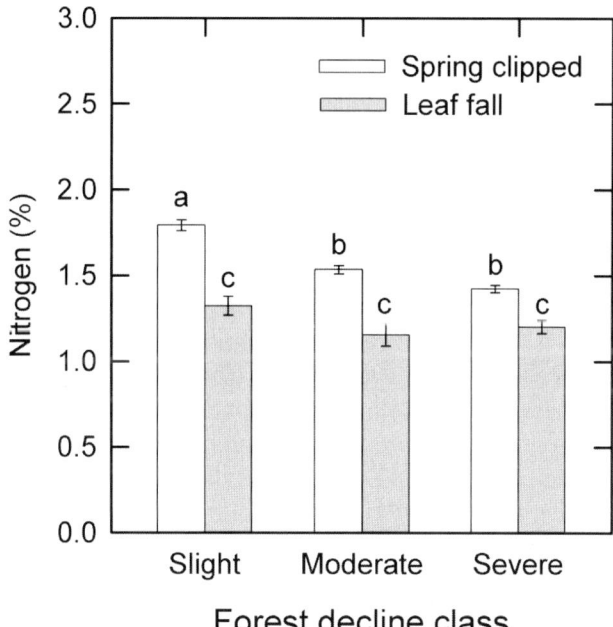

Fig. 13.9. Foliar nitrogen content of spring clipped leaves (white) and litterfall (patterned) of baldcypress and water tupelo trees across three forest conditions in 2002 and 2003 (*a priori* comparison of high vs. sparse using Fisher's LSD, $\alpha=0.05$, n=84 for each clipped bar, n=24 for litterfall, ±1 SE). Reprinted from Forest Ecology and Management, Vol 236, Effler et al. (2006) with permission from Elsevier.

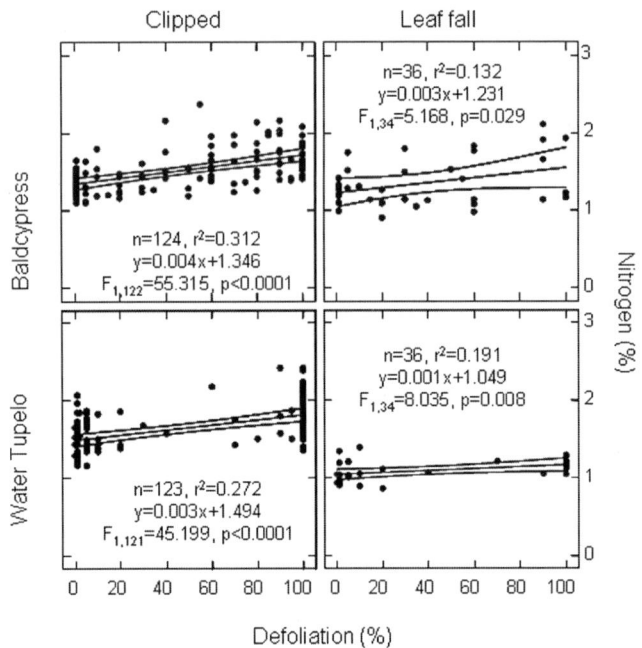

Fig. 13.10. Nitrogen content of clipped foliage (left) and litterfall (right) in 2002 and 2003 correlated with individual defoliation of baldcypress (top) and water tupelo (bottom) trees (95% C.I.). Modified from Effler et al. (2006).

Fertilizer applications increased foliar nitrogen levels across all habitat types, both species, and leaf types suggesting that the Lake Maurepas swamp is nitrogen limited in a wide array of stages of decline. Both predominant tree species therefore have the capacity to take up more nitrogen if available if these swamps are nutrient enhanced (via a diversion), with the N loading rate between of 43.4-86.8 m^3/s diversion rate. Further, a positive correlation between nitrogen levels in clipped leaves and forest health further supports the hypothesis that nitrogen is a limiting factor in this system.

Baldcypress litterfall is higher in nitrogen content and has a higher litter biomass than water tupelo, suggesting that baldcypress might be a better source of organic nutrients for aquatic food webs and other nutrient-cycling ecosystem processes. Nitrogen content in water tupelo frass was higher than in baldcypress frass, which could result in this being an important source of nitrogen input during the spring. However, these inputs must be considered in context with total biomass contributions. If baldcypress leafroller produces a larger quantity of frass than forest tent caterpillar, as suggested from the results, total overall nitrogen loading could be higher during years of peak defoliation. Interestingly, defoliation was positively correlated to the nitrogen content in litterfall, suggesting that defoliators not only increase organic nitrogen loading rates in the spring, but also could be causing changes in foliar chemistry that direct a higher nitrogen storage in the litterfall as opposed to utilizing resources for other storage organs and biochemical pathways (reproductive structures, roots, wood, defensive compounds, photosynthetic rates, etc.). In other swamps, nitrogen levels in clipped leaves for baldcypress were 1.43% (Schlesinger 1978) and 1.52-1.71% (Meeker 1992). The increased nutrient levels found in the canopy foliage following fertilization and corresponding lab assays showing these insects respond positively to nutrient addition to hosts (Souther-Effler 2004) suggest that defoliator populations could potentially increase.

13.5. Current and future research

13.5.1. Hurricane impacts

Baldcypress and water tupelo swamps historically comprised 90% of the wetlands in the Lake Pontchartrain Basin (Saucier 1963). This coverage has been radically reduced, and the swamps that remain have been classified as mostly non-sustainable (Chambers et al. 2005; Lake Pontchartrain Basin Foundation 2005). Several projects have been proposed to help restore these swamps, but hurricanes Katrina and Rita have altered the way we should think about wetland restoration (Boesch et al. 2006). In terms of flood- and wind-damage reduction, swamps may provide important protection from storms and surges (Williams et al. 1999; Danielsen et al. 2005).

Initial observations in the Maurepas swamp indicate minimal wind throw of overstory trees, but light to moderate crown damage from Hurricane Katrina (2005). In contrast, midstory wind throw was severe in certain areas and appears to be a function of overstory density. We need to quantify the relationship between wind throw and overstory tree density, which could be useful in determining optimal densities of artificial plantings. We also need to monitor post-hurricane tree and herbaceous ground cover production and compare it with that of normal- and drought-year conditions. These data could be very useful in refining Louisiana coast-wide landscape models to incorporate hurricane impacts on growth and survival of herbaceous and woody wetland species.

Hurricanes Katrina and Rita also impacted herbaceous vegetation in the Maurepas swamp and Manchac marsh. Complete mortality appears to have occurred for nine of 22 common species (Shaffer, unpublished data). In the future, we need to determine if these species re-establish from the seed-bank, or if the hurricanes produced a long-term reduction of species diversity. The hurricanes may have caused both increases and decreases in herbaceous production through deposition of re-suspended sediments and erosion, respectively. Pre-hurricane monitoring of herbaceous and woody production in the Maurepas swamp should be followed by post-hurricane studies to assess the impacts of the 2005 hurricanes.

Although cypress - tupelo swamps are extremely resistant to wind throw and deep flooding, they are less resistant to salt water intrusion and thus require a reliable source of freshwater for system flushing following tropical storm events. Swamps can survive short-term salinity pulses (Allen et al. 1994; Campo 1996; Conner et al. 1997); for example a cypress - ash nursery in St. Benard Parish was covered with 3 m of saline water from Hurricane Katrina for 3-4 days and survived; exact salinity levels were not recorded (Richard Goyer, personal observation). We plan to build a GIS containing all substantial point and non-point freshwater sources, including urban and agricultural runoff, storm water pumps, non-contact industrial cooling water, municipal wastewater treatment facilities, and potential Mississippi River diversion sites. At present, most of these sources are input to the basin to maximize drainage efficiency. Freshwater is routed into ditches and canals that carry it directly to the lakes, bypassing wetland contact. This creates a "lose-lose" situation as potential for eutrophication is maximized and the wetlands remain nutrient starved. In contrast, re-routing the water to maximize sheet flow will improve water quality, increase wetland net primary production, and decrease salt water intrusion. Furthermore, implementation of the proposed river diversions at Violet, Bonnet Carre, La Branche, and the Maurepas swamp (Coast 2050 1998) will greatly enhance restoration of historic salinity conditions. In addition to decreasing storm damage, increasing swamp acreage will improve water quality and may lead to net sediment accretion which will increase carbon sequestration (Trettin and Jorgensen 2003) and enhance several of the "multiple lines of defense" proposed by Lopez (2006). Ultimately, we must close the Mississippi River Gulf Outlet (MRGO) to help approximate historic salinity regimes and further assist in storm-damage reduction.

One concern that managers and the general public have with restoring repressed swamps is the amount of time required for swamp-like characteristics to emerge and manifest. Fortunately, given favorable hydrologic conditions, baldcypress and water tupelo seedlings can reach greater than

10-meter heights within one decade (Shaffer and Goyer, unpublished data). For example, a pilot planting of baldcypress seedlings at the Caernarvon diversion (Krauss et al. 2000) has yielded >10 m tall trees in a decade and all of these resisted wind throw during the hurricanes of 2005 (Rich Goyer, personable observation).

In summary, future research in the Maurepas swamp will seek to (1) determine whether the impacts of hurricanes Katrina and Rita were less severe in forested wetlands than other contiguous wetland habitat types, (2) follow the production and compositional response of the herbaceous and woody vegetation of the swamp to the hurricanes, (3) determine the relationship between density of canopy trees and frequencies of midstory wind throw, (4) isolate all substantial point and nonpoint sources of potential freshwater input in the basin, (5) build a GIS of these sources and the spatial extent of currently repressed swamp (swamp that has converted to marsh or open water) that could be restored by these sources, and (6) model the relationship of area of swamp and energy reduction of storms of varying strength, and compare this with other types of wetlands.

13.5.2. Sustainability and management

Current Maurepas swamp research projects largely focus on two related main questions: (1) to which degree are the Maurepas swamps self-sustaining in the face of continuing relative sea-level rise? and (2) what management or restoration actions need to be taken to maintain or improve current swamp conditions? The complexity of these questions dictates the use of a wide variety of scientific tools.

Researchers at Southeastern Louisiana University currently investigate the rates and variability of local subsidence and sediment deposition through the use of sediment elevation tables (SETs) and marker horizons (Cahoon et al. 1999). The high porosity of the organic swamp soils makes the application of commonly used feldspar marker horizons ineffectual, and thus new methods of measuring sediment accretion need to be developed. Similarly, the enormous water-holding capacity of the highly organic swamp soils has thus far resulted in highly variable and seasonally fluctuating subsidence measurements (Shaffer et al. 2003). Recently, Shaffer et al. installed a new array of deep SETs (Cahoon et al. 1999) throughout marsh and swamp habitats in the Maurepas basin to shed light on the complex interplay of soil compaction, expansion, and accretion. Once baseline measurements of local subsidence and accretion have been taken, future research plans include fertilization studies to investigate the role of belowground production in soil dynamics.

In an effort to understand the long-term implications of multiple stressors on the ecological health and maintenance of the existing swamps, researchers at Louisiana State University (Baton Rouge, Louisiana) and East Carolina University (Greenville, North Carolina) are in the process of developing computer simulation models. One of these modelling efforts seeks to predict habitat change on a landscape scale over 100 years (Enrique Reyes, East Carolina University, personal communication). The landscape model is a descendant of the Coastal Ecological Landscape Spatial Simulation (CELSS) model lineage, which has been developed over the last two decades to describe and predict wetland habitat change in coastal Louisiana under various restoration and climate change scenarios (Costanza et al. 1990; Martin et al. 2000; Reyes et al. 2000; Martin et al. 2002). The landscape model predicts habitat change and biomass production for a spatial scale of 3,640 km^2 at a spatial resolution of 1 km^2, and uses a mass-balance approach for water movement due to tidal exchange, wind shear, and riverine inputs. A related individual-based model (IBM) developed by researchers at Louisiana State University aims to investigate the effects of short-term and long-term environmental processes on individual trees as well as the overall swamp population dynamics on a spatial scale of 1 km^2 at a spatial resolution of 100 m^2. Swamp characteristics modelled in the IBM include stand density, basal area, species composition, recruitment, and productivity of the two dominant tree species, baldcypress and water tupelo. Ongoing research goals include the linkage of the coarser large-scale landscape model with the fine-scale swamp IBM. Future modelling goals may include the addition of nutrient cycling into both models. Models of nutrient loading rates of defoliation by-products may need to be established to describe nutrient flow in the system. Estimates of biomass and nutrient contributions could be made with regard to species composition and tree density/canopy closure. Biomass and nutrient contributions may be important in nutrient impoverished systems as well as for water quality.

The restoration potential of parts of the Maurepas swamp through tree plantings has become a focus of resource management concern as well as research interest. In general, planted baldcypress seedlings can grow and survive under the low salinity conditions typical of the Maurepas swamp (Myers et al. 1995). However, high mortality rates (~90%) of planted baldcypress seedlings have been observed periodically in the Maurepas swamp over the past 16 years (Shaffer, unpublished data). The large scale mortality periods generally occur during saltwater intrusion events, which are greatly exacerbated by regional droughts. In addition, a large restoration planting on Jones Island suffered 97% seedling mortality in 2005 (Effler, unpublished data) as a result of nutria damage despite herbivore control efforts (TreePro® tree shelters and Ropel® - mammalian chemical deterrent). Extensive seedling plantings with various types of tree shelters have occurred on Jones Island from 2000 to 2005, and have achieved limited success (Micheal Greene, personal communication, Southeastern Louisiana University, Hammond, LA). Over time, the local population of nutria may have become "tree-shelter smart," as evidenced by the wide-spread occurrence of lifted and destroyed tree-shelters in the frequently planted areas. Two other areas in the basin, which had not previously been planted, had few if any nutria related mortalities despite using similar herbivore protection devices. Further research into the effectiveness of herbivore protection devices may be necessary to enable the success of future restoration plantings.

13.6. Conclusion

The Maurepas swamps are characterized by nutrient poor waters, soils of extremely low strength indicative of stress, and salt water intrusions that occur during the late summer and fall seasons. Salinity and flooding interact as stressors throughout the Basin, but differ in importance in different areas. The Maurepas swamps are nitrogen limited, and nutrient stress is potentially as important as salt or flood stress, especially if accompanied with defoliation. Further, recruitment of saplings throughout the swamp is very low, probably not enough to sustain current forest. Most of the Maurepas swamp appears to be converting to marsh and open water primarily due to the lack of riverine input. Salt stress is killing trees proximal to the lake, whereas stagnant standing water and nutrient deprivation appear to be the largest stressors at interior sites.

Besides decreasing the detrimental effects of salinity throughout the Maurepas swamp, a proposed freshwater diversion from the Mississippi River also would increase the sediment load and nutrient supply to these wetlands. The potentially negative impacts of lake eutrophication due to the increase in nutrient loading to the swamp are unlikely to occur, as nutrient models indicate high nutrient retention in the swamp with nutrient removal efficiencies of 94-99%. Experimental nutrient augmentation enhanced biomass production of the herbaceous vegetation by up to 300%. This enhanced productivity is essential for subsiding coastal wetlands to offset relative sea-level rise. The exact duration and depth that cause the transition of swamp to marsh remains unknown. Without a diversion from the Mississippi River, however, the Maurepas swamp may soon resolve this issue all too clearly.

References

Abadie SW, Poirrier A (2000) Increased density of large rangia clams in Lake Pontchartrain after the cessation of shell dredging. J Shellfish Res 19:481-485

Abrahamson LP, Harper JD (1973) Microbial insecticides control forest tent caterpillar in southwestern Alabama. Research Note SO-157. U.S. Department of Agriculture, Forest Service, Southern Forest Experiment Station, New Orleans

Alfaro RI, MacDonald RN (1988) Effects of defoliation by the western false hemlock looper on Douglas-fir tree-ring chronologies. Tree-Ring Bull 48:3-11

Alfaro RI, Shepherd RF (1991) Tree-ring growth of interior Douglas-fir after one year's defoliation by Douglas-fir tussock moth. For Sci 37:959-964

Allen JA, Chambers JL, McKinney D (1994) Intraspecific variation in the response of *Taxodium distichum* seedlings to salinity. For Ecol Manage 70:203–214

Barras JA, Bourgeois PE, Handley LR (1994) Land loss in coastal Louisiana. Open File Report 94-01. National Biological Survey, National Wetlands Research Center, Lafayette

Boesch DF, Shabman L, Antle LG, Day JW, Dean RG, Galloway GE, Groat CG, Laska SB, Luettich RA, Mitsch WJ, Rabalais NN, Reed DJ, Simonstad CA, Streever BJ, Taylor RB, Twilley RR, Watson CC, Wells JT, Whigham DF (2006) A new framework for planning the future of coastal Louisiana after the hurricanes of 2005. Working Group for Post-Hurricane Planning for the Louisiana Coast

Bove MC, Zierden DF, O'Brien JJ (1998) Are Gulf landfalling hurricanes getting stronger? Bull Am Meteorol Soc 79:1327–1328

Braun DM, Goyer RA, Lenhard GJ (1989) Biology and mortality agents of the fruittree leafroller (Lepidoptera: Tortricidae) on baldcypress in Louisiana. J Entomol Sci 25:176-184

Brown CA, Montz GN (1986) Baldcypress and the tree unique, the wood eternal. Claitor's Publishing Division, Baton Rouge

Burns AC (1980) Frank B. Williams: Cypress lumber king. J For Hist 24:127-134

Cahoon DR, Day JW, Reed D (1999) The influences of surface and shallow subsurface soil processes on wetland elevation: A synthesis. Current Topics in Wetland Biogeochemistry 3:72-78

Campo FM (1996) Restoring a repressed swamp: The relative effects of a saltwater influx on an immature stand of baldcypress (*Taxodium distichum* (L.) Richard). M.S. thesis, Southeastern Louisiana University

Carter MR, Burns LA, Cavinder TR, Dugger KR, Fore PL, Hicks DB, Revells HL, Schmidt TW (1973) Ecosystems analysis of the Big Cypress Swamp and estuaries. U.S. Environmental Protection Agency, Region IV, South Florida Ecological Study

Chambers JL, Conner WH, Day JW, Faulkner SP, Gardiner ES, Hughes MS, Keim RF, King SL, McLeod KW, Miller CA, Nyman JA, Shaffer GP (2005) Conservation, protection and utilization of Louisiana's Coastal Wetland Forests. Final Report to the Governor of Louisiana from the Coastal Wetland Forest Conservation and Use Science Working Group. Louisiana Governor's Office of Coastal Activities, Baton Rouge

Coast 2050 (1998) Coast 2050: Toward a sustainable Louisiana. Louisiana Coastal Wetlands Conservation and Restoration Task Force, Louisiana Department of Natural Resources, Baton Rouge

Conner WH, Askew GR (1993) Impact of saltwater flooding on red maple, redbay, and Chinese tallow seedlings. Castanea 53(3):215-219

Conner WH, Day Jr JW (1976) Productivity and composition of a bald cypress-water tupelo site and a bottomland hardwood site in a Louisiana swamp. Am J Bot 63:1354-1364

Conner WH, Day Jr JW (1992) Water level variability and litterfall productivity of forested freshwater wetlands in Louisiana. Am Midl Nat 128(2):237-245

Conner WH, Gosselink JG, Parrondo RT (1981) Comparison of the vegetation of three Louisiana swamp sites with different flooding regimes. Am J Bot 68:320-331

Conner WH, McLeod KW, McCarron JK (1997) Flooding and salinity effects on growth and survival of four common forested wetland species. Wetl Ecol Manage 5:99-109

Costanza R, Sklar FH, White ML (1990) Modeling coastal landscape dynamics. BioScience 40:91-107

Danielsen F, Sorenson MK, Olwig MF, Selvam V, Parish F, Burgess ND, Hiraishi T, Karunagaran VM, Rasmussen MS, Hansen LB, Quarto A, Suryadiputan N (2005) The Asian tsunami: A protective role for vegetation. Science 310:643

Day Jr JW, Kemp GP, Mashriqui HS, Dartez D, Lane RR, Cunningham R (2004) Development plan for a diversion into the Maurepas swamp: Water quality and hydrologic modeling components. Final report. U.S. Environmental Protection Agency, Region 6, Dallas

Dicke SG, Toliver JR (1990) Growth and development of baldcypress/water-tupelo stands under continuous versus seasonal flooding. For Ecol Manage 33/34:523-530

Effler RS, Goyer RA, Lenhard GJ (2006) Baldcypress and water tupelo responses to insect defoliation and nutrient augmentation in Maurepas Swamp, Louisiana, USA. For Ecol Manage 236:295-304

Fitzgerald TD (1995) The tent caterpillars. Cornell University Press, Ithaca

Frazier DE (1967) Recent deltaic deposits of the Mississippi River – their development and chronology. Trans Gulf Coast Assoc Geol Soc 17:287-315

Goyer RA, Chambers J (1996) Evolution of insect defoliation in baldcypress and its relationship to flooding. Biological Science Report 8. U.S. Department of Interior, National Biological Service

Goyer RA, Lenhard GJ (1988) A new insect pest threatens baldcypress. LA Agriculture, LA Agricultural Experiment Station 31:16-17, 21

Hatton RS (1981) Aspects of marsh accretion and geochemistry: Barataria Basin, Louisiana. M.S. thesis, Louisiana State University

Hoover HT (1975) The Chitimacha People. The Indian Tribal Series, Phoenix

Johnson CW (2004) The performance of the baldcypress leafroller (*Archips goyerana*) Kruse, Lepidoptera: Tortricidae) in response to fertilization, thinning, and genetic variation in host baldcypress (*Taxodium distichum* L.Richard). M.S. thesis, Louisiana State University

Jones RT, Sharitz RR, McLeod KW (1989) Effects of flooding and root competition on growth of shaded bottomland hardwood seedlings. Am Midl Nat 121:165-175

Kemp PG, Conner WH, Day Jr JW (1985) Effects of flooding on decomposition and nutrient cycling in a Louisiana swamp forest. Wetlands 5:35-50

Klock GO, Wickman BE (1978) Ecosystems effects. In: Brookes MH, Stark RW, Campbell RW (eds) The Douglas-fir Tussock moth: A synthesis. Technical Bulletin 1585. U.S. Department of Agriculture, Forest Service, Washington, pp 90-94

Krauss KW, Chambers JL, Allen JA, Soileau Jr DM, DeBosier AS (2000) Growth and nutrition of baldcypress families planted under varying salinity regimes in Louisiana, USA. J Coast Res 16:153-163.

Lake Pontchartrain Basin Foundation (2005) Comprehensive Management Plan for the Lake Pontchartrain Basin. LPBF, New Orleans

Lane RR, Day JW, Thibodeaux B (1999) Water quality analysis of a freshwater diversion at Caernarvon, Louisiana. Estuaries 22:327-336

Lane RR, Mashriqui HS, Kemp GP, Day JW, Day JN, Hamilton A (2003) Potential nitrate removal from a river diversion into a Mississippi delta forested wetland. Ecol Eng 20:237-249

Lopez JA (2006) The multiple lines of defense strategy to sustain coastal Louisiana. White Paper. Lake Pontchartrain Basin Foundation, New Orleans

Lovett G, Tobiessen P (1993) Carbon and nitrogen assimilation in red oaks (*Quercus rubra* L.) subject to defoliation and nitrogen stress. Tree Physiol 12:259-269

Mancil E (1972) An historical geography of industrial cypress lumbering in Louisiana. Ph.D. thesis, Louisiana State University

Mancil E (1980) Pullboat logging. J For Hist 24:135-141

Martin JF, White ME, Reyes E, Kemp GP, Mashriqui H, Day JW Jr (2000) Evaluation of coastal management plans with a spatial model: Mississippi Delta, Louisiana, USA. Environ Manage 26:117-129

Martin JF, Reyes E, Kemp GP, Mashriqui H, Day Jr JW (2002) Landscape modeling of the Mississippi Delta. BioScience 52:357-365

Mattoon WR (1915) The southern cypress. Agriculture Bulletin 272. U.S. Department of Agriculture, Washington

Meeker JR (1992) Host quality of baldcypress and its influence of the fruit tree leafroller, *Archips agyrospila* (Walker) (Lepidoptera: Tortricidae), performance in forested wetlands of Louisiana. Ph.D. thesis, Louisiana State University

Megonigal JP, Conner WH, Kroeger S, Sharitz RR (1997) Aboveground production in southeastern floodplain forests: A test of the subsidy-stress hypothesis. Ecology 78(2):370-384

Mitsch WJ, Taylor JR, Benson KB (1991) Estimating primary productivity of forested wetland communities in different hydrologic landscapes. Landscape Ecol 5:75-92

Myers RS, Shaffer GP, Llwellyn DW (1995) Baldcypress (*Taxodium distichum* (L.) Rich.) restoration in southeast Louisiana: Relative effects of herbivory, flooding, competition, and macronutrients. Wetlands 15:141-148

Nachod LH (1977) Spring defoliation by forest insects in Louisiana. Insect and Disease Report. Louisiana Office of Forestry, Woodworth

Nachod LH, Kucera DR (1971) Observations of the forest tent caterpillar in south Louisiana. Insect and disease report. Louisiana Office of Forestry, Woodworth

NOAA (2001) National Oceanic and Atmospheric Administration (2001). Climate of 2000—Annual review, US summary. Available on the NOAA website at www.ncdc.noaa.gov/ol/climate/research/2000/ann/us_summary.html

Penland S, Ramsey KE (1990) Relative sea-level rise in Louisiana and the Gulf of Mexico: 1908-1988. J Coast Res 6:323-342

Progar RA, Schowalter TD, Morrell JJ, Freitag CM (2000) Respiration from coarse woody debris as affected by moisture and saprotroph functional diversity in western Oregon. Oecologia 124:426-431

Reyes E, White ML, Martin JF, Kemp GP, Day Jr JW, Aravamuthan V (2000) Landscape modeling of coastal habitat change in the Mississippi Delta. Ecology 81:2331-2349

Rheinhardt RD, Rheinhardt MC, Brinson MM, Faser K (1998) Forested wetlands of low order streams in the inner coastal plain of North Carolina, USA. Wetlands 18(3):365-378

Saucier RT (1963) Recent geomorphic history of the Pontchartrain Basin. Louisiana State Uuniversity Press, Baton Rouge

Schlesinger WH (1978) Community structure, dyes, and nutrient ecology in the Okenfenokee cypress swamp-forest. Ecol Monogr 48:43-65

Schowalter TD (1981) Insect herbivore relationship to the state of the host plant: Biotic relation of ecosystem nutrient cycling through ecosystem succession. Oikos 37:126-130

Schowalter TD (1992) Early decomposition and nutrient dynamics of oak (*Quercus*) logs at four sites across a North American gradient. Can J For Res 22:161-166

Schowalter TD, Crossley Jr DA (1983) Forest canopy arthropods as sodium, potassium, magnesium and calcium pools in forests. For Ecol Manage 7:143-148

Schowalter TD, Sabin TE, Stafford SG, Sexton JM (1991) Phytophage effects primary production, nutrient turnover, and litter decomposition on young Douglas-fir in western Oregon. For Ecol Manage 42:229-243

Schowalter TD, Caldwell BA, Carpenter SE, Griffiths RP, Harmon ME, Ingham ER, Kelsey RG, Lattin JD, Moldenke AR (1992) Decomposition of fallen trees: effects of initial conditions and heterotroph colonization rate. In: Singh KP (ed) Tropical ecosystems: Ecology and management. Wiley Eastern, Ltd., New Delhi, pp 371-381

Schowalter TD, Zhang YL, Sabin TE (1998) Decomposition and nutrient dynamics of oak (*Quercus* spp.) logs after five years of decomposition. Ecography 21:3-10

Scruton PC (1960) Delta building and deltaic sequence. In: Shepard FP, Phleger FB, van Andel THR (eds) Recent sediments, northwest Gulf of Mexico. American Association of Petroleum Geologists, Tulsa, pp 82-102

Seastedt TR, Crossley DA Jr (1984) The influence of arthropods on ecosystems. BioScience 34:157-161

Shaffer GP, Perkins TE, Hoeppner SS, Howell S, Beneard TH, Parsons AC (2003) Ecosystem health of the Maurepas Swamp: Feasibility and projected benefits of a freshwater diversion. Final Report. Environmental Protection Agency, Region 6, Dallas

Smith JD, Goyer RA (1986) Population fluctuations and causes of mortality for the forest tent caterpillar, *Malacosoma disstria* (Lepidoptera: Lasiocampidae), on three different sites in southern Louisiana. Environ Entomol 15:1184-1188

Souther-Effler RF (2004) Interactions of insect herbivory and multiple abiotic stress agents on two wetland tree species in southeast Louisiana swamps. Ph.D. thesis, Louisiana State University

Swank WT, Waide JB, Crossley Jr DA, Todd RL (1981) Insect defoliation enhances nitrate export from forest ecosystems. Oecologia 51:297-299

Taylor JR (1985) Community structure and primary productivity of forested wetlands in Western Kentucky. Ph.D. thesis, University of Louisville

Thomson DM, Shaffer GP, McCorquodale JA (2002) A potential interaction between sea level rise and global warming: Implications for coastal stability on the Mississippi River deltaic plain. Global Plan 32:49-59

Trettin CC, Jorgensen MF (2003) Carbon cycling in wetland forest soils. In: Kimble JM, Heath LS, Birdsey RA, Lal R (eds) The potential of U.S. forest soils to sequester carbon and mitigate the greenhouse effect. CRC Press, Boca Raton, pp 311-331

Trumble JT, Kolodny-Hirsch DM, Ting IP (1993) Plant compensation for arthropod herbivory. Annu Rev Entomol 38:93-119

Twilley RR, Barron EJ, Gohlz HL, Harwell MA, Millier RL, Reed DJ, Rose JB, Siemann EH, Wetzel RG, Zimmerman RJ (2001) Confronting climate change in the Gulf of Mexico coast region: Prospects for sustaining our ecological heritage. Union of Concerned Scientists, Cambridge, MA and The Ecological Society of America, Washington

Webb WL (1978) Effects of defoliation and tree energetics. In: Brookes MH, Stark RW, Campbell RW (eds) The Douglas-fir tussock moth: A synthesis. Technical Bulletin 1585. U.S. Department of Agriculture, Forest Service, pp 77-81

White DA (1983) Plant communities of the lower Pearl River Basin, Louisiana. Am Midl Nat 110(2):381-397

Williams K, Pinzon ZS, Stumpf RP, Raabe EA (1999) Sea-level rise and coastal forests on the Gulf of Mexico. Open-File Report 99-441. U.S. Geological Survey, St. Petersburg

Xu YJ, Wu K (2006) Seasonality and interannual variability of freshwater inflow to a large oligohaline estuary in the Northern Gulf of Mexico. Estuar Coastal Shelf Sci 68(3-4):619-626

Chapter 14 - Selection for Salt Tolerance in Tidal Freshwater Swamp Species: Advances Using Baldcypress as a Model for Restoration*

Ken W. Krauss[1], Jim L. Chambers[2], and David Creech[3]

[1] U.S. Geological Survey, National Wetlands Research Center, 700 Cajundome Blvd., Lafayette, LA 70506
[2] School of Renewable Natural Resources, Louisiana State University Ag Center, Renewable Natural Resources Building, Baton Rouge, LA 70803
[3] SFA Mast Arboretum, Stephen F. Austin State University, PO Box 13000, Nacogdoches, TX 75962

14.1. Introduction

Worldwide, the intrusion of salinity into irrigated and natural landscapes has major economic and cultural impacts and has resulted in large reductions in crop yields (Epstein et al. 1980; Flowers 2003). Losses have prompted wide-scale programs to improve the salt tolerance of many agronomic species or to identify crop species that can tolerate lands affected by low levels of salinity. Few historic research efforts have considered forest tree species in the United States, especially in nonurban areas.

Newer programs have focused on identifying salt tolerance in forest tree species but have mainly limited these efforts to compiling lists of salt tolerant species to be used in afforestation projects (Gogate et al. 1984; Shrivastava et al. 1988; Beckmann 1991; Bell 1999). Gogate et al. (1984), for instance, listed 26 potential species from Australia with silvicultural application to salt affected lands in India. More comprehensive efforts have considered species lists along with specific site requirements (Bell 1999); species tolerant to saline irrigation waters on dry land, for example, will not often be tolerant of salinity increases in wetland settings. Similar ideas

---

* The U.S. Government's right to retain a non-exclusive, royalty-free licence in and to any copyright is acknowledged.

have spawned field trials of native and nonnative tree species in India, Pakistan, Thailand, Australia, and the United States (Thomson 1988; Beckmann 1991; Krauss et al. 2000; Conner and Ozalp 2002; Marcar and Crawford 2004; Conner and Inabinette 2005).

Concerted attempts at salt tolerance improvement of forest tree species have been limited, owing in part to the diversity of regional issues that such programs must consider. Whereas food, fodder, and pulp yield may be the major improvement goal on salt affected lands in India (Mathur and Sharma 1984), identifying trees that can survive deicing salts (Townsend 1989), oil and gas brine discharges (Auchmoody and Walters 1988), or sea-level rise induced salinity changes (Pezeshki et al. 1987, 1990) are of greater interest to larger industrial nations. Nevertheless, salt tolerance research on a range of tree species has converged on one very important finding; among the mechanisms proposed for salt tolerance in nonhalophytes (Greenway and Munns 1980; Munns and Termaat 1986; Cheeseman 1988), ion exclusion from cellular processes, especially exclusion of Cl^-, ranks high (Townsend 1989). Identifying the principal mechanism and location of ion exclusion and determining the range of additive genetic variation available among physiological, morphological, and growth attributes for individual species have been the major elements of salt tolerance improvement programs for trees (Allen et al. 1994a).

14.1.1. Salinity and coastal swamp forests

Salinization is not uncommon to tidal freshwater swamps; fluxes of salts into these swamps can occur naturally under different sea-level rise or storm impact scenarios. William Bartram in 1791 was perhaps the first explorer to document a natural progression of swamp forest vegetation to marsh in the following excerpt:

> ". . . it seems evident, even to demonstration, that those salt marshes adjoining the coast of the main, and the reedy and grassy islands and marshes in the rivers, which are now overflowed at every tide, were formally high swamps of firm land, affording forests of Cypress, Tupilo, Magnolia grandiflora, Oak, Ash, Sweet Bay, and other timber trees, the same as are now growing on river swamps, whose surface is two feet or more above the spring tides that flow at this day; . . . when [planters] bank these grassy tide marshes for cultivation, they cannot sink their drains above three or four feet below the surface, before they come to strata of Cypress stumps and other trees, as close together as they now grow in the swamps." – p. 79 (Bartram 1928)

Bartram was traveling along the Atlantic coast of the Carolinas when he wrote this, and it was not until the early to mid 1900s that substantial detail was available on the condition and distribution of coastal swamp forests along the Gulf of Mexico Coast (Mattoon 1915; Penfound and Hathaway 1938; Penfound 1952; Chabreck 1972). By then, many of the swamp forests, at least in Louisiana, were logged, and humans were well into projects that would serve to exacerbate the areal extent and magnitude of salt water intrusion into tidal swamp forests. Hence, the periodic exposure of tidal freshwater swamp species to salinity is historically common but has been hastened and increased in extent by humans over the past century.

Increases in salinity have been associated with losses of large forested wetland areas in the southeastern United States, which contains 650,000 ha or more of baldcypress (*Taxodium distichum* [L.] L.C. Rich.) and water tupelo (*Nyssa aquatica* L.) communities (McWilliams and Rosson 1990). Salt water effects can range from total destruction and conversion of swamp forests to marsh and open water or can lead to reductions of standing timber. Basal area, for example, was reduced from 44-87 m^2/ha in forests with no salinity to 23-36 m^2/ha in forests with 1.3 to 3.0 ppt salinity in South Carolina and Louisiana (Chapter 9). Regeneration is also limited by increases in salinity and prolonged flooding (Chapter 16). Even small increases in salinity have considerable effect on both survival and conversion processes. Leveeing of rivers to prevent overbank flooding, dredging of canals, oil and gas exploration, and recent trends in sea-level rise continue to facilitate the intrusion of salt water into formerly freshwater swamp forests (Salinas et al. 1986; Pezeshki et al. 1987). Large tides and tropical storms are also important vectors of temporary or permanent saltwater intrusion. Storm surges can extend tens of kilometers inland and to at least a meter in depth in some coastal swamp forests. For example, porewater salinities of 0.2-1.4 ppt were replaced with salinities of 2.0-3.8 ppt just after the passage of Hurricane Rita (2005), which sent a 0.71 m storm surge over the soil surface of a swamp forest in Mandalay National Wildlife Refuge (Louisiana).

14.1.2. Acute and chronic susceptibility

The propensity for salt tolerance in forest tree species should be defined separately for acute and chronic scenarios. These are important distinctions for tree improvement programs because susceptibilities differ with these two exposures and potentially define different mechanisms for tolerance. On the one hand, storm surges, droughts, and even strong weather fronts can expose swamp forest species to rather high salinity for short periods of time. This is a common source of acute salinity pulses to swamp forests that are along the upper intertidal reaches of a river, closer to a freshwater source, or commonly hydrologically isolated. For these acute stress events, trees have to survive increases in salinity for shorter periods of time until rainfall and normal freshwater flushing restores porewater condition to freshwater levels. Tolerance thresholds for swamp species are higher under such flood scenarios; physiological and growth processes may be only temporarily stalled under this condition unless salinity is extremely high. For example, as many as 1.8 million ha of forests were affected by wind, overwash, and saltwater intrusion during Hurricane Hugo (1989) along the South Carolina coast (Hook et al. 1991); most have returned to normal, and salinity can still be detected on only a few sites today.

In parts of coastal Louisiana, it has been difficult to separate acute from chronic stress vectors because canal dredging has exposed large areas of swamp forest to salinity over short periods of time without provision for ameliorating porewater salinity condition. Mortality is quick, and the state change is permanent, often transitioning healthy swamp forests to a brackish or salt marsh in a matter of a decade (Figure 14.1). Moreover, forests that are hydrologically isolated have no way to dispose of excessive salts after single events such as storm surge and, therefore, become chronically stressed over time. Many former swamp forests and surrounding waterways in Louisiana currently have salinities ranging from 2–23 ppt (Wicker et al. 1982) and are surrounded by many hectares of declining or dying forest. Historically, coastal swamp forests were restricted to chronically imposed salinities below 2-3 ppt (Penfound and Hathaway 1938; Chabreck 1972). While survival of some species can be expected from acute pulses as high as 18.5 ppt (Conner and Inabinette 2005), persistence at those levels or higher for several weeks or greater leads to almost complete mortality (Conner and Askew 1992; McLeod et al. 1996; McCarron et al. 1998).

Sea-level rise, which may be as high as 0.09 to 0.88 m over the next 100 years (IPCC 2001), can cause a long-term, chronic shift in tidal prisms and salinity distributions in coastal wetlands. On the other hand, historical salinity conditions – acute or chronic – provide the very opportunity for salt tolerance selection of tidal freshwater tree species and genotypes. Accordingly, exacerbated salt water intrusion into freshwater swamps allows scientists to identify which species and genotypes are the most tolerant. As salinity enters large areas, only those species and genotypes (individuals) that are most tolerant to the chronic stress imposed by salinity will survive.

Chapter 14 - Selection of Salt Tolerance in Freshwater Species   389

Photo credit KW Krauss

Photo credit JA Allen

Fig. 14.1. Pristine (top) and degraded (bottom) coastal baldcypress sites in the Terrebonne Basin, Louisiana. The site on the bottom represents a state change from baldcypress-dominated swamp forest to cordgrass (*Spartina*) marsh after the construction of the Houma Navigation Canal. Coastal Louisiana has several constructed navigation corridors that vector salinity to otherwise freshwater swamps that originally looked similar to the top image.

14.2. Evidence of salt tolerance in baldcypress

Variation in the salt tolerance of natural tree populations can be altered over time through mutations, normal gene flow, natural selection, and genetic drift (Zobel and Talbert 1984). Tidal freshwater swamps may contain many tree species that segregate in clusters differentially along a floodplain in a way that affects those mechanisms and/or in a way that reflects the effects of those mechanisms. Chapter 12, for example, describes clumped distributions of tree and shrub species relative to hydroperiod in tidal forests along the lower Savannah River, Georgia. Accordingly, Rheinhardt and Hershner (1992) found that in areas with rapidly changing water table depths dominant forest vegetation segregates into primary species classes based upon mean water table depth in tidal forests of the Pamunkey River, Virginia.

Species segregation with salinity regime in tidal swamps can be even more distinctive. Tree diversity will decrease as salinity creeps from purely freshwater states upward to around 1.2 ppt salinity. At that concentration, species such as water tupelo, green ash (*Fraxinus pennsylvanica* Marsh.), and sweetgum (*Liquidambar styraciflua* L.) are eliminated from some forests (Chapter 9), while swamp tupelo and red maple appear to grow naturally at salinities up to 2.5 ppt. Only baldcypress typically remains as a dominant species at salinities above 1.8 ppt in natural swamp forests (Figure 14.2). Baldcypress, hence, appears not only to have the greatest flood tolerance but also the greatest tolerance to salinity among tidal swamp species in the southeastern United States. On the other hand, some species, such as water tupelo, maintain photosynthetic and growth reductions commensurate with those of baldcypress up to 3 ppt in seedling greenhouse studies (Pezeshki 1987, 1990; Pezeshki et al. 1989). However, these species tend not to be dominants along side baldcypress at elevated salinities in natural field settings.

Tolerance to salinity is most often measured by the growth and physiological performance of seedlings that result from seed or ramet collections from surviving trees. From these collections, some natural variation in the salt tolerance of baldcypress is evident, and the species in general appears to suppress reasonably the impacts of salinity up to approximately 3-4 ppt. This level of potential tolerance has prompted several studies on baldcypress.

Photo credit WH Conner

Photo credit KW Krauss

Photo credit WH Conner

Fig. 14.2. Tidal swamp forests along the lower Savannah River depicting (a) freshwater (0.1 ± 0.0 SE ppt), (b) moderate salinity (0.8 ± 0.1 SE ppt), and (c) high salinity states (3.1 ± 0.3 SE ppt). Tidal range along the lower Savannah River reaches to nearly 2 m and can impose chronic stress to forests downstream near tidal marshes. Note the dominance of baldcypress on the deteriorated site. These three sites span a river distance of approximately 12 km, though the tidal influence along the lower Savannah River can extend 72 km upstream (Way et al. 2003).

14.2.1. Salinity effects

What is most intriguing about baldcypress is that it can survive despite significant reductions in leaf gas exchange, water potential, and growth. Baldcypress is purportedly intolerant to waters harboring 0.89% (8.9 ppt) salt content (Hook 1984) yet can grow reasonably well in soils approaching a 0.38% (3.8 ppt) salt content in some settings (Xu and Long 1983). Even though the natural distribution of baldcypress may be limited to 3-4 ppt or less (Harlow and Harrar 1969; Chabreck 1972; Myers et al. 1995; Chapter 9), some seedlings and saplings appear more tolerant (Table 14.1).

While slight increases in salinity cause leaf water potential in baldcypress to drop as ion-induced uptake mediates changes in turgor, higher salinities cause foliar ions to increase and to disrupt cellular processes di-

Table 14.1. Percent reduction in leaf gas exchange (net photosynthesis, stomatal conductance) and growth (biomass, height) of baldcypress seedlings and saplings at different experimental floodwater salinities and exposure durations.

Salinity (ppt)	Leaf gas exchange	Growth	Reference
Chronic (30-90 d)			
2	32-58%	--	Pezeshki et al. 1986
	--	0%	Conner 1994
	~16-17%	~2%	Allen et al. 1997, 1994b
3	46%	~54%	Pezeshki 1990
4	67-74%	--	Pezeshki et al. 1986
	~13-22%	~20%	Allen et al. 1997, 1994b
	--	40%	Krauss et al. 1999
6	53%	~13%	Javanshir and Ewel 1993
	~43%	~39%	Allen et al. 1997, 1994b
	--	76%	Krauss et al. 1999
6-7	82-84%	--	Pezeshki et al. 1986
7	67-74%	--	Pezeshki et al. 1987
8	~39-52%	~25-38%	Pezeshki et al. 1995
	88%	--	Pezeshki et al. 1988
	~68-77%	~59%	Allen et al. 1997, 1994b
	--	~41%	Javanshir and Ewel 1993
10	--	~60%	Javanshir and Ewel 1993
	--	100%	Conner 1994
Acute (3-5 d)			
4	~12%	--	Krauss et al. 1996
6	~30-60%	--	Krauss et al. 1996
21	~90-100%	--	McLeod et al. 1996
23-30	--	100%	Conner and Askew 1992

rectly (see reviews by Greenway and Munns 1980; Hale and Orcutt 1987). For example, as leaf sodium concentrations increased from 0.11 mol/kg to 0.64 mol/kg in baldcypress, stomatal conductance decreased from 112 to about 15 mmol/m^2/s, and net assimilation decreased from 7.6 to 0.9 µmol/m^2/s (Pezeshki et al. 1988). In spite of reduced leaf gas exchange in old leaves of baldcypress under mild salinity stress, new leaf production may enhance gas exchange at these same salinities (Pezeshki et al. 1986). This demonstrates that baldcypress may be able to adjust osmotically to low salinity concentrations over long time periods (chronic) while flooded.

On the other hand, leaf gas exchange and leaf xylem water potential had different comparative responses in chronic versus acute stress simulations (McLeod et al. 1996). Net assimilation decreased significantly when inundated with chronic levels of 2 and 10 ppt salinity. However, predawn and midday leaf water potentials decreased only for the 10 ppt treatment. Following acute exposure to salinity ranging from 6 ppt (Krauss et al. 1996) to 21-30 ppt (Conner and Askew 1992; McLeod et al. 1996), all of these parameters decreased significantly and rapidly. These results are not uncommon; exposure to salinity of ≥10 ppt imposes either a toxicity or pressure imbalance from the external medium that is prohibitive to survival of baldcypress. At much lower salinities (< 3-6 ppt) baldcypress may only need to overcome disruptions in the osmotic balance. Though not necessarily causative (Termaat et al. 1985; Munns 1993), more negative leaf xylem water potentials and reduced net assimilation are correlated up to at least 6 ppt salinity in baldcypress (Figure 14.3).

14.3. Salt tolerance improvement in baldcypress

Certain genotypes of baldcypress may be more suited to the task of osmotic adjustment. Such phenomena might be best explained by a "biphasic model" proposed for cereal crops (Munns and Termaat 1986), whereby genotypic differentiation arises in the time vacuoles of mesophyll cells from different genotypes take to reach their maximum salinity-imposed ionic concentrations (Munns 1993). Therefore, water deficits dominate at low concentrations of salinity in phase 1 as a pressure imbalance results from the external salt medium, and ions begin to diffuse through the roots to the xylem (Figure 14.4). At higher salt concentrations in phase 2, the effects of accumulated ions start to dominate. The phase transition may be avoided during acute salinity pulses if salinity levels are low (<10 ppt) and exposure is short (1-2 d). If not, genotypic differentiation is likely to show up in phase 2, when genotypes vary in their abilities to maintain growth

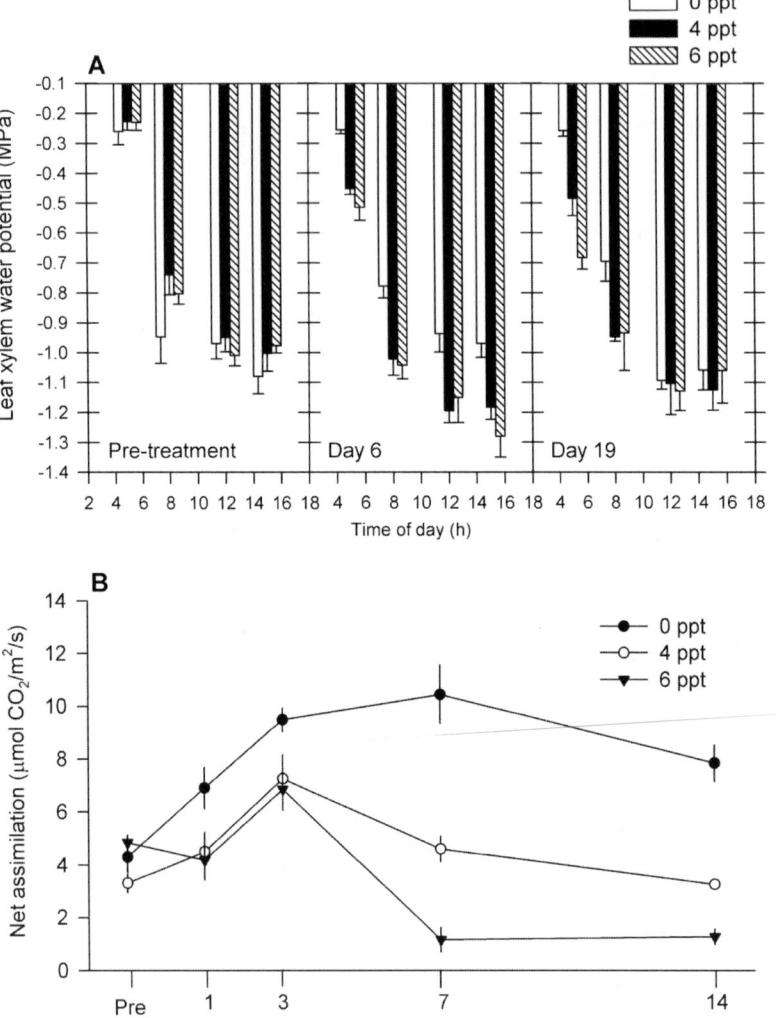

Fig. 14.3. Leaf xylem water potential (±1 SE) and net assimilation (±1 SE) of baldcypress seedlings growing in an environmental chamber at three concentrations of salinity. By day 6, water potential decreased significantly for seedlings treated with 4 and 6 ppt salinity, coincident with a drop in net assimilation for the same treatments by day 7 (after Krauss 1997).

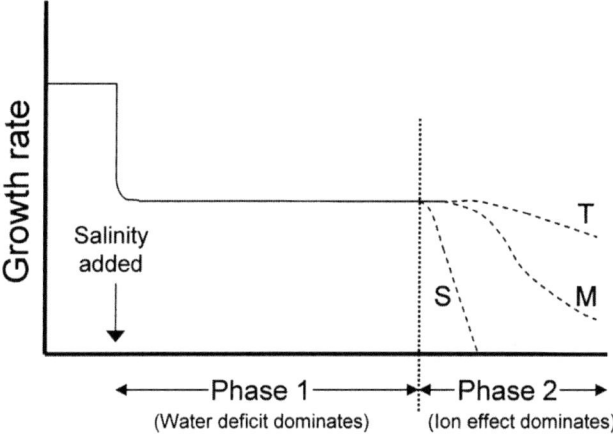

Fig. 14.4. Conceptual diagram of the biphasic stress response model of salt tolerance (after Munns 1993). During phase 1, growth rate decreases because of osmotic stress associated with salinity of the external rooting media. Phase 2 provides genotypic differentiation as ions are taken up faster in salt sensitive genotypes (S), slower for moderately tolerant genotypes (M), and slowest for tolerant genotypes (T). For baldcypress, the transition between phases 1 and 2 is probably around 3-6 ppt for greenhouse studies, but less in natural wetland settings. Note that differentiation in genotypes S, M, and T can also show up in phase 1.

and facilitate leaf gas exchange with slight increases in salinity beyond this threshold. For baldcypress, the transition between phases 1 and 2 is probably around 3-6 ppt salinity, depending upon genotype, under chronic, controlled conditions (Allen et al. 1994b); however, observations suggest a more restricted threshold in natural settings.

14.3.1. Selection of potential salt tolerant genotypes

Allen et al. (1994a) reviewed the prospects for the success of salt tolerance improvement programs in forest tree species. Several concerns were outlined, including contradictory results among past progeny tests and the long-term nature of using trees. More importantly, however, was the suggestion that progeny tests for baldcypress should be conducted under both flooding and salinity (Allen et al. 1996) since this combination defines the condition of prospective restoration sites in coastal areas likely to be subjected to increasing salinity in the southeastern United States.

As such, initial efforts began with mass selections of prospective genotypes from among degraded or dying, salt-impacted forest stands (Figures

14.1 and 14.2). The principle was simple: find trees that were living and in relatively healthy condition amongst dead or dying trees, collect seeds, and conduct experiments to evaluate genotypic variation among progeny. Intensive aerial and ground surveys of isolated stands along the Louisiana, Mississippi, and Alabama coasts gave rise to 15 prospective genotypes from estuarine and a few inland locations for initial studies (Allen et al. 1994b). Subsequent efforts evaluated seven (Krauss 1997; Krauss et al. 1999) and eight additional genotypes (Conner and Inabinette 2005).

14.3.2. Greenhouse progeny tests

For the first series of greenhouse progeny tests by Allen et al. (1994b), seeds were collected from individual parent trees from both brackish and freshwater sources. Hence, all members of a "family" were from open-pollinated cones from a single tree and were of at least half-sibling relation. Baldcypress progeny from 15 parent trees were subjected to 0, 2, 4, 6, and 8 ppt salinity after gradual acclimation to those levels for chronic stress assessment (Allen 1994). Survival, growth, and physiological proficiency were monitored for three months.

Salinity affected survival, with only 73% of seedlings surviving at 8 ppt salinity compared with 100% at 0 ppt. Progeny did not vary much in survival, with brackish water and freshwater sources maintaining similar potential for survival among the full range of salinity tested (Figure 14.5). Progeny did vary widely in leaf area and final biomass among all salinities; differences between brackish water and freshwater sources become accentuated at 6 and 8 ppt salinities (Allen et al. 1994b). For example, overall leaf area of progeny from brackish water sources was twice that of progeny of freshwater sources at 6 ppt salinity (Figure 14.5); progeny from brackish water sources also maintained 17% greater stem biomass and 55% greater root biomass (Allen et al. 1997). Allen et al. (1994b) developed tolerance indices to identify which genotypes performed better when a range of survival and growth criteria were included. Three genotypes (CB3, FA2, FA3), all from brackish water sources, were the best overall performers by retaining greater leaf area and appreciable biomass at 8 ppt salinity (Allen et al. 1994b).

Leaf gas exchange, on the other hand, varied so widely that sources (i.e., brackish water vs. freshwater) did not differ in response, even though individual families did (Allen et al. 1997). Salinity affected all progeny negatively and linearly from 0 to 8 ppt. Predawn leaf water potentials decreased approximately 0.07–0.15 MPa per 2 ppt increase in salinity and did vary appreciably by family at 6 and 8 ppt salinity. These differences in sensitiv-

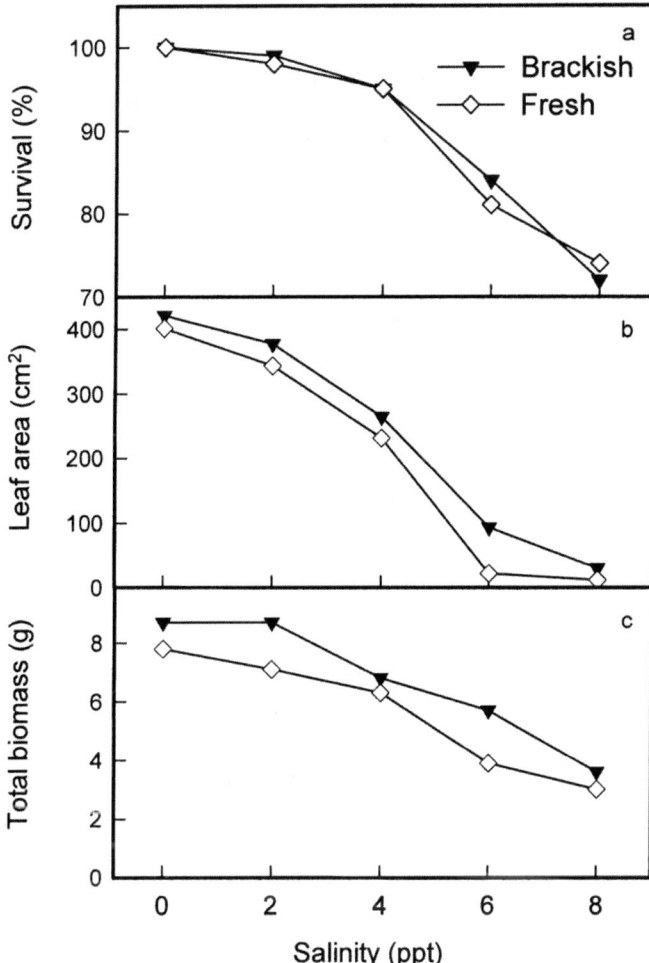

Fig. 14.5. Survival, leaf area, and total biomass for baldcypress seedlings collected from brackish water and freshwater sources and subjected to 0 to 8 ppt salinity in a greenhouse (after Allen et al. 1994b). Differences were not significant for survival; however, progeny from brackish water sources maintained greater leaf area and biomass than did progeny from freshwater sources.

ity correlated well with lower leaf tissue Na^+ and Cl^- concentrations. Exclusion of Na^+ and Cl^- must be important for the maintenance of turgor in more salt-tolerant baldcypress progeny (Allen et al. 1997).

Indeed, the greatest salt tolerance may not always be found in brackish water sources. In fact, what appears as salt tolerance may be an artifact of selection for an unrelated trait. Net assimilation, shoot growth, and root growth, for example, were higher for baldcypress seedlings grown from freshwater source collections than from brackish water sources (Pezeshki et al. 1995). Source assessments were also inconclusive in the field when genotypes were evaluated from among a range of salinities and site conditions common to south Louisiana restoration sites (Krauss et al. 2000). Difficulties in identifying potential sources of greater salt tolerance in baldcypress without extensive testing, hence, pose a slight dilemma. How do we search effectively for salt tolerance among genotypes from such a broad ranging tree species? Improvement efforts for additional tidal swamp tree species will pose similar difficulties. Limiting the search to coastal populations may provide the best technique for quick, initial gains, as has been demonstrated by Allen's research described above, but important genotypes will be overlooked.

Likewise, progeny tests under controlled settings reveal only a fraction of the story. It is difficult to maintain the combination of stressors in greenhouse designs that may interact greatly to influence the overall evaluation of salt tolerance. For baldcypress, the following important factors are often excluded from greenhouse designs: reduced soil oxidation-reduction potential, build-up of sulfides (Koch and Mendelssohn 1989), low soil fertility in some settings (Day et al. 2006), and competing vegetation (Myers et al. 1995). Perhaps these exclusions explain why so many forest tree improvement operations are designed as field trials from the start.

14.3.3. Field progeny tests

The potential for salt tolerance improvement was identified with greenhouse studies; however, salt tolerance improvement programs are predicated upon successful plantings of improved material on salt impacted field sites. To our knowledge, there have been only two field progeny tests for salt tolerance in baldcypress. One baldcypress progeny test used plant material collected from some of the same parent trees identified by Allen et al. (1994b) and was established on three salt-impacted field sites in coastal Louisiana (Krauss et al. 2000). The other progeny test was established on two abandoned, saline rice fields in South Carolina and included seed collections from eight states ranging from Virginia to Louisiana (Conner and Inabinette 2005). We should also mention a third progeny test that was established in South Carolina on sites impacted by a hurricane storm surge (Conner and Ozalp 2002); however, salinity was not detected on the site during the study and is thus excluded here as a test for salt tolerance. There have also been a number of trials using *Taxodium* hybrids in China, which we will briefly mention.

14.3.3.1. Louisiana plantations

Progeny from ten parent trees were collected from natural field sites in Louisiana and Alabama, grown for one full season in a greenhouse to a mean height of 87 cm, and planted on one of three field sites just southeast of New Orleans, Louisiana. Each site contained 40 individuals of each of ten genotypes arranged as 4-tree family rows in ten experimental blocks. Trees were planted on a 1-m grid in a study planned for two year duration (Krauss et al. 2000).

Salinity on the field sites were 0.1-0.8 ppt (low), 1.0-1.4 ppt (medium), and 1.5-3.0 ppt (high), depending upon location. The high salinity site experienced periodic (acute) salinity pulses to 15 ppt for short durations; however, salinity for the other two sites was relatively stable. Herbaceous vegetation differed widely among the three sites owing to their different salinity regimes.

Individual seedling survival, height, diameter, and leaf biomass differed significantly by site. Height increment, for example, averaged 109 cm after two years for the lowest salinity site and 33 cm for surviving seedlings on the highest salinity site. Foliar ion concentrations of Cl^- and Na^+ also increased with site salinity as baldcypress saplings became less adept at excluding both ions. Na/K ratios increased beyond what has been suggested as tolerable for nonhalophytes (> 1.0: Wyn Jones et al. 1979) on the high salinity site; mean Na/K ratios for the low, medium, and high salinity sites were 0.23, 0.54, and 1.14, respectively. The impact of salts in the soil, hence, was sufficiently high to warrant family-level differentiation.

Even though among-family survival ranged from 7% to 100% among different sites (Figure 14.6), height of surviving trees was the only significant intraspecific measured growth variable. In accordance with greenhouse studies, foliar ion composition differed among genotype and indicated that more sensitive genotypes may take up more Cl^- and cause a greater Na/K imbalance in the leaves. Moreover, foliar Na^+ content differed by family only on the high salinity site, suggesting that Na^+ has a role in conferring salt tolerance to seedlings under field settings. Of the genotypes identified by Allen et al. (1994b) as being more tolerant (CB3, FA2), neither ranked among the top three in height increment, diameter increment, or leaf biomass in the field. However, FA2 had the third lowest Na/K ratio, and both CB3 and FA2 tended to exclude more Na^+ from foliage. Survival of these genotypes was among the highest reported from the ten families evaluated across all sites, suggesting that slower growth and

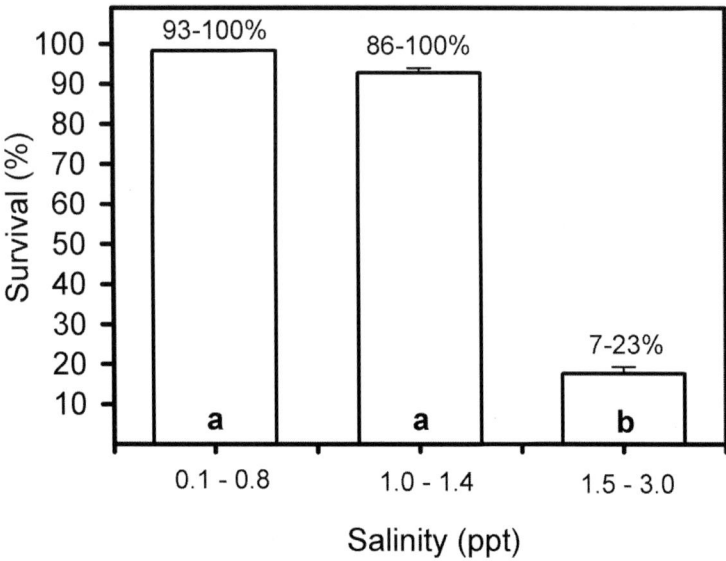

Fig. 14.6. Two-year survival of baldcypress progeny planted on three coastal field sites southeast of New Orleans, Louisiana ranging in salinity from 0.1 to 3.0 ppt. The highest salinity site experienced short-term pulses of up to 15 ppt. Ranges above the bars indicate survival percentages among progeny of 10 half-sibling families for a particular site.

ion exclusion may be part of a tolerance strategy for these genotypes (cf., Chapin 1991).

14.3.3.2. South Carolina plantations

Progeny from eight parent trees were collected from natural field sites in Virginia, North Carolina, South Carolina, Georgia, Florida, Alabama, Mississippi, and Louisiana; grown for two full seasons in nursery beds; and planted on one of two field sites at Hobcaw Barony roughly 7 km east of Georgetown, South Carolina (Conner and Inabinette 2005). Seedlings were root pruned at 23 cm and planted on two abandoned rice field sites as a multi-year evaluation.

Salinity on the sites ranged from 1-2 ppt just after planting to 7.5-12.2 ppt in late summer. This cycle of winter reductions to 1-3 ppt and late summer peaks continued throughout the five-year evaluation, reaching a peak salinity level of 18.5 ppt in November of 2001. Herbaceous vegetation was similar on both sites.

Survival differed significantly between sites. On one site, mortality was 100% shortly after planting while on the second site, mortality was low initially but reached 80% by the end of the first year. Genotypes collected from Louisiana, Alabama, and Florida had 73%, 22%, and 15% survival, respectively and were the only genotypes to survive to the end of the study (Figure 14.7). The genotype with the greatest survival and overall performance was collected from sites in Louisiana where recent salinization (since the early 1960s) caused large die-offs of coastal baldcypress forests. In this case, selection intensity associated with a large salinity treatment (3.5-12 ppt) helped to identify a good genotype amongst many dead or dying conspecifics.

14.3.3.3. Hybrid plantations

Since the 1960s, some Chinese scientists have been convinced that *Taxodium* hybridization promises to combine the best characteristics of superior parents for optimal plantation success on salt-affected lands in China (Wang 1995; Huang et al. 2000; Zhou et al. 2000; Chen et al. 2002; Yin et al. 2002; Lu et al. 2004; Chen Yong Hui, personal communication).

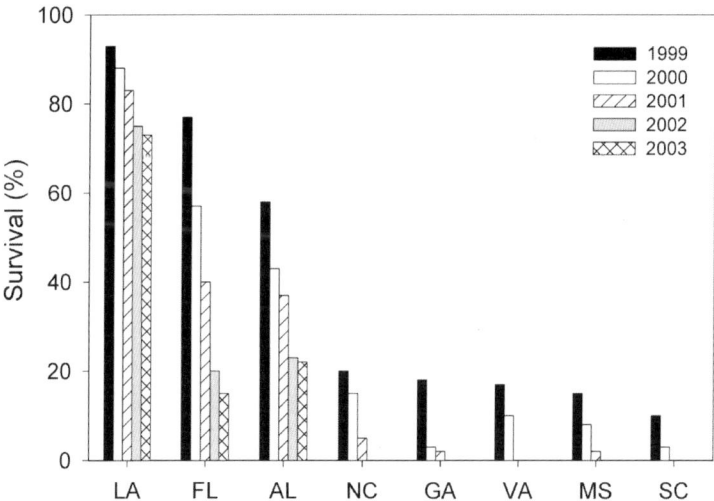

Fig. 14.7. Five-year survival of baldcypress progeny planted on two coastal field sites near Georgetown, South Carolina ranging to 18.5 ppt salinity for very short durations (after Conner and Inabinette 2005). State descriptors depict the location of parent trees from which seeds were collected. LA = Louisiana, FL = Florida, AL = Alabama, NC = North Carolina, GA = Georgia, VA = Virginia, MS = Mississippi, SC = South Carolina.

Three clones were isolated in China primarily for impressive growth and tolerance to alkaline and salt-rich coastal floodplains: a baldcypress × Montezuma baldcypress (*T. mucronatum* Ten.) hybrid and two pondcypress (*T. ascendens* Brongn.) × Montezuma baldcypress hybrids. After extensive trials, one (named "Nanjing Beauty": Creech and Yin 2003) has been recommended for soils with pH 8.0–8.5 and salt concentrations approaching 2 ppt. Other attributes of this genotype included 159% faster growth than baldcypress, good columnar form, longer foliage retention in fall and early winter, and no knees. This clone has been propagated in the United States since January 2002 and is currently under evaluation in over 30 locations in the South. In March 2005, the Stephen F. Austin State University (SFASU) Mast Arboretum received two new clones from China; both demonstrated strong salt tolerance in terms of improved growth and survival on field sites. These clones were selected from a field population of Nanjing Beauty × Montezuma baldcypress, but with improvements in form and vigor under salt stress.

Chinese scientists and nurserymen have integrated commercial production and research to scales not commonly seen in the west. In a nursery near Suzhou, D. Creech noted over 2.5 million cuttings of Nanjing Beauty, which are now in the rooting process (personal observation). In 2005, the central government of China committed two billion dollars ($USD) to initiate the planting of the coastal windbreak forest, an afforestation project on the mainland side of the coastal dike that runs along the sea, both north and south of the mouth of the Yangtze River. This is a region of China susceptible to annual and often devastating typhoons. This amazing project will ultimately involve the planting of billions of trees by the year 2038, and *Taxodium* hybrids, with their improved salt and alkalinity tolerance, are a significant part of the total mix of two hundred species planned for the project (D. Creech, personal observation).

14.4. Future prospects for salt tolerance improvement in tidal forest species

For baldcypress, genotypic variation in response to salinity discovered in field trials paralleled results from greenhouse progeny tests in some cases and suggests that screening progeny prior to outplanting can provide some assistance in conferring improved survival and growth to planting stock. Similar programs may also be feasible for other tidal freshwater swamp species. Several studies have indicated some variation in the leaf gas exchange and growth response of water tupelo, swamp tupelo, green ash, buttonbush (*Cephalanthus occidentalis* L.), eastern red-cedar (*Juniperus virginiana* L.), common honeylocust (*Gleditsia triacanthos* L.), and cabbage palm (*Sabal palmetto* (Walt.) Lodd. ex J.A. & J.H. Schultes) to a range of salinity levels (Table 14.2). Of these, water tupelo and green ash may have the greatest potential for inclusion in future salt tolerance improvement programs. On the other hand, many of these species are also very susceptible to both chronic and acute salinity exposure, indicating that field survival to the levels reported for baldcypress may not be easily attained. Nuttall oak (*Quercus texana* Buckl.), overcup oak (*Q. lyrata* Walt.), swamp chestnut oak (*Q. michauxii* Nutt.), and water oak (*Q. nigra* L.) are highly susceptible to salinity and have no apparent variation in their response to even low levels of salinity (Table 14.2; Conner et al. 1998).

Even for baldcypress, however, improvement programs have not advanced salt tolerance of the species appreciably. There have been few concerted efforts to improve the salt tolerance of selected material beyond initial half-sibling family collections. Within-family selection of clones from surviving genotypes outplanted to high salinity field sites is the next logical step. Clonal propagation of baldcypress can be high, i.e., 60% to 88% success by using young plant material (Copes and Randall 1993; Pezeshki and DeLaune 1994), so there are few operational barriers to advancing the salt tolerance of baldcypress in this way as long as sufficient natural genetic variation exists. For example, in one effort, cuttings from progeny of baldcypress from saline plantations (Krauss et al. 2000) were extracted from surviving trees after 10 years. Fifteen of the best formed trees were selected, and scion material was grafted onto nursery root stock in freshwater locations. The intent is that progeny from these clones will be outplanted to saline field sites after a series of controlled pollinations (R.A. Goyer, pers. comm.); the best progeny from those trials will again be selected.

It may also be important to screen from a wider range of genotypes before initiating large scale improvement efforts for any tidal swamp species in order to include as much genetic variation as possible. To date, fewer the 30 genotypes have been evaluated for baldcypress. Seed germination trials have been suggested as an acceptable surrogate for more intensive greenhouse-based progeny tests (Krauss et al. 1998). Germination trials can mimic greenhouse progeny tests in identifying greater salt tolerance among genotypes, can include hundreds of genotypes simultaneously, and are fairly inexpensive to conduct. On salt impacted field sites where hundreds of trees are surviving, screening prospective parent trees with foliar ion analysis may also be beneficial. Any protocol that hastens the identifi-

Table 14.2. Percent reduction in leaf gas exchange (net photosynthesis, stomatal conductance) and growth for seedlings and saplings of tidal swamp forest species of the southeastern United States at different experimental floodwater salinities and exposure durations.

Species/ Salinity (ppt)	Reduction in Leaf gas exchange	Reduction in Growth	Reference
Chronic (30-90 d)			
Water tupelo			
2	~0-5%	--	McLeod et al. 1996
3	24%	~53%	Pezeshki 1990
3	15-21%	~68%	Pezeshki et al. 1989
3	11-21%	--	Pezeshki 1987
10	~90-100%	--	McLeod et al. 1996
Swamp tupelo			
2	100%	100%	McCarron et al. 1998
10	100%	100%	McCarron et al. 1998
Green ash			
2	~27-50%	--	McLeod et al. 1996
4	--	0%	Monk and Peterson 1962
6	--	0%	Monk and Peterson 1962
8	--	0%	Monk and Peterson 1962
Nuttall, overcup, water, & swamp chestnut oaks			
2	--	100%	Conner et al. 1998
6	--	100%	Conner et al. 1998
Buttonbush			
2	~52-65%	~0-29%	McCarron et al. 1998
10	100%	100%	McCarron et al. 1998
Eastern red-cedar			
4	37%	--	Monk and Peterson 1962
6	35%	--	Monk and Peterson 1962
Common honeylocust			
4	--	0%	Monk and Peterson 1962
6	--	0%	Monk and Peterson 1962
8	--	15%	Monk and Peterson 1962
10	--	10%	Monk and Peterson 1962
Cabbage palm			
8	~34%	--	Perry and Williams 1996
15	~75%	--	Perry and Williams 1996
22	~76%	--	Perry and Williams 1996
Acute (3-5 d)			
Water tupelo			
21	~90-100%	--	McLeod et al. 1996
Swamp tupelo			
21	~43-50%	--	McCarron et al. 1998
Green ash			
21	~78-85%	--	McCarron et al. 1998
Buttonbush			
23	--	~56-75%	McCarron et al. 1998

cation of truly unique genotypes from the field will assist improvement programs for any tree species.

Salt tolerance improvement programs may be most effective at ameliorating against environmental extremes. In intertidal areas near the boundary between tidal freshwater swamp and marsh, periodic salt pulses cause incomplete mortality among trees (Figure 14.2). Extreme salinity pulses, as sea level rises or rivers are dredged, can kill remaining trees. Improved plant material can be used along this boundary. Often, this boundary shifts with anthropogenic activity. In South Carolina, activities include extensive rice culture and river dredging, while in Louisiana, activities include flood control programs and oil and gas mining operations (Chapter 1). There are even areas along the coast of the southeastern United States that are currently not regenerable either by natural or artificial means (Chapter 16). Salt tolerance improvement programs should focus on improving the success of plantings by providing the best material available to tolerate more extreme salinity conditions, especially as large-scale engineering approaches are implemented for some parts of the Gulf of Mexico Coast. Overall gains in salt tolerance improvement through conventional means for any tidal swamp species, however, are likely to be modest until newer, biotechnological techniques are included (e.g., identifying molecular markers, inserting gene sequences for salt tolerance) or until hybridization techniques and hybrid plantations gain wider acceptance. An appropriate role for the use of clones during ecological restoration activities will also need to be determined.

14.5. Acknowledgments

We would like to thank James A. Allen, Michael Stine, William H. Conner, Richard A. Goyer, Carroll L. Cordes, and Jarita Davis for providing text, peer, and editorial comments on this chapter. Support for this review was provided by the U.S. Geological Survey Global Climate Change Research program.

References

Allen JA (1994) Intraspecific variation in the response of baldcypress (*Taxodium distichum*) seedlings to salinity. Ph.D. thesis, Louisiana State University

Allen JA, Chambers JL, Stine M (1994a) Prospects for increasing the salt tolerance of forest trees: a review. Tree Physiol 14:843–853

Allen JA, Chambers JL, McKinney D (1994b) Intraspecific variation in the response of *Taxodium distichum* seedlings to salinity. For Ecol Manage 70:203–214

Allen JA, Pezeshki SR, Chambers JL (1996) Interaction of flooding and salinity stress on baldcypress (*Taxodium distichum*). Tree Physiol 16:307–313

Allen JA, Chambers JL, Pezeshki SR (1997) Effects of salinity on baldcypress seedlings: physiological responses and their relation to salinity tolerance. Wetlands 17:310–320

Auchmoody LR, Walters RS (1988) Revegetation of a brine-killed forest site. Soil Sci Soc Am J 52:277–280

Bartram W (1928) Travels of William Bartram. In: van Doren M (ed.) Reprint edition, Dover Publications, New York

Beckmann R (1991) Trees for salty land. ECOS 67:20–23

Bell DT (1999) Australian trees for the rehabilitation of waterlogged and salinity-damaged landscapes. Aust J Bot 47:697–716

Chabreck RH (1972) Vegetation, water and soil characteristics of the Louisiana coastal region. Agricultural Experimental Station Technical Bulletin No. 664. Louisiana State University, Baton Rouge

Chapin FS III (1991) Integrated response of plants to stress. BioScience 41:29–36

Cheeseman JM (1988) Mechanisms of salinity tolerance in plants. Plant Physiol 87:547–550

Chen Y, Pan S, Zhang J, Zhu W, Niu H, Qu Z, Wang J, Shen D, Ye Z (2002) RAPD analysis of genetic relationships among natural populations of hybrid *Taxodium mucronatum* Tenore. J Fudan Univ (Nat Sci) 41:641–645

Conner WH (1994) The effect of salinity and waterlogging on growth and survival of baldcypress and Chinese tallow seedlings. J Coast Res 10:1045–1049

Conner WH, Askew GR (1992) Response of baldcypress and loblolly pine seedlings to short-term saltwater flooding. Wetlands 12:230–233

Conner WH, Inabinette LW (2005) Identification of salt tolerant baldcypress (*Taxodium distichum* (L.) Rich) for planting in coastal areas. New For 29:305–312

Conner WH, Ozalp M (2002) Baldcypress restoration in a saltwater damaged area of South Carolina. In: Proceedings of the 11th Biennial Southern Silvicultural Research Conference. SRS-48. U.S. Department of Agriculture, Forest Service, Asheville, pp 357–361

Conner WH, McLeod KW, McCarron JK (1998) Survival and growth of seedlings of four bottomland oak species in response to increases in flooding and salinity. For Sci 44:618–624

Copes DL, Randall WK (1993) Rooting of baldcypress stem cuttings. Tree Planters Notes 44:125–127

Creech D, Yin Y (2003) *Taxodium* × 'Nanjing Beauty': a new landscape plant for the south. HortScience 38:1292–1293

Day RH, Doyle TW, Draugelis-Dale RO (2006) Interactive effects of substrate, hydroperiod, and nutrients on seedling growth of *Salix nigra* and *Taxodium distichum*. Environ Exp Bot 55:163–174

Epstein E, Norlyn JD, Rush DW, Kingsbury RW, Kelley DB, Cunningham GA, Wrona AF (1980) Saline culture of crops: a genetic approach. Science 210:399–404

Flowers TJ (2003) Improving crop salt tolerance. J Exp Bot 55:307–319

Gogate MG, Mittal RC, Pyarelal (1984) Screening through germination trials – Australian species for saline areas. Indian For 110:982–988

Greenway H, Munns R (1980) Mechanisms of salt tolerance in nonhalophytes. Ann Rev Plant Physiol 31:149–190

Hale MG, Orcutt DM (1987) The physiology of plants under stress. John Wiley & Sons, New York

Harlow WM, Harrar ES (1969) Textbook of dendrology, 5th edn. McGraw Hill Book Company, New York

Hook DD (1984) Waterlogging tolerance of lowland tree species of the south. Southern J Appl For 8:136–149

Hook DD, Buford MA, Williams TM (1991) Impact of Hurricane Hugo on the South Carolina coastal plain forest. J Coast Res SI-8:291–300

Huang L, Wang Q, Li X, Wang W, Chen Z, Jiang Z (2000) Studies on the variation of provenances and families in Genus *Taxodium*: variation on shoot cutting rooting ability. J Jiangsu Forestry Sci Tech 27:1–6

IPCC (2001) Changes in sea level. In: Houghton JT, Ding Y, Griggs DJ, Noguer M, van der Linden PJ, Dai X, Maskell K, Johnson CA (eds) Climate change 2001: The scientific basis. Cambridge University Press, UK, pp 639–693

Javanshir K, Ewel KC (1993) Salt resistance of bald cypress. In: Leith H, Al Masoom A (eds) Towards the rational use of high tolerant plants, vol 2. Kluwer Academic Publishers, The Netherlands, pp 285–291

Koch MS, Mendelssohn IA (1989) Sulphide as a soil phytotoxin: differential response in two marsh plant species. J Ecol 77:565–578

Krauss KW (1997) Intraspecific variation in baldcypress (*Taxodium distichum* (L.) Rich): Responses to salinity and potential for restoration of wetlands impacted by saltwater intrusion. M.S. thesis, Louisiana State University

Krauss KW, Chambers JL, Allen JA (1996) Intraspecific variation in the physiological response of baldcypress (*Taxodium distichum* (L.) Rich.) to a rapid influx of saltwater. In: Flynn KM (ed) Proceedings of the southern forested wetlands ecology and management conference. Consortium for Research on Southern Forested Wetlands, Clemson, pp 183–189

Krauss KW, Chambers JL, Allen JA (1998) Salinity effects and differential germination of several half-sib families of baldcypress from different seed sources. New For 15:53–68

Krauss KW, Chambers JL, Allen JA, Luse BP, DeBosier AS (1999) Root and shoot responses of *Taxodium distichum* seedlings subjected to saline flooding. Environ Exp Bot 41:15–23

Krauss KW, Chambers JL, Allen JA, Soileau DM Jr, DeBosier AS (2000) Growth and nutrition of baldcypress families planted under varying salinity regimes in Louisiana, USA. J Coast Res 16:153–163

Lu X, Mao Z, Chen Y, Huang L (2004) Propagation technology of *Taxodium* 'zhongshansha'. J Jiangsu Forestry Sci Tech 31:38–42

Marcar NE, Crawford DF (2004) Trees for saline landscapes. CSIRO Publishing, Australia.

Mathur NK, Sharma AK (1984) *Eucalyptus* in reclamation of saline and alkaline soils in India. Indian For 110:9–15

Mattoon WR (1915) The southern cypress. Bulletin 272. U.S. Department of Agriculture, Forest Service, Washington

McCarron JK, McLeod KW, Conner WH (1998) Flood and salinity stress of wetland woody species, buttonbush (*Cephalanthus occidentalis*) and swamp tupelo (*Nyssa sylvatica* var. *biflora*). Wetlands 18:165–175

McLeod KW, McCarron JK, Conner WH (1996) Effects of flooding and salinity on photosynthesis and water relations of four Southeastern Coastal Plain forest species. Wetl Ecol Manage 4:31–42

McWilliams WH, Rosson Jr JF (1990) Composition and vulnerability of bottomland hardwood forests of the coastal plain province in the south central United States. For Ecol Manage 33/34:485–501

Monk R, Peterson HB (1962) Tolerance of some trees and shrubs to saline conditions. Proc Am Soc Hort Sci 81:556–561

Munns R (1993) Physiological processes limiting plant growth in saline soils: some dogmas and hypotheses. Plant Cell Environ 16:15–24

Munns R, Termaat A (1986) Whole-plant responses to salinity. Aust J Plant Physiol 13:143–160

Myers RS, Shaffer GP, Llewellyn DW (1995) Baldcypress (*Taxodium distichum* (L.) Rich) restoration in southeast Louisiana: the relative effects of herbivory, flooding, competition, and macronutrients. Wetlands 15:141–148

Penfound WT (1952) Southern swamps and marshes. Bot Rev 18:413–445

Penfound WT, Hathaway ES (1938) Plant communities in the marshlands of southeastern Louisiana. Ecol Monogr 8:1–56

Perry L, Williams K (1996) Effects of salinity and flooding on seedlings of cabbage palm (*Sabal palmetto*). Oecologia 105:428–434

Pezeshki SR (1987) Gas exchange response of tupelo-gum (*Nyssa aquatica* L.) to flooding and salinity. Photosynthetica 21:489–493

Pezeshki SR (1990) A comparative study of the response of *Taxodium distichum* and *Nyssa aquatica* seedlings to soil anaerobiosis and salinity. For Ecol Manage 33/34:531–541

Pezeshki SR, DeLaune RD (1994) Rooting of baldcypress cuttings. New For 8:381–386

Pezeshki SR, DeLaune RD, Patrick Jr WH (1986) Gas exchange characteristics of bald cypress (*Taxodium distichum* L.): evaluation of responses to leaf aging, flooding, and salinity. Can J For Res 16:1394–1397

Pezeshki SR, DeLaune RD, Patrick Jr WH (1987) Response of baldcypress (*Taxodium distichum* L. var. *distichum*) to increases in flooding salinity in Louisiana's Mississippi River Deltaic Plain. Wetlands 7:1–10

Pezeshki SR, DeLaune RD, Patrick Jr WH (1988) Effect of salinity on leaf ionic content and photosynthesis of *Taxodium distichum* L. Am Midl Nat 119:185–192

Pezeshki SR, Patrick Jr WH, DeLaune RD, Moser ED (1989) Effects of waterlogging and salinity interactions on *Nyssa aquatica* seedlings. For Ecol Manage 27:41–51

Pezeshki SR, DeLaune RD, Patrick Jr WH (1990) Flooding and saltwater intrusion: potential effects on survival and productivity of wetland forests along the U.S. Gulf Coast. For Ecol Manage 33/34:287–301

Pezeshki SR, DeLaune RD, Choi HS (1995) Gas exchange and growth of bald cypress seedlings from selected U.S. Gulf Coast populations: responses to elevated salinities. Can J For Res 25:1409–1415

Rheinhardt RD, Hershner C (1992) The relationship of below-ground hydrology to canopy composition if five tidal freshwater swamps. Wetlands 12:208–216

Salinas LM, DeLaune RD, Patrick Jr WH (1986) Changes occurring along a rapidly subsiding coastal area: Louisiana, USA. J Coast Res 2:269–284

Shrivastava MB, Tewari KN, Shrivastava M (1988) Afforestation on salt affected soils in India. Indian J For 11:1–12

Termaat A, Passioura JB, Munns R (1985) Shoot turgor does not limit shoot growth of NaCl-affected wheat and barley. Plant Physiol 77:869–872

Thomson LAJ (1988) Salt tolerance in *Eucalyptus camaldulensis* and related species. Ph.D. thesis, University of Melbourne

Townsend AM (1989) The search for salt tolerant trees. Arboricultural J 13:67–73

Wang Q (1995) Studies on the variation of provenances and families in Genus *Taxodium*: introduction to the Genus. J Jiangsu Forestry Sci Tech 22:15–20

Way F, Goodrich M, Liu H, Mendelsohn DL (2003) 3-D hydrodynamic, salinity, dissolved oxygen and sediment transport modeling of the Savannah River Estuary. In: Coastal Zone 03, proceedings of the 13th biennial coastal zone conference. NOAA/OSC/20322-CD. NOAA Coastal Services Center, Charleston

Wicker KM, Castille GC, Davis DJ, Gagliano SM, Roberts DW, Sabins DS, Weinstein RA (1982) St Bernard Parish: A study in wetland management. Coastal Environments, Inc., Baton Rouge

Wyn Jones RG, Brady CJ, Speirs J (1979) Ionic and osmotic regulation in plant cells. In: Laidman DL, Wyn Jones RG (eds) Recent advances in biochemistry of cereals. Academic Press, London, pp 63–103

Xu YQ, Long WB (1983) The adaptive character and species choice of main planting trees of farmland shelterbelt in the Pearl River Delta. Scientia Silvae Sinicae 19:225–234

Yin X, Yin Y, Chen Y (2002) Isozyme analysis of *Taxodium* 'Zhongshansha 302', *T. mucronatum* Tenore and their hybrids. J Plant Resour Environ 11:59–61

Zhou K, Jia C, Chen Y, Yin Y, Sun Z (2000) Analysis of growth of *Taxodium distichum* and *T*. 'zhongshansa 302' on the alkaline lowland. J Jiangsu Forestry Sci Tech 27:16–18

Zobel B, Talbert J (1984) Applied forest tree improvement. Waveland Press, Prospect Heights

Chapter 15 - Assessing the Impact of Tidal Flooding and Salinity on Long-term Growth of Baldcypress Under Changing Climate and Riverflow*

Thomas W. Doyle[1], William H. Conner[2], Marceau Ratard[3], and L. Wayne Inabinette[2]

[1]*U.S. Geological Survey, National Wetlands Research Center, 700 Cajundome Blvd., Lafayette, LA 70506*
[2]*Baruch Institute of Coastal Ecology and Forest Science, Clemson University, Box 596, Georgetown, SC 29442*
[3]*Delgado Community College, 615 City Park Ave., New Orleans, LA 70119*

15.1. Introduction

Successful restoration of coastal wetlands depends mostly on the proper understanding of the physical environment and the ecological requirements and tolerances of eligible and expected biota, *in situ* or introduced. Tidal freshwater forested wetlands present a challenging coastal habitat that undergoes daily, seasonal, decadal, and even millennial cycles of environmental change that dictate the distribution and migration of this forest type over time and space. The potential extent of tidal freshwater forested wetlands within any given estuary is related to river gradient and flow coupled with tidal range and forcing that increase river stage and surface water salinities many km upstream. The response of individual trees to changing hydroperiod and salinity affected by tides is related to their species tolerance and to the degree of acute or chronic exposure. Tidal freshwater forested wetlands are composed of tree species that are nonhalophytic, unlike

---

* The U.S. Government's right to retain a non-exclusive, royalty-free licence in and to any copyright is acknowledged.

mangroves, but can survive low salt concentrations for long periods or modest salt levels for short durations to varying degrees. Improving our understanding of physical processes of these coastal systems and biological attributes of associated species will improve our ability to plan and predict the outcome of protected and restored wetlands along our coasts.

The cumulative impact of a subsiding coast, rising sea level, recurring hurricanes, and reduced freshwater flow will result in forest dieback and coastal retreat as the intertidal zone migrates upslope. The ability to predict when and where changing climate or freshwater flow will negatively or positively impact the structure, function, and distribution of tidal freshwater systems in the future may be aided most by reviewing the past. Tree-ring analysis provides a unique tool and metric for investigating patterns of tree and forest productivity in relation to historic climate and riverflow. The long-lived attribute of tree species has spurred many new discoveries of factors that influence tree growth relations. Tree-ring chronologies from coastal settings are few but have provided new insights on drought frequencies (Stahle et al. 1988, 1998), saltwater intrusion (Yanosky et al. 1995), and hurricane history and impact (Doyle and Gorham 1996). Identifying how tree species respond to elevated tidal flooding and salinity exposure during low flow years and droughts or from hurricane surge will provide the empirical basis for developing predictive models under climate change for tidal freshwater forests.

The most ubiquitous tree species in tidal freshwater swamps across the Southeast is baldcypress (*Taxodium distichum* [L.] L.C. Rich.). Baldcypress can be found in association with a host of hardwood species throughout tidal swamps and depressional wetlands of all kinds within the southeastern Coastal Plain. It is well known for its flooding tolerance and ability to withstand deepwater environments, and it has a reputation for having a life span that can exceed 1000 years. Tree cores and growth analysis of baldcypress in the Southeast have yielded millennial-aged climate reconstructions (Stahle et al. 1985b, 1988) and have been applied to explain the plausible demise of colonial settlements in Virginia and South Carolina (Stahle et al. 1988). Few tree-ring studies have been conducted specifically in tidal swamps at the estuarine interface (Yanosky et al. 1995) to relate salinity effects on tree growth for gauging the vulnerability and sensitivity of these systems to changing climate and riverflow.

In this chapter we investigate the effects of tidal flooding, river source, and salinity on the long-term growth response of baldcypress trees and forests in the tidal freshwater zone of the Pee Dee and Waccamaw Rivers near Georgetown, South Carolina. Paired sites of tidal freshwater forest dominated by baldcypress were established on the western and eastern shore of Sandy Island which is regularly flooded by the Pee Dee and Waccamaw Rivers, respectively. Additional paired sites below Sandy Island along the Waccamaw River are closer to the estuary and are thought to be more prone to salinity incursions.

15.1.1. Factors influencing growth variation in baldcypress in tidal wetlands

A complexity of factors may influence growth variation in baldcypress in a tidally driven system that may act to compound or impede growth response. First, tidal systems would be expected to be in flood conditions more often and to a greater degree than would nontidal forests of comparable elevation which share the same river water but are beyond the influence of tide. In this case, increasing hydroperiod or waterlogging is generally considered a stress on plants caused by prevailing anaerobic conditions yielding reduced growth potential (Harms et al. 1980; Pezeshki et al. 1987; Pezeshki and Santos 1998; Young et al. 1995). While tidal systems are in flood during high tide, they also drain during low tide cycles, allowing for some aeration and water exchange. Moreover, baldcypress also has been shown to grow more favorably in flooded situations when the water source is flowing rather than stagnant (Brown 1981; Lugo and Brown 1984), or periodically flooded versus continuously flooded (Conner and Flynn 1989; Dicke and Toliver 1990; Megonigal and Day 1992). Tidal forests flooded by redwater rivers that carry higher nutrient and sediment loads should experience greater growth potential than blackwater and spring-fed rivers that are typically nutrient poor (Brown 1981; Mitsch and Ewel 1979). Baldcypress forests receiving nutrient-enriched wastewater and seedling studies in fertilized hydroponic experiments show greater growth potential with increased nutrient availability (Mitsch and Ewel 1979; Hesse et al. 1998; Souther and Shaffer 2000; Day et al. 2006).

The major constituent of concern in tidal wetlands is the impact that salinity pulsing and concentrations have on tree growth and survival. It is expected that forests more proximal to the estuary interface endure the higher stage fluctuations and salinity levels. By definition, freshwater tree species, like baldcypress, are not adapted to survive in saltwater environments, but can tolerate low chronic exposure and acute episodes of modest concentrations if subsequently refreshed by rain or riverflow (Conner 1994; Allen et al. 1996; Krauss et al. 1998, 2000; Williams et al. 1999; Day et al. 2006). Any extended periods of chronic or elevated exposure to saltwater intrusion without freshwater recharge will ultimately lead to tree death and forest dieback. In all cases, the presence of saltwater has been shown to compromise tree health and growth potential such that exposed sites and forests near the estuary interface are expected to grow more slowly than are tidal forests without salinity pulsing.

The timing of flood or drought events may be an important determinant that can explain the growth variation between years or sites within a floodplain or tidal reach. The dynamics of tide and riverflow are such that they vary daily, monthly, seasonally, and annually as a function of climate variability. While tides are fairly predictable with respect to astronomical forcing, riverflow is much more variable and depends on the specific drainage basin and rainfall tendencies. During extended droughts, riverflows can remain near base flow, allowing greater tidal pulsing and salinity intrusions further upstream than normal. During extended wet seasons and with active tropical storm seasons, riverflows can remain high most of the growing season and abate to some degree salinity incursions and even refresh zones that salinize during drought periods.

15.1.2. Tree-ring studies with baldcypress

Baldcypress has many features which make it suitable for tree-ring analysis (Stahle et al. 1985b; Bowers et al. 1990). Baldcypress is a very long-lived species; ages as great as 1700 years old have been verified for some trees (Stahle et al. 1988). These trees are also sensitive to climatic changes, and this sensitivity is expressed in their ring widths (Stahle et al. 1985b). In addition, baldcypress lends itself very well to cross-dating both locally and regionally (Stahle et al. 1985b). Another feature especially useful for long-term studies is that baldcypress timber has been used for a long time, and sub-fossil logs are often found in very good condition because of the timber's inherent decay resistance (Stahle et al. 1985b). Because of these properties, baldcypress has been used for a number of tree-ring studies to examine drought severity going back hundreds of years (Stahle et al. 1985a; Stahle et al. 1988, 1998), hurricane signals (Reams and Van Deusen 1998), heavy metals in the environment (Latimer et al. 1996), wastewater amendment (Hesse et al. 1998), and even day length (Stahle et al. 1991).

15.2. River and estuary characteristics

This study was conducted in tidal freshwater forest around Sandy Island, South Carolina, upriver from the coast and harbor town of Georgetown (Figure 15.1). Sandy Island is a large (3,700 ha) island that includes

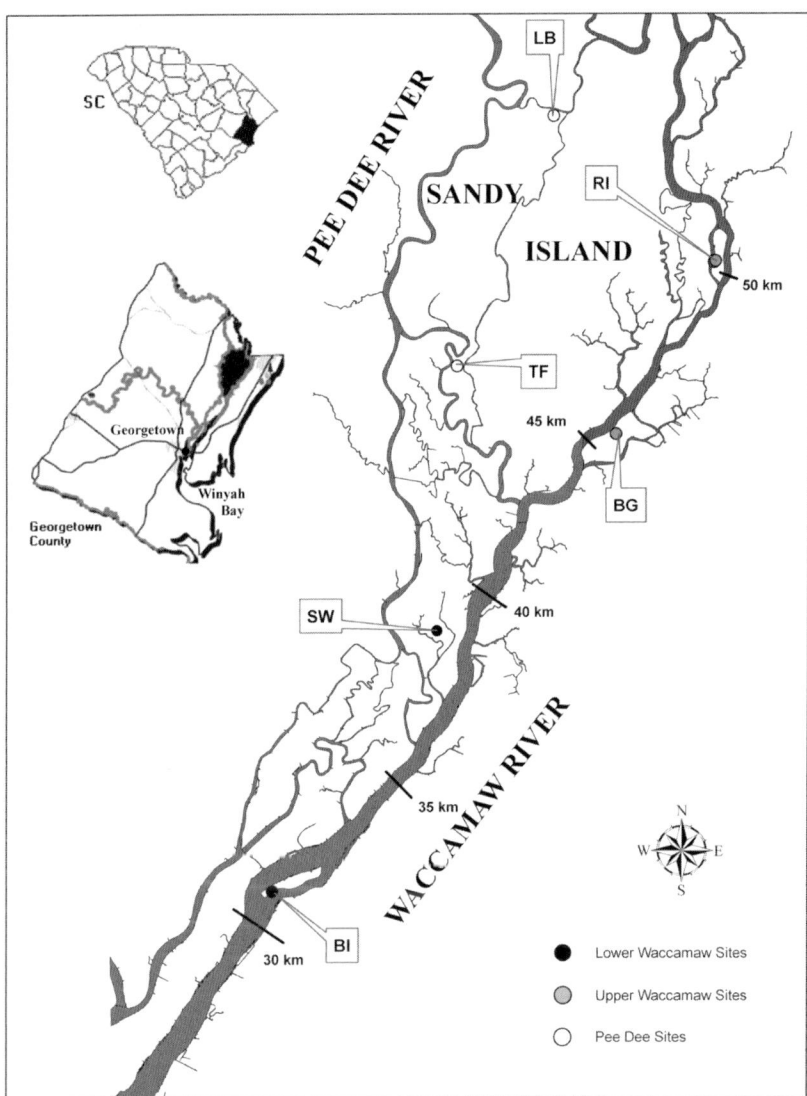

Fig. 15.1. Map of forest study sites in tidal freshwater forested wetlands of the Pee Dee and Waccamaw River drainages of Georgetown County, South Carolina. Locations of study sites and designated river groupings around and below Sandy Island with river distances (km) from the Atlantic Ocean are shown. The Pee Dee River grouping is composed of Little Bull Creek (LB) and Thorofare Creek (TF) forest sites on the western flank of Sandy Island. The upper Waccamaw River grouping includes Richmond Island (RI) and Brookgreen (BG) forest sites on the eastern flank of Sandy Island. The lower Waccamaw River group included Samworth (SW) and Bull Island (BI) forest sites downriver of Sandy Island 6-15 km.

the Waccamaw National Wildlife Refuge and is shaped by the confluence of the Pee Dee and Waccamaw Rivers. The Pee Dee and Waccamaw Rivers flow along the western and eastern shores of the island, respectively. Interdistribuary creeks connect the two rivers on the northern and southern ends of the island where some mixing occurs, depending on flow rates. Freshwater tidal forests occur on the margins of the island and are subject to daily flooding caused by tidal action, although there is normally enough river flow to prevent saltwater intrusion.

15.2.1. Pee Dee River drainage

The Pee Dee watershed and headwaters flow from the Blue Ridge Mountains in western North Carolina, draining more than 22,880 km^2 (at PeeDee, SC) of the Piedmont and Coastal Plain of the Carolinas. More than half of the basin is in agriculture or unforested, which contributes to the river's "redwater" nutrient and silt loads. A series of impoundments in the Piedmont sections of the river regulate its hydrology. The Pee Dee eventually enters the upper estuary of Winyah Bay where it joins the Waccamaw, Black, and Sampit Rivers. Water quality analyses of surface water from the Pee Dee River obtained from previous U.S. Geological Survey (USGS) studies demonstrated higher conductivity, turbidity, pH, and nitrate concentrations in comparison with the Waccamaw River (Table 15.1).

15.2.2. Waccamaw River drainage

In contrast, the Waccamaw River is a blackwater stream that originates and flows throughout the Coastal Plain of North Carolina and South Carolina. The USGS gage near Sandy Island at Hagleys Landing on the Waccamaw River experiences tidal fluxes and daily flow rates that can range from zero to over 1700 m^{3}s^{-1} at flood stage. Periods of high flow are usually generated by excessive rainfall associated with tropical storms. Water quality data for the Waccamaw River near Sandy Island verified a low nutrient, more humic and acidic stream typical of blackwater streams (Table 15.1).

15.2.3. Winyah Bay estuary

Freshwater input into Winyah Bay estuary from several South Carolina rivers, most notably, the Waccamaw and Pee Dee drainages, ranges up to 30,000 m^{3}s^{-1} at flood stage with normal mean runoff around 450 m^{3}s^{-1}. The

Table 15.1. Comparison of water quality measures between sample sets taken from the Pee Dee and Waccamaw Rivers respectively contrasting redwater and blackwater constituent concentrations. Data were extracted from U.S. Geological Survey surface water studies conducted from 1951-2000.

Constituent	Pee Dee River	Waccamaw River
River type	Redwater	Blackwater
Sample size	n=15	n=12
Conductivity (ms)	75.2	43.7
Turbidity	3	0
pH, field	6.64	6.03
pH, lab	6.36	5.87
Nitrate, filtered (mg/ml)	1.078	0.011
Organic carbon, H_2O (mg/	0.052	0.035
DOC, suspended (mg/l)	0.470	0.021

tides of the South Carolina coast are mesotidal and semi-diurnal with a diurnal range up to 2.2 m. Mean tidal amplitude is on the order of 1.4 m within Winyah Bay and near 1 m along the Waccamaw River at Hagleys Landing downstream of Sandy Island. In Winyah Bay, a salt wedge effect occurs as denser saltwater moves up the estuary along the bottom with the flooding tide, even though the overlying fresh water may be flowing toward the ocean. During periods of low riverflow, flooding tides push saltwater 32 km or more upstream of the ocean, but under average riverflow, the penetration is usually less than 5 km. Differences between surface and bottom salinities during these periods may be more than 20 ppt. A synoptic salinity survey of the Waccamaw River conducted by the authors in the fall of 2001 under drought conditions recorded salinity ranges above 3 ppt at the southern end of Sandy Island and as much as 20 ppt in the lower reaches of the river above the estuary (Figure 15.2).

15.3. Study design and site description

The purpose of this study was to investigate if river source and flow, tidal flooding, and salinity intrusions affect the stemwood production of dominant baldcypress trees in tidal freshwater swamps of a coastal riverine system in South Carolina. Six study sites consisting of 0.01 ha plots of baldcypress-water tupelo (*Nyssa aquatica* L.) forest were inventoried, and trees were cored to extract long-term growth records to correlate with environmental data of riverflow and climate (Ratard 2004). Forest plots were established in paired sites for each river and section of the tidal reach to contrast river source and tidal influence (distance from ocean) (Figure

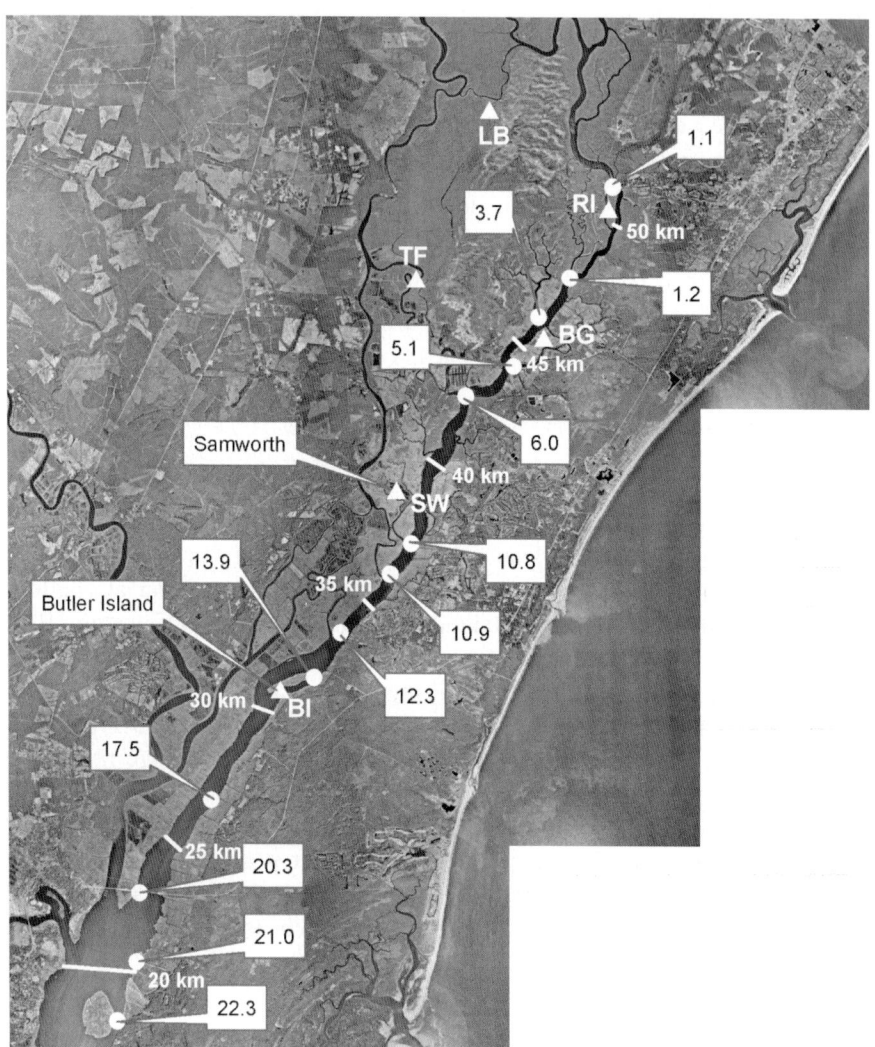

Fig. 15.2. Salinity values (ppt) observed at the surface along the lower Waccamaw River overlapping forest study sites from the upper estuary to Sandy Island during a low flow and drought period on November 30, 2001.

15.1). Little Bull Creek (LB) and Thorofare Creek (TF) are paired sites on the western flank of Sandy Island that are flooded routinely by the Pee Dee River and are called the PDR grouping. Richmond Island (RI) and Brookgreen (BG) are complimentary sites to LB and TF on the eastern flank of the island flooded daily by the Upper Waccamaw River and are labeled the UWR grouping. Butler Island (BI) and Samworth (SW) are paired sites downriver from Sandy Island along the Lower Waccamaw River expected to experience some saltwater impacts in low flow periods and are referred to as the LWR grouping. The LWR sites, SW and BI, are approximately 31 and 39 km upriver from the ocean front at Winyah Bay, respectively. The PDR and UWR sites around Sandy Island are collectively 46-50 km upstream from the Atlantic Ocean. River discharge records for both Waccamaw and Pee Dee were obtained from USGS river gages along the respective rivers.

15.3.1. Tree core collections and processing

At each sampling site, up to 20 dominant canopy baldcypress trees were cored with a 5-mm increment borer. Two cores were taken from each tree, and the height, crown ratio, diameter at breast height (DBH), and core azimuth were recorded. The cores were stored in acrylic tubes for transport back to the lab where they were placed into drying racks to air dry for several days. The dried cores were glued into grooved mounting blocks with proper trachied alignment and sanded with progressively finer grades of sand paper to a glossy, polished finish. The core samples were dated with the aid of a disecting microscope and by assigning calendar years to each growth ring while accounting for the possibility of missing rings. The cores were measured to the nearest 0.01 mm by using a mechanical stage with an optically gaged screw and digital encoder for automatic data recording. Basic core preparation techniques followed Stokes and Smiley (1968). Age was estimated by counting rings until the pith was reached or by estimating the pith date from the curvature and growth of the innermost rings. Core dates were verified with cross-correlation analysis accomplished with the Cofecha software developed and distributed by the Tree-Ring Laboratory, University of Arizona, Tucson.

Tree core measurements from opposite radii were summed to establish yearly diameter increment, an equivalent metric to annual remeasurement of tree diameter. Diameter increment was then used with field measurements of tree diameter at core height to calculate basal area increment by decrementing successive yearly measurement and diameter from bark to pith. Basal area increment equals the cross-sectional area of stemwood growth that accounts for tree size and provides a more comprehensive estimate of actual stemwood production on an annual basis. Two trees from the same site might share similar ring width or diameter increment for a given calendar year, but as tree size and age at the time the ring was formed differ, so does the realized growth in stemwood area. Generally, diameter increment and basal area increment will be highly correlated by derivation, but basal area increment is a more absolute growth estimate for making comparisons of growth potential between trees and populations (Phipps 1985).

15.3.2. Tree-ring analysis procedures

Tree populations were compared between sample sites to determine whether they were of near equal age and size to minimize confounding differences related more to stand history and development than to prevailing river and climate conditions. A univariate analysis of variance was used to test for differences between mean age and size of sampled trees, applying Tukey's HSD test to determine significance.

To simplify growth comparisons for patterns associated with site-level and river group responses, decadal growth increments were calculated by collapsing yearly variation into decadal periods to test and visualize temporal trends not confused by climate variability and individual sensitivity. The basal increments of individual trees were summed for even decadal periods in a balanced design where all trees had measured growth going back 40 years, thus generating four decadal periods for comparisons. These data would be equivalent to diameter remeasurement data gathered every tenth year and converted to stemwood area based on tree size and increment. These decadal summaries were analyzed by using a repeated measures ANOVA procedure in SAS version 8.2 which tested for differences between sites and river groups. It was assumed that the climatic variation within a decade and between years would be minimized, and we tested if growth departed from potential, based on stand maturity. We expected that climate and riverflow variation might be somewhat equalized over a ten year period and that growth differences, if any, might relate only to site condition as influenced by fertility and river nutrient input from redwater and blackwater loadings between PDR and UWR groups with a potentially negative effect of saltwater exposure in LWR group.

Annual growth variation between sites and river groups was investigated to determine whether given climate and riverflow years resulted in greater or lesser stemwood production. A repeated measures ANOVA was used to distinguish differences in growth performance for select high and low growth years successively from 1960 to 2000. Mean basal area increment (BAI) chronologies were used to contrast growth performance and variation between sites and river groups and relate them to environmental conditions. Mean BAI chronologies were also used to filter growth variability between paired sites and river groups to distinguish yearly departures and temporal trends.

15.4. Study results

15.4.1. Forest history and development

Tidal freshwater forests investigated in this study were of comparable age from 60-75 years; this similarity resulted from wide-area logging in the 1930s and 1940s (Table 15.2). Mean age of most trees and stands was 65 years which made growth contrasts compatible between sites at similar stages of stand development. There were no apparent growth releases evident for any individual growth history; an observation which suggests that all sites were clear cuts and resulted in second growth. Mean diameter of trees sampled for each site also was fairly similar, ranging from 37-47 cm across all sites (Table 14.2). Differences in mean size for sites of similar age indicate that average growth rate may be greater or lesser by site. The PDR sites (LB, TF) are influenced by redwater nutrient loading and were expected to exhibit greater growth than were trees flooded by low nutrient, blackwater riverflow from UWR sites (RI, BG). Differences in tree size (less than 10 cm) and age (slightly more than a decade) between sites were significant between the larger LB and BG populations and the younger RI population. Overall, the sites are strikingly similar, akin to a single cohort, and these similarities aided in direct comparisons of decadal and yearly growth between sites and river groups.

Table 15.2. Mean diameter at breast height (DBH) and age comparisons of cored baldcypress trees in a South Carolina tidal freshwater system. Different letter designations represent significantly different groups according to Tukey HSD test (p < 0.05).

Groups and Sites	No. of trees	Mean DBH (cm)	Mean Age (yrs)
Pee Dee River (PDR)			
Little Bull (LB)	13	47.077 a	68 ab
Thorofare (TF)	11	41.735 ab	66 ab
Upper Waccamaw River (UWR)			
Richmond Island (RI)	13	37.000 b	62 a
Brookgreen (BG)	12	46.159 a	75 ab
Lower Waccamaw River (LWR)			
Samworth (SW)	10	39.421 b	65 ab
Butler Island (BI)	14	39.380 b	63 ab

15.4.2. Decadal growth analysis

The strength of tree-ring analysis lies in the year to year variation of low and high growth performance that may be correlated with environmental observations such as climate and riverflow. Because all these forest sites are within 25 km of each other and are within the same river reach and county, it is expected that common growth variation is controlled by the regional macroclimate. The process of dating the cores and rings affirmed this assumption with shared marker years of low or high annual growth among all sites. Aggregating the annual growth response into a decadal measurement serves the advantages of minimizing the signal noise of yearly environmental change and accentuating growth trends related to site and water quality by river source or disturbance. For all sites, growth behaviour by tree or stand basis did not show any suppression or release episodes that might be related to a major disturbance. While there were relatively few missing rings, most absent and micro rings were associated with 1990 and 1991, years following a major hurricane strike from Hurricane Hugo (1989). Despite passing nearly 100 miles south of the study area, Hugo was a powerful category 4 storm with estimated winds of a category 2 storm at Sandy Island sufficient to down inferior trees, fell branches, and strip leaves to some degree.

Decadal growth patterns varied by river group and time period (Figure 15.3). The Pee Dee River (PDR) group outperformed both the upper and lower Waccamaw River groups for every decade over the last 40 years. This result corresponds with the slightly larger diameter sizes of the trees of this river group for the same relative age. This finding suggests that the greater nutrient concentrations of redwater rivers like the Pee Dee may contribute to greater soil fertility and growth potential than do blackwater rivers and swamps. There are numerous tree-ring and experimental studies that show that elevated nutrients in soil or aqueous solution can enhance tree growth and forest productivity (Hesse et al. 1998; Day et al. 2006). The fact that the growth enhancement is very uniform for all decades lends support to a chronic condition such as river source and water quality rather than to random or acute phenomena such as disturbance effects.

From the 1970s on, the rank order of growth potential by river group and decade from largest to smallest remains PDR, LWR, UWR; although there is no statistical difference among Waccamaw sites/groups. The robust growth response of the LWR group in the 1970s exceeds the expected trajectory particularly for sites where salinity intrusion in low flow years might compromise growth potential. Field surveys conducted in 2005 for a complimentary research study (unpublished data) reported maximum observed soil salinities up to 3.6 ppt at the Butler Island site (LWR group) in

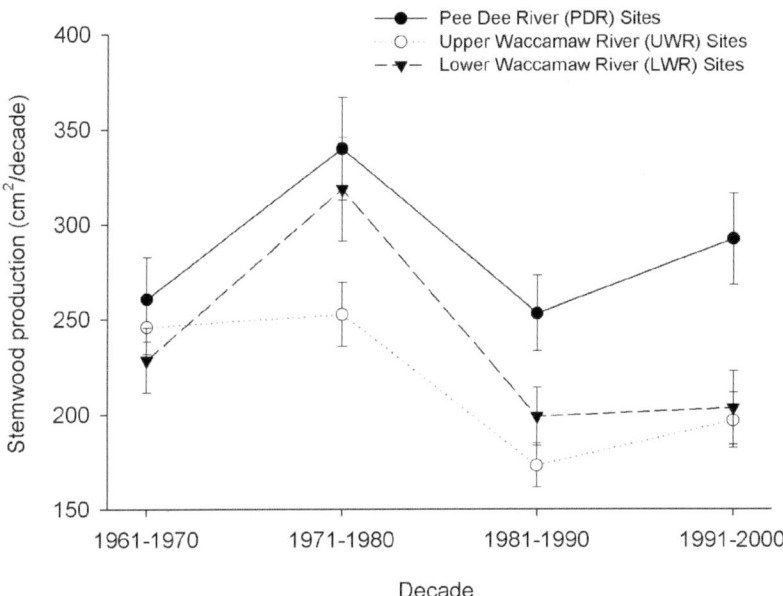

Fig. 15.3. Mean stemwood production for four successive decadal periods by river group.

contrast to 0.2 ppt at the Richmond Island site (UWR group) upstream (K. Krauss, personal communication). The elevated salinity reading at BI appears to correspond to storm tides from Hurricane Katrina which highlights the role that hurricanes can play in chronically and acutely raising soil and surface water salinities in interior tidal freshwater zones. The LWR group sustained greater stemwood production in the 1980s but lost its growth margin during the 1990s.

One explanation for the higher stemwood production in the 1970s of the LWR group may be the higher than normal riverflow for that same period. The 1970s decade experienced a series of years with record riverflows that in combination produced the highest flow for a given decade over the period of record from 1939-2000 (Figure 15.4). It is possible that the degree of surplus freshwater flow also may be indicative of climate conditions that promote tree growth and elevate riverflow, such as higher and more frequent precipitation events. Climate and streamflow are more often correlated such that droughts contribute to low flow records depending on how local the drought might be and how large an area the river basin may include. Average decadal riverflow in the 1980s exceeded the 1990s only slightly but may account for a concomitant drop in growth performance.

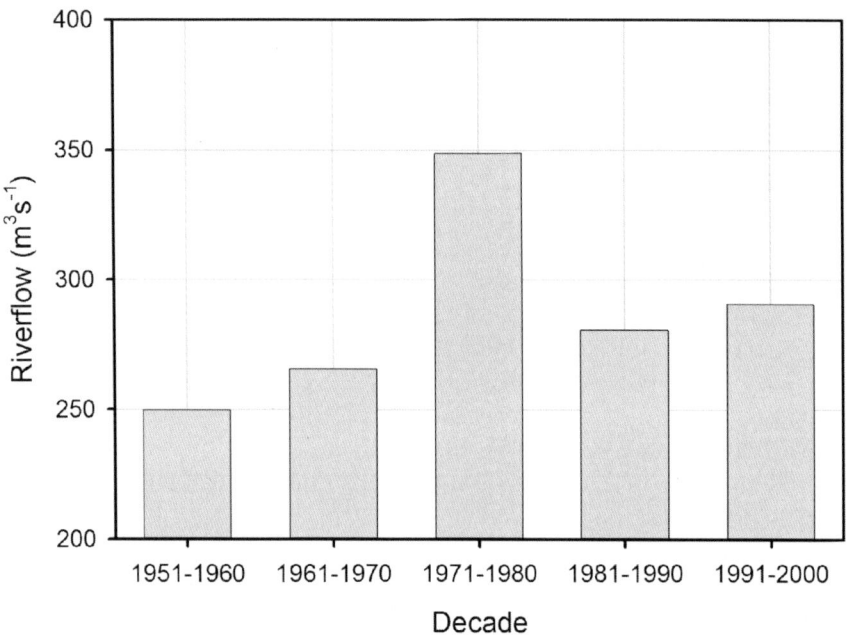

Fig. 15.4. Streamflow average of the Pee Dee River at Pee Dee, South Carolina by decade from 1951 to 2000.

Floods and high flow years of the 1970s may have flushed salt concentrations from the soils along the lower reaches of the Waccamaw River and allowed the trees to rebound a robust growth potential with freshwater recharge. Consequently, it also may be expected that growth reductions should be the result in low flow years or extended drought periods which may explain the waning growth performance in the 1980s and 1990s. The sea-level trend for this coastal reach of South Carolina is 5.17 mm/yr (Springmaid Pier, NOAA website). Accounting for historic eustatic sea-level rise of 1.8 mm/yr during the 20th century, the remaining 3.37 mm/yr rate of the sea-level trend is attributed to local subsidence which is greater than other locales along the Atlantic and Gulf of Mexico Coast, excluding the Mississippi Delta region. It is plausible that sea-level rise has risen sufficiently over the last 40 years to reach a point where trees and forests in the lower reaches of coastal rivers are responding to increased salinity exposure and may undergo dieback in the near future under this pace.

15.4.3. Yearly growth analysis

Mean BAI chronologies were developed for each site and river grouping by averaging the annual increment for sampled trees. Baldcypress trees are commonly known for producing tight or narrow rings when stressed by standing in deep-water, permanently flooded sites. Tree growth in these second growth forests was fairly robust, however, exceeding 1 cm for a maximum ring width or approaching 3 cm diameter growth per annum. Trees with high growth rates usually produce fairly complacent ring series such that the variation in width from year to year is nearly the same and often complicates dating verification and environmental analysis. All of the trees used in this study showed similar growth patterns, sufficient ring size variation, and marker years of lower and higher growth for solid chronology development.

Mean growth chronologies of stemwood production were constructed from basal area increment (BAI) calculations of all sampled trees from a given site or paired sites by river group and by year. Mean BAI chronologies were compared between paired sites by river group to show relative magnitude, degree of synchronicity and uncommon variance (Figure 15.5). It was expected that sites nearest each other sharing the same river water should be more similar. Results indicate that sites within each river group express synchronous behavior and common patterns of high and low growth. Noticeable contrasts include the convergence behavior of the PDR sites around 1970 and thereafter are almost identical (Figure 15.5a). The UWR sites exhibit a similar chronology and track each other well over the entire period of record (Figure 15.5b). The LWR groups also share a common magnitude and synchronized annual growth (Figure 15.5c). The one exception is the peak growth performance from the Butler Island collection and trees in the 1970s that were also detected in the decadal analysis.

Contrasting the annual stemwood production between river groups PDR and UWR provided a more detailed comparison of year to year variation between a redwater and blackwater influence, equally fresh and equally flooded. Results show a remarkable parallel behavior of PDR outperforming UWR from year to year, similar to results in the decadal analysis (Figure 15.6a). Both chronologies show almost identical year to year variation of highs and lows that demonstrates that the two groups share the same flooding pattern and/or climate response but the added nutrient availability from the Pee Dee River and soils may be augmenting forest growth in relation to the Waccamaw River sites. On average, growth performance was 10-20% greater for PDR trees and sites than for the UWR group. On a

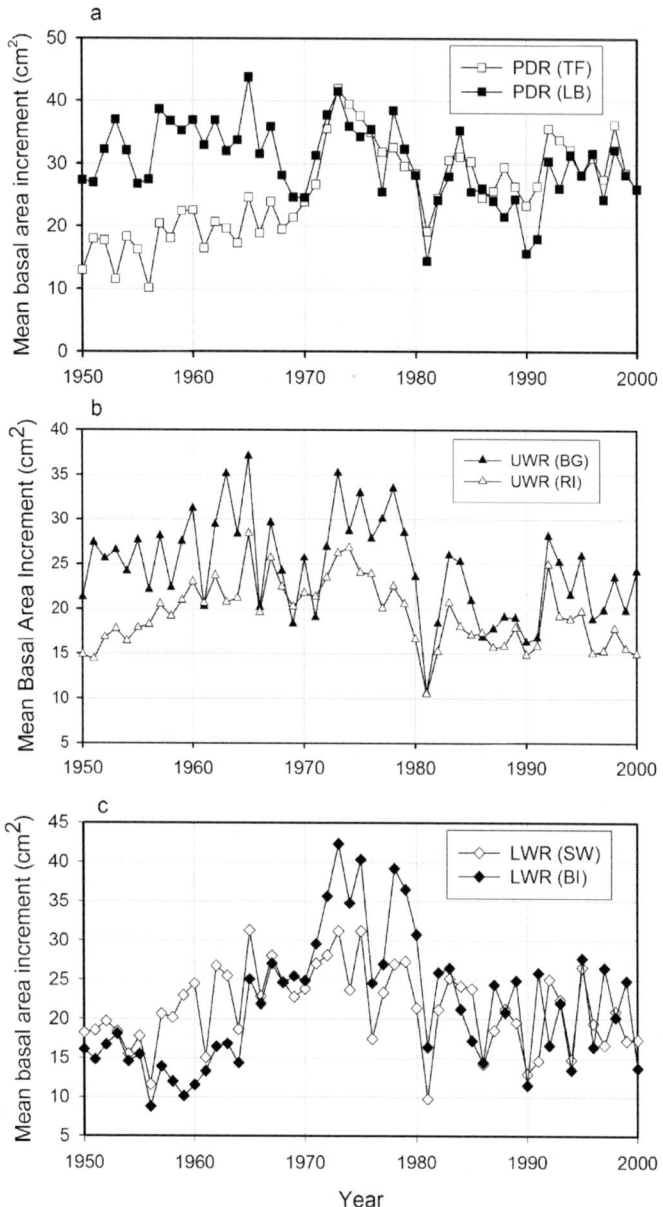

Fig. 15.5. Mean chronology of annual basal area increment for each site by river group: (a) Pee Dee River (PDR), fresh, redwater source; (b) upper Waccamaw River (UWR), fresh, blackwater source; and (c) lower Waccamaw River (LWR), saline, blackwater source. Paired data show within site variation. TF=Thorofare Creek, LB=Little Bull Creek, BG=Brookgreen, RI=Richmond Island, SW=Samworth, and BI=Butler Island.

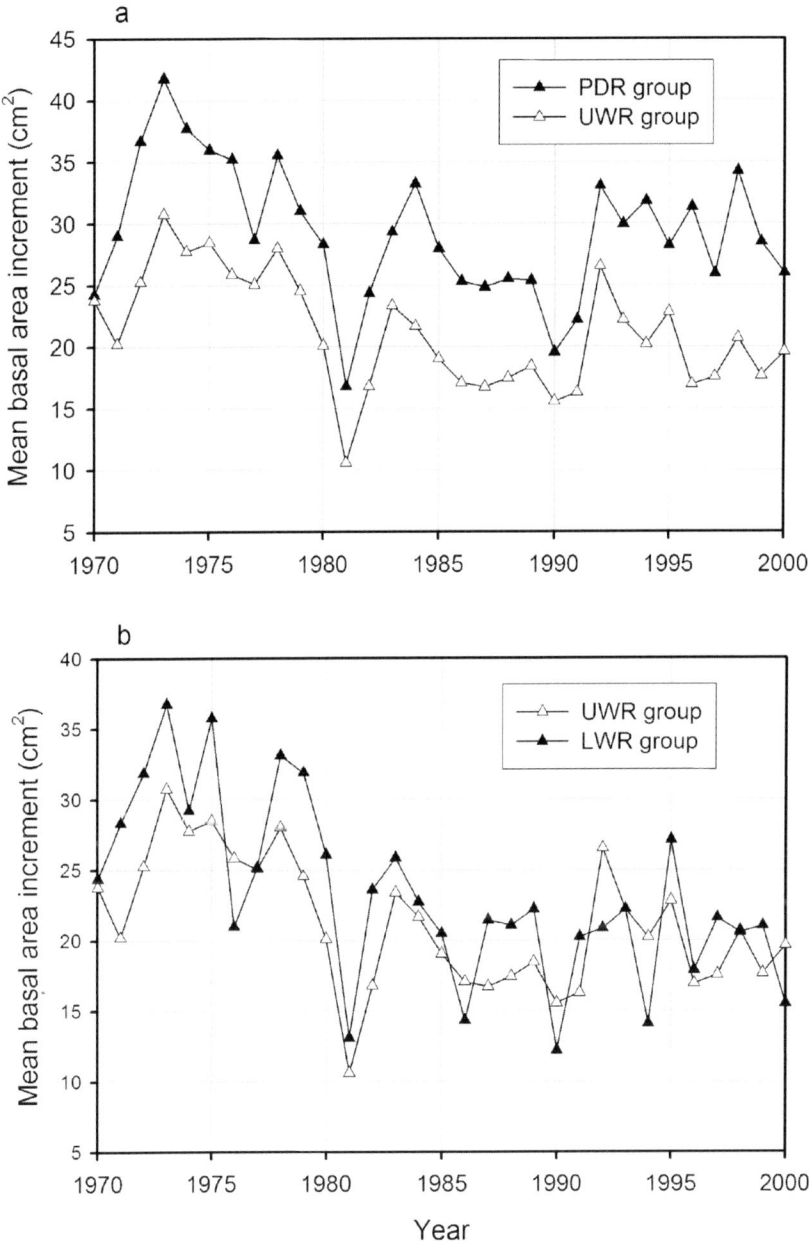

Fig. 15.6. Mean annual basal area increment by river group contrasted between (a) fresh, redwater Pee Dee River (PDR) and fresh, blackwater upper Waccamaw River (UWR) and (b) fresh, blackwater upper Waccamaw River (UWR) and saline, blackwater lower Waccamaw River (LWR) swamp forest.

yearly basis, PDR stemwood production was on average 8.1 cm^2/yr more than the UWR group over the 31 year period.

The comparison of mean BAI growth between the UWR and LWR by year demonstrates a similar growth trend from 1970 to 2000, excepting greater yearly sensitivity and some sharp growth reductions in the LWR group for known drought years (Figure 15.6b). LWR outperforms UWR in nondrought years when river flow may be higher than normal, ameliorating elevated salinity concentrations accumulated in low flow years when tidal influence is more prevalent. The overall growth trend over the last 30 years is comparatively similar among all river groups and shows a decreasing monotonic slope between 30 and 20 cm^2/year basal area growth.

15.4.4. Normalized growth differences between river group and forest sites

A common technique for tree-ring studies is to convert mean growth chronologies to indexed ratios based on various curve fitting methods designed to remove low-frequency change and to enhance high-frequency differences of year to year variation for climate analysis. The compatibility of tree age, size, and growth potential among all sites allowed direct comparison of annual growth between site and group chronologies. A simple and effective normalization method for comparing growth chronologies is to contrast series differences in relation to series sums. The resultant index is constrained to values 0.0 to ±1.0 such that the sign indicates which series is greater or dominant. Resultant values at or near zero indicate little or no difference between series, and values at ±1.0 reflect complete dominance by one series and no growth in the other. Index values of 0.325 and 0.5 are indicative of a two and threefold difference in growth values, respectively. Normalized indexed values for tree-ring data commonly range from 0 to 0.1 which equates to a 20% difference or more in relative growth performance. This methodology assumes that trees within and between sites that are proximal to one another also share the same macroclimate which should produce similar high and low growth responses to climate; any significant departures are the result of other tree or site-specific influences. This methodology has been used effectively for detection of hurricane impact on an individual tree and stand basis (Doyle and Gorham 1996). The normalized series should show relative patterns of dominance by year in response to climate or other local environmental factors and indicate the direction of changing dominance over time, if present.

The normalized series contrasting the PDR (redwater) and UWR (blackwater) groups showed an expectant pattern of dominant growth of the PDR group over the UWR between 0.1 and 0.25 (Figure 15.7a). The PDR growth exceeded UWR for every year during the 1970-2000 period with an average stemwood production difference of 8.1 cm^2/yr. In years when both sample groups responded to drought or temperature effects (i.e., 1994, 1996), the indexed value raised not so much on absolute change in stemwood production, but more so relative to decreased growth performance found in both sample groups. Contrasting the normalized series from the UWR (fresh) and LWR (saline) groups, however, demonstrates stark growth departures during drought years (Figure 15.7b). Normal climate and riverflow years actually favored the LWR group by slight and insignificant margins, but nonetheless resulted in consistently greater stemwood production in all but 6 of the 31 years. The upper and lower Waccamaw River groups demonstrated the same growth potential overall excepting certain drought and low-flow years. However, the indexed growth ratio of UWR and LWR groups demonstrates synchronous growth departures for the 1981 drought and differential sensitivity to subsequent droughts in the 1980s and 1990s.

Drought severity and/or tide variation had some effect on whether the negative impact was expressed at fresh (UWR) and saline (LWR) river sites. The severe drought of 1981 was a marker year for dating all cores at all sites. Likewise, both the UWR and LWR sites and river groups were negatively impacted. Other drought years more prominently impacted the LWR group, however, and may have related to saltwater incursions and distance of upstream distribution. Because the forest sites are distributed at different distances from the estuary, it might be expected that the closest site to ocean tides and saltwater, Butler Island (BI), may be most sensitive to both drought and low-flow years and to high tide events and years. Therefore, we propose that drought and low-flow years are expected to promote saltwater incursions and elevate surface water and soil salinities that negatively impact tree and forest growth depending on the degree, timing, and constancy.

The growth departure of the UWR (fresh) group and the LWR (saline) forest sites was indexed separately at Samworth (SW) and Butler Island (BI) to determine if the same years impacted growth and if varying degrees related to site proximity and propensity for salinity overwash. Both sites respond similarly for drought and low-flow years, but the Butler Island site shows a greater degree of response both positively and negatively (Figure 15.8). The Butler Island site is expected to be more susceptible to salinity intrusion by virtue of closer proximity to Winyah Bay and more responsive

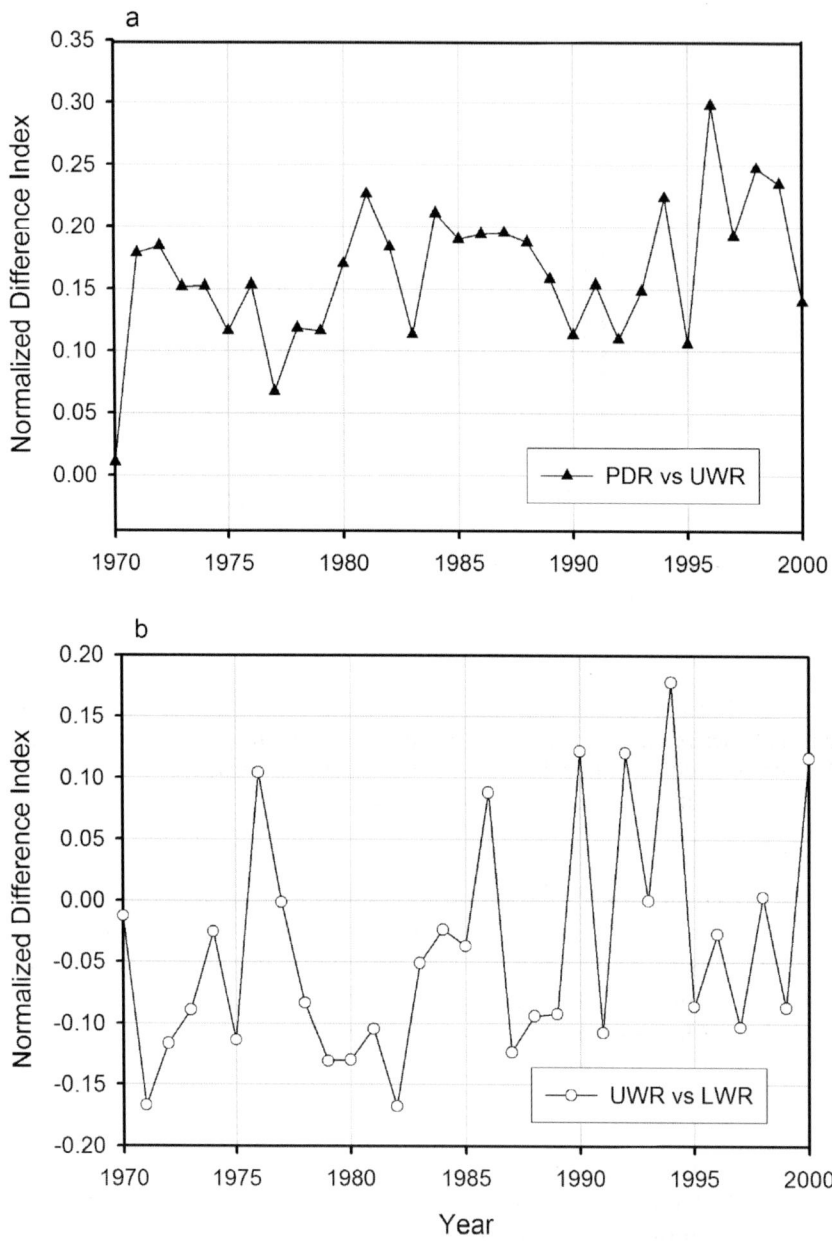

Fig. 15.7. Normalized growth series of mean stemwood production between (a) fresh, redwater Pee Dee River (PDR) and fresh, blackwater upper Waccamaw River (UWR) sites and (b) comparing upstream, fresh upper Waccamaw River (UWR) with downstream, saline lower Waccamaw River (LWR) sites.

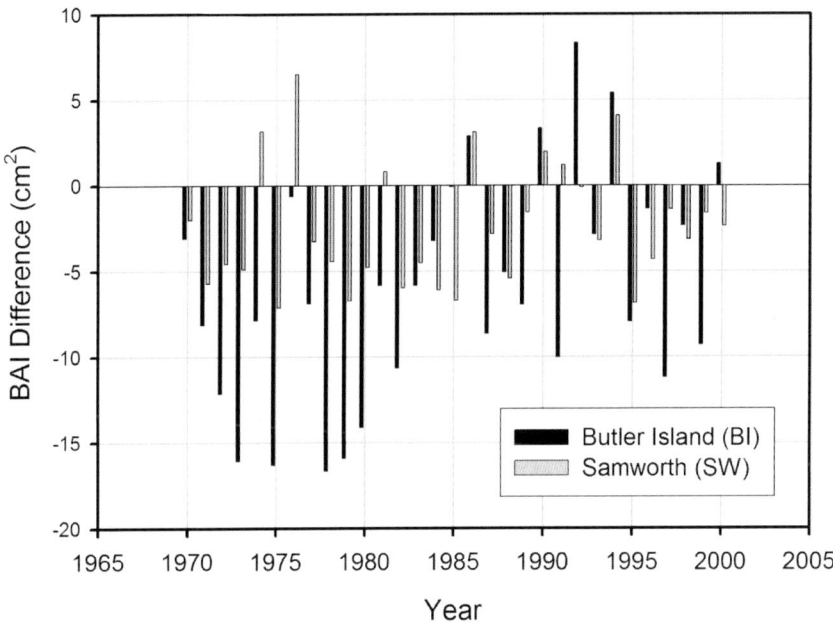

Fig. 15.8. Difference in annual basal area increment (BAI) of Samworth and Butler Island sites relative to upper Waccamaw River (UWR) group chronology from 1970-2000. Positive values indicate greater stemwood production by the UWR group, and negative values indicate greater SW or BI stemwood production.

to freshwater recharge in high-flow years when soil salinities are flushed, stimulating root regrowth. Soil salinities at Butler Island in recent years (Conner, unpublished data) have exceeded 3 ppt during the summer months, and one would expect these levels to impede tree growth and physiology. The 1970s had some of the highest flow rates for the period of record that may have been sufficient and continuous enough to flush the Butler Island site of any salinity. The Butler Island stand is also less dense than other sites and have higher crown ratio and spacing per individual tree which may explain the degree of rebound growth during high-flow periods. The Butler Island site also shows greater sensitivity to drought years than do the Samworth trees during the 1980s and 1990s.

15.5. Discussion

15.5.1. Stand age and history

The tree ages of about 65 years which were observed at most sites were consistent with forest clearing that took place in and around Sandy Island in the 1930s and 1940s (USFWS 1997). Tree stands of the same cohort class and species such as these can be expected to be about the same size, assuming the same growth rate under similar climate and flooding conditions. The size of sampled trees for the cored trees and stand composition (Ratard 2004) are relatively similar and differ less than 10 cm in diameter between sites (Table 15.2). The fairly narrow range of variation in tree size by site did account for statistical differences in larger and smaller groupings. River source and differences in stand density can at least partially explain the variation in tree size and growth by site. Pee Dee River sites, Little Bull and Thorofare, are redwater sites with higher nutrients in water and soils that also can contribute to greater stem growth. Samworth has the highest stand density (1690 stems/hectare) and the smallest average tree diameter (38 cm dbh), while Little Bull had the largest trees (49 cm dbh) and an intermediate density (1100 stems/hectare) (Ratard 2004). The Brookgreen (BG) site was slightly older than other sites examined in this study, and the added years of stand development may account for their larger stem diameters. Given the relatively young age of the trees, growth has been generally robust over the last 50 years. Observed mean basal area growth for baldcypress trees in southwestern Louisiana ranged from about 20.0-35.0 cm^2/yr (Hesse et al. 1998). This is comparable to tree growth in this study, where mean basal increments ranged from 18.0-30.0 cm^2/yr. Species composition at the SW and BI sites in the lower Waccamaw River were composed of more than 90% baldcypress in contrast to 60% or less for all upstream sites (Ratard 2004). The fact that the less salt-tolerant water tupelo is absent from these salt-prone sites may be an additional indicator of saltwater impact.

15.5.2. Water source and quality

Redwater rivers typically have higher nutrient loads which could potentially contribute to increased growth relative to trees along more nutrient-poor blackwater rivers (Kellison et al. 1998). The Pee Dee River sites had larger size trees and greater annual growth than any of the upper or lower Waccamaw River sites in this study. Stemwood production was uniformly greater on an annual basis by 8.1 cm^2/yr, exhibiting the same annual growth variation over the last 30 years (Figure 15.6). Controlled fertilizer studies and wastewater applications into baldcypress swamps have been shown to increase growth potential (Mitsch and Ewel 1979; Hesse et al. 1998; Souther and Shaffer 2000; Day et al. 2006). Site quality differences may account for higher growth potential at Pee Dee sites by virtue of the mineral soils accumulated over the years of sediment deposition from Piedmont drainage as compared to little or no sediment transport in the Waccamaw River system. Hesse et al. (1998) showed a similar increase in growth potential over a 40-year period from wastewater effluent discharge into a backswamp baldcypress forest. The change in growth potential in their study was evident in the treated stand pre- and post-effluent discharge and in contrast to a nearby unaffected stand above the effluent outfall. The fact that the growth pattern and difference have been so uniform for the last 30 years argues for a difference in site quality and the absence of any disturbance phenomenon or significant change in stand development. Day et al. (2006) conducted an experimental study of Atchafalaya River water with and without nutrient amendments to determine if nutrient-rich river water would be sufficient to maximize sapling growth of baldcypress. Their results showed that baldcypress growth was enhanced with fertilizer amendments in aqueous application, which shows that relative growth rates can be controlled by relative nutrient concentrations of water source. The threefold or greater levels of nitrogen and phosphorous concentrations in the Pee Dee waters above the nutrient-poor Waccamaw River (Table 15.1) may be adequate to promote and sustain greater tree growth as observed between Pee Dee and Waccamaw chronologies.

15.5.3. Flooding regimes

Tidal fluctuations produce daily draining and flooding that also act to synchronize flooding among sites. Flooding depth or range generally increases within the river channel and in forest sites closer to the estuary than sites upriver where tide influence eventually wanes. Based on monitored water level observations, Ratard (2004) showed that the downstream Samworth (SW) site flooded above soil elevation longer and to a greater degree by 33% or more than did the upstream Brookgreen (BG) site. Hourly readings of water level at the Richmond Island (RI), Samworth (SW), and Butler Island (BI) sites in the spring of 2005 demonstrated increased flood range and duration with downriver sites and proximity to the estuary (K. Krauss, pers. comm.) The flood range for the Richmond Island site for this synoptic period was 42.77 cm compared with 49.45 cm and

63.11 cm for the downstream SW and BI sites, respectively. These sites also drain with the tide and to a greater degree with downriver position and proximity to the estuary. The drain events appear to be adequate and frequent enough to support productive forests. The tidal effect may behave more like an irrigation system, and the exchange of waters on each tide cycle also may benefit tree growth. Several studies on baldcypress growth have shown that flowing and fluctuating water regimes stimulate more growth and healthier trees than do stagnant and continuously flooded swamps (Brown 1981; Lugo and Brown 1984; Conner and Flynn 1989; Dicke and Toliver 1990; Megonigal and Day 1992).

Abrupt changes in the hydroperiod of a swamp environment can lead to abrupt changes in growth response in baldcypress and other flood tolerant species (Keeland 1994; Young et al. 1995). None of the Sandy Island chronologies investigated in this study, however, demonstrated any abrupt or incremental changes in growth pattern that can be associated with altered hydrology or other exogenous disturbances.

15.5.4. Saltwater intrusion

Saltwater intrusion may be the more important factor or process that can reduce growth potential in tidal freshwater swamps dominated by baldcypress. The degree to which the hydrology of a given forest stand is coupled with the local tides and river stage will determine the frequency and concentration of salt exposure. Obviously, sites closer to the estuary are more likely to incur salinity pulses than are upstream sites. Upstream sites may still experience an increase in stage and flooding related to tide effects but may only contend with salinity during severe droughts or with hurricane surge. Drought severity and hurricanes are climate phenomena that can exacerbate salinity excursions beyond normal tides. Baldcypress has been shown to tolerate moderate salinity exposure of low chronic conditions or short acute pulses, but not sustained levels beyond modest concentrations (Conner 1994; Allen et al. 1996; Krauss et al. 1998; Williams et al. 1999; Krauss et al. 2000; Day et al. 2006).

No salinity measurements or monitoring stations are available in the lower Waccamaw River or other tidal rivers in the area to correlate with these tree-ring chronologies. Except for a few synoptic surveys (Figure 15.2), we have limited understanding of the potential for salinity incursions during severe or prolonged drought periods. Both the river and tides are dynamic on a daily, seasonal, interannual basis, and this characteristic emphasizes the need for continuous salinity monitoring to understand the actual frequency, distribution, and duration of salinity concentrations in surface waters and soils of these tidal freshwater swamps. Results of a tree band study by Ratard (2004) on some of these same forest sites showed retarded growth behavior from the Samworth (SW) site compared to the Brookgreen (BG) site that was attributed to saltwater intrusion. Without a yearly data set and a corresponding salinity record, it is difficult to determine whether there have been salinity incursions. The soils and tree cores themselves could be tested for chloride concentration to at least establish relative exposure. Yanosky et al. (1995) demonstrated the use of proton-induced X-ray emission (PIXE) technology to estimate chloride concentrations in tree rings of baldcypress from the Cape Fear River system in North Carolina. Thus comparing stemwood chloride concentrations among the sites and growth rings could provide evidence for salinity incursion and, possibly, timing of exposure. Yanosky et al. (1995) cautioned that trans-mobilization of chloride ions from outer sapwood rings to inner heartwood zones may confound dating of salinity uptake.

15.5.5. Timing of riverflow and tides

Baldcypress growth chronologies across the Southeast have been used to reconstruct historic climate, determine drought frequencies and surplus water availability, and evaluate effects of hydrologic alterations. Several baldcypress tree-ring studies conducted over a wide area of the Southeast have shown a positive growth correlation with spring precipitation (Stahle and Cleaveland 1992, 1994). Still others have found that growing season droughts, which are also often associated with low streamflow, negatively impact baldcypress tree growth (Stahle et al. 1988; Cleaveland 2000; Fye et al. 2003). Generally, there is a close association between the Palmer Drought Severity Index and riverflow in summer months for this coastal region (Figure 15.9). The intercorrelation of climate and riverflow can confuse the analysis and interpretation of factors and processes controlling tree growth, particularly in coastal rivers and tidal freshwater swamps where tides and salinity add another complicating factor.

During low flow periods, it is expected that normal and supranormal tides can raise saltwater concentrations and distribution into freshwater zones. Even under moderate river discharge, tide stage more than river stage dictates flooding levels and saltwater pulsing. The conditions that elevate salinity intrusions upstream as previously shown in a November 2001 synoptic survey of the lower Waccamaw River (Figure 15.2) seem related to a combination of low riverflow and high tides. The monthly riverflow and tide readings for 2001 illustrate a case where tides increased in stage throughout the year while riverflow over the same period was declin-

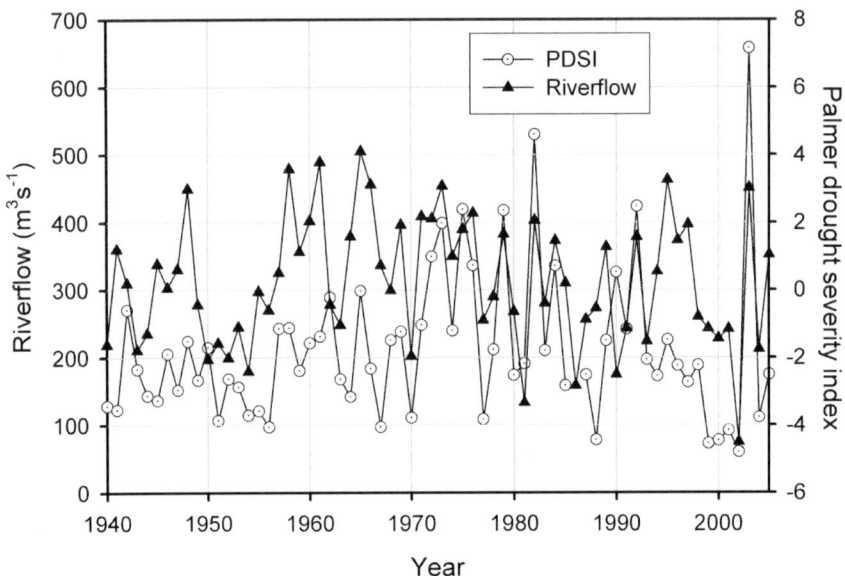

Fig. 15.9. Association of Palmer Drought Severity Index (PDSI) with Pee Dee riverflow for this coastal region during the growing season month of June over a 60-year period.

ing, thus elevating salinity concentrations and incursions upstream (Figure 15.10). The long-term pattern of riverflow and tides during the growing season does not, however, indicate any dependent relation. The mean Pee Dee riverflow and tide stage during the growing season month of June, for the common period of record 1977-2005, illustrates the various uncorrelated patterns of high and low riverflow associated with high and low tide stages (Figure 15.11).

While droughts and reduced riverflow increase the potential for salinity incursions of higher concentration and further upstream, tides are also highly variable on a monthly and annual basis, such that the timing and periodicity of elevated tides probably play a larger role in controlling growth and forest dieback than has been examined to date. Not all years with significantly reduced growth rates can be attributed to drought alone. The Samworth (SW) and Butler Island (BI) growth chronologies definitely exhibit more interannual variation in contrast to upper Waccamaw River sites that includes drought years and nondrought years. Assuming that salinity incursions may be more deleterious to growth during the growing season than during the dormant season, years of supranormal tides in late spring and early summer may explain some of the relative growth departures be-

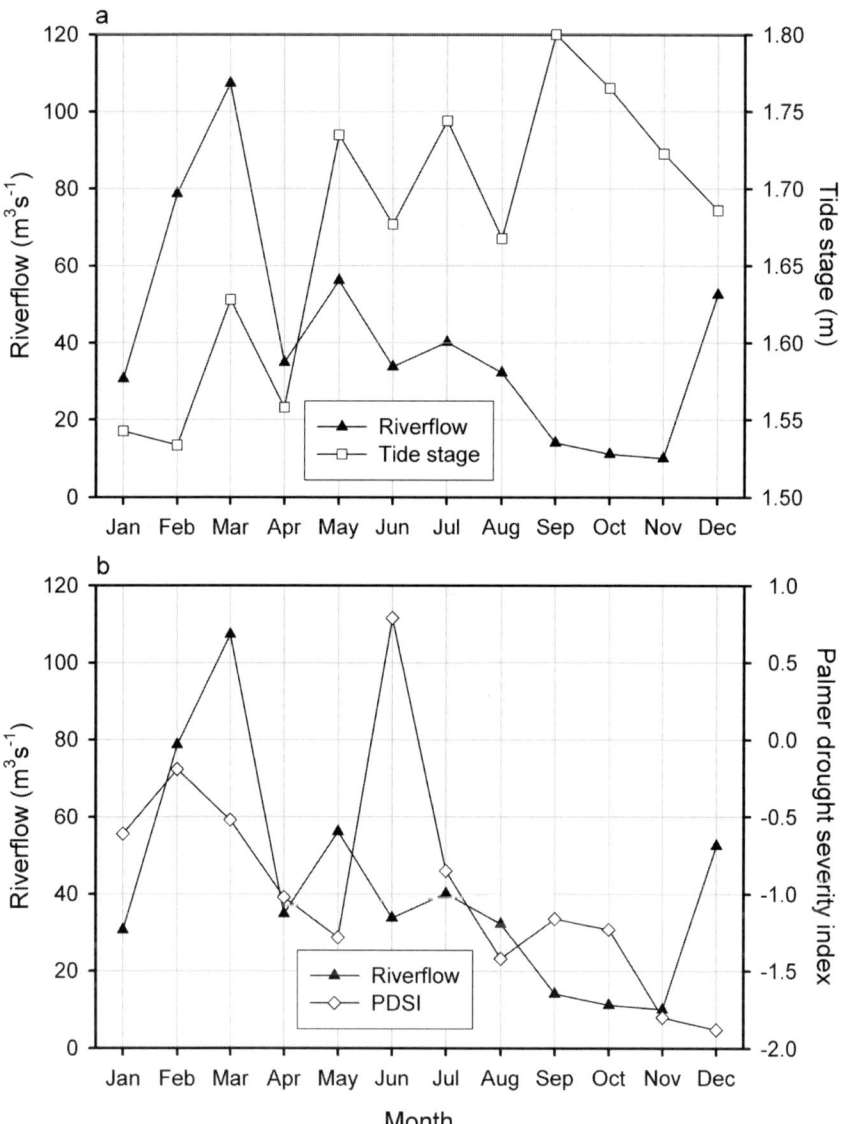

Fig. 15.10. Pattern of (a) Pee Dee riverflow and tide stage and (b) Palmer Drought Severity Index (PDSI) for the year 2001, illustrating a declining trend in both climate and hydrology that contributed to elevated saltwater intrusion in a November synoptic survey of upriver surface water salinities.

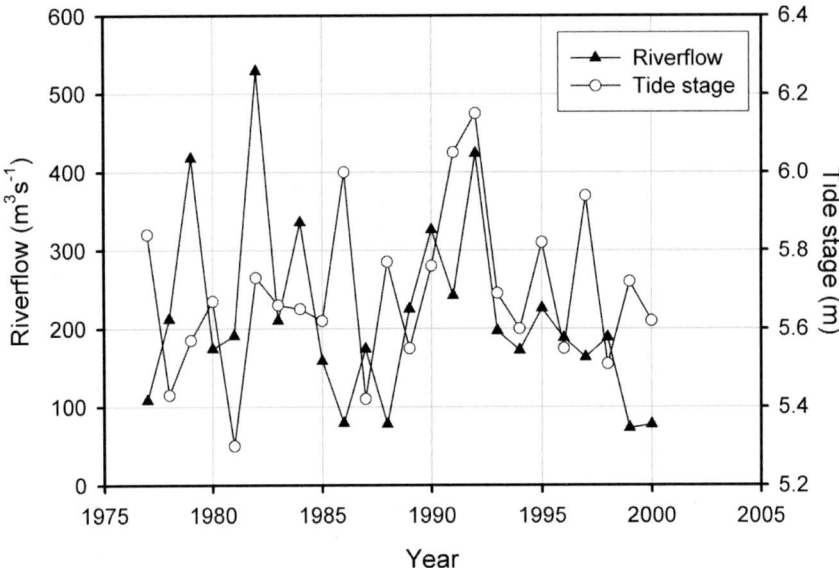

Fig. 15.11. Mean monthly pattern of Pee Dee riverflow and Springmaid Pier tide stage during the growing season month of June for the period 1977-2000, showing uncorrelated patterns of high and low riverflow associated with high and low tide stages for the same month and year.

tween the lower and upper Waccamaw sites and chronologies. Years 1976, 1986, 1990, 1992, 1994, and 2000 stand out where growth response favored the upper Waccamaw sites but dramatically inhibited growth at SW and BI (Figures 15.7 and 15.8). Most were drought years, excluding 1992 and 1994 which were preceeded by drought conditions that may have incrementally raised soil salinities without sufficient freshwater recharge to refresh the sites before their growing season. The highest June tides for the period of record (1977-2005) occurred in 1992, and the next highest were in 1991 and 1986, with mean high water averages near or above 1.8 m msl. The most severe drought year of recent history was 1981 which affected all sites somewhat equally but also had the lowest mean tide which may have spared the SW and BI sites of additional salinity stress as experienced in other drought episodes. A pattern may be emerging wherein the SW and BI populations are increasingly falling behind in relative growth performance in latter years, perhaps because of sea-level rise.

15.5.6. Climate change and sea-level rise

Sea level has been rising on a global scale during the 20th century at a rate of 1.8 mm yr^{-1} (Douglas 1991, 1997). Sea level has risen and fallen over eons mostly in response to the various warming and cooling periods that have preceeded the current threat of fossil fuel consumption and elevated atmospheric CO_2 concentrations that has ushered in a "greenhouse effect" referred to as "climate change." The northeastern coast of South Carolina is currently experiencing sea-level rise at 5.2 mm/yr which equates to a regional subsidence rate around 3.4 mm yr^{-1} (Figure 15.12). This means that the tidal freshwater swamps of this coastal region are undergoing moderate geologic submergence that over time will account for sinking land elevation, greater flooding and higher salinity concentrations that eventually will determine the fate for freshwater species. Evidence exists that coastal forests across the southeastern United States already may be undergoing decline from saltwater impacts related to sea-level rise (Ross et al. 1994; Williams et al. 1999; Doyle et al. 2003). While these studies document changes in tree species demographics and mortality, none have examined changes in growth rates or growth potential to project the process of gradual saltwater exposure and decline in tree health.

The evolving pattern of growth potential for the lower Waccamaw River forests is that drought years in concert with low riverflow and high tides increase saltwater exposure sufficiently to cause relatively poor growth performance compared to that of the trees and sites upstream. Periods of high riverflow and rainfall may be adequate to flush or ameliorate saltwater concentrations and stimulate substantial rebound of growth potential. The frequency and trajectory of growth impact has increased at SW and BI in recent decades along with sea level which might suggest that a threshold may soon be exceeded whereby former freshwater zones will be become more estuarine and unsuitable for freshwater tree species. There are no substantial baldcypress stands below the Butler Island location that are not either highly reduced in stem density or do not show many dead snags and stumps of former trees and forests. By the year 2100, sea-level changes are expected to rise by at least 0.5 m without any accelerated changes in eustatic rates from climate change. This rise in sea level and the tidal plane will eventually compromise the health of lower treeline sites within the lower Waccamaw River and overtake existing sites at Butler Island and further upstream with a concomitant rise in salinity concentration and distribution.

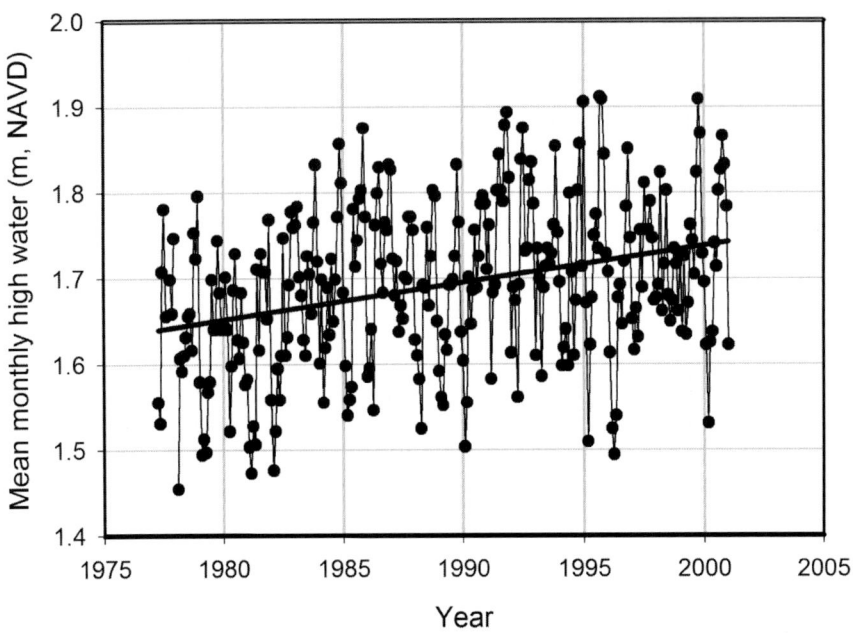

Fig. 15.12. Mean monthly tide level for northeastern South Carolina NOAA gage at Springsmaid Pier in Myrtle Beach, South Carolina for the period of record 1977 –2001.

15.6. Conclusions

Successful restoration of coastal wetlands depends mostly on the proper understanding of the physical environment and the ecological requirements and tolerances of eligible and expected biota, *in situ* or introduced. The potential extent of tidal freshwater swamps within any given estuary is related to river gradient and flow coupled with tidal range and forcing that increase river stage and surface water salinities many kilometers upstream. Indicators of tidal impact in these forests include the change in tree species distribution particularly in relation to elevated salinity concentrations in soils and surface water. Many studies have shown that salt-intolerant tree species will either die or fail to regenerate in tidal swamps when and where salinity concentrations intrude and persist. In the lower reaches of the Waccamaw and Pee Dee Rivers, baldcypress is the dominant tree species while water tupelo and swamp tupelo are equally prominent upstream, but water tupelo and swamp tupelo (*Nyssa biflora* Walt.) become noticeably absent closer to the estuary. In this study, we investigated the growth history and variation in baldcypress stands of similar age and size in two different coastal rivers of different water quality and in forests of varying distances from direct tidal influence to determine factors of flooding and salinity acting to control tree growth.

Tree-ring analysis was used to review growth history and relation to climate, riverflow, and tides in tidal freshwater swamps along the lower reaches of the Pee Dee and Waccamaw Rivers around and below Sandy Island, South Carolina. Tidal freshwater swamps around Sandy Island were found to be second growth forests as a result of wide-area logging conducted in the early part of the 20th century. Baldcypress tree growth has been robust for the last 50 years and seems to tolerate and thrive under the ebb and flood cycle of tides. Tree growth during the 1970s exceeded all other decades and years at all forest sites concomitant with higher than normal river flooding. Paired sites along the Pee Dee River demonstrated significantly greater stemwood production than sites located in the upper and lower Waccamaw River because of soil differences and higher nutrient concentrations between a redwater and blackwater river, respectively. Conversely, forest sites closer to the estuary, SW and BI, showed lower stemwood growth and greater sensitivity to drought episodes more so than did upstream forests for years when tide levels were above normal on a seasonal and annual basis. Findings indicate that high riverflow for sustained periods may be sufficient to refresh forest soils of elevated salinities and promote robust growth in wet years. Tree growth suppression during drought years and above normal tides may be an indicator of an eventual threat to tree health and potential for forest dieback from salinity intrusion.

Rising sea level of 20 cm or more over the last 65 years of the life of these trees and forests may already be affecting sites of the lower Waccamaw River with increasing incidence of suppressed growth in the 1980s and 1990s. The implications for restoration suggest that salinity intrusions will increase with frequency and intensity as sea levels continue to rise whether under accelerated rates from climate change or not. As sea level continues to rise over the next century from regional subsidence and eustatic changes, it is expected that salinity intrusions will lead to higher salinity concentrations and circulation upstream compromising forest growth and function. Sites already experiencing elevated soil salinities and showing signs of growth suppression or tree mortality should be considered a low prospect for restoration of freshwater species in the absence of any managed freshwater flows. Under changing climate, it would be possible to calculate the range of river slope likely to be impacted by salinity incursions with a projected change in sea level. Improving our understanding of physical processes of coastal systems and biological attributes of associated species will improve our ability to plan and predict the outcome of protected and restored wetlands along our coasts.

15.7. Acknowledgments

This material is based upon work supported by the USGS Global Climate Change Program and by the CSREES/USDA, under project number SC-1700271. Technical Contribution No. 5281 of the Clemson University Experiment Station. The authors thank Drs. Dave Stahle and Gary Shaffer for their helpful comments on this manuscript.

References

Allen JA, Pezeshki SR, Chambers JL (1996) Interaction of flooding and salinity stress on baldcypress (*Taxodium distichum*). Tree Physiol 16:307-313

Bowers LJ, Gosselink JG, Patrick Jr WH, Choong BT (1990) Investigation of six anatomical and four statistical features of bald-cypress (*Taxodium distichum*) tree rings. For Ecol Manage 33/34:503-508

Brown S (1981) A comparison of the structure, primary productivity, and transpiration of cypress ecosystems in Florida. Ecol Monogr 51:403-427

Cleaveland MK (2000) A 963-year reconstruction of summer (JJA) streamflow in the White River, Arkansas, USA, from tree-rings. Holocene 10:33-41

Conner WH (1994) Effect of forest management practices on southern forested wetland productivity. Wetlands 14:27-40

Conner WH, Flynn K (1989) Growth and survival of baldcypress (*Taxodium distichum*) planted across a flooding gradient in a Louisiana bottomland hardwood forest. Wetlands 9:207-217

Day RH, Doyle TW, Draugelis-Dale RO (2006) Interactive effects of substrate, hydroperiod, and nutrients on seedling growth of *Salix nigra* and *Taxodium distichum*. Environ Exp Bot 55:163–174

Dicke SG, Toliver JR (1990) Growth and development of baldcypress/water tupelo stands under continuous versus seasonal flooding. For Ecol Manage 33/34:523-530

Douglas BC (1991) Global sea level rise. J Geophys Res 96:6981–92

Douglas BC (1997) Global sea rise: A redetermination. Surveys in Geophysics 18:279-292

Doyle TW, Gorham LE (1996) Detecting hurricane impacts and recovery from tree rings. In: Dean JS, Meko DM, Swetnam TW (eds) Tree rings, environment, and humanity. Radiocarbon 1996, Department of Geosciences, The University of Arizona, Tucson, pp 405-412

Doyle TW, Day RH, Biagas JM (2003) Predicting coastal retreat in the Florida Big Bend region of the Gulf Coast under climate change induced sea-level rise. In: Ning ZH, Turner RE, Doyle T, Abdollahi K (eds) Integrated assessment of the climate change impacts on the Gulf Coast Region. GRCCC and LSU Graphic Services, Baton Rouge, pp 201-210

Fye FK, Stahle DW, Cook ER (2003) Paleoclimatic analogs to twentieth-century moisture regimes across the United States. Am Meteorological Soc 84:901-909

Harms WR, Schreuder HT, Hook DD, Brown CL, Shropshire FW (1980) The effects of flooding on the swamp forest in Lake Ocklawaha, Florida. Ecology 61:1412-1421

Hesse ID, Day Jr JW, Doyle TW (1998) Long-term growth enhancement of baldcypress (*Taxodium distichum*) from municipal wastewater application. Environ Manage 22:119-127

Keeland BD (1994) The effects of hydrologic regime on diameter growth of three wetland tree species of the southeastern USA. Ph.D. thesis, University of Georgia

Kellison RC, Young MJ, Braham RR, Jones EJ (1998) Major alluvial floodplains. In: Messina MG, Conner WH (eds) Southern forested wetlands ecology and management. Lewis Publishers, Boca Raton, pp 291-319

Krauss KW, Chambers JL, Allen JA (1998) Salinity effects and differential germination of several half-sib families of baldcypress from different seed sources. New Forests 15:53-68

Krauss KW, Chambers JL, Allen JA, Soileau Jr DM, DeBosier AS (2000) Growth and nutrition of baldcypress families planted under varying salinity regimes in Louisiana, USA. J Coastal Res 16:153–163

Latimer SD, Devall MS, Thomas C, Ellgaard EG, Satish KD, Thien LB (1996) Dendrochronology and heavy metal deposition in tree rings of baldcypress. J Environ Qual 25:1411-1419

Lugo AE, Brown SL (1984) The Oklawaha River forested wetlands and their response to chronic flooding. In: Ewel KC, Odum HT (eds) Cypress swamps. University of Florida Press, Gainesville, pp 365-373

Megonigal PJ, Day FP (1992) Effects of flooding on root and shoot production of bald cypress in large experimental enclosures. Ecology 73:1182-1193

Mitsch WJ, Ewel KC (1979) Comparative biomass and growth of cypress in Florida wetlands. Am Midl Nat 101:417-426

Pezeshki SR, Santos MI (1998) Relationships among rhizosphere oxygen deficiency, root restriction, photosynthesis, and growth in baldcypress (*Taxodium distichum* L.) seedlings. Photosynthetica 35:381-390

Pezeshki SR, DeLaune RD, Patrick Jr WH (1987) Response of baldcypress (*Taxodium distichum* L. var. *distichum*) to increases in flooding salinity in Louisiana's Mississippi River Deltaic Plain. Wetlands 7:1–10

Phipps RL (1985) Collecting, preparing, cross-dating, and measuring tree increment cores. Water Resources Investigations Rep 85-4148. U.S. Geological Survey, Lakewood

Ratard MAG (2004) Factors affecting growth and regeneration of baldcypress in a South Carolina tidal freshwater swamp. Ph.D. thesis, Clemson University

Reams GA, Van Deusen PC (1998) Detecting and predicting climatic variation from old-growth baldcypress. In: Mickler RA, Fox S (eds) The productivity and sustainability of southern forest ecosystems in a changing environment. Springer-Verlag, New York, pp 701-716

Ross M, O'Brien J, Sternberg L (1994) Sea level rise and the decline of Pine Rockland forests in the Lower Florida Keys. Ecol Appl 4(1):144-156

Souther RF, Shaffer GP (2000) The effects of submergence and light on two age classes of baldcypress (*Taxodium distichum* (L.) Richard) seedlings. Wetlands 20:697-706

Stahle DW, Cleaveland MK (1992) Reconstruction and analysis of spring rainfall over the southeastern United-States for the past 1000 years. Bull Am Meteorol Soc 73:1947-1961

Stahle DW, Cleaveland MK (1994) Tree-ring reconstructed rainfall over the southeastern USA during the Medieval warm period and Little Ice-Age. Climatic Change 26:199-212

Stahle DW, Cleaveland MK, Hehr JG (1985a) A 450-year drought reconstruction for Arkansas, United States. Nature 316:530-532

Stahle DW, Cook ER, White JW(1985b) Tree-ring dating of baldcypress and the potential for millenna-long chronologies in the southeast. Am Antiquity 50(4):796-802

Stahle DW, Cleaveland MK, Hehr JG (1988) North Carolina climate changes reconstructed from tree Rings - AD 372 to 1985. Science 240:1517-1519

Stahle DW, Cleaveland MK, Cerveny RS (1991) Tree-ring reconstructed sunshine duration over central USA. International J Climatol 11:285-295

Stahle DW, Cleaveland MK, Blanton DB, Therrell MD, Gay DA (1998) The Lost Colony and Jamestown droughts. Science 280:564-567

Stokes MA, Smiley TL (1968) An introduction to tree ring dating. University of Chicago Press, Chicago

U.S. Fish and Wildlife Service (1997) Proposed establishment of Waccamaw National Wildlife Refuge. U.S. Department of the Interior, U.S. Fish and Wildlife Service, Southeastern Region, Atlanta

Williams K, Ewel KC, Stumpf RP, Putz FE, Workman TW (1999) Sea-level rise and coastal forest retreat on the west coast of Florida, USA. Ecology 80:2045-2063

Yanosky TM, Hupp CR, Hackney CT (1995) Chloride concentrations in growth rings of *Taxodium distichum* in a saltwater-intruded estuary. Ecol Appl 5:785-792

Young PJ, Keeland BD, Sharitz RR (1995) Growth response of baldcypress [*Taxodium distichum* (L.) Rich.] to an altered hydrologic regime. Am Midl Nat 133:206-21

Chapter 16 - Conservation and Use of Coastal Wetland Forests in Louisiana*

Stephen P. Faulkner[1], Jim L. Chambers[2], William H. Conner[3], Richard F. Keim[2], John W. Day[4], Emile S. Gardiner[5], Melinda S. Hughes[2], Sammy L. King[6], Kenneth W. McLeod[7], Craig A. Miller[2], J. Andrew Nyman[2], and Gary P. Shaffer[8]

[1]*U.S. Geological Survey, National Wetlands Research Center, 700 Cajundome Blvd., Lafayette, LA 70506*
[2]*School of Renewable Natural Resources, Louisiana State University Ag Center, Renewable Natural Resources Building, Baton Rouge, LA 70803*
[3]*Baruch Institute of Coastal Ecology and Forest Science, Clemson University, Department of Forestry and Natural Resources, PO Box 596, Georgetown, SC 29442*
[4]*Department of Oceanography and Coastal Sciences and Coastal Ecology Institute, School of the Coast & Environment, Louisiana State University, Baton Rouge, LA 70803*
[5]*Center for Bottomland Hardwoods Research, USDA-Forest Service Southern Hardwoods Laboratory, PO Box 227, Stoneville, MS 38776*
[6]*USGS Louisiana Cooperative Fish and Wildlife Research Unit, Louisiana State University Ag Center, School of Renewable Natural Resources, Baton Rouge, LA 70803*
[7]*Savannah River Ecology Laboratory, PO Drawer E, Aiken, SC 29802*
[8]*Department of Biological Sciences, Southeastern Louisiana University, Box 10736, Hammond, LA 70402*

16.1. Introduction

The natural ecosystems of coastal Louisiana reflect the underlying geomorphic processes responsible for their formation. The majority of Louisiana's wetland forests are found in the lower reaches of the Mississippi

---

* The U.S. Government's right to retain a non-exclusive, royalty-free licence in and to any copyright is acknowledged.

Alluvial Valley and the Deltaic Plain. The sediments, water, and energy of the Mississippi River have shaped the Deltaic Plain as natural deltas have been formed and abandoned over the last 5,000 years (Coleman et al. 1998). During the regressive or constructional phase of the delta cycle, the system is dominated by freshwater riverine inputs with the formation of corresponding freshwater marshes and swamps, which then deteriorate during the marine-dominated transgressive phase (Roberts 1997). These processes have resulted in the current coastal landscape of bottomland hardwood forests on the remnant natural levees of the distributary channels with swamps dominated by baldcypress (*Taxodium distichum* [L.] L.C. Rich.), pondcypress (*Taxodium ascendens* Brongn.) and water tupelo (*Nyssa aquatica* L.) occupying lower elevations (Penfound 1952; Mitsch and Gosselink 2003).

Historically, the coastal wetland forests in the Deltaic Plain were intimately connected to the Mississippi River and its distributaries. Annual pulses of freshwater, sediments, and nutrients that sustained these forests were dispersed during flood events. The construction of flood-control levees, however, has isolated these forests from these flood pulses. The cumulative effects of human activity (e.g., levees), eustatic (actual) sea-level rise, and tectonic activity, have resulted in high rates of subsidence that dominate the surface elevation and geomorphology of the Deltaic Plain (Saucier 1994; DeLaune et al. 2004). Deltaic Plain subsidence rates range from 0.3 to 0.9 cm yr^{-1} (Gagliano 1998) with relative (eustatic + subsidence) sea-level rise predicted to range from 50 to >100 cm over the next 100 years (Twilley et al. 2001). Since over half of the 809,000 ha of Louisiana's forested wetlands occur in the coastal parishes (Figure 16.1), these natural and anthropogenic changes in hydrology and geomorphology have reduced productivity in many coastal wetland forest areas and have caused the complete loss of forest cover in some places.

Nearly all of the cypress-tupelo forest in Louisiana today are second growth, originating as natural regeneration after logging about 100 years ago (Norgress 1947; Mancil 1972). The area of greatest commercial timber production included all of the alluvial floodplain of the Mississippi River, but was mainly concentrated in the area south of Baton Rouge. The period of maximum harvest was the 1890s through the 1930s (Figure 16.2) and many of these forests have not regenerated with cypress-tupelo (Mattoon 1915; Norgress 1947; Mancil 1972).

Little is known about the present state of cypress ecosystems at the scale of the entire coastal Louisiana region. This knowledge gap has developed because of physical inaccessibility, lack of active forest management following the period of intense logging in the early 20th century, and the

Chapter 16 - Conservation and Use of Coastal Forests in Louisiana 449

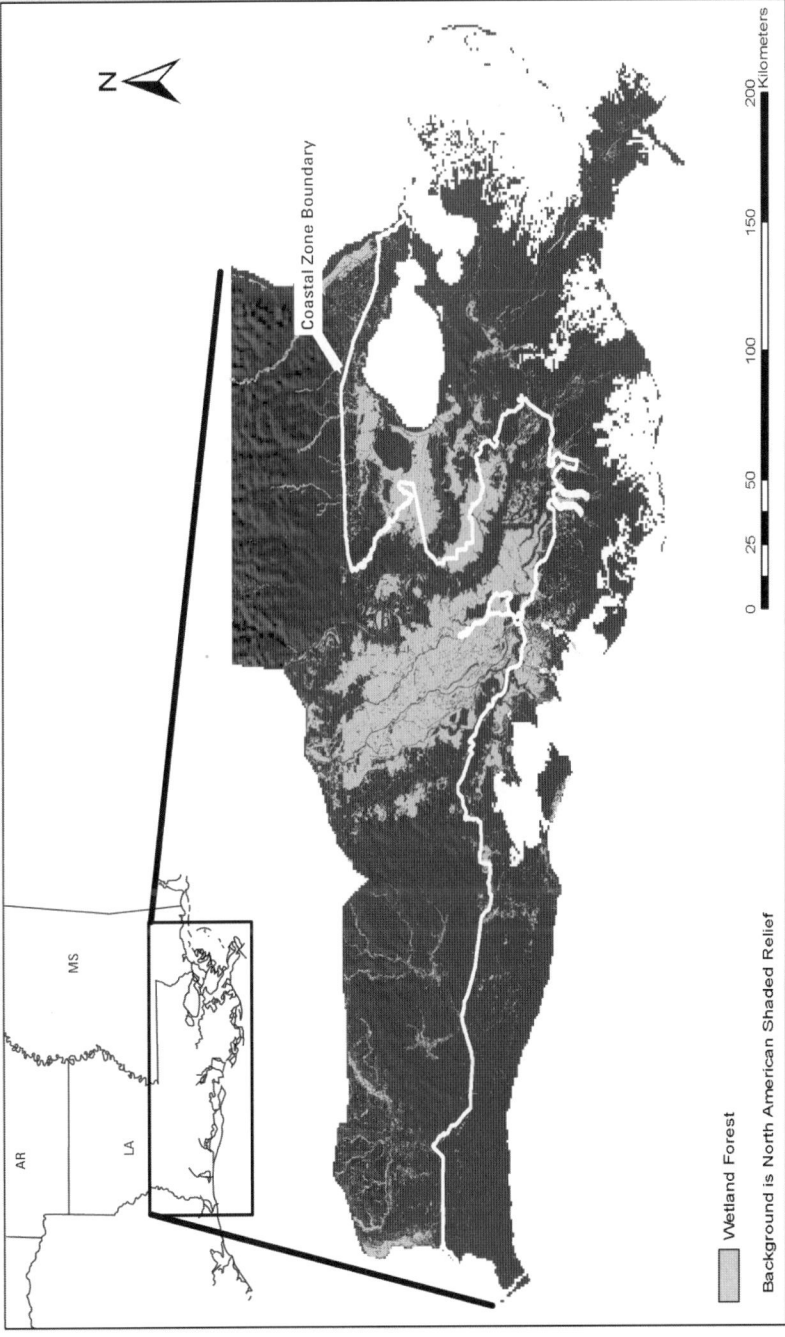

Fig. 16.1. Map of forested wetlands (shown in lighter shading) in southern Louisiana. The white line indicates the approximate northern boundary of the coastal zone.

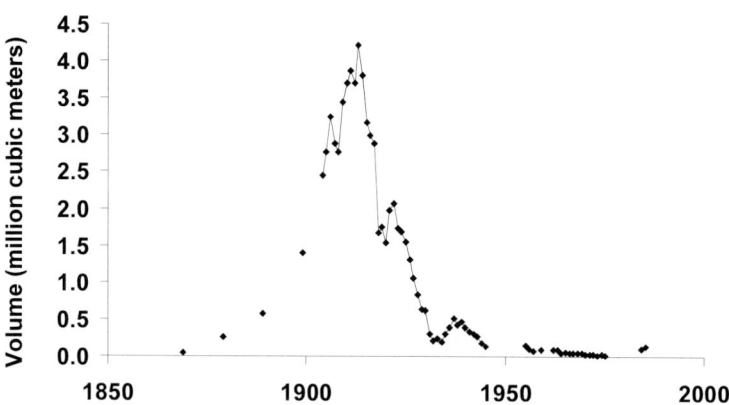

Fig. 16.2. Volume of baldcypress cut in Louisiana from the late 1870's to the 1980's (Data from Louisiana Department of Conservation 1943; Steer 1948; Louisiana Forestry Commission 1957; Louisiana Forestry Commission Progress Reports 1956-76; Mistretta and Bylin 1987).

emphasis on coastal marshes in coastal restoration planning. Conner and Toliver (1990) reported that cypress-dominated ecosystems of coastal Louisiana have experienced widespread hydrological, biogeochemical, and biological changes over the past century, and declines in some populations have been apparent. Numerous scientific reports of cypress-tupelo forest death and decline beginning in the 1980s have raised concerns regarding the long-term viability of Louisiana's coastal wetland forests (Brinson et al. 1985; DeLaune et al. 1987; Pezeshki et al. 1990; Allen 1992; Conner 1993; Gresham 1993; Williams 1993; Krauss et al. 2000; Conner and Inabinette 2003).

Consequently, concern over the decline of swamp forests in south Louisiana has increased in recent years and scientists are examining if harvesting under inappropriate conditions can result in nonviable regeneration preventing long-term establishment of new forests. Demand for forest products such as cypress mulch has spurred new investments and interest in harvesting cypress swamps. Approximately 60% of the landscape mulch sold in Florida is made from cypress (Duryea 2001). Quantitative data on cypress logging are not readily available for Louisiana; however, a new cypress sawmill was opened in Roseland, Louisiana in 2005 (Louisiana Department of Agriculture and Forestry 2005). Both the U.S. Army Corps of Engineers (ACE) and the U.S. Environmental Protection Agency (EPA) have reported increased interest in cypress logging in south Louisiana including a proposed 20,000 ha cypress mulching operation (Bruza 2006; Ettinger 2006). The Louisiana Coastal Wetlands Conservation and Restoration Task Force (1998) concluded that up to 93,000 ha of swamps in coastal Louisiana would be lost by 2050, despite current restoration efforts. This figure does not include additional losses that may occur with the renewed interest in harvesting existing baldcypress forest in south Louisiana or losses to development and agriculture.

A multidisciplinary approach summarizing the available science and providing management recommendations was needed to address these issues. In response, the Louisiana Governor's office formed the Coastal Wetland Forest Conservation and Use Science Working Group (SWG) and an associated Advisory Panel in early 2004 to provide the Governor with information and suggestions of strategies for environmental and economic utilization, conservation, and protection of Louisiana's coastal wetland forest ecosystem. The SWG was comprised of coastal scientists representing a variety of disciplines and expertise in coastal wetland forest issues.

The mission of the SWG was to provide information and guidelines for the long-term utilization, conservation, and protection of Louisiana's coastal wetland forest ecosystem, from both environmental and economic perspectives. To accomplish this mission, the following objectives were developed by the SWG (Chambers et al. 2005):

(1) Gather and synthesize scientific information available on regeneration, growth, and potential harvesting effects on coastal wetland forests.
(2) Gather and summarize field information on general characteristics of previously harvested cypress and tupelo forest stands to evaluate their potential to regenerate, become established, and remain vigorous.
(3) Review existing laws, regulations, policy, and guidelines affecting coastal forestry activities (and current forest conditions).
(4) Develop science-based, interim guidelines for the conservation and utilization of coastal wetland forests.
(5) Identify critical areas of priority research needed to refine these interim guidelines.

A policy oriented Advisory Panel on Coastal Wetland Conservation and Use was also established by the Governor's Office to advise the SWG. The SWG was able to hear concerns and needs of various interest groups within the Advisory Panel, including private landowners, environmental groups, forest industry, nongovernmental organizations, and state and federal agencies through regular input from the Advisory Panel. Regular interaction between the SWG and the Advisory Panel encouraged a balance between conservation, restoration, and use of coastal renewable natural forest resources.

The SWG published a draft report and solicited comments from the Governor's Advisory Panel and the public. The comments submitted to the SWG and to the Governor's office were reviewed by the SWG, and a revised final report was submitted to the Governor's office in 2005 (Chambers et al. 2005). This chapter summarizes the primary issues and findings of the SWG and those readers wanting a more complete understanding of these issues are referred to the full report (Chambers et al. 2005). The efforts and findings of the SWG represent a case study in applying scientific data and ecological concepts to help resolve complex resource management issues in landscapes that include tidal freshwater swamp ecosystems in Louisiana.

16.2. Science Working Group findings

Louisiana's coastal wetland forests are of tremendous economic, ecological, cultural, and recreational value to the people of Louisiana and the United States. They provide flood protection, water quality improvement, storm protection, and mitigate greenhouse gas emissions through carbon sequestration. The coastal wetland forests are habitat for threatened species such as the Louisiana black bear (*Ursus americanus luteolus*) and the bald eagle (*Haliaeetus leucocephalus*). Virtually all of the eastern neotropical migrant land bird species in the United States and numerous species from the western United States migrate through the coastal forests of Louisiana during spring and fall migration (Lowery 1974; Barrow et al. 2000). These forests also support billions of dollars of economic benefits from fishing, crawfishing, hunting, timber harvesting, and ecotourism. Based on current stumpage volumes and rates, the value of the standing cypress-tupelo timber in the area delineated by the SWG has been estimated by the Louisiana Department of Agriculture and Forestry to be $3.3 billion.

The functions and ecosystem services of Louisiana's coastal wetland forests are threatened by both large- and small-scale hydrologic and geomorphic alterations and by conversion of these forests to other uses. Subsidence, sea-level rise, canal dredging, and levee construction are the principal large-scale hydrologic and geomorphic alterations responsible for the loss of Louisiana's coastal wetland ecosystems including coastal wetland forests. The cumulative effects of small-scale or local factors can be of equal or greater importance in coastal wetland forest loss and degradation than large-scale alterations. These factors include increased depth and duration of flooding, saltwater intrusion, nutrient and sediment deprivation, herbivory, invasive species, and direct loss caused by conversion. Under less severe impacts, many of the important functions and ecosystem services are lost or degraded even though the trees may be intact and the forest may appear unaffected. Without appropriate human intervention to alleviate the factors causing degradation, most of coastal Louisiana will inevitably experience the loss of coastal wetland forest functions and ecosystem services through conversion to open water, marsh, or other land uses.

Spatially explicit data of coastal wetland forest conditions necessary to guide restoration, regulatory, and management efforts are scarce. While there are several inventory and classification programs [e.g., U.S. Forest Service (FS) Forest Inventory and Analysis, National Land Cover and Data, U.S. Fish and Wildlife Service (FWS) National Wetlands Inventory), there is no single, spatially explicit database that provides an accurate assessment of historic and current coastal wetland forest conditions. The actual acreage and amount of cypress-tupelo forests loss through time in south Louisiana has not been well documented mainly because of the various ways the resource has been measured in the past (Norgress 1947; Mancil 1972; Conner and Toliver 1990). Some of the limitations of the currently available datasets include changing definitions of forest cover type through time and lack of adequate sample points throughout the entire range of the cypress-tupelo forest cover type. The current Louisiana Coastal Zone Boundary, which is a programmatic boundary established by legal statute, does not accurately reflect the full extent of Louisiana's coastal wetland forests. As a result, these forest areas are more vulnerable to loss and degradation from detrimental impacts since large-scale restoration and protection activities are focused on those areas inside the Louisiana Coastal Zone.

Regeneration is a critical process of specific concern in maintaining coastal wetland forest resources. Successful natural regeneration of this resource following harvesting in the early 1900's was due to more favorable conditions existing at that time (primarily a drier hydrologic regime since 100 years of subsidence had not lowered the soil elevation). The hydrologic and geomorphic alterations identified above have led to increased flooding depths and durations under current conditions resulting in a lack of regeneration following harvesting in coastal cypress-tupelo forests today. In those areas where flooding prevents or limits the natural regeneration of the cypress-tupelo forest, artificial regeneration through tree planting is the only currently viable mechanism to regenerate the forest. Some swamps are altered to such a significant extent that even artificial regeneration is not possible. A review of the available scientific literature and a field survey of eighteen previously harvested sites indicate that coppice or stump sprouting does not provide sufficient numbers of viable trees to reliably regenerate the forest, even under optimum conditions, e.g., trees less than 60 years old harvested during the dormant season (Mattoon 1915; Williston et al. 1980; Kennedy 1982; Conner et al. 1986; Gardiner et al. 2000; Chambers et al. 2005).

Conditions affecting the potential for forest regeneration and establishment are recognizable based upon existing site-based biological and physical factors. The SWG developed a set of condition classes for the dominant wetland forest type in Louisiana's coastal cypress-tupelo forests. Assuming average rainfall conditions and no extreme or unusual events, the SWG set the general description of the three condition classes for cypress-tupelo as follows (see Chambers et al. 2005 for more details and discussion):

SWG Condition Class I: Sites with Potential for Natural Regeneration
> Sites that are generally connected to a source of fresh surface or ground water and flooded or ponded periodically on an annual basis. These sites must have seasonal flooding and dry cycles (pulsing), have both sediment and nutrient inputs, and are not subsiding.

SWG Condition Class II: Sites with Potential for Artificial Regeneration Only
> These sites may have overstory trees with full crowns and few signs of canopy deterioration but are either permanently flooded (which prevents seed germination and seedling establishment of cypress and tupelo) or are flooded deeply enough that when natural regeneration does occur during low water, seedlings cannot grow tall enough between flood events for at least 50% of their crown to remain above the high water level during the growing season. Water depth is restricted to a maximum of two feet for practical reasons related to planting and production of tree seedlings. These conditions require artificial regeneration (i.e., planting of tree seedlings).

SWG Condition Class III: Sites with No Potential for Either Natural or Artificial Regeneration
> These sites are either flooded for periods long enough to prevent natural regeneration and practical artificial regeneration, or they are subject to saltwater intrusion with salinity levels that are toxic to

cypress-tupelo forests (two ppt for tupelo; four ppt for cypress). Two trajectories are possible for these two conditions: (1) freshwater forests transitioning to either floating marsh or open fresh water, or (2) forested areas with saltwater intrusion that are transitioning to open brackish or saltwater (marsh may be an intermediate condition).

16.3. Science Working Group recommendations

The SWG made recommendations to the Governor's Office regarding actions that the state could take to conserve and protect these forests (Chambers et al. 2005). In general terms, the SWG recommended the following:
- Make conserving, restoring, and managing these coastal wetland forests a state priority.
- Recognize the set of three condition classes outlined by the SWG relative to regeneration of specific sites.
- Place priority on maintaining the hydrological regime of the most productive sites while avoiding loss of the more sensitive sites.
- Place a moratorium on harvesting state-owned Condition Class III forests, and seek ways to delay harvesting those private forest lands not likely to regenerate until site environmental conditions are changed (consider use of incentives).
- Help to ensure proper management and regeneration through written forest management plans with specific actions to ensure regeneration of cypress-tupelo forest.
- Develop spatially explicit databases, and establish and maintain long-term monitoring efforts to guide management decisions.
- Recognize an expanded area for coastal forest conservation beyond the current Louisiana Coastal Zone Boundary.
- Ensure that all agencies and organizations share and coordinate information, develop practices

to prevent coastal wetland forest loss, and actively pursue restoration of degraded forests.
- Enhance ecosystem functions through hydrological management decisions related to construction and other activities in wetland areas.

16.4. Summary and conclusions

Establishment of the SWG is an important example of developing a multidisciplinary, science-based solution to address resource management problems. The SWG consolidated the scientific knowledge related to the functions, ecosystem services, and underlying processes affecting the condition of Louisiana's coastal wetland forests and made recommendations based on that science-based assessment. The collaboration between scientists, resource managers, and policy makers continues in the effort to implement the SWG recommendations. An outside group composed of members of the Advisory Panel is currently working on a set of Interim Forest Practices Guidelines relative to harvesting and regeneration of coastal wetland forests in Louisiana. A coastal forest regeneration research meeting was held in November 2005 to prioritize research related to regeneration of coastal wetland forests. The state of Louisiana has conducted public hearings and taken public comments on the SWG's final report. Future actions include recommendations from the Advisory Panel regarding the SWG report and balancing conservation, restoration, and use of coastal wetland forest resources; developing methods to identify the SWG Condition Classes on the ground and map them using remotely sensed data; providing a set of incentives to encourage landowner conservation of coastal wetland forests; and developing and implementing methods to restore coastal wetland forests.

Hydrology in much of south Louisiana has been altered over the years by several factors including levee construction, pipeline operations, oil and natural gas exploration, shipping concerns, and subsidence. Current forestry best management practices have failed to take into account the impact of these changes on the regeneration and productivity of coastal wetland forests in Louisiana. Established forestry activities in wetlands (including harvesting) are generally exempt from permit requirements under Section 404 of the Clean Water Act, however, concerns over the viable regeneration of cypress-tupelo forests have arisen regarding these exemptions. Based in part on the findings of the SWG, the EPA has recently ruled that a landowner in Louisiana would not be exempt from the Section

404 permit requirement unless he could establish that the cypress forest would successfully regenerate (EPA 2006; Mississippi Interstate Cooperative Resource Association 2006). Sustainable and long-term use of coastal wetland forests will require major changes in the state's efforts to restore ecosystem processes (e.g. sediment and nutrient input, flood pulsing) and the careful coordination among agencies (local, state, and federal), landowners, and industries working in coastal areas to ensure that the best scientific knowledge is used to develop management practices that result in the viable regeneration of these forest ecosystems. A concerted effort expanding sound research knowledge of these forests and their restoration is also critical to their future.

References

Allen JA (1992) Cypress-tupelo swamp restoration in southern Louisiana. Restor Manage Notes 10:188-189

Barrow WC, Hamilton RB, Powell MA, Ouchley K (2000) Contribution of landbird migration to the biological diversity of the northwest gulf coastal plain. Texas J Sci 52:151-172

Brinson MM, Bradshaw HD, Jones MN (1985) Transitions in forested wetlands along gradients of salinity and hydroperiod. J Elisha Mitchell Sci Soc 101(2):76-94

Bruza, JD (2006) Personal Communication. US Army Corps of Engineers, New Orleans District Regulatory Branch

Chambers JL, Conner WH, Day JW, Faulkner SP, Gardiner ES, Hughes MS, Keim RF, King SL, McLeod KW, Miller CA, Nyman JA, Shaffer GP (2005) Conservation, protection and utilization of Louisiana's coastal wetland forests. Final Report to the Governor of Louisiana from the Coastal Wetland Forest Conservation and Use Science Working Group. Louisiana Governor's Office of Coastal Activities, Baton Rouge

Coleman JM, Roberts HH, Stone GW (1998) Mississippi River Delta: An overview. J Coast Restor 14:698-717

Conner WH (1993) Artificial regeneration of baldcypress in three South Carolina forested wetland areas after Hurricane Hugo. In: Brissette JC (ed) Proceedings of the seventh biennial southern silvicultural research conference. General Technical Report SO-93. U.S. Department of Agriculture, Forest Service, Southern Forest Experiment Station, New Orleans, pp 185-188

Conner WH, Inabinette LW (2003) Tree growth in three South Carolina (USA) swamps after Hurricane Hugo: 1991-2001. For Ecol Manage 182:371-380

Conner WH, Toliver JR (1990) Long-term trends in the baldcypress (*Taxodium distichum* (L.) Rich.) resource in Louisiana. For Ecol Manage 33/34:543-557

Conner WH, Toliver JR, Sklar FH (1986) Natural regeneration of baldcypress (Taxodium distichum (L.) Rich.) in a Louisiana swamp. For Ecol Manage 14:305-317

DeLaune RD, Pezeshki SR, Patrick Jr WH (1987) Response of coastal plants to increase in submergence and salinity. J Coast Res 3(4):535-546

DeLaune RD, Callaway JC, Patrick Jr WH, Nyman JA (2004) An analysis of marsh accretionary processes in Louisiana coastal wetlands. In: Davis DW, Richardson M (eds) The coastal zone: Papers in honor of H. Jesse Walker. Geoscience Publications, Department of Geography and Anthropology, Louisiana State University, Baton Rouge, pp 113-130

Duryea ML (2001) Landscape mulches: What are the choices in Florida? Florida Cooperative Extension Service Publication FOR 80. Institute of Food and Agricultural Sciences, University of Florida, Gainesville

Ettinger J (2006) Personal Communication. USEPA Region 6 Water Quality Protection Division

Gagliano SM (1998) Faulting, subsidence, and land loss in coastal Louisiana. Coastal Environments Inc., Baton Rouge

Gardiner ES, Russell Jr DR, Hodges JD, Fristoe TC (2000) Impacts of mechanical tree felling on development of water tupelo regeneration in the Mobile Delta, Alabama. South J Appl For 24:65-69

Gresham CA (1993) Changes in baldcypress-swamp tupelo wetland soil chemistry caused by Hurricane Hugo induced saltwater inundation. In: Brissette JC (ed) Proceedings of the seventh biennial southern Silvicultural research conference. General Technical Report SO-93. U.S. Department of Agriculture, Forest Service, Southern Research Station, New Orleans, pp. 171-175

Kennedy Jr HE (1982) Growth and survival of water tupelo coppice regeneration after six growing seasons. South J Appl For 6:133-135

Krauss KW, Chambers JL, Allen JA, Soileau Jr DM, DeBosier AS (2000) Growth and nutrition of baldcypress families planted under varying salinity regimes in Louisiana, USA. J Coastal Res 16:153-163

Louisiana Coastal Wetlands Conservation and Restoration Task Force (1998) Coast 2050: Toward a sustainable coastal Louisiana. Louisiana Department of Natural Resources, Baton Rouge

Louisiana Department of Agriculture and Forestry (2005) Cypress sawmill opens today in Roseland. Press Release July 8, 2005

Louisiana Department of Conservation (1943) Report on timber production in Louisiana, 1939-1942. Division of Forestry, New Orleans

Louisiana Forestry Commission (1957) 1956 timber production in Louisiana. Louisiana Department of Conservation, Baton Rouge

Louisiana Forestry Commission (1956-1976) Biennial progress reports. Louisiana Department of Conservation, Baton Rouge

Lowery Jr GH (1974) Louisiana birds. Louisiana State University Press, Baton Rouge

Mancil E (1972) A historical geography of industrial cypress lumbering in Louisiana. Ph.D. thesis, Louisiana State University

Mattoon WR (1915) The southern cypress. Bulletin No. 272. U.S. Department of Agriculture, Washington

Mississippi Interstate Cooperative Resource Association (2006). River Crossings 15:9

Mistretta PA, Bylin CV (1987) Incidence and impact of damage to Louisiana's timber, 1985. Research Bulletin SO-117. U.S. Department of Agriculture, Forest Service, Washington

Mitsch WJ, Gosselink JG (2003) Wetlands, 3rd edn. Van Nostrand Reinhold, New York

Norgress RE (1947) The history of the cypress lumber industry in Louisiana. La Hist Quart 30:979-1059

Penfound WT (1952) Southern swamps and marshes. Bot Rev 18:413-446

Pezeshki SR, DeLaune RD, Patrick Jr WH (1990) Flooding and saltwater intrusion: potential effects on survival and productivity of wetland forests along the U.S. Gulf Coast. For Ecol Manage 33/34:287-301

Roberts HH (1997) Dynamic changes of the holocene Mississippi River delta plain: the delta cycle. J Coast Res 13:605-627

Saucier RT (1994) Geomorphology and quaternary geologic history of the Lower Mississippi Valley. Volume 1 (Text). US Army Corps of Engineers, Waterways Experiment Station, Vicksburg

Steer HB (1948) Lumber production in the United States, 1799-1946. Miscellaneous Publication No. 669. U.S. Department of Agriculture, Washington

Twilley RR, Barron EJ, Gholz HL, Harwell MA, Miller RL, Reed DJ, Rose JB, Siemann EH, Wetzel RG, Zimmerman RJ (2001) Confronting climate change in the Gulf Coast region: Prospects for sustaining our ecological heritage. Union of Concerned Scientists, Cambridge, and Ecological Society of America, Washington

U.S. Environmental Protection Agency (2006) Letter from Miguel Flores, Chief, Region 6 Water Quality Protection Division to Col. Richard Wagenaar, USACE New Orleans District, dated June 6, 2006

Williams TM (1993) Salt water movement within the water table aquifer following Hurricane Hugo. In: Brissette JC (ed) Proceedings of the seventh biennial southern silvicultural research conference. General Technical Report SO-93. U.S. Department of Agriculture, Forest Service, Southern Forest Experiment Station, New Orleans, pp 177-183

Williston HL, Shropshire FW, Balmer WE (1980) Cypress management: a forgotten opportunity. Forestry Report SA-FR 8. U.S. Department of Agriculture, Forest Service, Southeastern Area, Atlanta

Chapter 17 - Tidal Freshwater Forested Wetlands: Future Research Needs and an Overview of Restoration*

William H. Conner[1], Courtney T. Hackney[2], Ken W. Krauss[3], and John W. Day, Jr.[4]

[1]*Baruch Institute of Coastal Ecology and Forest Science, Clemson University, Box 596, Georgetown, SC 29442*
[2]*Department of Biology and Marine Biology, University of North Carolina at Wilmington, Wilmington, NC 28403*
[3]*U.S. Geological Survey, National Wetlands Research Center, 700 Cajundome Blvd., Lafayette, LA 70506*
[4]*Department of Oceanography and Coastal Sciences and Coastal Ecology Institute, School of the Coast & Environment, Louisiana State University, Baton Rouge, LA 70803*

17.1. Introduction

Studies of tidal freshwater forested wetlands are few in contrast to the diversity of conditions and information needs that exist for this ecosystem type. Basic information is lacking on the physiological ecology of major wetland tree species under natural settings, the structure and dynamics of pure and mixed species communities, soil-plant interactions, biogeochemistry, hydrology, soils, wildlife habitat, primary biotic and abiotic functions, and the response of these systems to natural and human-caused disruptions. Existing information is often not in a form that can be applied to ecosystem problems, especially those related to management, restoration, or creation of tidal swamps. Accordingly, there is a critical need for research on fundamental biotic and abiotic processes and functions in tidal forested wetland landscapes on a local and regional scale. In this chapter, we detail those research needs, and we highlight some restoration ideas for tidal freshwater forested wetlands with the hope that much additional research will follow.

---

* The U.S. Government's right to retain a non-exclusive, royalty-free licence in and to any copyright is acknowledged.

17.2. Hydrogeomorphic setting

Tidal swamps exist in a dynamic equilibrium in which the influence of tides diminishes as waterborne sediments and nutrients either are added to the wetland surface with each tide or are eroded or entrained by ebbing tides. In addition, autochthonous organic inputs from vegetation, which are not removed by tides, may also raise the surface elevation closer to a point where flooding by tides does not occur. For tidal swamps to maintain themselves, relative sea-level rise must equal the accumulation of inorganic and organic inputs. If relative sea level remains static, tidal swamps would eventually move toward bottomland hardwood/softwood forests as allochthonous inorganic, river-borne sediments, accumulated decaying leaves and stems, and in situ root production, elevate the swamp floor above the tides (Figure 17.1). Alternatively, when sea-level rise exceeds the input of sediments (allochthonous and autochthonous), tidal swamps

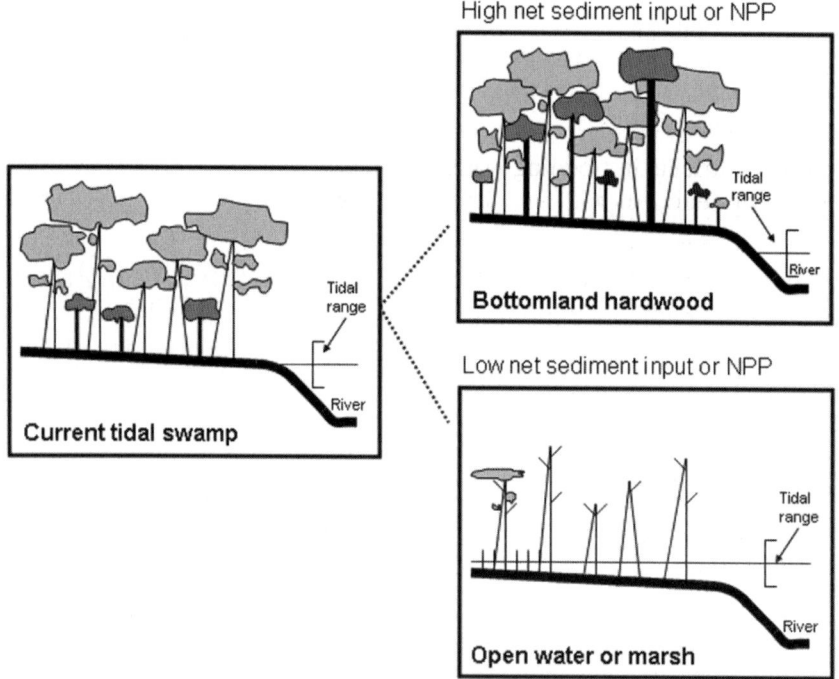

Fig. 17.1. Tidal swamps exist in a dynamic equilibrium in which sediment accumulation increases the surface elevation at the same rate as sea level rises. When sediment additions exceed sea-level rise, the swamp moves toward a bottomland hardwood system, while the opposite scenario ultimately drowns the swamp, resulting in marsh or open water. NPP = Net Primary Production.

are increasingly flooded. In this scenario, swamp forest would gradually shift to herbaceous marsh and ultimately become open water in a process called reverse succession (Isaak et al. 1959). During this transition, the swamp surface becomes increasingly vertically heterogeneous, with woody vegetation growing almost exclusively on hummocks derived from fallen trees and stumps (Rheinhardt and Hershner 1992). The broad, flat coastal plains along the Gulf of Mexico and Atlantic coasts of the southeastern United States contain extensive tidal swamps in this stage of change.

Tidal swamps periodically connected to redwater rivers and streams high in suspended materials, during overbank flooding, would be more likely to maintain their elevation relative to sea level than would those associated with slow moving, blackwater streams, which are common on the outer coastal plain of the southeastern United States (Smock and Gilinsky 1992). Indeed, sediments in tidal swamps adjacent to redwater rivers in North Carolina contained more inorganic sediments than did those swamps adjacent to blackwater streams (see Chapter 8). Tidal swamps exposed to eutrophic conditions, either from natural or human activity, would likely accumulate organic material more rapidly and reach a condition above tides faster than tidal swamps in a more oligotrophic setting and would be more capable of maintaining their position relative to sea level than those with few nutrients in flooding water. However, if relative sea level exceeds the rate of sediment and autochthonous organic accumulation, then tidal flooding gradually becomes longer with deeper floodwaters in any setting. Human activities or natural changes that alter any of the forcing functions can potentially alter the equilibrium that produces and maintains tidal swamps.

A basic conceptual model suggests that there are several natural and unnatural ways by which tidal swamps can be altered or disturbed indirectly by human activity to the extent that they are converted to open water or bottomland hardwood/softwood forest (Figure 17.2). A more comprehensive conceptual model for tidal freshwater forested wetlands might look much like those developed for mangroves (cf., Davis et al. 2005) or for tidal freshwater marshes (cf., Simpson et al. 1983) because disturbance events, relative sea-level rise, freshwater flow volume, and water distribution also influence tidal freshwater swamps. Any change that decreases sediment input (e.g., local levees or dams upriver) or increases suspended sediment load of flood waters (e.g., development, boat traffic, fishing activities, etc.) would generally change the rate of accumulation on the swamp surface. Increased nutrients in flood water from sewage treatment plants, fish farms, animal operations, etc., typically lead to higher primary

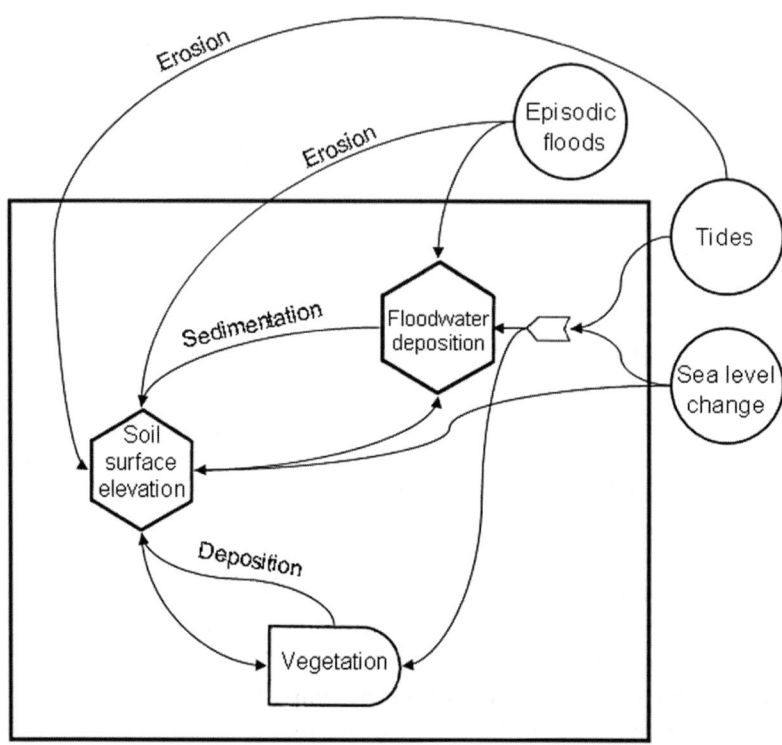

Fig. 17.2. The key element controlling the maintenance of tidal swamps is the soil surface elevation. Inorganic sediment deposition is largely controlled by episodic flooding (i.e., upstream floods) and tidal flooding. Increased flooding induced by a rise in sea level leads to increased deposition when suspended sediments are abundant in floodwater (e.g., in redwater rivers). Soil surface elevation is also related to the surface input of organic materials (leaves, twigs, stems, etc.) from tidal swamp vegetation. Note that roots and bulk stems may be added in situ by tree growth and thus increase surface elevation. Disturbances that lead to increased or decreased mineralization (e.g., change in flooding frequency or duration) can alter the rate of decomposition or organic matter accumulation. Not shown in this model is the importance of dissolved nutrients, which can alter both the growth rate and subsequent deposition of organic materials and the mineralization of organic soils.

production and thus more potential for organic matter accumulation on the wetland surface. This condition is especially true under flooded states where mineralization of organic material proceeds slowly and typically leads to the accumulation of peat. Inputs of nitrates, sulphates, or other oxidants favored by anaerobic soil microorganisms, however, can accelerate the mineralization of organic matter and affect soil surface elevation. Finally, tidal flooding is the product of the ocean tide, which varies greatly around the world, and is affected strongly by the geomorphology of the estuary, stream channel configuration, tidal range, and seasonal patterns of freshwater flow. Freshwater release from upstream dams and channel modification dramatically influence tidal flooding frequency, duration, and depth, which in turn can have dramatic influence on soil and vegetation of tidal freshwater forested wetlands.

17.3. Disturbance

Wetlands are affected by environmental change to a greater extent than upland ecosystems (Burkett and Kusler 2000). Wetland biota is especially sensitive to small changes in water tables and fragmentation by dams, highways, dikes, drainages, and other changes that restrict movement of organisms to areas more favorable to survival (Brinson 2006). As seen in previous chapters, tidal freshwater forested wetlands have been, and continue to be, subjected to a wide variety of impacts. Many wetlands, including those associated with rivers and tides, do not exist in a steady state condition. Instead, they are pulsed by important inputs, most notably flooding (Odum et al. 1995). Their landscape position in the upper intertidal zone adjacent to nontidal forests or uplands (Figure 17.3) makes them particularly susceptible to disturbances that exceed the normal pulsing, especially with respect to water level and saltwater intrusion. At some point, normal feedback mechanisms that allow tidal swamps to maintain position relative to sea level may cease to function (Figure 17.2). Continued sea-level rise combined with reduced freshwater input caused by upstream dams and land-use changes, as well as channel construction and dredging also results in saltwater intrusion and mortality of forests (Pezeshki et al. 1990; Allen et al. 1994; Chapter 1). Salinity stress exacerbates the stress of increased flooding (Allen et al. 1996; McLeod et al. 1996; Mendelssohn and Morris 2000), and multiple biotic and abiotic stressors combine to degrade ecosystem functions and services in coastal forests (Allen et al. 1998; Effler et al. 2006).

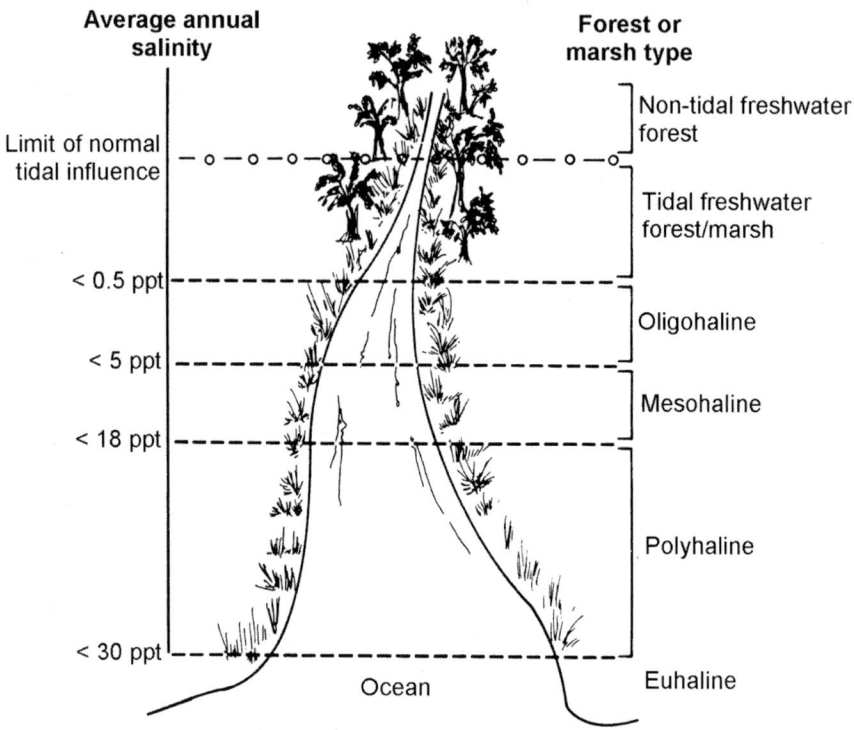

Fig. 17.3. The location of tidal freshwater forested wetlands along an idealized riverine landscape (after Odum et al. 1984). Although salinity in tidal freshwater forested wetlands is not normally expected to rise above 0.5 ppt, many chapters in this book indicate situations in which global climate change, drought, and/or human factors exacerbate salinities beyond 0.5 ppt in normally freshwater forested wetlands.

17.3.1. Salinity

Coordinated research over the past two decades has provided scientists and managers with a good knowledge base on species' tolerances to salinity (Chapter 14). Additional experimental studies may benefit specific tidal freshwater forested wetlands where dominant tree species have not been subjected to rigorous salt tolerance studies. For example, tidal Atlantic white cedar (*Chamaecyparis thyoides* [L.] B.S.P.) wetlands in Mississippi appear to be very susceptible to salinity, but it is uncertain whether prolonged sea-level rise is limiting Atlantic white cedar seedling regeneration or whether chronic or pulsed salinity encroachment is killing seedlings and mature trees (Chapter 4). Even among those species rigorously tested, assumptions are used to associate seedling, sapling, and mature tree responses. Future studies should consider principal field components that rate mature tree response. There are some discrepancies between how some tidal freshwater swamp species perform in greenhouse settings versus how they fare in the field. For example, water tupelo (*Nyssa aquatica* L.) has been documented as fairly tolerant to salinities up to 3 ppt in greenhouse experiments (Pezeshki 1987, 1990; Pezeshki et al. 1989), but it is difficult to find mature individuals at less than half of that salinity concentration in the field (Chapter 9). Determining how salinity affects reproduction in tidal freshwater swamp forests by influencing seed production and viability, or by influencing the long-term genetic structure of certain populations, offer important avenues for future research.

Furthermore, the frequency, duration, and magnitude of salinization events within tidal freshwater forested wetlands have important consequences for how effectively vegetation and soils can tolerate the perturbation. What is almost completely lacking in the literature on tidal freshwater forested wetlands is a description of whether salt enters as surface water storm surge, as groundwater storm surge, or during abnormal surface and groundwater pulses from spring tides. Determining the interrelatedness of site hydrology and salinity to define appropriate hydrological targets would be important in any effort to restore or enhance the ecological processes of tidal forests (cf., Lewis 2005). Hydrological evaluations are an important first step (e.g., Rheinhardt 1992; Rheinhardt and Hershner 1992; Chapter 2).

17.3.2. Flooding

Species' tolerances to flooding have been established from a fairly comprehensive network of field observations and greenhouse tests that were conducted to produce reasonable flood tolerance rankings for use by foresters, landscape modelers, and resource managers (McKnight et al. 1981; Hook 1984). Rankings do not consider tide, but tides not only can influence the physical environment of coastal wetlands but also can supply nutrients and oxygen while flushing soils of phytotoxic respiratory byproducts (Odum et al. 1983). The net effect might be larger or smaller species' pools for the colonization of tidally flooded sites than in permanently, or even seasonally, flooded sites with identical annual flood durations.

Important research contributions along these lines might include determining the short-term and long-term effects of tidal flooding on ecosystem dynamics, such as mature tree gas exchange, belowground production, soil carbon dioxide (CO_2), methane (CH_4), and nitrous oxide (N_2O) efflux, and stand-level atmospheric and hydrologic carbon exchange. It is certainly important to understand whether mature trees experience the same ecophysiological responses as do seedlings while flooded, such as changes in stomatal closure, and whether these changes differ with landscape position. Being able to extrapolate seedling response to trees would be useful, and knowing whether previous exposure to different frequencies and durations of tidal flooding (akin to landscape position) influences ecophysiological processes would define an overall ecosystem capacity for resilience to global climate change.

The response to tidal flooding can also determine aspects of stand water use and can influence coastal water conservation. Water use by tidal mangroves, for example, is conservative at both the leaf level (Ball 1986, 1988) and the individual tree level (Krauss et al. 2007). Water transiting mangrove stems, hence, contributes a relatively small fraction to overall wetland water loss through evapotranspiration, and thus stands reduce atmospheric water losses as intact, coastal forests. Low rates of water use by mangrove trees have been attributed to strategies for avoiding the uptake of excessive salts. On the other hand, cypress domes (dominated by pond cypress, *Taxodium ascendens* Brongn.) in south Florida demonstrate a similar capacity for water conservation even without the influence of salinity (Brown 1981; Brown et al. 1984). Water and phosphorus limitations influence water conservation in cypress domes (Brown 1981). Defining water-use strategies of tidal freshwater forested wetland stands in relation to a diversity of hydroperiod, salinity, and nutrient regimes would be a fruitful avenue of research and would define an often overlooked service provided by tidal swamps. It may be important to restore and/or preserve tidal swamp forests based on water conservation alone; however, emergent properties associated with water filtration, nutrient use, and carbon sequestration in forests with different hydrological signatures, for example, may be just as important.

17.3.3 Tropical storms

Coastal forested wetlands of the southern United States have developed with hurricanes as normal aperiodic events (Lugo et al. 1983; Conner et al. 1989; Weaver 1989; Boucher 1990; Michener et al. 1997). Because of their intermittent nature and unpredictable behavior, however, hurricane influences are difficult to study, and data concerning before and after responses of these ecosystems to storms are rare (Spiller et al. 1998). Although hurricane winds can cause extensive tree mortality in upland forests (Hedlund 1969; Gresham et al. 1991; Sharitz et al. 1993; Doyle et al. 1995; Rybczyk et al. 1995; Nix et al. 1996), wetland forests populated with baldcypress (*Taxodium distichum* [L.] L.C. Rich.), water tupelo, and swamp tupelo (*Nyssa biflora* Walt.) are generally very stable and suffer little damage or mortality because of wind (Gresham et al. 1991; Hook et al. 1991; Putz and Sharitz 1991; Duever and McCollom 1993; Sharitz et al. 1993; Loope et al. 1994; Doyle et al. 1995). Trees are toppled by hurricane winds, and shrubs are pushed over by subsequent flood waters associated with these events (Hackney, unpublished data). Downed trees and their root balls form the basis for hummocks, where most woody, less water tolerant species are found in tidal swamps (Rheinhardt and Hershner 1992).

Wind effects are only a portion of the concern. Heavy rains associated with some hurricanes can cause extreme flooding of coastal systems with fresh water. However, certain storm trajectories can vector floodwater salinity far upriver and impact even those tidal freshwater forested wetlands far removed from major centers of storm activity. For example, as Hurricane Katrina (2005) made landfall in southeast Florida on its trajectory from the Atlantic Ocean toward the Gulf of Mexico, appreciable storm surge and porewater salinity pulses were noted within tidal swamps along the Savannah River in Georgia and the Waccamaw River in South Carolina, hundreds of kilometers away.

Little is understood about the relationship between river salinity and porewater salinity in tidal freshwater forested wetlands. For example, one location along the main channel of the Savannah River often has salinities of up to 4 ppt but with no apparent influence on adjacent tidal freshwater forested wetlands, which have chronically low salinity levels (<0.1 ppt) (Chapter 2). Nutrient dynamics from surface water contribution during storm surge undoubtedly follow a similar pattern; high levels of salinity and nutrients in-channel may not equate as appreciably to tidal swamp condition as off-channel or surface water concentrations. Future research should attempt to link in-channel and swamp salinity and nutrient concentrations synoptically with different tidal ranges, volume of freshwater discharge, variable stage, and intensity of frontal passage in tidal swamps positioned along coastal rivers, tidal lakes, and embayments.

17.4. Biogeochemistry

The biogeochemistry of estuaries is complex and is dominated by strikingly different processes depending on landscape position (Bianchi et al. 1999). Landscape position places tidal freshwater forested wetlands at the cusp of many chemical transformations in the water column. Chemical transformations, along with hydroperiod and geomorphology, contribute to the nutrient load of flooding waters and can affect site fertility (G. Noe and C. Hupp, unpublished data). We address the basic needs for biogeochemistry research in tidal freshwater forested wetlands around three themes: nitrogen and phosphorus, remineralization, and nutrient uptake. We recognize, however, that biogeochemistry is much more than just nutrient cycling.

17.4.1 Nitrogen and phosphorus

Studies conducted in tidal saltwater forests (mangroves) have described important links to soil nitrogen (N) and phosphorus (P); however, very little of the literature associated with tidal freshwater forested wetlands has described soil nutrients, porewater characteristics, or floodwater chemistry. Research from the Pocomoke River (Chapter 5) indicates that P has a tight association with clay and mineral sediments while N is associated with organic matter (Noe and Hupp 2005). Evidence from Noe and Hupp (2005) also suggests that P dynamics are linked to hydrological connectivity in tidal swamps. Similar processes have been described for mangroves: the relative concentration of soil N:P, which is greatly affected by the frequency of tidal floods (McKee et al. 2002), has been shown to control forest productivity (Chen and Twilley 1998, 1999) and species composition (Sherman et al. 1998). These nutrients, and others, are important to consider in future studies; their role in controlling site quality and microbial processes in tidal freshwater forested wetlands will be critical for evaluating and predicting restoration success.

17.4.2. Remineralization

Microbial remineralization processes that convert organic carbon to gaseous forms of CO_2 and CH_4 are dominant in flooded, organic soils. While soils are flooded, oxygen diffusion is limited, and oxic respiration provides only a relatively small contribution to soil respiration via CO_2 efflux. The relative production of CO_2, hence, is directly related to soil oxygen state, so on many tidal freshwater forested wetland sites during years of typical rainfall depth, much remineralization is expected to occur via anaerobic methanogenesis or through other biochemical pathways (Wolfe and Higgins 1979; Chapter 8). As salinity enters the system, microbes associated with sulfate reduction begin to outcompete methanogens and eventually shut down CH_4 production (Martens and Berner 1974). Future research in tidal freshwater forested wetlands should establish several trajectories associated with aerobic and anaerobic soil remineralization: (1) determine the relative contributions of CO_2, CH_4, and hydrogen sulfide (HS) efflux to the atmosphere and to porewater with landscape forest position and/or community composition (proxy for relative stand degradation), (2) determine the role of pH in mediating these shifts, (3) determine the influence of drought in controlling the form of gaseous carbon loss to the atmosphere and porewater, and (4) develop multiple temporal profiles for describing CO_2, CH_4, and HS production with tidal flows and ebbs, diurnally and seasonally as soil temperatures change and aperiodically as normal rainfall years transition to drought years.

17.4.3. Nutrient uptake

Collectively, the accumulation of soil N and soil P in tidal freshwater forested wetlands of one watershed has been shown to range from 0.2 to 13.4 $g\ m^{-2}\ yr^{-1}$ (Noe and Hupp 2005); however, understanding how the soil surface interacts with flood waters to facilitate this deposition or transform nutrients would define a larger functional capacity of tidal swamp forests to provide important ecosystem services as restoration and protection of these systems are considered. Nutrient flumes provide a good technique for determining concentrations of dissolved inorganic N (NH_4^+, NO_2^-, NO_3^-), dissolved organic N, particulate N, and total suspended sediments entering but not necessarily leaving wetlands with each tidal event (Childers and Day 1988). This technique has been useful in defining different mangrove forests as sources or sinks of nutrients (Rivera-Monroy et al. 1995; Childers et al. 2000); application to tidal freshwater forested wetlands would be important for defining their role as sources, sinks, or conveyors of different human-derived and natural forms of nutrients.

17.5. Crabs

Fauna can have important influences on ecological processes in tidal forested wetlands. Research from mangrove ecosystems have certainly provided examples of how fish, infauna, bacteria, and macroinvertebrates can affect growth and nutrient cycling (Hogarth 1999). Transient faunal communities are especially difficult to study, but a prominent, yet unstudied, component of tidal freshwater forested wetlands is the crab community. In particular, what effect are crabs likely to have on regeneration, carbon storage, and biogeochemistry of these wetlands? The crab community, especially small grapsid crabs (family Grapsidae), has received much attention in the mangrove literature in their capacity to limit recruitment of tree seedlings (cf., Allen et al. 2003). The literature on these crabs, as well as their insect counterparts, has provided several intriguing hypotheses defining the mechanisms of tree species dominance and colonization (e.g., dominance-predation: Smith 1987; gap refugia: Sousa et al. 2003). Further, the literature defines additional roles for crabs: crabs alter community dynamics by facilitating ammonification (Alongi et al. 1992), by promoting organic matter decomposition (Robertson and Daniel 1989; Micheli et al. 1991), by grazing on leaf material (Onuf et al. 1977; Beever et al. 1979), by aerating anoxic soils through burrowing (Smith et al. 1991), and by altering soil microtopography by building mounds (Warren and Underwood 1986; Minchinton 2001). Intriguing recent studies have even shown that crabs influence carbon losses in salt marshes by maintaining less carbon in surficial sediments through burrowing activity and reducing the amount of carbon available for tidal export (Gutiérrez et al. 2006). Many of these processes might be applicable to tidal freshwater forested wetlands as well.

Understanding the ecological processes of tidal freshwater forested wetlands may be greatly enhanced by focused study on the crab community. Studies should determine major functional guilds of crabs and identify the intensity and differential predation that crabs have on seeds from tidal forested wetland tree species. Biogeochemically, it might be important to define the role of crab burrows in oxygenating the soil, how crabs influence mineralization of N at an ecosystem level, and how crabs influence overall carbon degradation, storage, and efflux (especially as CO_2) along tidal hydroperiod gradients. Practically any study of the crab community, and other faunal constituents, would make a primary research contribution to the study of tidal freshwater forested wetlands.

17.6. Elevation change and sedimentation

The high rate of subsidence in the Mississippi River Delta (10 mm yr^{-1}), and in many other coastal wetland settings, along with the possibility of an acceleration of sea-level rise caused by global warming (Penland et al. 1988; IPCC 2001), will lead to stresses on tidal freshwater forested wetlands in the southeastern United States. If coastal wetland soils do not achieve elevation gain at a rate equal to the rate of relative sea-level rise (RSLR = eustatic sea-level rise plus subsidence) then they will become stressed, because of either waterlogging or salt, and ultimately disappear

(Conner and Day 1988; Mendelssohn and McKee 1988; Webb et al. 1995; Mendelssohn and Morris 2000).

Only a couple of studies have attempted to explain sedimentation processes in tidal freshwater forested wetlands. One study found that vertical accretion ranged widely, from 2 to 6 mm yr^{-1}, depending upon the location of the tidal swamp in relation to the distance from a main river channel in the Chesapeake Bay area (Chapter 5). In tidal swamps and marsh along the Cape Fear River in North Carolina accumulation rates varied between 3 and 9.5 mm yr^{-1} (Chapter 8). Another study by Rybczyk et al. (2002), in which vertical accretion ranged from 1.4–11 mm yr^{-1}, is discussed below (Section 17.7.1) as part of an investigation into the effects of sewage effluent on vertical accretion in a degraded tidal swamp in Louisiana.

Sedimentation processes are complex and cannot be fully assessed by one technique alone. Whereas marker horizons measure accreted sediments and thus serve as a baseline for determining the degree of mineral sediment deposition, they ignore the actual elevation change of the wetland. Soil elevation is a combination of vertical accretion, erosion, belowground root production, and subsidence (Cahoon et al. 1999). A combined approach that uses sedimentation-erosion tables (Boumans and Day 1993) has been implemented in many coastal wetlands throughout the Southeastern United States and the world (Cahoon et al. 2002, 2006). Although wetlands do not stay above the submergence threshold through one process alone, often one process does dominate the propensity for elevation gain in a specific location.

Future research on tidal freshwater forested wetland sedimentation should attempt to define the processes, as well as the rates, associated with elevation change in vulnerable coastal settings. During these studies, microtopography needs to be considered. Elevation processes associated with hummocks (topographic highs) are likely to differ considerably from those associated with hollows (topographic lows), for example. Sedimentation processes are also likely to differ considerably in areas with organic sediments (e.g., Louisiana) from processes along the black-, and redwater rivers that define tidal freshwater forested wetlands along the Atlantic coast of the southeastern United States (Chapter 1). It is also very important to understand how elevation change processes might be modified from current mechanisms with future sea-level changes, river flow patterns, and coastal modifications.

17.7. Restoration

Restoration is defined as the process of assisting recovery of an ecosystem that has been degraded, damaged, or destroyed (Society for Ecological Restoration 2002). Unfortunately, there are many unknowns and uncertainties in attempting to reassemble nature (Zedler 2006). In many cases, we are still learning as we go, producing data that can be used to refine future restoration attempts.

17.7.1. Use of treated wastewater as a restoration tool in Louisiana

Coastal wetlands can be used to treat wastewater and in turn the effluent may serve as a wetland restoration or enhancement tool. The discharge of secondarily treated effluent into wetlands can stimulate biomass production and enhance sediment accretion rates (Rybczyk et al. 2002).

Secondarily treated effluent delivers nutrient-rich water to wetlands, stimulating vegetative productivity. In coastal Louisiana, both short-term and long-term applications of wastewater have enhanced productivity of forested wetland sites in some cases. One such wetland, located near Breaux Bridge, has received effluent discharge for over 50 years. A dendroecological analysis of stem wood growth from 1920 to 1992 found that tree growth was significantly higher in the control site compared to the treatment site prior to the introduction of wastewater, but after the onset of wastewater application, tree growth was significantly higher in the treatment site than in the control site (Hesse et al. 1998). Short-term records at this site also show generally higher stemwood and leaf productivity in treatment areas (Table 17.1). Similar results have been reported for two additional coastal sites in Louisiana: Amelia, a site that has received effluent for more than 25 years (Day et al. 2006), and Mandeville, where secondarily treated effluent has been discharged into a forested wetland for 15 years (Brantley 2005).

Rybczyk et al. (2002) used marker horizons to measure accretion rates at treatment and control sites at another effluent study location near Thibodeaux, Louisiana. Pre-effluent accretion rates averaged 7.8 mm yr^{-1} in the treatment site and 5.2 mm yr^{-1} in the control site, and they were not significantly different. After wastewater application began, accretion rates in the treatment site (11 mm yr^{-1}) were significantly higher than accretion rates measured at the control site (1.4 mm yr^{-1}). The new accretion rate in the treatment site fell within one standard deviation of the estimated rate of RSLR in the region (Rybczyk et al. 2002).

Table 17.1. Aboveground productivity (g dry wt $m^{-2}yr^{-1}$) at three Louisiana forested wetland sites receiving municipal effluent. Reference sites receive no effluent (Delgado-Sanchez 1995; Day et al. 2004; Brantley 2005).

Site/Date	Stemwood Production	Leaf Production	Total Aboveground Production
Breaux Bridge			
Reference (1993)	677.9	514.0	1191.9
Treatment (1993)	780.0	420.0	1200.0
Reference (1994)	593.2	547.3	1140.5
Treatment (1994)	1383.4	745.8	2129.2
Reference (1995)	574.8	705.2	1280.0
Treatment (1995)	847.7	763.6	1611.3
Mandeville (1998-2002)			
Reference	217.0	649.0	866.0
Treatment Upstream	462.0	498.0	960.0
Treatment Downstream	456.0	746.0	1202.0
Amelia (1995-1996)			
Reference	302.4	412.4	714.8
Treatment (near)	750.8	716.7	1467.5
Treatment (far)	776.0	666.4	1442.4
Treatment (far)	638.5	546.1	1184.6

A before-after-control-impact statistical analysis revealed that neither leaf-litter decomposition rates nor initial leaf-litter N and P concentrations were affected by wastewater effluent (Rybczyk et al. 2002). A similar analysis revealed that final N and P leaf litter concentrations did increase significantly in the treatment site relative to the control site after effluent was applied. A wetland elevation-sediment dynamics model developed for this system revealed that changes in wetland elevation were much more responsive, or sensitive, to changes in primary production than to changes in rates of decomposition (Rybczyk et al. 1998). Trends would suggest that increased organic matter production (including root production) and accretion would offset increased decomposition rates.

Based upon the above examples, it seems natural that wetland assimilation of nutrients and carbon may be used as a coastal restoration tool. Planning is currently underway to implement a large-scale project to restore coastal wetlands southeast of New Orleans by using treated effluent. The area contains over 4,000 ha of historically tidal baldcypress wetlands that were killed or degraded after the construction of the Mississippi River-Gulf Outlet navigation canal connected the high-salinity water of the Gulf of Mexico to the low-salinity water of the wetland forests. To restore the forested wetland, effluent will be pumped into the area, and baldcypress will be planted. Over 375,000 m^3 day^{-1} of treated effluent from New Orleans and from St. Bernard Parish (just downstream of New Orleans) will be pumped into the area, facilitating habitat restoration, improving water quality, and enhancing storm protection. The effluent will lead to freshwater conditions and moderate levels of nutrients. The project in New Orleans will be the largest restoration effort to date that uses treated effluent. Additional research along these lines is needed, especially in areas where close associations exist between swamp forests and human development.

17.7.2. Freshwater diversions

In a lot of ways, diverting sewage effluent from municipalities is a scaled-down version of a freshwater diversion, which is designed to accomplish similar restoration goals on a much broader scale. Freshwater diversions from large rivers, such as the Mississippi River, have the purpose of diverting freshwater and sediment from main, secondary, or tertiary river channels to bypass levees and flow directly into coastal wetlands (Figure 17.4). The primary intent of this management strategy is to replicate natural hydrologic cycles of overbank flooding to produce new wet-

Fig. 17.4. Freshwater diversion at Caernarvon designed to divert Mississippi River water and sediment to receiving marshes in southeastern Louisiana in order to offset the high rate of coastal subsidence and marsh loss in the region. Similar diversions are being proposed for tidal freshwater forested wetland rehabilitation (e.g., Maurepas Swamp, Chapter 13).

land areas or rehabilitate existing wetlands by promoting positive elevation gain through sedimentation and reduced salinity (Turner and Boyer 1997). Diversions at this scale have principally focused on marshes, where nutrient assimilation by vegetation and reductions in salinity (cf., Lane et al. 1999) have prompted wide-scale interest in expanding these diversion to include both tidal and non-tidal freshwater forested wetlands. Future research should identify the effects that diverted river water has on ecosystem properties as diversions are considered for tidal freshwater forested wetlands (Chapter 13). These structures have the potential to affect very large tidal forested wetland areas.

17.7.3. Planting efforts

We have less experience with restoring coastal forested wetlands than with restoring bottomland hardwoods or pine forests (Mitsch and Gosselink 1993). Although there has been little success in planting tupelo (DeBell et al. 1982), better results have been obtained with baldcypress. Planting of baldcypress began in the 1950s with good success (Peters and Holcombe 1951). Rathborne Lumber Company planted nearly 1 million baldcypress seedlings on cutover land in Louisiana with 80–95 percent survival (Rathborne 1951). The Soil Conservation Service, however, experienced severe herbivory problems, and they recommended suspension of planting until some means of controlling nutria (*Myocastor coypus* [Molina, 1782]) was developed (Blair and Langlinais 1960). Nutria remain a serious problem to newly planted seedlings (Conner 1988; Brantley and Platt 1992; Myers et al. 1995) in coastal areas of Louisiana, and they may even damage mature trees (Hesse et al. 1996).

Planting seedlings may be necessary to restore swamps because natural regeneration is unreliable (Hook et al. 1967; Kennedy 1982; Hamilton 1984; Conner et al. 1986; Smith 1995). Planting 1-year-old baldcypress seedlings that are at least 1 m tall and larger than 1.25 cm at the root collar improves early survival and growth (Faulkner et al. 1986). Planting is recommended in the late fall and in winter so that seedlings become established during low water periods (Mattoon 1915). Even when planted in permanent standing water, height growth averages 20–30 cm per year for baldcypress when there are no herbivory problems (Conner 1988; Conner and Flynn 1989). Tree shelters generally increase the chances of survival of planted seedlings, but they do not prevent all herbivory (McLeod 2000; Chapter 13).

A simple planting technique has been successfully tested for planting seedlings in standing water (Conner and Flynn 1989; Conner 1995; Funderburk 1995; McLeod et al. 1996). Root pruning, or trimming off the lateral roots and cutting the taproot to approximately 20 cm, allows planters to grasp each seedling at the root collar and push it into the sediment until their hand hits the sediment. This method has worked well in trials with baldcypress and water tupelo but not as well with green ash (*Fraxinus pennsylvanica* Marsh.) and swamp tupelo.

Establishing effective and reasonable planting techniques for additional tidal freshwater forested wetland tree species would be a good contribution to restoration. Too often planting efforts rate only seedling, sapling, and tree growth and mortality, but little is understood about the underlying individual tree physiological mechanisms that cause growth successes or failures (J.L. Chambers, personal communication). Many greenhouse studies, on the other hand, do investigate causal mechanisms. For example, studies may determine whether a reduction in seedling growth is associated with reduced leaf xylem water potential or with shifts in foliar Na:K ratios. However, natural field settings provide the true test for what scientists infer from controlled investigations. Since seedlings typically establish well on suitable plantation sites, and adequately on marginal sites, and ecophysiological equipment has become easily transportable to the field, there are few operational barriers to conducting future ecophysiological investigations on tidal freshwater swamp seedlings and saplings in natural settings.

17.8. Future of tidal forests

Forests in general are considered to be resilient: after disturbance, they typically revert to a similar forest if given enough time. Many forests are considered to be self-renewing, and recovery may occur in a relatively short period; however, tidal freshwater forested wetlands that occur in a narrow fringe between fresh and salt water are frequently at the tipping point and influenced by multiple, simultaneous stressors (Figure 17.5). The ability to repair the habitat and to restore forest structure is complicated by diverse ecological and societal conditions. The time required for a severely degraded site to recover is unknown, but it may take decades to centuries (Stanturf 2005).

Chapter 17 - Future Research Needs and Restoration Overview   479

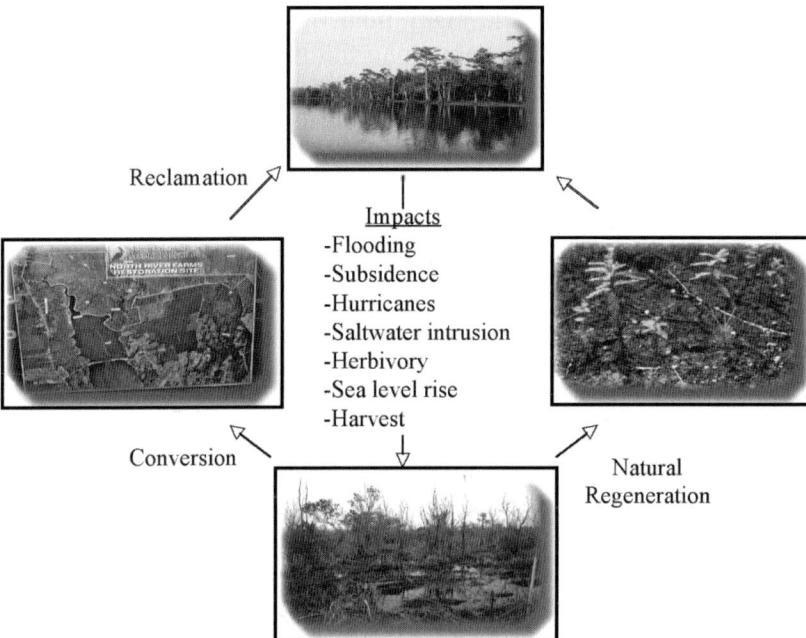

Fig. 17.5. Tidal freshwater forested wetlands are subject to a wide variety of natural and human-caused stresses. Impacted forests will naturally regenerate to their former composition in time if increased flooding and salt water intrusion are prevented. Otherwise, forests convert to marsh or open water. In some cases, extensive drainage and impoundment networks lead to wetland forests that are being converted to other uses (e.g., agriculture). Projects for reclaiming these converted wetlands are proposed for sites in North Carolina.

17.9. Conclusion

"Restoration" is a term that implies returning a landscape to a former condition, but there is no way to determine whether the newly established forest is as complex and diverse as the one displaced (Sauer 1998). Future ecological studies should attempt to define specific geomorphological, biogeochemical, and disturbance-abatement roles of tidal freshwater forested wetlands in order to provide a better link to coastal protection and to define specific ecological services provided by this wetland type. Then, restoration can take on several trajectories, ranging from direct planting of seedlings to inclusion of tidal forests in wetland treatment systems. For the latter, research in Louisiana indicates that simultaneous ecological goals of enhancing water quality, stimulating vertical accretion, and increasing productivity can be achieved while providing a benefit to society.

Accordingly, restoration of the many tidal freshwater forested wetlands that were converted to rice culture along the Atlantic coast of the southeastern United States (Chapter 1) should meet with good success, as long as tidal hydrological signatures were reestablished and maintained.

17.10. Acknowledgments

We thank Jim L. Chambers, Richard H. Day, Thomas McGinnis II, Thomas W. Doyle, and Victoria Jenkins for critical reviews of this chapter. This material is based upon work supported by the Cooperative State Research, Education, and Extension Service of the U.S. Department of Agriculture (CSREES/USDA), under project number SC-1700271, and by the Global Change Research Program of the U.S. Geological Survey. Technical Contribution No. 5287 of the Clemson University Experiment Station.

References

Allen JA, Chambers JL, McKinney D (1994) Intraspecific variation in the response of *Taxodium distichum* seedlings to salinity. For Ecol Manage 70:203-214

Allen JA, Pezeshki SR, Chambers JL (1996) Interaction of flooding and salinity stress on baldcypress (*Taxodium distichum*). Tree Physiol 16:307-313

Allen JA, Conner WH, Goyer RA, Chambers JL, Krauss KW (1998) Freshwater forested wetlands and global climate change. In: Guntenspergen GR, Vairin BA (eds) Vulnerability of coastal wetlands in the southeastern United States: Climate change research results, 1992-97. Biological Science Report USGS/BRD/BSR-1998-0002. U.S. Geological Survey, Biological Resources Division, Lafayette, pp 33-44

Allen JA, Krauss KW, Hauff RD (2003) Factors influencing the intertidal distribution of the mangrove species *Xylocarpus granatum*. Oecologia 135:110-121

Alongi DM, Boto KG, Robertson AI (1992) Nitrogen and phosphorus cycles. In: Robertson AI, Alongi DM (eds) Tropical mangrove ecosystems, Coastal and Estuarine Studies 41. American Geophysical Union, Washington, pp 251-292

Ball MC (1986) Photosynthesis in mangroves. Wetlands (Australia) 6: 12-22

Ball MC (1988) Ecophysiology of mangroves. Trees 2:129-142

Beever JW III, Simberloff D, King LL (1979) Herbivory and predation by the mangrove tree crab *Aratus pisonii*. Oecologia 43:317-328

Bianchi TS, Pennock JR, Twilley RR eds (1999) Biogeochemistry of Gulf of Mexico Estuaries. John Wiley and Sons Inc, New York

Blair RM, Langlinais MJ (1960) Nutria and swamp rabbits damage baldcypress plantings. J For 58:388-389

Boucher DH (1990) Growing back after hurricanes. BioScience 40:163-166

Boumans R, Day Jr JW (1993) High precision measurements of sediment elevation in shallow coastal areas using a sedimentation-erosion table. Estuaries 16:375-380

Brantley CG (2005) Nutrient interactions, plant productivity, soil accretion, and policy implications of wetland enhancements in coastal Louisiana. Ph.D. thesis, Louisiana State University

Brantley CG, Platt SG (1992) Experimental evaluation of nutria herbivory on baldcypress. Proc La Acad Sci 55:21-25

Brinson M (2006) Consequences for wetlands of a changing global environment. In: Batzer DP, Sharitz RR (2006) Ecology of freshwater and estuarine wetlands. University of California Press, Berkeley, pp 436-461

Brown SL (1981) A comparison of the structure, primary productivity, and transpiration of cypress ecosystems in Florida. Ecol Monogr 51:403-427

Brown SL, Cowles SW, Odum HT (1984) Metabolism and transpiration of cypress domes in north-central Florida. In: Ewel KC, Odum HT (eds) Cypress swamps. University of Florida Press, Gainesville, pp 145-163

Burkett V, Kusler J (2000) Climate change: potential impacts and interactions in wetlands of the United States. J Am Water Resour Assoc 36:313-320

Cahoon DR, Day Jr JW, Reed DJ (1999) The influence of surface and shallow subsurface processes on wetland elevation: a synthesis. Current Topics in Wetland Biogeochemistry 3:72-88

Cahoon DR, Lynch JC, Hensel P, Boumans R, Perez BC, Segura B, Day Jr JW (2002) High precision measurements of wetland sediment elevation: I. Recent improvements to the sedimentation-erosion table. J Sedimentary Res 72:730-733

Cahoon DR, Hensel P, Spencer T, Reed D, McKee KL, Saintilan N (2006) Coastal wetland vulnerability to relative sea-level rise: Wetland elevation trends and process controls. In: Verhoeven J, Whigham DF (eds) Wetlands and natural resource management. Springer, New York, pp 272-292

Chen R, Twilley RR (1998) A gap dynamic model of mangrove forest development along gradients of soil salinity and nutrient resources. J Ecol 86:37-51

Chen R, Twilley RR (1999) Patterns of mangrove forest structure and soil nutrient dynamics along the Shark River Estuary, Florida. Estuaries 22:955-970

Childers DL, Day Jr JW (1988) A flow-through flume technique for quantifying nutrient and material fluxes in microtidal estuaries. Estuar Coast Shelf Sci 27(5):483-494

Childers DL, Day Jr JW, McKellar Jr HN (2000) Twenty more years of marsh and estuarine flux studies: Revisiting Nixon (1980). In: Weinstein MP, Kreeger DQ (eds) Concepts and controversies in tidal marsh ecology. Kluwer Academic Press, Nordrecht, The Netherlands, pp 385-414

Conner WH (1988) Natural and artificial regeneration of baldcypress (*Taxodium distichum* [L.] Rich.) in the Barataria Basins of Louisiana. Ph.D. thesis, Louisiana State University

Conner WH (1995) Baldcypress seedlings for planting in flooded sites. In: Edwards MB (comp) Proceedings of the eighth biennial southern silvicultural research conference. General Technical Report SRS-1. U.S. Department of Agriculture, Forest Service, Southern Research Station, Asheville, pp 430-434

Conner WH, Day Jr JW (1988) Rising water levels in coastal Louisiana: Implications for two forested wetland areas in Louisiana. J Coast Res 4:589-596

Conner WH, Flynn K (1989) Growth and survival of baldcypress (*Taxodium distichum* [L.] Rich.) planted across a flooding gradient in a Louisiana bottomland forest. Wetlands 9:207-217

Conner WH, Toliver JR, Sklar FH (1986) Natural regeneration of cypress in a Louisiana swamp. For Ecol Manage 14:305-317

Conner WH, Day Jr JW, Baumann RH, Randall J (1989) Influence of hurricanes on coastal ecosystems along the northern Gulf of Mexico. Wetl Ecol Manage 1:45-56

Davis SM, Childers DL, Lorenz JJ, Wanless HR, Hopkins TE (2005) A conceptual model of ecological interactions in the mangrove estuaries of the Florida Everglades. Wetlands 25:832-842

Day Jr JW, Templet PH, Ko J-Y (2004) The Mississippi delta: System functioning, environmental impacts, and sustainable management. In: Caso M (ed) Environmental diagnosis of the Gulf of Mexico, vol. 2. Mexican National Institute of Ecology, Mexico City, pp 851–880

Day Jr JW, Westphal A, Pratt R, Hyfield E, Rybczyk J, Kemp GP, Day JN, Marx B (2006) Effects of long-term municipal effluent discharge on the nutrient dynamics, productivity, and benthic community structure of a tidal freshwater forested wetland in Louisiana. Ecol Eng 27:242-257

DeBell DS, Askew GR, Hook DD, Stubbs J, Owens EG (1982) Species suitability on a lowland site altered by drainage. South J Appl For 6:2-9

Delgado-Sanchez P (1995) Effects of long-term wastewater discharge in the Cypriere Perdue forested wetland at Breaux Bridge, Louisiana. M.S. thesis, Louisiana State University

Doyle TW, Keeland BD, Gorham LE, Johnson DJ (1995) Structural impact of Hurricane Andrew on forested wetlands of the Atchafalaya Basin in coastal Louisiana. J Coast Res 18:354-364

Duever MJ, McCollom JM (1993) Hurricane Hugo effects on old-growth floodplain forests communities at Four Hole Swamp, South Carolina. In: Brissette JC (ed) Proceedings of the seventh biennial southern silvicultural research conference. General Technical Report SO-93. U.S. Department of Agriculture, Forest Service, Southern Forest Experiment Station, New Orleans, pp 197-202

Effler RS, Goyer RA, Lenhard GJ (2006) Baldcypress and water tupelo responses to insect defoliation and nutrient augmentation in Maurepas Swamp, Louisiana, USA. For Ecol Manage 236:295-304

Faulkner PL, Zeringue F, Toliver JR (1986) Genetic variation among open-pollinated families of baldcypress seedlings planted on two different sites. In: Proceedings of the eighteenth southern forest tree improvement conference. Sponsored Publication No. 40. Southern Forest Tree Improvement Committee, pp 267-272

Funderburk EL (1995) Growth and survival of four forested wetland species planted on Pen Branch delta, Savannah River Site, SC. M.S. thesis, Clemson University

Gresham CA, Williams TM, Lipscomb DJ (1991) Hurricane Hugo wind damage to southeastern U. S. coastal forest tree species. Biotropica 23(4):420-426

Gutiérrez JL, Jones CG, Groffman PM, Findlay SEG, Iribarne OO, Ribeiro PD, Bruschetti CM (2006) The contribution of crab burrow excavation to carbon availability in surficial salt-marsh sediment. Ecosystems 9:647-658

Hamilton DB (1984) Plant succession and the influence of disturbance in Okefenokee Swamp. In: Cohen AD, Casagrande DJ, Andrejko MJ,

Best GR (eds) The Okefenokee Swamp: Its natural history, geology, and geochemistry. Wetland Surveys, Los Alamos, pp 86-111

Hedlund A (1969) Hurricane Camille's impact on Mississippi timber. Southern Lumberman 219(2728):191-192

Hesse I, Doyle TW, Day Jr JW (1998) Long-term growth enhancement of baldcypress (*Taxodium distichum*) from municipal wastewater application. Environ Manage 22(1):119-127

Hesse ID, Conner WH, Day Jr JW (1996) Herbivory impacts on the regeneration of forested wetlands. In: Flynn KM (ed) Proceedings of the southern forested wetland management and ecology conference. Clemson University, Clemson, pp 23-28

Hogarth PJ (1999) The biology of mangroves. Oxford University Press, Oxford

Hook DD (1984) Waterlogging tolerance of lowland tree species of the south. South J Appl For 8:136-149

Hook DD, LeGrande WW, Langdon OG (1967) Stump sprouts on water tupelo. Southern Lumberman 215(2680):111-112

Hook DD, Buford MA, Williams TM (1991) Impact of Hurricane Hugo on the South Carolina coastal plain forest. J Coast Res SI-8:291-300

IPCC (Intergovernmental Panel on Climate Change) (2001) Climate change 2001: The scientific basis contribution of Working Group 1 to the Third Assessment Report. Cambridge University Press, Cambridge

Isaak D, Marshall W, Buell M (1959) A record of reverse plant succession in a tamarack bog. Ecology 40:317-320

Kennedy Jr HE (1982) Growth and survival of water tupelo coppice regeneration after six growing seasons. South J Appl For 6:133-135

Krauss KW, Young PJ, Chambers JL, Doyle TW, Twilley RR (2007) Sap flow characteristics of neotropical mangroves in flooded and drained soils. Tree Physiol 27:775-783

Lane RR, Day Jr JW, Thibodeaux B (1999) Water quality analysis of a freshwater diversion at Caernarvon, Louisiana. Estuaries 22:327-336

Lewis RR III (2005) Ecological engineering for successful management and restoration of mangrove forests. Ecol Engineer 24:403-418

Loope L, Duever M, Herndon A, Snyder J, Jansen D (1994) Hurricane impact on uplands and freshwater swamp forest. BioScience 44:238-246

Lugo AE, Applefield M, Pool DJ, McDonald RB (1983) The impact of Hurricane David on the forests of Dominica. Can J For Res 13:201-211

Martens CS, Berner RA (1974) Methane production in the interstitial waters of sulfate depleted marine sediments. Science 185:1167-1169

Mattoon WR (1915) The southern cypress. Agriculture Bulletin 272. U.S. Department of Agriculture, Washington

McKee KL, Feller IC, Popp M, Wanek W (2002) Mangrove isotopic ($\delta^{15}N$ and $\delta^{13}C$) fractionation across a nitrogen vs. phosphorus limitation gradient. Ecology 83:1065-1075

McKnight JS, Hook DD, Langdon OG, Johnson RL (1981) Flood tolerance and related characteristics of trees of the bottomland forests of the southern United States. In: Clark JR, Benforado J (eds) Wetlands of bottomland hardwood forests. Elsevier, The Netherlands, pp 26-69

McLeod KW (2000) Species selection trials and silvicultural techniques for the restoration of bottomland hardwood forests. Ecol Engineer 15:S35-S46

McLeod KW, Reed MR, Ciravolo TG (1996) Reforesting a damaged stream delta. Land and Water 40:11-13

Mendelssohn I, McKee K (1988) *Spartina alterniflora* die-back in Louisiana: time-course investigation of soil waterlogging effects. J Ecol 76:509-521

Mendelssohn I, Morris J (2000) Eco-physiological controls on the productivity of *Spartina alterniflora* Loisel. In: Weinstein M, Kreeger D (eds) Concepts and controversies in salt marsh ecology. Kluwer Academic Publishers, Dordrecht, The Netherlands, pp 59-80

Micheli F, Gherardi F, Vannini M (1991) Feeding and burrowing ecology of two East African mangrove crabs. Mar Biol 111:247-254

Michener WK, Blood ER, Bildstein KL, Brinson MM, Gardner LR (1997) Climate change, hurricanes and tropical storms, and rising sea level in coastal wetlands. Ecol Appl 7:770 801

Minchinton TE (2001) Canopy and substratum heterogeneity influence recruitment of the mangrove *Avicennia marina*. J Ecol 89:888-902

Mitsch WJ, Gosselink JG (1993) Wetlands, 3rd Edition. John Wiley, New York

Myers RS, Shaffer GP, Llewellyn DW (1995) Baldcypress (*Taxodium distichum* (L.) Rich.) restoration in southeast Louisiana: the relative effects of herbivory, flooding, competition, and macronutrients. Wetlands 15(2):141-148

Nix LE, Hook DD, Williams JG, Van Blaricom D (1996) Assessment of hurricane damage to the Santee Experimental Forest and the Francis Marion National Forest with a geographic information System. In: Haymond JL, Hook DD, Harms WR (eds) Hurricane Hugo: South Carolina forest land research and management related to the storm. General Technical Report SRS-5. U.S. Department of Agriculture, Forest Service, Southern Research Station, Asheville, pp 44-51

Noe GB, Hupp CR (2005) Carbon, nitrogen, and phosphorus accumulation in floodplains of Atlantic coastal plain rivers, USA. Ecol Appl 15:1178-1190

Odum EP, Birch JB, Cooley JL (1983) Comparison of giant cutgrass productivity in tidal and impounded marshes with special reference to tidal subsidy and waste assimilation. Estuaries 6:88-94

Odum WE, Smith III TJ, Hoover JK, McIvor CC (1984) The ecology of tidal freshwater marshes of the United States east coast: A community profile. FWS/OBS-87/17. U.S. Fish and Wildlife Service, Washington

Odum WE, Odum EP, Odum HT (1995) Nature's pulsing paradigm. Estuaries 18:547-555

Onuf CP, Teal TM, Valiela I (1977) Interactions of nutrients, plant growth and herbivory in a mangrove ecosystem. Ecology 58:514-526

Penland S, Ramsey KE, McBride RA, Mestayer JT, Westphal KA (1988) Relative sea level rise and delta-plain development in the Terrebonne Parish Region. Coastal Geology Technical Report No. 4. Louisiana Geological Survey, Baton Rouge

Peters MA, Holcombe E (1951) Bottomland cypress planting recommended for flooded areas by soil conservationists. Forests and People 1(2):18, 32-33

Pezeshki SR (1987) Gas exchange response of tupelo-gum (*Nyssa aquatica* L.) to flooding and salinity. Photosynthetica 21:489-493

Pezeshki SR (1990) A comparative study of the response of *Taxodium distichum* and *Nyssa aquatica* seedlings to soil anaerobiosis and salinity. For Ecol Manage 33/34:531-541

Pezeshki SR, Patrick Jr WH, DeLaune RD, Moser ED (1989) Effects of waterlogging and salinity interactions on *Nyssa aquatica* seedlings. For Ecol Manage 27:41-51

Pezeshki SR, DeLaune RD, Patrick Jr WH (1990) Flooding and saltwater intrusion: potential effects on survival and productivity of wetland forests along the U.S. Gulf Coast. For Ecol Manage 33/34:287-301

Putz FE, Sharitz RR (1991) Hurricane damage to old-growth forest in Congaree Swamp National Monument, South Carolina, U.S.A. Can J For Res 21:1765-1770

Rathborne JC (1951) Cypress reforestation. Southern Lumberman 183(2297):239-240

Rheinhardt RD (1992) A multivariate analysis of vegetation patterns in tidal freshwater swamps of lower Chesapeake Bay, U.S.A. Bull Torrey Bot Club 119:192-207

Rheinhardt RD, Hershner C (1992) The relationship of below-ground hydrology to canopy composition in five tidal freshwater swamps. Wetlands 12:208-216

Rivera-Monroy VH, Day Jr JW, Twilley RR, Vera-Herrera F, Coronado-Molina C (1995) Flux of nitrogen and sediment in a fringe mangrove forest in Terminos Lagoon, Mexico. Estuar Coast Shelf Sci 40:139-160

Robertson AI, Daniel PA (1989) The influence of crabs on litter processing in high intertidal mangrove forests in tropical Australia. Oecologia 78:191-198

Rybczyk JM, Zhang XW, Day Jr JW, Feagley S (1995) The impact of Hurricane Andrew on tree mortality, litterfall, nutrient flux, and water quality in a Louisiana coastal swamp forest. J Coast Res SI-21:340-353

Rybczyk JM, Callaway JC, Day Jr JW (1998) A relative elevation model (REM) for a subsiding coastal forested wetland receiving wastewater effluent. Ecol Modelling 112:23-44

Rybczyk JM, Day Jr JW, Conner WH (2002) The impact of wastewater effluent on accretion and decomposition in a subsiding forested wetland. Wetlands 22(1):18-32

Sauer LJ (1998) The once and future forest: A guide to forest restoration strategies. Island Press, Washington

Sharitz RR, Vaitkkus MR, Cook AE (1993) Hurricane damage to an old-growth floodplain forest in the southeast. In: Brissette, JC (ed) Proceedings of the seventh biennial southern silvicultural research conference. General Technical Report SO-93. U.S. Department of Agriculture, Forest Service, Southern Forest Experiment Station, New Orleans, pp 203-210

Sherman RE, Fahey TJ, Howarth RW (1998) Soil-plant interactions in a neotropical mangrove forest: iron, phosphorus and sulfur dynamics. Oecologia 115:553-563

Simpson RL, Good RE, Leck MA, Whigham DF (1983) The ecology of freshwater tidal wetlands. BioScience 33:255-259

Smith II LE (1995) Regeneration of Atlantic white-cedar at the Alligator River National Wildlife Refuge and Dare County Air Force Bombing Range. M.S. thesis, North Carolina State University

Smith TJ III (1987) Seed predation in relation to dominance and distribution in mangrove forests. Ecology 68:266-273

Smith TJ III, Boto KG, Frusher SD, Giddins RL (1991) Keystone species and mangrove forest dynamics: the influence of burrowing by crabs on soil nutrient status and forest productivity. Estuar Coast Shelf Sci 33:419-432

Smock L, Gilinsky E (1992) Coastal plain blackwater streams. In: Hackney C, Adams S, Martin W (eds) Biodiversity of the Southeastern United States: Aquatic communities, pp 271-314

Society for Ecological Restoration (2002) Science and Policy Working Group primer on ecological restoration, www.ser.org/content/ecological_restoration_primer.asp

Sousa WP, Quek SP, Mitchell BJ (2003) Regeneration of *Rhizophora mangle* in a Caribbean mangrove forest: interacting effects of canopy disturbance and a stem-boring beetle. Oecologia 137:436-445

Spiller DA, Losos JB, Schoener TW (1998) Impact of a catastrophic hurricane on island populations. Science 281:695-697

Stanturf JA (2005) What is forest restoration? In: Stanturf JA, Madsen P (eds) Restoration of boreal and temperate forests. CRC Press, Boca Raton, pp 3-11

Turner RE, Boyer ME (1997) Mississippi River diversions, coastal wetland restoration/creation and an economy of scale. Ecol Engineering 8:117-128

Warren JH, Underwood A (1986) Effects of burrowing crabs on the topography of mangrove swamps in New South Wales. J Exp Mar Biol Ecol 102:223-235

Weaver PL (1989) Forest changes after hurricanes in Puerto Rico's Luquillo mountains. Interciencia 14:181-192

Webb EC, Mendelssohn IA, Wilsey BJ (1995) Causes for vegetation dieback in a Louisiana salt marsh: a bioassay approach. Aq Bot 51:281-289

Wolfe RS, Higgins IJ (1979) Microbial biochemistry of methane-a study in contrast. Int Rev Biochem 21:267-353

Zedler JB (2006) Wetland restoration. In: Batzer DP, Sharitz RR (eds) Ecology of freshwater and estuarine wetlands. University of California Press, Berkeley, pp 348-435

Appendix 1

Scientific and common names of plant species found in the text are standardized based on the Integrated Taxonomic Information System found at http://www.itis.gov.

Scientific name	Common name
Acer barbatum Michx.	florida maple
Acer rubrum L.	red maple
Acer rubrum var. *drummondii* (Hook. & Arn. ex Nutt.) Sarg.	drummond maple
Alnus serrulata (Ait.) Willd.	alder
Alternanthera philoxeroides (Mart.) Griseb.	alligatorweed
Amaranthus australis (Gray) Sauer	southern amaranth
Amaranthus cannabinus (L.) Sauer	tidalmarsh amaranth
Amblystegium Schimp. in B.S.G.	amblystegium moss
Amblystegium serpens (Hedw.) Schimp. in B.S.G.	amblystegium moss
Amblystegium varium (Hedw.) Lindb.	amblystegium moss
Amelanchier canadensis (L.) Medik.	Canadian serviceberry
Anomodon attenuatus (Hedw.) Hüb.	anomodon moss
Apios americana Medik.	groundnut
Arisaema triphylum (L.) Schott	Jack in the pulpit
Aristida L.	threeawn
Aster L.	aster
Athyrium filix-femina spp. *asplenioides* (Michx.) Hulten	asplenium ladyfern
Atrichum crispum (James) Sull.	atrichum moss
Baccharis halimifolia L.	eastern baccharis
Bazzania trilobata (L.) S. Gray	threelobed bazzania
Betula nigra L.	river birch
Bidens connata Muhl. ex Willd.	purple-stem beggarticks
Bidens coronata (L.) Britt.	crowned beggarticks
Bidens laevis (L.) B.S.P.	smooth beggarticks
Bignonia capreolata L.	crossvine
Boehmeria cylindrica (L.) Sw.	small-spike false nettle
Boltonia asteroides (L.) L'Hér.	white doll's-daisy
Bostrychia rivularis Harvey	

Brachythecium acuminatum (Hedw.) Aust.	acuminate brachythecium moss
Bryoandersonia illecebra (Hedw.) Robins.	bryoandersonia moss
Bryohaplocladium microphyllum (Hedw.) Wat. & Iwats.	bryohaplocladium moss
Bryum Hedw.	bryum moss
Caloglossa leprieurii Montagne	
Carex L.	sedge
Carex bromoides Schkuhr ex Willd.	bromelike sedge
Carex crinita Lam.	fringed sedge
Carex crinita var. *brevicrinis* Fern.	fringed sedge
Carex gracillima Schwein.	graceful sedge
Carex hyalinolepis Steud.	shoreline sedge
Carex lupulina Muhl. ex Willd.	hop sedge
Carpinus caroliniana Walt.	American hornbeam
Carya aquatica (Michx. f.) Nutt.	water hickory
Carya cordiformis (Wangenh.) K. Koch	bitternut hickory
Carya glabra (P. Mill.) Sweet	pignut hickory
Catenella caespitosa	
Celtis laevigata Willd.	sugarberry
Cenchrus myosuroides Kunth	big sandbur
Cephalanthus occidentalis L.	buttonbush
Cephalozia lunulifolia (Dum.) Dum.	cephalozia
Cercis canadensis L.	eastern redbud
Chamaecyparis thyoides (L.) B.S.P.	Atlantic white cedar
Chasmanthium latifolium (Michx.) Yates	broadleaf uniola
Chasmanthium laxum (L.) Yates	slender woodoats
Chasmanthium sessiliflorum [Poir.] Yates	longleaf spikegrass
Chelone glabra L.	white turtlehead
Chiloscyphus polyanthos (L.) Corda	
Cicuta maculata L.	common water hemlock
Cinna arundinacea L.	stout woodreed
Cinnamomum camphora (L.) J. Presl	camphor tree
Cladium mariscus ssp. *jamaicense* (Crantz) Kükenth.	jamaica sawgrass
Clasmatodon parvulus (Hampe) Hook. & Wils. ex Sull. in Gray	clasmatodon moss
Clethra alnifolia L.	coastal sweetpepperbush
Cliftonia monophylla (Lam.) Britt. ex Sarg.	buckwheat tree
Climacium americanum Brid.	American climacium moss
Climacium kindbergii (Ren. & Card.) Grout	Kindberg's climacium moss
Clinopodium coccineum (Nutt. ex Hook.) Kuntze	scarlet calamint
Cololejeunea minutissima (Sm.) Schiffn.	
Cololejeunea setiloba Evans	
Commelina virginica L.	Virginia dayflower
Cornus foemina P. Mill.	stiff dogwood

Crataegus L.	hawthorn
Crataegus virdis L.	green hawthorn
Crinum americanum L.	seven sisters
Cuscuta L.	dodder
Cyclospermum leptophyllum (Pers.) Sprague ex Britt. & Wilson	marsh parsley
Cyperus odoratus L.	fragrant flatsedge
Cyrilla racemiflora L.	swamp titi
Decodon verticillatus (L.) Ell.	swamp-loosestrife
Desmodium laevigatum (Nutt.) DC.	smooth tickclover
Dichanthelium (A.S. Hitchc. & Chase) Gould	rosette grass
Dichanthelium clandestinum (L.) Gould	deertongue
Dichanthelium commutatum (J.A. Schultes) Gould	variable panicgrass
Dioscorea villosa L.	wild yam
Diospyros virginiana L.	common persimmon
Dulichium arundinaceum (L.) Britt.	threeway sedge
Echinochloa walteri (Pursh) Heller	coast cockspur
Echinodorus cordifolius (L.) Griseb.	creeping burhead
Eichhornia crassipes (Mart.) Solms	water hyacinth
Eleocharis R. Br.	spikerush
Entodon seductrix (Hedw.) C. Müll.	seductive entodon moss
Eryngium aquaticum L.	rattlesnakemaster
Eurhynchium hians (Hedw.) Sande Lac.	eurhynchium moss
Eurhynchium pulchellum (Hedw.) Jenn.	eurhynchium moss
Fagus grandifolia Ehrh.	American beech
Fissidens fontanus (B. Pyl.) Steud.	fissidens moss
Fontinalis sullivantii Lindb.	Sullivant's fontinalis moss
Forestiera acuminata (Michx.) Poir.	swamp privet
Fraxinus caroliniana P. Mill.	Carolina ash
Fraxinus pennsylvanica Marsh.	green ash
Fraxinus profunda (Bush) Bush	pumpkin ash
Frullania kunzei Lehm. and Lindenb.	
Galium tinctorium L.	stiff marsh bedstraw
Gleditsia aquatica Marsh.	water locust
Gleditsia triacanthos L.	common honeylocust
Glyceria striata (Lam.) A.S. Hitchc.	fowl manna grass
Hamamelis virginiana L.	American witchhazel
Hydrocotyle L.	hydrocotyle
Hymenocallis caroliniana (L.) Herbert	Carolina spiderlily
Hymenocallis floridana (Raf.) Morton	Florida spiderlily
Hypericum galioides Lam.	bedstraw St. Johnswort
Hypericum hypericoides (L.) Crantz	St. Andrew's cross
Hypericum mutilum L.	dwarf St. Johnswort
Hypnum lindbergii Mitt.	Lindberg's hypnum moss
Hypolepis repens (L.) K. Presl.	bramblefern

Ilex cassine L.	dahoon
Ilex coriacea (Pursh) Chapman	large gallberry
Ilex decidua Walt.	possumhaw
Ilex glabra (L.) Gray	inkberry
Ilex longipes Chapman ex Trel.	Georgia holly
Ilex opaca Ait.	American holly
Ilex verticillata L.	common winterberry
Ilex vomitoria Ait.	yaupon
Impatiens capensis Meerb.	jewelweed
Imperata cylindrica (L.) Beauv.	cogon grass
Iris virginica L.	Virginia iris
Isoetes flaccida Shuttlw. ex A. Braun	southern quillwort
Isopterygium tenerum (Sw.) Mitt.	isopterygium moss
Itea virginica L.	Virginia sweetspire
Jamesoniella autumnalis (Decandolle) Steph.	
Juncus roemerianus Scheele	needlegrass rush
Juniperus virginiana L.	eastern red-cedar
Juniperus virginiana var. *silicicola* (Small) J. Silba	southern redcedar
Lejeunea flava (Sw.) Nees	
Lejeunea laetivirens Nees et Mont.	
Leptodictyum humile (P. Beauv.) Ochyra	eptodictyum moss
Leptodictyum riparium (Hedw.) Warnst.	streamside leptodictyum moss
Leucobyrum albidum (Brid.) Lindb. (not in ITIS)	
Leucodon brachypus Brid.	leucodon moss
Leucodon julaceus (Hedw.) Sull.	leucodon moss
Leucolejeunea clypeata (Schwein.) Evans	
Leucolejeunea conchifolia Evans	
Leucothoe racemosa (L.) Gray	swamp doghobble
Lilaeopsis chinensis (L.) Kuntze	eastern grasswort
Lindbergia brachyptera (Mitt.) Kindb.	lindbergia moss
Lindera Thunb.	spicebush
Lindera benzoin (L.) Blume	northen spicebush
Liquidambar styraciflua L.	sweetgum
Ludwigia palustris (L.) Ell.	marsh primrose-willow
Ludwigia uruguayensis (Camb.) Hara	water primrose
Lycopus virginicus L.	Virginia water horehound
Lyonia ligustrina (L.) DC.	maleberry
Lyonia lucida (Lam.) K. Koch	fetterbush lyonia
Magnolia grandiflora L.	southern magnolia
Magnolia virginiana L.	sweetbay
Metzgeria Raddi sp.	
Mikania scandens (L.) Willd.	climbing hempvine
Mitchella repens L.	partridgeberry
Morella cerifera (L.) Small	wax myrtle
Morus rubra L.	red mulberry

Murdannia keisak (Hassk.) Hand.-Maz. — Asian spiderwort
Nuphar luteum (L.) J. E. Smith — pond-lily
Nyssa aquatica L. — water tupelo
Nyssa biflora Walt. — swamp tupelo
Nyssa sylvatica Marsh. — black tupelo
Odontoschisma prostratum (Sw.) Trev.
Orontium aquaticum L. — goldenclub
Osmunda cinnamomea L. — cinnamon fern
Osmunda regalis L. — royal fern
Packera aurea (L.) A. & D. Love — golden ragwort
Pallavicinia lyellii (Hook.) Carruth.
Panicum dichotomiflorum Michx. — fall panicgrass
Panicum hemitomon J.A. Schultes — maidencane
Panicum repens L. — torpedo grass
Peltandra virginica (L.) Schott — green arrow arum
Persea borbonia (L.) Spreng. — redbay
Persea palustris (Raf.) Sarg. — swamp bay
Phanopyrum gymnocarpon (Ell.) Nash — Savannah-panicgrass
Pinus palustris P. Mill. — longleaf pine
Pinus taeda L. — loblolly pine
Plagiomnium cuspidatum (Hedw.) T. Kop. — toothed plagiomnium moss
Planera aquatica J.F. Gmel. — water-elm
Pluchea odorata (L.) Cass. — sweetscent
Polygonum arifolium L. — halberdleaf tearthumb
Polygonum hydropiper L. — mild water-pepper
Polygonum punctatum Ell. — dotted smartweed
Polygonum virginianum L. — jumpseed
Pontederia cordata L. — pickerelweed
Populus deltoides Bartr. ex Marsh. — eastern cottonwood
Populus fremontii S. Wats. — Freemont cottonwood
Porella pinnata L.
Ptelea trifoliata L. — hoptree
Quercus falcata Michx. — southern red oak
Quercus geminata Small — sand live oak
Quercus hemisphaerica Bartr. ex Willd. — Darlington's oak
Quercus laurifolia Michx. — laurel oak
Quercus lyrata Walt. — overcup oak
Quercus michauxii Nutt. — swamp chestnut oak
Quercus nigra L. — water oak
Quercus phellos L. — willow oak
Quercus texana Buckl. — nuttall oak
Quercus virginiana P. Mill. — live oak
Racomitrium Brid. — racomitrium moss
Radula complanata (L.) Dum.
Rhododendron canescens (Michx.) Sweet — mountain azalea
Rhododendron viscosum (L.) Torr. — swamp azalea

Rhus copallina L. — dwarf sumac
Rhus copallinum L. — flameleaf sumac
Rhynchospora corniculata (Lam.) Gray — shortbristle horned beaksedge
Rhynchospora inundata (Oakes) Fern. — narrowfruit horned beaksedge
Riccardia multifida (L.) Gray
Rosa palustris Marsh. — swamp rose
Rumex verticillatus L. — swamp dock
Sabal minor (Jacq.) Pers. — dwarf palmetto
Sabal palmetto (Walt.) Lodd. ex J.A. & J.H. Schultes — cabbage palm
Saccharum giganteum (Walt.) Pers. — sugarcane plumegrass
Sagittaria lancifolia L. — bulltongue arrowhead
Salix caroliniana Michx. — coastal plain willow
Salix nigra Marsh. — black willow
Salvinia Seguier — watermoss
Sambucus nigra spp. *canadensis* (L.) R. Bolli — elderberry
Saururus cernuus L. — lizard's tail
Schinus terebinthifolius Raddi — Brazilian peppertree
Schlotheimia rugifolia (Hook.) Schwaegr. — ruggedleaf schlotheimia moss
Schoenoplectus americanus (Pers.) Volk. ex Schinz & R. Keller — American bulrush
Scutellaria lateriflora L. — mad dog skullcap
Selaginella apoda (L.) Spring — meadow spike-moss
Sematophyllum adnatum (Michx.) Britt. — sematophyllum moss
Serenoa repens (Bartr.) Small — saw palmetto
Sium suave Walt. — common waterparsnip
Smilax bona-nox L. — saw greenbrier
Smilax laurifolia L. — laurel greenbrier
Smilax rotundifolia L. — common greenbriar
Solidago sempervirens L. — seaside goldenrod
Spartina cynosuroides (L.) Roth — big cordgrass
Sphagnum affine Ren. & Card. — sphagnum
Sphagnum palustre L. — prairie sphagnum
Sphagnum recurvum P. Beauv. — recurved sphagnum
Steerecleus serrulatus (Hedw.) Robins. — steerecleus moss
Symphyotrichum elliotii (Torr. & Gray) Nesom — Elliott's aster
Symphyotrichum subulatum (Michx.) Nesom — eastern annual saltmarsh aster
Syrrhopodon texanus Sull. — Texan syrrhopodon moss
Taxiphyllum deplanatum (Bruch & Schimp. ex Sull.) Fleisch. — taxiphyllum moss
Taxiphyllum taxirameum (Mitt.) Fleisch. — taxiphyllum moss
Taxodium ascendens Brongn. — pondcypress
Taxodium distichum (L.) L.C. Rich. — baldcypress
Taxodium mucronatum Ten. — Montezuma bald cypress

Telaranea nematodes (Gott. Ex Aust.)

M. A. Howe
Thalictrum pubescens Pursh — king of the meadow
Thuidium allenii Aust. — Allen's thuidium moss
Thuidium delicatulum (Hedw.) Schimp. in B.S.G. — delicate thuidium moss
Toxicodendron radicans (L.) Kuntze — poison ivy
Trachelospermum difforme (Walt.) Gray — climbing dogbane
Triadenum walteri (J.G. Gmel.) Gleason — greater marsh St. Johnswort
Triadica sebifera (L.) Small — Chinese tallow tree
Typha latifolia L. — broadleaf cattail
Ulmus americana L. — American elm
Ulmus crassifolia Nutt. — cedar elm
Ulmus rubra Muhl. — slippery elm
Vaccinium arboreum Marsh. — farkleberry
Vaccinium corymbosum L. — highbush blueberry
Vaccinium elliottii Chapman — Elliott's blueberry
Vaccinium stamineum L. — deerberry
Vaccinium virgatum Ait. — smallflower blueberry
Viburnum dentatum L. — arrowwood
Viburnum nudum L. — possumhaw
Vigna luteola (Jacq.) Benth. — deer pea
Viola affinis Le Conte — Arizona bog violet
Viola cucullata Ait. — marsh blue violet
Vitis L. — grape
Woodwardia areolata (L.) T. Moore — netted chainfern
Woodwardia virginica (L.) Sm. — Virginia chainfern
Zizania aquatica L. — annual wildrice
Zizaniopsis miliacea (Michx.) Doell & Aschers. — giant cutgrass

Index

A

Accreting 20, 43
Accretion 143, 154, 167, 297, 375, 473-475
Age (tree) 77, 236, 278, 419-421, 432
Agriculture 12, 69, 100, 115, 141, 184, 224, 260, 298, 385
Alabama 4, 7, 9, 89-90, 105, 297, 324, 400-401
Alluvial 6, 56, 124-126, 129, 133, 255, 324
Amite River 353
Ammonium 358
Anadromous 161
Anaerobic 66, 69, 200, 465
Anaerobic condition 55-56, 470-471
Apalachicola River 4, 50, 52, 65, 68, 70-71, 77, 81, 262, 307, 324
Astronomical tide 6
Atlantic white cedar 49, 89-106, 147, 466

B

Backswamp 3, 91, 125, 131, 331, 344
Baldcypress 12, 15, 49, 53, 55, 57, 96, 100-101, 107, 117, 130, 165, 206, 212, 223-245, 249, 255, 294, 307, 315, 322, 324, 340, 344, 390-401, 412, 448, 469

Basal area 77-79, 96, 98, 123, 130, 149, 303, 307, 309, 330, 338, 340, 359, 368-369, 419, 426-427
Basal area increment 387, 419, 426-427
Big Bend region 258
Biodiversity 15, 149, 157
Biomass 55, 79, 166, 234-235, 360, 368
Biphasic model 393, 395
Blackwater river 5, 71-72, 79, 81, 113, 121, 187, 224, 413, 416, 463, 473
Bottomland hardwood forest 46, 183, 404, 462-463
Brackish 103-104, 143, 152, 184, 197, 200, 236, 297, 327, 396-398, 466
Bryophytes 209

C

Caernarvon diversion 375
Canopy 76, 95, 99, 103-107, 145, 166, 224, 239, 241, 257, 276, 292, 295, 302, 307, 324, 331, 340, 343, 361
Cape Fear River 4, 55, 66, 75, 77, 183-218, 473
Carbon (soils, cycling) 5, 74, 122, 133, 200, 243, 244, 452, 472
Carbon dioxide (CO_2) 394, 468, 470-472

497

Cedar Key 259
CH_4 74-76, 468, 470-471
Chassahowitzka River 279
Chesapeake Bay 50, 53, 66, 113, 116, 120, 122, 128, 139-143, 152, 161-162, 473
Chickahominy River 170
Chloride 154, 190, 330, 386
Climate 18, 19, 65, 74, 91, 108, 184, 186, 230, 244, 262, 298, 302, 326, 345, 355, 423
Climate change 18, 19, 158, 238, 376, 439, 466
Coarse woody debris 463, 469
Coastal 1, 2, 6, 67
    beaches 32, 46
    margins 30, 255
    plain 67-68, 113, 123, 129, 141, 183, 192, 224, 229, 291, 322
    reach 40, 81
    retreat 412
    water levels 33, 36, 40
    zone 68, 229
Cogon grass 275
Competition 200, 275, 368
Composition 55, 89-90, 96, 101, 105, 108, 123, 130, 140, 147, 188, 214, 239, 241, 245, 276, 291, 303, 314, 329, 333, 344
Conservation 2, 67, 142, 244, 447
Conversion (forest) 55, 141, 199, 217, 245, 266-267, 280-282, 453, 462-465
Cousiac Swamp 163
Crabs 82, 214, 312-313, 471-472
Creeks, tidal 49, 52, 55, 58, 68, 79, 295, 305, 313, 331, 341, 343
Crown damage 12
Cyclones 33
Cypress (see Baldcypress)

D

Dams 23, 40, 58, 67, 69, 184, 187, 245, 344, 463
DCA ordination 167, 175
Decay rate (k) 82
Decline, forest 20, 227, 238, 350, 360, 367, 369-371, 439, 450
Decomposition 56, 70, 72, 79, 82-83, 243, 329, 366, 464, 475
Delmarva Peninsula 113, 115-116, 139, 141
Dendrochronology 282
Denitrification 66, 81
Deposition 20, 56, 67, 83, 122-125, 145, 187, 225, 273, 326, 472-473, 477
Diameter 95, 241
    at breast height 123, 147, 303, 309, 329, 419, 421
    growth 234
    increment 419
Dieback 15, 177, 233, 238, 366, 412-413
Dieoff 15
Digital elevation model 280-281
Dismal Swamp 113
Disturbance 12, 275, 341, 422, 465
Ditches 45, 356, 374
Diversions 365, 367, 374
Diversity 55, 57, 70, 101, 145, 151, 191, 212, 214, 329, 340, 342
Drainage 30, 33 51, 103, 184, 188, 226, 229, 291, 295, 329, 331, 339, 342, 353, 414
    basin 3, 114, 120, 123, 185, 292, 298, 315
Dredging 12, 33, 39, 58, 188, 191, 218, 325-326, 354
Drought 6, 42-45, 76, 139, 149, 152, 156, 191, 194, 207-208, 239, 244, 262, 266, 283, 327, 345, 357, 362, 387, 412, 414, 424, 429, 439, 466, 471

E

Early settlement
　European 10, 141, 165, 184, 226, 258, 352, 386
　Spanish 225
Ecological 82, 120, 245, 327, 452
　indicators 314
　restoration 385, 402-405, 474
　succession 462-465
Ecosystem management 474-477, 452
Ecotone 83
Edaphic conditions 65, 69, 154, 331, 462
Elevation 31, 46, 50-51, 65, 69, 71, 74, 120, 140, 145, 147, 157, 169-170, 187, 195, 209, 232, 258, 268, 281, 295, 305, 324, 439, 448, 462-463, 472-473
Elsing Swamp 168
Environmental 147, 152, 158, 333
　condition 56, 241, 243, 269
　gradients 166
Erosion 52, 116, 133, 164, 224, 273-274, 354, 473
Escatawpa River 94
Estuary 20, 42, 67, 70, 139, 143, 154, 161, 186, 191, 206, 217, 239, 297, 313, 327, 411, 434, 466
Eutrophication 141, 374, 377
Evaporation 31, 49, 306
Evapotranspiration 33, 183, 203, 230, 264, 468
Everglades 9

F

Facultative
　species 212
　wetlands
Fauna 188, 291, 471
Fertility 398, 470
Fertilization 365, 370, 372, 433, 474
Fire 224, 256

Flatwoods 297
Flood 413-414, 424, 448
　control 3, 387
　duration 69, 72, 192, 236, 467
　event 12, 100, 145, 183, 238, 299, 306, 354
　protection 452
　stage 8, 126
　tide 36, 326, 345
　tolerant 76, 165, 234, 467
Flooding 12, 36, 40, 45, 50, 52, 55, 57, 76, 81, 90, 99, 100, 103, 107, 123-124, 143, 157, 173, 179, 184, 188, 192, 196, 224, 233, 262, 291, 299, 306, 327, 344, 433-434, 467-468
Floodplain 4, 50, 52, 66-71, 77, 80-81, 91, 113, 117, 124, 187, 224, 232, 292, 297, 303, 308, 330, 448
Floridian aquifer 258, 264
Florida 4, 7, 9, 32-33, 50, 52, 65, 68-69, 81-82, 89, 89, 93, 105, 107, 230, 258, 297, 468
Florida Panhandle 257
Forest 65, 68, 71, 77, 103, 106, 120, 141, 156, 178, 266, 278, 321
　biomass 55
　composition 46, 55, 268
　cover 4, 115, 117, 280
　defoliation 366
　development 52
　dieback 15, 238
　ghost 15, 389, 391
　health 57, 278
　history 421
　inventory 9, 243
　management 227
　productivity 55-57, 79
　regeneration 2, 244
　soils 71, 77, 81, 464
　structure 149, 243, 282, 387
　succession 463

Fragmentation 465
Frass 370-372
Freshwater 1, 139, 143, 212, 261, 413, 424, 448
   diversion 476-477
   marsh 9, 12, 30, 49, 68, 82, 139, 225, 312, 326, 466
   recharge 431
   zone 29, 143
Frontal passage 57, 230

G
Genotype 56, 393, 395-401
Geologic history 66, 115, 258, 332, 351
Geology 186, 295
Geomorphic setting 72
Geomorphology 2, 36, 57, 66, 74, 232, 448, 465
Geophysical processes
Georgia 33, 40, 297, 321, 324, 327
Global change 238, 280
Global climate change 158, 238-239
Global warming 19
Groundwater 81, 102, 114, 132, 143, 243, 262, 265-266, 297, 345
   discharge 120
   flushing 77
   inputs 79
   salinity 116, 467
   seepage 82, 422
   sources 65, 102
   withdrawal 283
Growth 4, 8, 77, 147, 207, 226, 234, 243, 367, 422
   chronologies 425, 428
   rate 55, 157
   ring 206, 419, 425

H
Habitat loss
Hammocks 46, 49, 52, 57, 294, 297, 305, 311
Harvest impacts 141, 260-261, 344, 354, 450
Headwaters 58, 79, 114, 184, 228, 292
Herbivory 366, 377
Hollows 50-52, 58, 69, 100, 120, 130, 140, 145, 147, 151, 166-169, 340, 344
Holocene 66, 141, 186, 322
Houma Navigation Canal 389
Human 66, 77
   exploitation 23, 260
   impacts 387, 463
   intervention 2
Hummocks 50-52, 58, 69, 100, 120, 130, 140, 145, 147, 151, 166-169, 206, 303, 305, 311, 315, 324, 337, 340-341, 463, 469, 473
HURASIM 17, 262
Hurricanes 49, 76, 91, 116, 203, 230, 236, 262, 280, 355, 373, 398, 468-469
   Agnes 299
   Alma 299
   Floyd 116
   Hugo 230, 238, 388
   Josephine 306
   Katrina, 94, 103-105, 230, 243, 373-374, 423, 469
   Rita 94, 103, 230, 242, 243, 373-374, 387
Hybrids 401-402, 405
Hydraulics 29, 56
Hydrogen sulfide (HS) 200, 218, 471
Hydrology 2, 6, 30, 55, 57, 69, 72, 75, 83, 90, 93-94, 102, 108, 188, 191-192, 195, 231, 262-266, 291, 298, 303, 315, 327, 356, 448, 467-468
Hydroperiod 45, 50, 55, 56, 113, 120, 122, 125, 143, 147, 152, 157, 223, 256, 321, 390, 413, 434, 465, 468, 470

Hydrophytic 170
Hypoxic conditions 467-468

I
Impoundments 11, 12, 14
Insect defoliation 365-366, 370
Insects 212, 214, 472
Intertidal 9, 42, 139, 142, 190, 278, 297, 388, 405, 412
Inundation 10, 31, 45, 51, 55, 66, 82, 114, 124, 127, 143, 157, 176, 195, 206, 217, 238, 262
Invertebrates 313, 471-472
Iron 82
Isolated wetlands 388, 396

J
James River 170
Juncus marshes 261

K
Knees 45

L
Lake Maurepaus 353
Lake Pontchartrain 349, 352
Land clearing 353
Land use 4, 10, 115, 141, 164-165, 224, 226, 258-261, 295, 322-323, 351-353
Landscape 2, 55, 71, 108, 141, 158, 229, 232, 241, 295, 302, 306, 331, 335, 343, 376, 452, 466, 469
Leaf 233, 392-398
Leaf Area Index (LAI) 149
Levee 14, 32, 43, 53, 83, 91, 96, 99-100, 103, 117, 124, 130, 145, 225, 227, 245, 305, 308, 331, 387, 448, 476
Limestone 5, 77, 81, 229, 255, 258, 295, 297, 307
Litter 81, 98, 104, 123, 125
  biomass 82
  production 367, 370

Litterfall 83, 372
Logging 10-11, 170, 226, 323-324, 344, 354, 387, 421, 450
Louisiana 4, 7, 9, 32, 36, 40-43, 49, 53, 67, 75, 223-245, 387-389, 396, 398-400, 447, 449, 473, 474-472

M
Major alluvial rivers 4
Management 2, 11, 67, 77, 227, 244, 282, 326, 375, 448, 456
Mangrove 9, 31, 68, 79, 82-83, 184, 312, 411, 463, 470-472
Mapping 10, 70, 72, 303
Maritime forest 46, 49, 297
Marsh 1, 9, 29-30, 49, 52, 55, 67, 82, 116-117, 120, 139, 144-145, 166-167, 179, 184, 187, 190, 195, 223, 228, 283, 322, 326, 359, 377, 462-463, 466, 472, 477
Maryland 4, 7, 9, 36, 45-46, 50, 113, 115, 139, 143
Maurepaus Swamp 349
Mean high water 31, 39-40, 46, 48-49, 57
Mean low water 50, 169
Meteorological 57
  conditions 6, 8
  tides 349
Microtopography 65, 76, 82, 140, 145, 303, 344, 472
Migration 178, 411
Mineralization 66, 75, 83, 125, 324
Mississippi 89, 103-105
Mississippi River Gulf Outlet 354
Mobile Delta 77-79
Models 76, 224, 314, 337, 463
Moisture 102, 230, 330
  soil 314
  stress 77
Mortality 17, 233, 236, 238, 242, 245, 273, 277-278, 359, 374, 376, 387-388, 439

Muck 71-72, 297, 307, 315
Mulch 228, 255, 450

N

Nanticoke River 46, 139-158
National Wetlands Inventory (NWI) 9, 15, 321, 453
National Wildlife Refuge (NWR) 231, 242, 243, 387
Navigation canals 16, 227, 245, 325, 387
Nitrate 298, 358, 367-368, 471
Nitrification 472
Nitrogen 56, 82-83, 122, 125, 365, 367-368, 371-372, 433, 469-472
Nutria 355, 377, 477
Nutrient 3, 56, 72, 83, 122, 125, 358, 368, 377, 413, 432, 458, 468, 469
    availability 130, 132
    concentrations 79, 125, 134, 331, 470
    cycling 82-83, 470
    load 124, 463, 474-476
    remineralization 470-471
    trapping 56, 133, 477
    uptake, 471

O

Okefenokee Swamp 291
Organic 71-72, 123, 133, 306
    matter 51, 55-56, 66, 69, 71, 72, 79, 122, 125, 154, 187, 200, 256, 330, 332, 336, 344, 473
    soils 79, 113, 462-465
Overbank flooding 20, 476
Overstory 239, 241
Overwash 15
Oxidation 56, 74, 76
Oxygen 200, 313, 467, 470

P

Palmer Drought Severity Index (PDSI) 436-437

Palmetto
    cabbage 49, 256, 260, 279, 296, 308, 311, 315
    dwarf 224
    saw 99, 101
Palustrine 42, 49
Pamunkey River 46, 50, 66, 68-69, 71-72, 83, 130, 161-162, 390
Peat 72, 164, 255
Pee Dee River 4, 70, 224, 228-229, 414-416
pH 55, 81, 332, 358, 471
Phosphorus 56, 66, 82-83, 122, 125, 330, 358, 367-368, 433, 470-471
Photosynthesis 234-236, 392-394, 404
Photosynthetically Active Radiation 149
Physiological ecology 468, 478
Piedmont 5, 71, 81, 162, 185-187, 224, 324
Pine, 117, 130, 223, 256, 311, 315
    flatwoods 282
    forests 261
    savannah 91, 99, 104
Plantations 10-11, 164, 225, 228, 257, 260-261, 352, 399-403
Pleistocene 66, 164, 186, 258, 297, 322
Pocomoke River 45-46, 53, 113-117, 120, 125, 130, 470
Population 298, 313, 315
Precipitation 31, 33, 43, 67, 82, 91, 116, 142, 163, 186, 230, 262, 280, 302, 355
Predation 472
Pre-settlement 258, 386-387
Principal Components Analysis 209
Productivity 55-57, 166, 224, 243, 245, 312, 327, 360, 363-364, 419, 430

Q

Quartz 186

Quaternary 115, 141, 322

R
Rail 10
Railroad 245, 356, 366
Railway 12, 226, 228
Ramsar wetlands 31
Recruitment 190, 238, 274, 360
Redwater rivers 5, 72, 81, 162, 187, 224, 413, 416, 463-464, 473
Reforestation 2, 385, 477-478
Regeneration 2, 57, 105, 108, 238, 244, 269-275, 387, 405, 453-454, 471
Remineralization 200, 203, 470-471
Respiration 200, 470, 472
Restoration 23, 365, 373, 376, 385, 411, 440-441, 450, 461
Rice 164, 224, 228, 323, 327
Riparian 121, 133
Riverflow 23, 299, 302, 306, 313, 414, 435-438
Rivers 2-4, 33, 38, 42, 57, 65, 68, 70, 141, 184, 192, 224, 228, 245, 291, 331, 387, 390-391, 469, 473, 475-477
Roads 10, 225
Roots 69, 77, 120, 145, 167, 206, 212, 243, 396, 464, 473
Runoff 3, 29, 32, 43, 141, 154, 197, 201, 357

S
Salinity 8, 29, 41-44, 53, 55-56, 68, 71, 76-77, 79, 94, 101, 103-104, 120, 152-153, 163, 184, 188, 192, 195, 231-241, 269, 291, 302, 305, 308, 326, 328, 356, 386-389, 466-467, 479
Salt-pulsed 157, 405, 413, 434
Sandy Island 415-416
Saturated 50-52, 58, 132, 183, 195, 217, 236, 238, 306, 315, 327, 345

Savannah River 66, 224, 228, 321-345, 390, 391, 469
Scale 228, 245, 331
  landscape 2, 147, 468
  stand 468
Sea level 19, 30, 40, 66-67, 115, 120, 129, 141, 143, 156, 225, 259, 297, 303, 315, 424
Sea-level rise 41, 58, 66, 116, 141, 144, 157, 161, 177-179, 187, 239, 258, 280, 314, 356, 386, 388, 439, 448, 462-463, 472, 479
Sediment Elevation Table (SET) 133, 375, 473
Sediments 3, 56, 66, 81, 115, 120, 123, 125, 141, 154, 164, 186-187, 200, 212, 223, 244, 258, 323, 344, 413, 448, 463
Seed bank 374
Seed germination 403
Seedling 39, 55-56, 90, 99, 103, 105, 149, 151, 232-239, 244, 269-275, 303, 308, 374, 390, 392, 396-398, 404, 477-478
Seepage 255, 264
Sewage treatment 463, 474-476
Shell middens 258
Sink hole 295
Sinks 124, 367, 471
Slope 49, 57
Snags 389, 391
Soil 5, 12, 30, 32, 45, 50-52, 55-56, 91, 93, 100, 102-104, 108, 123, 154, 158, 187, 225, 236, 241, 265, 278, 295, 302, 306, 322, 327, 329, 331, 357, 422, 431
South Carolina 33, 40, 42, 66, 82, 223-245, 297, 302, 306, 321, 324, 469, 400-401
Spartina marshes 389

Species 42, 46, 49, 52, 89, 95-96,
    99-108, 117, 132, 184,
    223-224, 233-237, 291,
    295, 303, 307, 321, 326,
    329, 341
    composition 55, 90, 96, 101,
        123, 130, 147, 241, 257,
        332, 390
    distribution 17, 29, 467
    diversity 55, 57, 330
    richness 173, 191, 209, 214
Springs 5, 255, 264, 295
St. Marks River 257
Stage 51, 192, 299, 315
State change 388
Stomatal conductance 392-393, 404
Storm surge 15, 17, 44, 46, 77, 95,
    116, 231, 236-238, 241,
    257, 262-263, 291, 299,
    302, 314, 387, 469
Stream 36, 38, 40, 68, 128, 187,
    229, 343
Streamflow 117
Stress 142, 144, 165, 173, 209, 233,
    236, 238, 333, 335, 364,
    377, 387
Subcanopy 224, 292, 302, 309, 331,
    341, 343
Subsidence 19-20, 41, 68, 116, 133,
    144, 156, 232, 244, 375,
    424, 448, 472-473
Succession 224
Sulfide 154, 200, 203, 218, 471
Surface water 122, 144, 195, 232,
    243, 467
Survival 56, 234, 244, 396-401
Suwanee River 33, 46, 51-52, 65,
    68, 71, 77, 261, 291-316

T

Temperature 142, 190, 230, 262,
    280, 302, 314, 326, 355
Terraces 297
Tertiary 115, 141, 258, 476
Tidal 46, 115, 163, 261, 356, 414

    amplitude 122, 178, 231
    creeks 49, 52, 58, 68, 167, 294,
        305, 323, 331, 341
    cycles 143, 166, 326, 433
    flooding 172, 188, 268, 271-
        272, 283, 412, 467-468
    flushing 49, 470
    inundation 31
    marsh 139, 184, 197, 462-463,
        472
    patterns 191
    range 68-69, 76, 142, 298,
        327, 465
    reaches 113, 294, 299
    surge 76, 373
    wave 46
Tidal freshwater forested wetlands
    definition 31, 42
    distribution 29, 43
Tide 6, 8, 29, 33-40, 129, 163, 166,
    228, 258, 264-265, 326,
    414, 434-438
    astronomical 29, 36, 57, 467-468
    meteorological 57, 265, 356
    storm 57, 262, 265, 423
    surge 44, 46
Timber 2, 10, 141, 165, 178, 226,
    228, 298, 352-353, 414, 448
Tolerance 206, 219, 359, 411
    flood 50, 56, 165, 173, 237, 434
    salinity 56, 157, 237, 272, 282,
        308, 385-405, 466-467
    shade 173
Topography 32, 42, 48, 50, 120,
    130, 161, 258-259, 291,
    303, 324, 340, 344
Transpiration 468
Transport 67, 76
Tree rings 206, 282, 412, 414
Tributaries 10, 50, 53, 66, 114, 141,
    149, 164, 178, 184, 186,
    191, 224, 295, 356
Tropical storms 57, 116, 230-231,
    299, 387
Turbidity 141, 157, 186, 354

U

Understory 147, 173, 241
Upland 18, 30, 45, 71, 83, 165, 188, 197, 201, 297, 359

V

Value 152, 157, 245, 313, 330, 333
Vegetation 50, 52, 55-56, 58, 141, 147, 184, 190, 207, 241-242, 292, 302, 312, 315, 326
Vertebrates 312, 461
Virginia 46, 66, 75, 113, 115, 140, 390

W

Waccamaw River 42-43, 224, 228-229, 414-418, 469
Waccassasa Bay 267, 269, 279
Wastewater 374, 413, 433
Water 5, 116-117, 122, 127, 297, 326
    balance 393-397
    depth 30, 465
    quality 5, 298, 314, 354, 417, 432
    table 179, 307, 344
Waterlogging 157, 235, 413, 467-468
Watershed 17, 79, 116, 126, 133, 139, 141, 152, 184-185, 211, 229, 238, 243
Weather 245
Wind 31, 33, 36, 45, 52-54, 57, 65, 116, 127, 133, 142, 241, 245, 299, 314
Winyah Bay 416-417
Wood 12, 167, 260
Woody debris (see Coarse woody debris)
Wrack 273-275

Z

Zonation 145, 200, 267-269